Engineering Advanced Materials for Manufacturing, Energy, and Smart Systems

Deepanraj Balakrishnan
https://orcid.org/0000-0001-8986-6692
Prince Mohammad Bin Fahd University, Saudi Arabia

Ratnasunil Buradagunta
https://orcid.org/0000-0001-9855-7808
Prince Mohammad Bin Fahd University, Saudi Arabia

Wattala Fernando
University of Dundee, UK

IGI Global
Scientific Publishing
Publishing Tomorrow's Research Today

Vice President of Editorial — Melissa Wagner
Director of Acquisitions — Mikaela Felty
Director of Book Development — Jocelynn Hessler
Production Manager — Mike Brehm
Cover Design — Jose Rosado

Published in the United States of America by
IGI Global Scientific Publishing
701 East Chocolate Avenue
Hershey, PA, 17033, USA
Tel: 717-533-8845 | Fax: 717-533-7115
Website: https://www.igi-global.com E-mail: cust@igi-global.com

Library of Congress Cataloging-in-Publication Data

LCCN: (CIP Data Pending)
ISBN13: 9798337394541
Isbn13Softcover: 9798337394558
EISBN13: 9798337394565
British Cataloguing in Publication Data
A Cataloguing in Publication record for this book is available from the British Library.

Table of Contents

Detailed Table of Contents

The pursuit of enhanced lightweight materials is intensifying, with stakeholders in metallurgy, thermal/mechanical treatments, and materials fabrication converging to develop high-performance alloys. Aluminum (Al), Titanium (Ti), and Magnesium (Mg) based alloys are at the forefront, drawing significant attention. Among their critical properties, fatigue performance—alongside corrosion resistance is crucial. This review aims to unify perspectives from diverse domains on factors affecting fatigue behavior, fostering collaborative progress in material and process development. The literature on Al, Ti, and Mg-based alloys was systematically reviewed, categorized, and presented for easy access. The influence of alloying elements, phases, and microstructures on fatigue properties is examined. Additionally, the impacts of thermal and thermo-mechanical treatments—including roller burnishing, peening, deep rolling, severe plastic deformation, hydrostatic extrusion, and friction stir processing on microstructural evolution and fatigue enhancement are highlighted.

methods like rapid solidification, powder metallurgy, and additive manufacturing create refined microstructures and improve mechanical performance. Advanced joining techniques such as laser beam and friction stir welding enhance damage tolerance and manufacturability. Future trends emphasize digitalization, sustainable production, multifunctional hybrid materials, and tailored alloy compositions to meet next-generation aerospace demands. Overall, Al-Li alloys offer vital weight savings, durability, and cost-effectiveness for high-speed aircraft structures.

It is known that the aluminum and aluminum alloys are materials frequently preferred for the aerospatial equipment due to their properties to ensure a certain strength and a low weight of the equipment components. For such reasons, the aluminum and aluminum alloys were utilized just from the first use of metallic materials in the aerospatial equipment. The abrasive jet machining is a machining method based on the effects generated on the workpiece surface by the abrasive particles placed in a high-speed gas stream. Essentially, these effects are materialized by a material removal process and/or a change of the surface aspect. The abrasive jet machining is applied as a subtractive machining method. The objective of the proposed chapter was to offer a general image concerning the theoretical principles and applications of using the abrasive jet machining in the manufacturing of aluminum and aluminum alloys parts included in the aerospatial equipment. A synthesis concerning the main aspects specific to the abrasive jet machining of the aluminum and aluminum alloys was achieved.

Smart materials that can sense and respond to environmental stimuli such as temperature, stress, magnetic fields, and electric currents are transforming the design and functionality of engineering systems across diverse sectors. From aerospace actuators to biomedical implants, these materials offer tailored performance and adaptability, often mimicking biological responsiveness. Concurrently, metal additive manufacturing, or metal 3D printing, is redefining possibilities in fabrication, offering geometric complexity, customization, and material efficiency that traditional methods cannot match. The intersection of metal Additive Manufacturing and smart materials presents a powerful opportunity: manufacturing functional, responsive structures with tailored properties, spatially varying functionalities, and integrated sensors or actuators from the ground up. This chapter proposes to explore the current landscape, challenges, and prospects of fabricating smart materials using Additive Manufacturing.

Uzwalkiran Rokkala, Aditya University, Surampalem, India
Faseeulla Khan Mohammad, King Faisal University, Saudi Arabia
Saad Patel Md, Vellore Institute of Technology, Vellore, India
Akash Biradar, Dr. D.Y. Patil Institute of Technology, India
Nooruddin Ansari, Oryx Universal College, Liverpool John Moores University, Doha, Qatar
Prudvireddy Paresi, Federation University, Ballarat, Australia
Raffi Mohammed, National Institute of Technology Andhra Pradesh, India

Wire Arc Additive Manufacturing (WAAM) is a novel metal additive manufacturing process based on direct energy deposition destined to revolutionize the manufacturing sector in the next few decades. This process enables sequential deposition of metal on a substrate with the capability of creating intricate geometries with enhanced mechanical properties.This chapter discusses the basic concepts of WAAM, its categorizations, process planning strategies, control approaches, toolpath generation, material choice, defect avoidance and monitoring techniques. Recent innovations, such as intelligent path optimization, real-time process monitoring, and simulation-driven control, highlight the rapid development of the technology. Industries like aerospace, automotive, and tooling are examined for critical applications, while future developments include hybrid production system integration, sustainability-driven practices, and the development of industry standards. WAAM presents an attractive solution for future manufacturing with its flexibility in design, cost-effectiveness and scalability.

Wirecut Electrical Discharge Machining (WEDM) is a cost-effective and productive method of cutting electrically conductive hard materials. WEDM has the added benefit of being more accurate when it comes to manufacturing component forms. In this research, the performance of reciprocated WEDM (RWEDM) on a titanium grade 2 alloy with a molybdenum wire electrode is evaluated by measuring the residual stresses present on the machined part for different conditions. Observations indicate that considerable residual stresses are seen in the specimen as watched from the X-ray diffraction test through measuring the residual stress, shear stress, D-spacing, full width at half maximum height, intensity, and breadth of stresses. Observations show that when feed rate of wire electrode in RWEDM and flow rate of dielectric is higher and higher variable frequency and higher current, residual stress tends to rise due to higher generation of spark density, which correspondingly removes higher quantity of material from the workpiece

3D printing has transformed the manufacturing of medical implants by enabling precise and customized designs. This review explores key powder-based additive manufacturing (AM) methods, including Powder Bed Fusion (PBF) and Directed Energy Deposition (DED), used to create titanium-based implants. It also discusses titanium powder production techniques such as gas atomization (GA), plasma atomization (PA), and the plasma rotating electrode process (PREP). The study highlights the benefits of additive manufacturing (AM), including its cost-effectiveness and design flexibility. However, it addresses significant challenges like powder quality, porosity, and contamination. The increasing adoption of titanium implants

However, traditional electronics have been made of hard materials which have strong limitations on sensations of feeling, flexibility, durability whereas all of those are required for wearable devices. With the de velopment of IoTdriven wearables, flexible and stretchable material integration into the device could be a new opportunity to break through these obstacles. These advanced materials, which include conductive polymers, graphene, and elastomers, can be used to fabricate wearable electronics for curving, stretching and matching the human body. In this chapter, we review the use of such materials for designing and developing the ubiquitous wearable IoT devices described above by discussing their characteristics, benefits, and processing methods. Issues such as flexible electronics integration, sensor accuracy, power management, and long-term durability are also addressed.

Safety and service life of infrastructures including bridges, buildings, and industrial plants can be validated by the structural health monitoring (SHM) techniques. Combining Internet of Things (IoT) technology with advanced image processing can present a solution for real-time, non-destructive, automated structure health monitor. In this chapter, we review smart solutions that integrate IoT-enabled sensor networks with image-based data acquisition for structural defect detection, classification, and analysis. A recent overview for the developments in IoT and image processing for SHM is summarized, such as machine learning assisted crack detection, and cloud-edge based computation model. Case studies show the functionality of these integrated systems allows for improved accuracy of monitoring, maintenance expenditures, and the use of prognostic maintenance tactics. Challenges such as environmental variability, data handling, and computation limitations are addressed, and the future direction of more reliable and scalable SHM is presented.

The natural composites are widely preferred nowadays in versatile application due to easily degradable and moderate mechanical properties when compare to polymer matrix composites. The kenaf fiber are used as reinforcement in composites wherever high flexural strength and modulus is required whereas sisal fiber is preferred for tensile strength, impact and wear resistance. In our work hybrid composite comprising kenaf and sisal as reinforcement and epoxy as matrix is fabricated and its mechanical properties are explored and compared with the conventional composite fabricated using only kenaf or sisal as reinforcement. The hybrid composite shows excellent flexural strength but inferior tensile strength when compare to conventional composite. The failure of the specimens are investigated using scanning electron microscopy (SEM). In this chapter detailed insight is presented for emphasizing the need for tailored fiber selection and processing to achieve optimal mechanical properties.

The advancement of sustainable energy systems requires the development of high-performance photovoltaic materials that meet the demands of efficiency and scalability. Thin-film photovoltaic technologies have emerged as a promising solution. This chapter provides a review of second generation thin-film PV technologies, including amorphous silicon (a-Si), cadmium telluride (CdTe), copper indium selenide (CIS) and copper indium gallium selenide (CIGS), as well as emerging technologies including perovskites, dye-sensitized solar cells (DSSCs), and organic solar cells (OSCs). The chapter addresses material properties, device architectures, deposition processes, and the current performance trends. It also discusses challenges such as material scarcity, toxicity, and long-term stability. Case studies are provided on building-integrated photovoltaic (BIPV) systems for practical applications and performance outcomes. The chapter offers technical insights for researchers,

Preface

The rapid evolution of engineering systems in manufacturing, energy, aerospace, biomedical, and smart technologies has placed unprecedented demands on materials performance, functionality, and sustainability. Advanced materials now sit at the core of innovation, enabling lighter structures, higher efficiencies, smarter responses, and longer service lifetimes under increasingly complex operating conditions. In this context, the ability to engineer materials—rather than merely select them—has become a defining capability of modern engineering practice. This edited volume, *Engineering Advanced Materials for Manufacturing, Energy, and Smart Systems*, is conceived as a timely and comprehensive reference that captures this shift toward materials-by-design.

As editors, we recognize that contemporary challenges cannot be addressed within isolated disciplinary boundaries. Progress in advanced materials requires the integration of materials science, manufacturing processes, digital tools, and application-driven design philosophies. This book therefore brings together contributions from researchers and practitioners across academia and industry, representing diverse geographical regions and technical specializations. The chapters collectively reflect how advanced materials are being tailored through innovative processing routes, characterization techniques, and system-level integration to meet the demands of next-generation engineering applications.

The overarching objective of this volume is to bridge fundamental understanding with practical implementation. Rather than focusing solely on material properties in isolation, the chapters emphasize processing–structure–property–performance relationships, scalability of manufacturing routes, and real-world constraints such as cost, reliability, sustainability, and regulatory considerations. By doing so, the book serves not only as a scholarly reference for researchers, but also as a practical guide for engineers, designers, and decision-makers engaged in deploying advanced materials in manufacturing, energy systems, and smart infrastructures.

CHAPTER OVERVIEW

The book opens with **Chapter 1**, which presents a state-of-the-art review on fatigue resistance in lightweight aluminum, titanium, and magnesium alloys processed through advanced manufacturing techniques. By synthesizing insights on alloying, microstructural evolution, and thermo-mechanical treatments, this chapter establishes a strong foundation for understanding durability and performance in lightweight structural applications.

Chapter 2 focuses on machining science, employing numerical simulation and optimization techniques to study coating thickness and tool geometry during the turning of nano-hybrid aluminum composites. The chapter highlights how process parameter optimization can significantly influence cutting forces and thermal behavior, contributing to improved machinability and tool life.

In **Chapter 3**, the attention shifts to aluminum–lithium alloys for high-speed aircraft. This chapter examines alloy development, advanced manufacturing routes, and joining techniques that enable lightweight, high-strength components capable of withstanding extreme aerodynamic and thermal loads, offering insights critical to next-generation aerospace systems.

Chapter 4 provides a comprehensive overview of abrasive jet machining of aluminum and aluminum alloys, particularly for aerospace applications. The chapter synthesizes theoretical principles and practical considerations, emphasizing surface integrity and material removal mechanisms in nontraditional machining processes.

Additive manufacturing emerges as a central theme in **Chapters 5 and 6. Chapter 5** explores the convergence of metal additive manufacturing and smart materials, discussing design-to-deployment challenges and opportunities for creating responsive, multifunctional structures. **Chapter 6** delves into wire arc additive manufacturing (WAAM), covering process fundamentals, control strategies, industrial applications, and future perspectives for scalable metal AM.

Chapter 7 investigates residual stress development in commercially pure titanium machined using reciprocated WEDM. Through detailed experimental analysis, the chapter underscores the importance of process control in ensuring dimensional accuracy and structural integrity of hard-to-machine materials.

Biomedical applications are addressed in **Chapters 8 and 9. Chapter 8** reviews titanium-based implants produced via additive manufacturing, examining powder production, processing challenges, and future research directions. **Chapter 9** broadens the scope to polymers—natural, synthetic, and biodegradable—highlighting their roles in tissue engineering, drug delivery, and regenerative medicine.

The integration of advanced materials with digital and smart systems is the focus of **Chapters 10 and 11. Chapter 10** reviews flexible and stretchable materials for IoT-driven wearable devices, addressing material selection, processing, and system-level challenges. **Chapter 11** examines smart materials and IoT-enabled structural health monitoring using image processing, demonstrating how materials, sensors, and data analytics converge to enhance infrastructure safety.

Sustainability and renewable energy applications are emphasized in **Chapter 12** and **Chapter 13. Chapter 12** explores natural fiber hybrid composites, offering insights into tailoring mechanical behavior through fiber hybridization. **Chapter 13** provides a detailed review of thin-film photovoltaic materials, deposition techniques, and engineering applications, highlighting their role in scalable and sustainable energy systems.

The volume concludes its technical journey with **Chapter 14**, which reviews RF-sputtered wide bandgap semiconductors for next-generation electronic and optoelectronic devices. This chapter outlines synthesis challenges, optimization strategies, and application potential, reinforcing the importance of advanced functional materials in future technologies.

CONCLUSION

This edited volume reflects the dynamic and multidisciplinary nature of advanced materials engineering. Across its chapters, a unifying theme emerges: the necessity of integrating materials development with manufacturing processes, application requirements, and system-level considerations. Whether addressing fatigue resistance in lightweight alloys, scalability of additive manufacturing, or functionality in smart and energy systems, the contributions collectively demonstrate how materials innovation underpins technological progress.

We believe this book offers lasting value as a reference for researchers seeking consolidated insights, for graduate students building foundational and applied knowledge, and for practicing engineers navigating the complexities of advanced materials selection and processing. The diversity of topics and approaches ensures that readers gain both breadth and depth, enabling cross-fertilization of ideas across traditional domain boundaries.

As editors, we are grateful to the contributing authors for their scholarly rigor and commitment, and to the reviewers and publishers for their support in shaping this volume. It is our hope that *Engineering Advanced Materials for Manufacturing, Energy, and Smart Systems* will stimulate further research, foster interdisciplinary collaboration, and contribute meaningfully to the advancement of sustainable, intelligent, and high-performance engineering solutions for the future.

Chapter 1
State-of-the-Art Review on Fatigue Resistance of Lightweight Alloys Through Advanced Manufacturing Techniques

Namrata Gangil
Ajay Kumar Garg Engineering College, India

Arshad Noor Siddiquee
Jamia Millia Islamia, India

Faseeulla Khan Mohammad
https://orcid.org/0000-0002-5193-3260
King Faisal University, Saudi Arabia

Mohammed E. Ali Mohsin
King Faisal University, Saudi Arabia

Syed Quiadir Moinuddin
King Faisal University, Saudi Arabia

Aarti Singh
Ajay Kumar Garg Engineering College, India

ABSTRACT

The pursuit of enhanced lightweight materials is intensifying, with stakeholders in metallurgy, thermal/mechanical treatments, and materials fabrication converging to develop high-performance alloys. Aluminum (Al), Titanium (Ti), and Magnesium (Mg) based alloys are at the forefront, drawing significant attention. Among their critical properties, fatigue performance—alongside corrosion resistance is crucial. This review aims to unify perspectives from diverse domains on factors affecting fatigue behavior, fostering collaborative progress in material and process develop-

DOI: 10.4018/979-8-3373-9454-1.ch001

ment. The literature on Al, Ti, and Mg-based alloys was systematically reviewed, categorized, and presented for easy access. The influence of alloying elements, phases, and microstructures on fatigue properties is examined. Additionally, the impacts of thermal and thermo-mechanical treatments—including roller burnishing, peening, deep rolling, severe plastic deformation, hydrostatic extrusion, and friction stir processing on microstructural evolution and fatigue enhancement are highlighted.

1. INTRODUCTION

The weight reduction is rigorously adopted with a view to cutting down on energy consumption. This has given fillip to the deployment of light materials and several applications involving conventional materials are being subsumed by light metals. The mechanical performance of materials greatly depends on the structural integrity under applied forces, type of load and importantly cyclic loading. Cyclically varying loads aggravate micro/macro-structural irregularities to eventually develop into gross damage. Such damage results in fatigue failure, which occurs without warning and at much below the material's static strength (Moshiri et al., 2021). This necessitates periodic inspection and involves service breaks and associated inspection cost. The rate and degree of damage largely depends on microstructure and surface conditions. Fatigue of metallic materials is associated with microscopic crack initiation and crack's prorogation. Other fatigue phenomenon includes fretting and roller contact fatigue, creep fatigue at elevated temperature or corrosion fatigue associated with corrosive environmental damage (Prasad et al., 2014; W. Song et al., 2021; Y. Song et al., 2021; Teng et al., 2021; Z. Wang et al., 2021). Materials endurance to fatigue is very sensitive to the microstructural state, surface condition and stability under the action of environment and load (Garbacz et al., 2011). In the pursuit of weight saving, reducing operational energy consumption without compromising safety, interest in high strength lightweight materials is increasing tremendously. Among available options, high strength Al, Mg and Ti alloys are the front runners, whilst the former has cost advantage over the latter, but it suffers from poor performance at high temperature (Dwivedi et al., 2021; Gangil et al., 2020). Of late, Mg has also found increased interest in the light metals category after issues of inflammability settled down. High strength Al alloys have a great advantage due to high specific strength, and corrosion resistance yet their fatigue related properties are greatly important to technological readiness for strategic applications in aviation, space and aerospace (Khan et al., 2017).

The processing routes, and post processing treatments are known to alter microstructure and enhance the mechanical properties including fatigue performance. The present paper is conceived with the objective of portraying the properties, microstructural characteristics and in service performance of Al, Mg and Ti based light alloys. The paper outlines the micro/macrostructural features obtained after various treatments and mechanical properties including fatigue behavior. Current state of development of alloys, especially in Mg, and Al systems, has been included and their characteristics presented so that the stakeholder can identify and advance upon the evolved characteristics. The thermal/thermomechanical processing routes that alter microstructure to favor mechanical properties and performance under fatigue are also presented.

2. LIGHT WEIGHT ALLOYS AND THEIR PROPERTIES

Among light materials, Al, Ti and Mg based alloys have a distinction over ferrous materials because of high specific strength and ease of manufacturability. Al alloys are large in number which are generally categorized as cast alloys and wrought alloys. The 2XXX and 7XXX series Al alloys are age-hardenable which possess very high specific strength and find major applications in aircraft and aerospace (as typically depicted in Figure 1(Guru et al., 2015; Jaseem et al., 2018; Meenia & Khan, 2016). However, Ti alloys are costly in comparison to Al and Mg based alloys, but possess excellent specific strength, high temperature strength and corrosion resistance (Yang et al., 2014). These alloys find applications in aircraft and aerospace components employed in high temperature regions. The Mg based alloys, on the other hand, are the lightest among these but have poor corrosion resistance and toughness. Newer rare earth- based Mg alloys from WE-series possess enhanced corrosion resistance and toughness. The Li based Al alloys are among the lightest Al alloys and are now being developed for the gauge-to-gauge replacement of high strength age-hardened Al alloys. Consequently, Al-Li alloys system is covered in this work.

2.1 Aluminum-Lithium Alloys

The alloying of Al with ultralight Li significantly reduces the density of Al-Li alloys. Among Al-Li alloys, the AA2020 has long been in use for several decades but did not find encouragement due to anisotropy, poor toughness and poor manufacturability. Recently, developments in the 2nd and 3rd generation Al-Li alloys have overcome almost all limitations of earlier alloys. There is renewed interest in the further development and improvement of ultra-light weight alloys which apart from Al-Li also include Al-Be and Al-Be-Mg alloys etc. All these alloys find critical applications in space, aerospace and aviation. Li being the lightest metal, its alloying with Al reduces density significantly and improves the stiffness of resultant alloy. The Al-Li alloys from 2XXX series such as AA2020, AA2050, AA2055, AA2060,

AA2065, AA2076, AA2090, AA2091, AA2096, AA2098, 2099 and 8XXX series such as AA8090, etc. are widely used in aircraft applications.

The Al-Li alloy system contains Li up to 3% and addition of Li even in small amounts significantly improves the properties. Typically, 1% addition of Li reduces the density by 3% and enhances the modulus by 6% (Rioja & Liu, 2012). These alloys are 7-10% lighter as compared to traditional high-strength Al alloys and possess excellent resistance to corrosion and high stiffness. Despite ultra-light weight, these alloys suffer from low ductility, issues with fracture toughness, poor short-transverse properties and significant fatigue crack growth rates (FCGR) especially when cracks are microscopic, (Deschamps et al., 2013; L. Wu et al., 2012). The weight savings due to high specific strength motivated the systematic replacement of other high strength alloys with the Al-Li alloys (e.g., AA2090-T81, T83, T84 and T86 were comparable and suitable to replace 7075- T6; likewise, AA2091-T8X and AA8090-T81AA are paired with 2024-T3).

The journey of high strength Al alloys began with Al-Cu system (Figure1(a)) which formed the T1(Al_2CuLi) and θ'(Al_2Cu) strengthening precipitates (T1 being the most prominent); whereas controlled amount of precipitate T2(Al_6CuLi_3) improves toughness. The T1was found to nucleate at sub-grain boundaries and dislocation networks, whereas T2 at grain boundaries. The addition of Li in Al-Cu reduces the density and produces δ'(Al_3Li) which produces an effect like T1 and θ'. Manganese (Mn), in the alloy, forms an incoherent dispersoid ($Al_2oCu_2Mn_3$) which improves fracture toughness and fatigue performance (Li et al., 2019; S. C. Wang & Starink, 2005; Yoshimura et al., 2003a, 2003b).

Figure 1. (a) Historical development of high strength Al-alloys(Runge, 2018) . Application of Al-Li alloy AA8090 in E101 Helicopter(Y. Wang & Zhao, 2020) ; (b) in Fuselage and (c) in Lift frame.

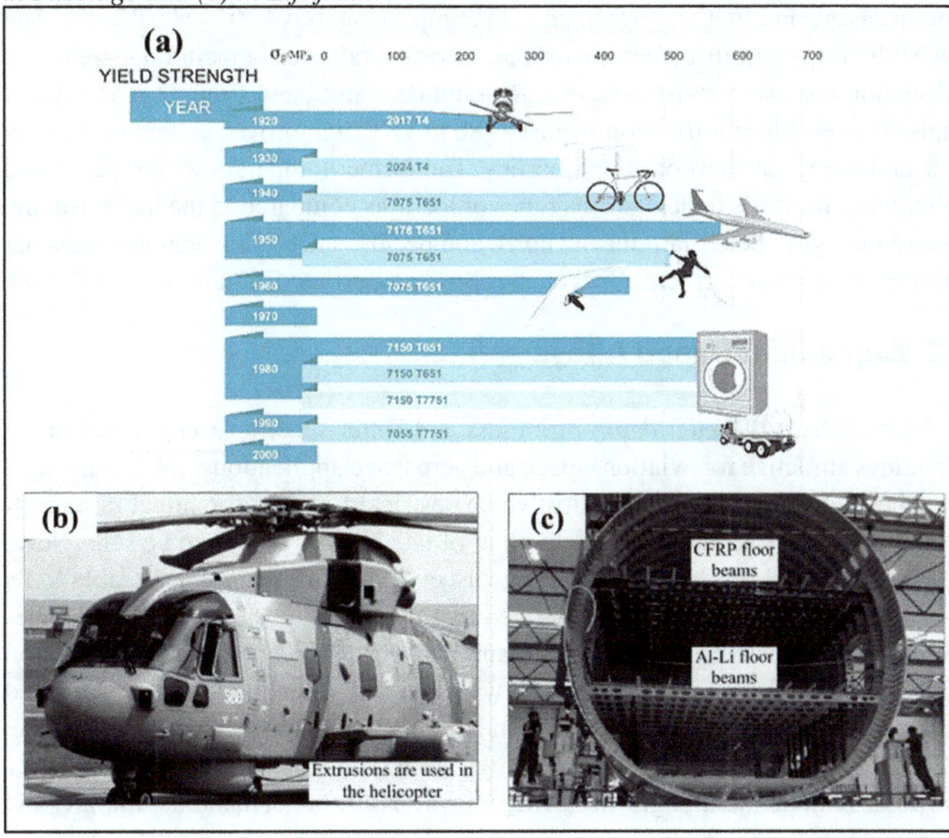

Addition of Mg in Al-Li reduces density and contributes to strength via solid solution and precipitation mechanisms. The addition of Ag mainly contributes to strength via solid solution mechanism. The Zn present as solute, is heavier but, contributes to solid solution strengthening and improves corrosion resistance as well. Mn controls grainsize, microstructural texture evolution and improves fracture toughness during thermo-mechanical processing.

Crystallographic an isotropy and texture evolution are one of the limitations of some Al-Li alloys. The anisotropy along L, LT, ST (i.e., in plane and through thickness) directions is affected by all factors i.e., grain size and grain morphology, texture, type and amount of cold work, and aging treatment. The AA8090-T86 (2nd generation alloy) alloy shows significantly high anisotropy in comparison to AA2199 (3rd generation alloy) alloy (Rioja & Liu, 2012). The anisotropy is attributed to the presence of grain boundary precipitates, clustering of precipitates, and presence of high amount of insoluble second phases. Anisotropy leads to unpredictable FCGR and thus creates design difficulty.

Investigations demonstrate that AA2199-T8E80 alloy has resulted in 8% greater weight saving and FGCR performance in comparison to AA2024 HDT-T351. The AA2199 alloy exhibits better performance on several criteria including stress and exfoliation corrosion resistance, FCGR, toughness and yield strength under direct loads. The temper modification from T8T80 to T8T79 improved weight savings by 15% and improved most of the properties. The former temper, however, possessed better ultimate stress under various types of loads as compared to the latter. But the later temper gave better fatigue and corrosion performance. This demonstrates that aging treatment has a critical effect on the alloys properties (Giummarra et al., 2007).

2.2 Magnesium Based Alloys

Lighter than Al, better damp-proofness and better specific strength makes the Mg alloys attractive for aviation, space and aerospace applications. It has long been apprehended for its inflammability which restricted its use for in-cabin aircraft applications such as chair-frames etc.; but, of late Mg has proven to be safe (Neve, 2019). Almost all reports of its burning susceptibility are mainly applicable to its melt condition. It has been demonstrated that the addition of rare earth (RE) elements in Mg, which forms the "WEXX and EVXX" series Mg alloys, have ignition temperature range comparable to Al alloys (Ravi Kumar et al., 2003). This opened new avenues for the use of Mg alloys in aviation and aerospace. The AZXX (i.e. AZ61, AZ80, ZK30, ZMC711 etc. are high strength alloys; whereas AZ31, ZM21 etc. possess moderate strength) series Mg alloys possess better manufacturability and have lowering it ion and concerns of corrosion. The strengthening, manufacturability and corrosion responses of the Mg alloys have been adequately addressed in the WE and EV series alloys(F Khan & Rokkala, 2023; Kumar et al., 2021).

Furthermore, fatigue behavior greatly differs from usual mechanical response and the material fails under fatigue at considerably lower stresses. Typically, fatigue failure occurs in three steps (Figure 2): (i) crack initiation at a microstructural site acting as stress raiser, (ii) crack propagation under variable stresses and (iii) eventual fatigue failure. In case of Mg alloys, the pre-crack initiation is associated with microstructural changes including dislocation generation, movement and arrangement of dislocation in structural pattern. These pre-crack-initiation changes sometimes make the affected regions soft or hard depending on age-hardening condition and type of precipitation. The precipitates in Mg alloy systems may be either coherent or incoherent. Former possesses properties like the Mg matrix, whereas in the case of latter the properties differ significantly. The dislocations, during movement, easily shear through the coherent precipitates and further refine them. The dislocations cannot cross through the in-coherent precipitates and thus they create the pinning effect, and the affected regions harden (Potzies & Kainer, 2004). The RE-based Mg systems produce coherent precipitates; whereas precipitates in Mg-Al, Mg-Li based systems are generally incoherent (NIE, 2012). The deformed microstructure possesses slip bands which also act as crack initiation sites in Mg alloys. At or near room temperature, the dislocation movement about the basal slip planes is greatly favored, whereas at elevated temperatures the prismatic slip planes become favorable. The presence of defects like voids, inclusions, shrinkages cracks or even particulate phases act as the initiation site.

Figure 2. (a) The three stages of Fatigue (Gao et al., 2021), (b) The three stages on fractured cross section (Zhu et al., 2023), (c) Schematic of stress corrosion crack (SCC) fatigue; crack initiation and growth.

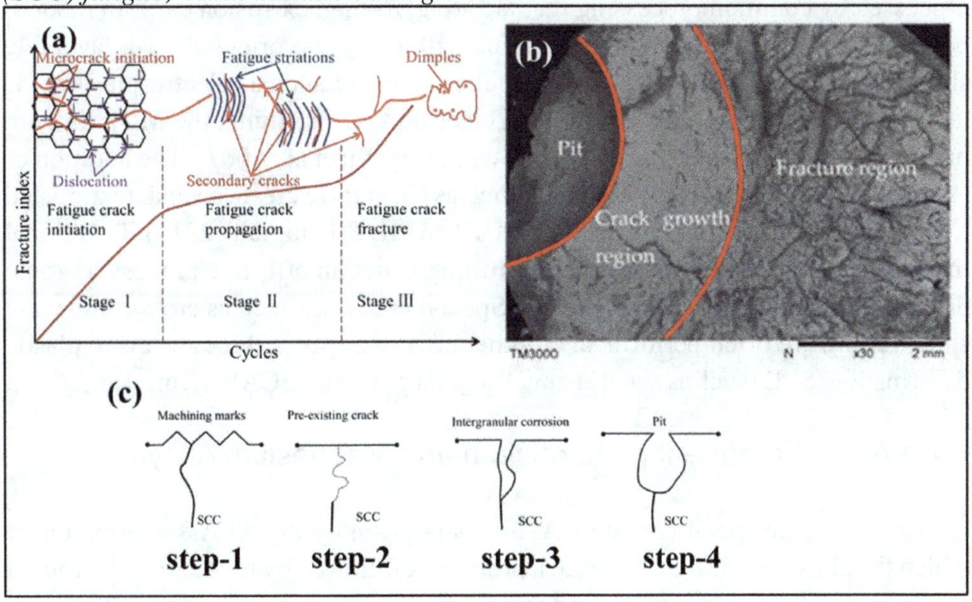

Thus, microstructural integrity, alloy composition, age-hardening and strain-hardening treatments, all affect the fatigue response of Mg alloys. The response of alloy under cyclic loading is the most critical selection criteria for key applications. Mn in small concentration, for example 0.4%, improves the strength and hardness of Mg alloys and up to 0.58% and alloys fatigue properties are also improved. Excess Mn, beyond 0.58%, forms large intermetallic which act as tiny stress raiser and preferred crack initiation site; consequently, reduces fatigue performance. The Mg alloys are found to exhibit high fatigue life sensitivity and smooth transition from low to high cyclic fatigue. The Mg alloys are prone to early crack initiation at stress levels higher than endurance limit. Such a response of alloys under fatigue conditions continues to be investigated for the prevalent fatigue mechanisms. Fatigue investigations on wrought AZ31 (Chamos et al., 2008) demonstrated trans-granular crack initiation followed by intergranular crack propagation leading to eventual fatigue failure. Such behavior is attributed to lack of plastic deformation on account of hcp structure and availability of lesser number of slip systems.

The layered basal planes and lesser slip systems in the HCP Mg make critical shear 40 times less than the prismatic slip. Anisotropy makes room temperature (and warm forming below 250°C) difficult in Mg alloys. The room temperature anisotropy of the Mg alloys can be reduced by adding suitable alloying elements such as Ca, Zn, Ba, Y, La and Sr(Mishra et al., 2022; Sahoo & Khan, 2018). Anisotropy can be tackled in the polycrystalline Mg alloys through crystallographic texture. A strong texture in Mg means strong basal texture which is the case after the Mg alloys get processed by common processing methods (e.g. rolling, extrusion etc.). In random polycrystalline weak texture, some grains will always be oriented for a favorable slip for every deforming load. Mg's peculiar behavior such as electro-plasticity is tapped in alternate forming methods. In electro-plastic forming, the Mg alloys are processed while high electric current flows past it (Xu et al., 2007). The addition of RE elements and some other elements such as Ca and Be create a weak texture and improve formability (F Khan et al., 2019; F Khan MD & Panigrahi, 2018). Traditional thermo-mechanical processing such as rolling is not an effective processing route due to presence of strong basal texture. Special processes such as cross rolling and shear rolling are often performed. Another effective way is through severe plastic deformation (SPD) such as equal channel angular pressing (ECAP) (Kim et al., 2003).

2.2.1 Aging Treatment of Aluminium and Magnesium Alloys

Most versatile specification of Al alloys is given by the Al Association under which the alloys are designated with a prefix AA followed by four digits in wrought alloys (and a dot between third- and fourth-digit specifier in cast alloys). The alloy series which respond to heat treatment are termed as age-hardening or precipitation hardening. Age-hardening enhances the strength of alloy immensely; however, the strengthening effect is meta-stable. The treatment involves sequential heating and cooling cycles: (i) heating and soaking the alloys at a high temperature at which the meta-stable phase dissolves in the primary α matrix, and this operation is known as solutionizing. Solutionizing creates super saturated solid solution (SSSS). (ii) Quenching the alloys after solutionizing, followed by (iii) aging at room temperature (known as natural aging) or artificially reheating the quenched alloys at a temperature significantly lower than solutionizing temperature. The aging process transforms the SSSS and produces a sequence of metastable ultrafine precipitates which grow with time in density and morphology. These precipitates contribute to strengthening through pinning effect. While natural aging requires several days or even months for aging to complete; artificial aging, on the other hand, is complete in several hours. Aging treatment enhances the strength of alloys by several folds, it is also found to affect corrosion and fatigue properties. The precipitate's morphology, density and distribution of precipitation mainly control strengthening, fatigue and corrosion behavior of age-hardened alloys. The response of aging treatment of different alloy specifications is different and standard procedures are available.

The precipitates in Mg alloys comprise of finely distributed plate or lath-rod shaped fine equilibrium phase intermetallic which are formed parallel and normal to basal planes. The Mg-alloys in as-cast structure are designated (Refer to Table.1) as "F" i.e., as-fabricated. The wrought alloys in the annealed state are designated by "O" which is the softest temper condition. Wrought alloys after strain-hardening by cold work (with or without thermal treatment) are specified by "H" followed by one or two digits. Such as H1, which indicates alloy being strain hardened to impart strengthening without thermal stress relieving. The digit following H1 indicates the degree of straining. The designation H2 specifies strain-hardening coupled with partial annealing. The digit following H2 indicates the degree of straining after partial annealing. The temper specified by H3 treatment indicates strain hardening followed by low temperature stabilization which may slightly lower the strength and improve the toughness. The temper designated with "W" indicates solution heat-treatment. The numeric post-specifier indicates the time for natural aging. During "W", the alloy attains an unstable temper by spontaneous natural aging from solutionizing temperature for a short duration. Finally, when the alloy is artificially aged to attain a stable temper, then its condition is specified at "T". The subdivision of temper "T" from 1 to 10 specifies specific thermal and straining treatment given to the alloy to bring about stable temper.

Table 1. Magnesium alloys temper conditions and designations (Westengen & Rashed, 2016)

Designation	Treatment
F	As fabricated
H	Strain hardened (wrought products only)
H1X	Strain hardened only. X is a number denoting degree of hardening
H2X	Strain hardened and partially annealed.
H3X	Strain hardened and stabilized.
O	Soft annealed, recrystallized (wrought products only)

continued on following page

Table 1. Continued

Designation	Treatment
T	Heat treated to produce tempers other than F, O, and H
T1	Cooled and naturally aged
T3	Solution heat treated and then coldworked
T5	Cooled and artificially aged
T4	Solution heat treated
T6	Solution heat treated and artificially aged
T7	Solution heat treated and stabilized
T8	Solution heat treated, coldworked, and artificially aged
T9	Solution heat treated, artificially aged, and coldworked
T10	Cooled from thermo- mechanical treatment, artificially aged and cold worked
W	Solution heat treatment and room temperature aging

2.3 Titanium Based Alloys

Ti-based alloys are widely used in aircraft, aerospace, formula cars and bio-medical applications, due to their high specific strength, low specific gravity, high thermal stability and excellent corrosion resistance. The crystal structure of Ti-alloys changes with temperature at 883°C from alpha (α) hcp structure to beta (β) bcc structure. Ti-alloys are classified as α, near α or super α, α/β, β and γ alloys (Boyer, 1996; Ghonem, 2010). Al is the principal alloying element in α- alloys; on adding β stabilizing elements (Mo and V), the alloys respond well to thermo-mechanical treatments and the alloys are categorized as super α alloys. They are further processed to the α/β and β phases to attain either equi-axed α, acicular transformed β microstructure, or duplex microstructures in which α platelets (equi-axed) are embedded in the colonies of lamellar β (acicular transformed) matrix (Boyer, 1996; Weiss & Semiatin, 1999). Distinctive microstructures witnessed in near-α and α/β Ti alloys are: bimodal, lamellar, and acicular comprise of bimodal distribution of α grains, lamellar α grains and acicular α grains, respectively as typically depicted in Figure 3(a1) - (a3). Recrystallization of α-phase just below β transfuse gives bi-modal microstructure which contains primary equi-axed α grains and lamellar α+β from the prior β grains. As shown in Figure 3, slow cooling from β phase gives the lamellar microstructure and this phase contains no primary α. The rapid cooling from β phase, on the other hand results in acicular microstructure that consists of martensite (Verdhan et al., 2015). The α-platelets are generally found in two basic arrangements one is Widmanstatten formation, also called basket weave pattern, or in fully lamellar microstructure in similarly aligned patterns (Ghonem, 2010).

Lamellar microstructure, bimodal microstructure, and acicular microstructures are generally found in rolled, forged, and welded regions, respectively.

Figure 3. Types of microstructures (a1) lamellar, (a2) bimodal, and (a3) acicular(Akram et al., 2019) and (b) a1 Time-Temperature-Transformation for β phase(Yan & Yu, 2015)

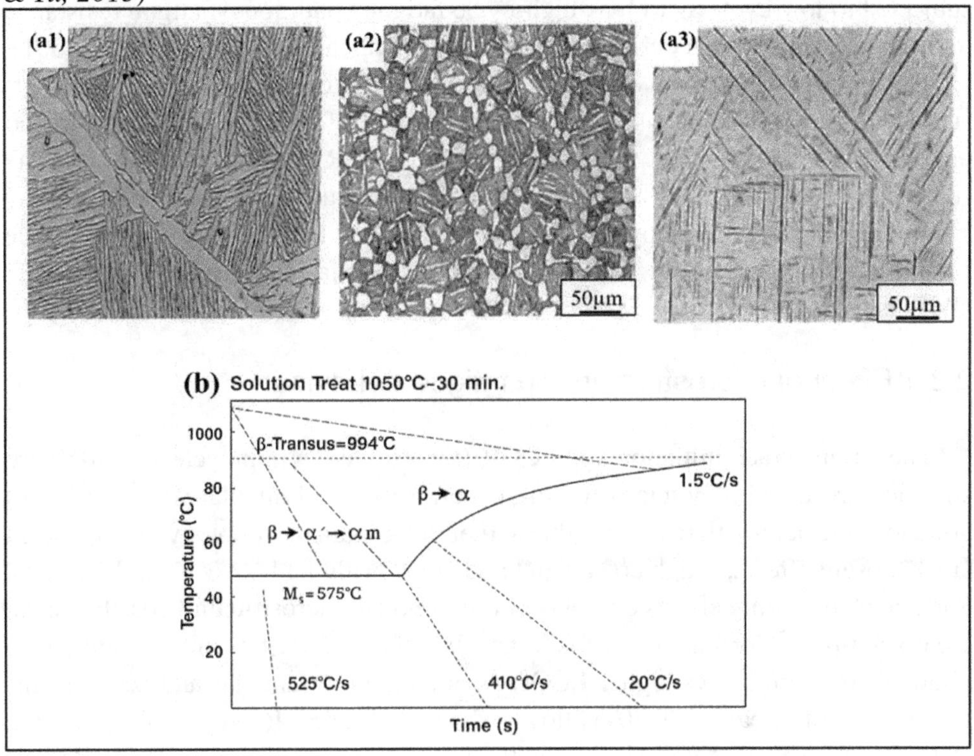

Studies show that these distinct microstructural characteristics significantly affect the fatigue performance of Ti alloys. The α-alloys offer superior creep resistance and good low-cycle fatigue (LCF) resistance, whereas, β-alloys displays superior fatigue strength, higher ductility, fracture toughness and crack initiation resistance (Lutjering, 1999; Polmear, 1989). The commonly used near-α alloys are VT-20 (Verdhan et al., 2015). Timetal-1100 (Ti-6Al-2.7Sn-4Zr-0.4Mo-0.45Si) and Ti6242 (Ti-6Al-2Sn-4Zr-2Mo) alloys (Ghonem, 2010), α/β alloys are Ti-6Al-4V, Ti-6Al-6V-2Sn and Ti-6Al-2Sn-4Zr–6Mo alloys, and near-β alloy is TB6 (Delai, 2011). VT-20 alloy is used in load carrying bulk frame of supersonic fighter aircraft (Verdhan et al., 2015). Timetal-1100 and Ti6242 are widely utilized in high strength and high temperature applications (as high as 760°C) automobile engine valves and compressor disc and impellers, respectively (Ghonem, 2010). TB6 is used in some latest aircraft landing gear structures, helicopter horizontal rotor hub system, support beams due to its better mechanical properties and forging abilities than any other Ti alloy (Delai, 2011). Light weight structural applications such as in aerospace, marine and defense sectors widely employ Ti-6Al-4V alloy due to its structural property benefits (Baragetti et al., 2011). Aircraft parts such as turbine engine blades and discs are made up of γ-TiAl-based alloys (Semenova et al., 2016). These parts are subjected to low-cycle as well as high-cycle fatigue conditions. Fatigue resistance of these materials is a very critical issue for such applications (Lanning et al., 2001).

The damage mechanism of FCGR largely depends on the microstructural characteristics of all these alloys. A significant amount of research has been focused on the relationship between microstructure and FCGR behavior of Ti alloys. The subsequent section presents the effect of microstructure type, grain size, colony size the orientation of colony, loading parameters and holding time on the fatigue performance of these alloys. The summary of fatigue failure of some important Ti alloys is presented in Table 2.

2.3.1 Effect of microstructure on fatigue behaviour

The fatigue crack initiation and FCGR (represented in mm/cycle as da/dN) are significantly affected by microstructural characteristics. Lamellar structure shows good long crack growth resistance due to their long grain morphology that results in crack closure effects, rough crack front and crack bridging between nearby grains. Such crack impeding effects do not occur in bimodal microstructure possibly since the crack tip size for long cracks is larger than their average grain size and hence bimodal structure shows higher FCGR as compared to lamellar and acicular microstructure of a near-α VT-20Ti alloy as shown in Figure. 4(Cvijovic-Alagic et al., 2014; Verdhan et al., 2015) The plot in the figure shows that the FCGR (i.e. da/dN) is higher for bimodal microstructure and the other two microstructure (lamellar and acicular) overlay on each other. However, bimodal structure shows good small crack growth resistance than lamellar structure(Benedetti & Fontanari, 2004).

Figure 4. (a) FCGR (da/dN) vs. the applied stress intensity factor range ΔK, (b) FCG profile for different microstructures (Liang et al., 2016).

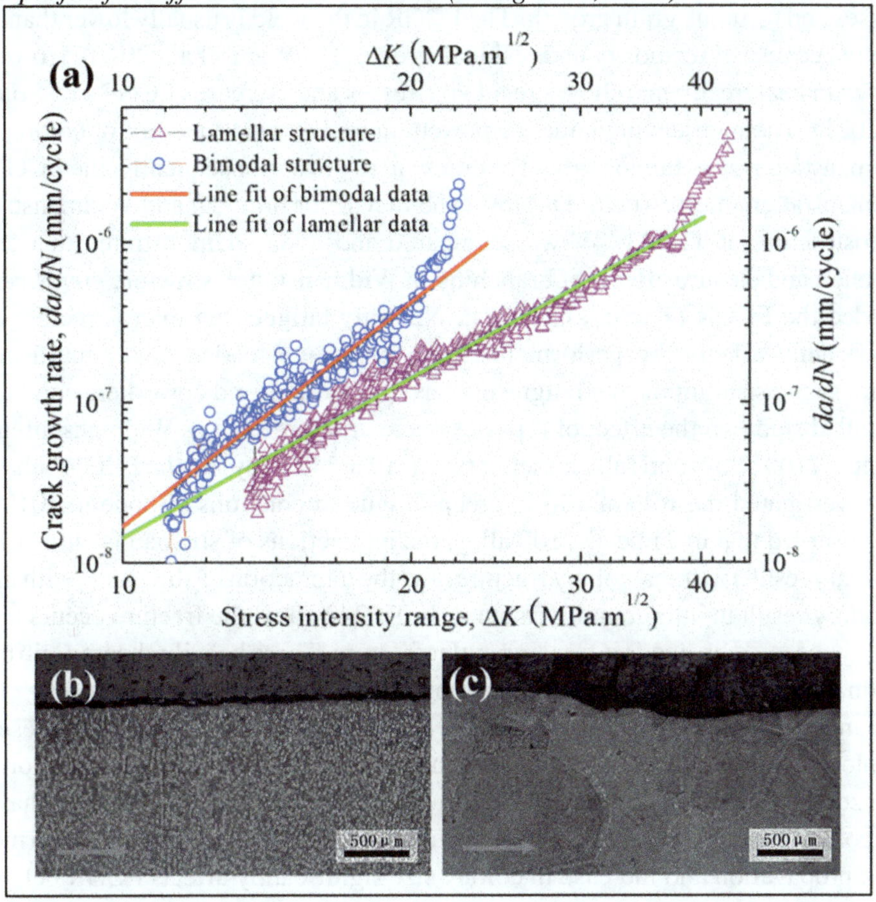

The FCGR follows characteristically different patterns which closely depend on microstructure (typically FCGR pattern for VT-20 alloy is shown in Figure 4(c)). In lamellar structure, the crack propagates and changes its direction from one colony to another with minor deflections striated (as evident from Figure 4(b)). In acicular microstructure, the α grains, which are randomly oriented with in prior β grains, influence the crack path (Figure 4(c)). The fatigue crack grows by striated and intergranular fracture of α grains in lamellar and a circular microstructure, respectively. Both microstructures undergo tortuous FCGR path. However, fatigue crack grows by the cleaving of large primary α grain and FCGR pattern seems to be unaffected to the microstructure in bimodal type and shows a quite straight crack (Verdhan et al., 2015).

In case of fusion welds, the cast microstructure significantly differs from base material (BM) microstructure. The weld structure is marked with high residual stresses and tortuous grain growth. The FCGR in the weld is usually lower than BM mainly because of tortuosity and residual stresses (X. Wang et al., 2010). However, post weld heat treatment relieves residual stresses and increases FCGR (W. Song et al., 2021). The interleaving lamellae present in basket weave pattern of colonies of Widmanstatten structure obstruct the crack tip to grow; consequently, the FCGR of Widmanstatten microstructure is slow. The FGCR for bimodal and Widmanstatten microstructure of Ti-6Al- 4V was compared and it was demonstrated that crack diversion and closure effects of branching in Widmanstatten structure significantly retarded the FCGR (Zuo et al., 2008) . Not only fatigue, the microstructure also significantly affects the performance of components loaded under multi-mode failure such as simultaneous fatigue and creep. Ghomen and coworkers conducted a detailed study on the effect of types of microstructure such as Widmanstatten (in Timetal-1100 alloy) and fully lamellar type (in Ti6242 alloy) on the FCGR behavior and investigated the role of slip in creep-fatigue interactions (Ghonem, 2010). It was observed that in Timetal-1100 alloy, the interactions of slip bands with colony boundary results in trans-colony fracture, and the interaction of slip bands with grain boundary result in intergranular fracture. In Ti6242 alloy, the fracture occurs along heavily sheared slip bands is trans-granular fracture. These heavily sheared slip band within a single colony are characterized by angle between the corresponding crack path and the long axis direction of lamella (Ghonem, 2010). Evidently, the presence of colonies of typical microstructural feature certainly affects fatigue behavior, but the size of colony boundary also found to have significant impact on FCGR behavior. The colony boundaries in duplex microstructure of α/β alloys act as obstacle to micro crack propagation and increase in colony size significantly affects FCGR behavior of these alloys as it increases the effective slip length and increases the micro crack propagation rate. Likewise, grain size also has a significant influence on FCGR, and it is observed that the coarse grains give better FCGR resistance even in crack closure-free locations (W. Song et al., 2021) It was observed that the fine-grained microstructure retards initiation of cracks, whereas coarse-grained microstructure inhibits crack propagation (Ghonem, 2010).

Regardless of FCGR behavior, the fatigue life in case of high-cycle fatigue is majorly influenced by crack initiation. It is reported that porosity and inclusions do not influence fatigue strength; rather, grain structure anomalies cause crack initiation (Takeuchi et al., 2008). It is demonstrated that, in bimodal microstructure, primary α grains with large boundaries provide crack initiation sites and for basket-weave microstructure the sites are long interfaces of α–β lamellae. The fatigue strength of bimodal structure is higher than that of basket weave structure and both the structures show a decreasing S-N curve (Figure 5(a)) (Crupi et al., 2017).

Figure 5. (a) S-N curve during very high cycle fatigue test of Ti-6Al-4V alloy(Y. Wu et al., 2024) (b) Fracture morphology of lug and bolts (Kong & Liu, 2018). Crack depth (mm) vs. Number of cycles since occurrence of initiation of crack in (c) laboratory air and (d) NaCl solution (Baragetti et al., 2011)

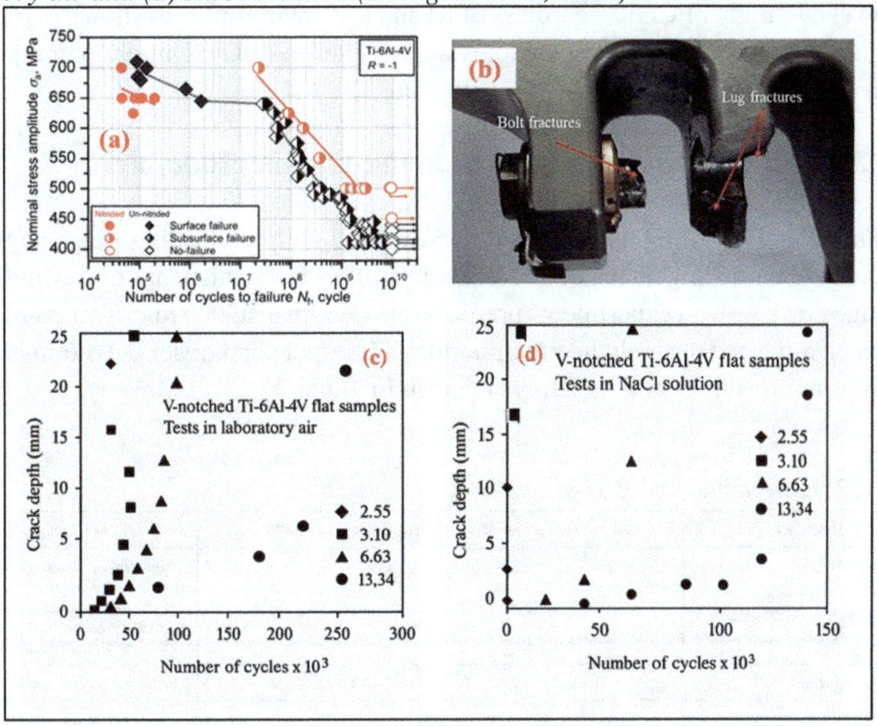

Fretting damage sensitivity is another very important characteristic for the safety of aircraft. High fretting wear and micro-dynamic damage sensitivity may reduce fatigue life by as much as 20-50% (Delai, 2011; Xiao-Bing et al., 2010). The fretting of Ti alloys generally occurs by two mechanisms: (i) fretting wear and (ii) fretting fatigue. A study of fatigue teston the "lug" (TB6 alloy) and "bolts" (steel) of helicopter structural component (Figure 5(b) showed a typical fretting fatigue failure (Kong & Liu, 2018). It was observed that the fretting wear damage of inner surface of hole in the lug results in the initiation of fatigue crack. Ti is highly susceptible to fretting fatigue damage because of its high reactivity, six square structure and low thermal conductivity. During the wear process the Ti easily combines with other elements and gets oxidizes easily too. The adhesion failure occurs because of easy movement of its {1 0 0} slip system (Collins & Tovey, 1972). Studies have reported presence of 32% "O" in the source area of fatigue crack which confirmed that the fretting wear had occurred and has also caused the initiation of fatigue crack (Kong & Liu, 2018).

Wide application of Ti-6Al-4V in aircraft and offshore applications require careful understanding of this alloy in corrosive environment. The corrosion fatigue (its stages as shown in Figure 2(c) behavior of this alloy in air and recirculated 3.5 wt.% NaCl solution with stress concentration factors (Kt) of 2.55, 3.10, 6.63 and 13.34 shows higher FCGR regardless of Kt and test environment as shown in Figure 5(c) and (d).The FCGR of test samples were higher in NaCl solution as compared in air at all the K_t investigated (Baragetti et al., 2011).

2.3.2 Effect of loading frequency on the fatigue properties

The selectivity of either boundary is correlated with the loading frequency and the ensuing hardening on the active slip planes. This mechanism has been validated through crack growth experiments preformed on Widmanstätten microstructure that has been modified through the precipitation of internal slip barriers. A summary of fatigues failures in Ti and its Alloys is given in Table 2.

Table 2. Fatigue failure of Ti alloys

Ti Alloy (Reference)	Test location/material	Fatigue test and Failure type	Failure Location
Ti-4.5Al-2.5Cr-1.2Fe-0.1C (Ti531C) (Nakai et al., 2019)	FSW Weld Joint	High-cycle fatigue failure	Isothermal ω phase in SZ and discontinuous microstructure at the boundary between the SZ and BM were potential Stress concentration sites
TB6 (Kong & Liu, 2018)	Lug and bolts	Fretting fatigue	Fretting wear damage of inner
Ti–5Al– 5Mo–5V–3Cr–1Zr (Huang et al., 2017)	530 mm diameter hot-forged bar of Ti-55531 was heat treated as: (1) 860°C/2 h, air cooling (AC)+600°C/6h, AC to get LM;(2)790°C/2h, AC+600°C/6h, AC to get BM	High-cycle fatigue failure	Lamellar microstructure(LM) and bimodal microstructure (BM) were compared and BM showed higher strength, lower ductility and higher HCF strength.
Ti–6Al–2Zr–1.5Mo–1.5V (VT-20) Near-α (Verdhan et al., 2015)	Rolled, Forged and TIG welded condition, Different microstructure: lamellar, bimodal, acicular	Decreasing and increasing applied stress intensity factor range ΔK Fatigue crack growth tests at 90 HZ.	Grain boundary (GB) α promotes crack initiation in LM than that in BM. Bimodal microstructure showed higher FCGR due to the cleavage of primary α.
Ti-6Al-4V (Crupi et al., 2017)	Different microstructure: basket weave and bimodal	Very High-cycle fatigue failure	Bimodal microstructure showed higher fatigue strength than basket weaving microstructure.

continued on following page

Table 2. Continued

Ti Alloy (Reference)	Test location/material	Fatigue test and Failure type	Failure Location
Timetal-1100 alloy (Ghonem, 2010)	Timetal-1100 alloy with Widmenstatten microstructure under varying temperatures and loading frequencies	Influence of Widmanstätten and fully lamellar microstructures on the crack growth behavior under varying loading frequency and temperature	In Widmanstätten microstructure, the interactions of slip bands with colony boundary and grain boundary result in trans colony and intergranular fracture, respectively. The predominant trans-granular fracture in lamellar structure, occurs along shear slip bands is categorized in terms of angle between the crack path and the lamella. The parallel- and transverse- to-lamella crack directions were found to be governed by slip along the a 1 and a 2- prism directions of the α phase
TB6 high strength titanium alloy (Kong & Liu, 2018)	Effect of surface fretting wear damage on the lug's performance.	Fatigue crack growth	Fretting wear damage to the inner surface of lug hole was the main reason for the fatigue crack initiation
Ti-6Al-4V (Baragetti et al., 2011)	Dog bone with V- notched flat	Step-loading fatigue test air and re-circulated 3.5 wt.% NaCl solution	High crack growth rates were observed regardless of K_t and test environment Crack growth rates in NaCl solution were higher than they were in air A fatigue limit reduced to - 28% for the samples tested in NaCl solution at K_t lower than 6.63
Grade 4 commercial purity (Semenova et al., 2016)	Hot-rolled 12 mm diameter rods subjected to ECAP-C followed by drawing	Rotating bend fatigue tests at up to $N=10^7$ cycles at constant stress amplitude and symmetrical loading (R=-1), at a frequency, f = 50 Hz	The fatigue resistance of UFG Ti increased to 610MPa due to high strength and ductility (UTS = 1290 MPa with elongation = 13%). This value is comparable with the fatigue resistance of the titanium alloy Ti–6Al– 4V.

On a concluding note, among light metals and their alloy systems, there often remains competing interest among the Ti, Al and Mg alloys from the application point of view. Among all three, Ti possesses excellent corrosion resistance, but structurally Ti and Mg are both hcp, and the Ti alloys have far superior high temperature performance, and it is by far very costly among the three. The alloys of Al and Mg, on the other hand share similar technical and economic features. The FCGR and fracture toughness of Mg alloys are lower than Al alloys, yet some Mg alloys exhibit comparable fatigue life to that of Al alloys.

2.4 Modeling and Simulation for Fatigue Analysis of Light Alloys

The design against fatigue is the most critical factor in safety of equipment and systems mainly so because the fatigue failure stress is several order lower than the classing design stress. This fact makes the design conservative and enhances the weight. The fatigue at riveted and bolted joints, dovetail and mortise-tenon joints in BLISKS used in gas turbine, fan and compressor place in aero-engine are critical example for typical fatigue failure. More so, the analysis of fatigue failure location is complex due to sharp cross-sectional variations, contact stress at the interface and environmental issues. Also, the segregation and separate analyses of the two stages of fatigue failure i.e. crack initiation and crack propagation is and challenging (Golden & Grandt, 2004). The modern ICT tools especially modeling, analysis and simulation platforms such as ABACUS, AnSYS, MATLAB, etc. provide opportunity to visualize the factors affecting fatigue failure in various stages before failure. The outcome such analyses not only save time, effort, material and cost but also generate better understanding the factors that are critical to various stage of fatigue failure. (Fadag et al., 2008)(Verma et al., 2022).

These systems though have limitations mainly due to the conservative assumptions made during synthesis of the studies, selection of shape and size of elements, governing equations etc. The shape and size of element significantly affect the computation time and computing system configuration. These aspect need to be critically examined by effective experimental validation. In any case, an analysis by modeling and simulations before actual design significantly improves the systems life, safety, saves on materials and cost. Important fatigue factors and parameters such as the stress intensity factor, crack growth rate, prediction of fatigue life of the components, can be computed by modelling and simulations with ease and in time/effective manners. Studies have reported the effects of contact load, cyclic stress, and effect of initial crack length on crack propagation behavior for Ti-alloys used in aircraft and aerospace (Fadag et al., 2008).

3. METHODS FOR IMPROVING MECHANICAL PROPERTIES AND FATIGUE LIFE OF THE LIGHTWEIGHT ALLOYS

The damage due to fatigue grows in sequential stages involving (Frunz verde et al., 2010): microstructural asperities/existing tiny crevice acting as crack initiation site and nucleate microscopic crack; Progression of crack and its coalescence with enroute microscopic asperities develop into a macro-crack. The macro-cracker undergoes a stable crack growth up to a critical size which causes structural instability leading to eventual fracture. Each stage of crack growth, and the rate of its advance, can be strongly affected by several mechanical, microstructural and environmental factors. This fact indicates that all factors responsible for each stage, if controlled, can in turn be used to control fatigue performance. Evidently, the routes through which a material is processed can induce favorable microstructural changes can improve the fatigue performance of materials (Bhuiyan et al., 2012). Literature reports of considerable improvement in fatigue behavior of Ti alloys (including Cp-Ti and Ti-6Al-4V) through microstructural modification by SPD such as ECAP, multistep forging, hydrostatic extrusion etc. (Bagherpour et al., 2019; EC et al., 2019). The forthcoming sections outline the effects of important processing routes on the fatigue performance of materials.

3.1 Surface Treatment

Notably, the cyclic loading conditions in fatigue often subject to the maximal stress at the surface of the component, consequently, the surface imperfections initiate the cracks at surface under the working stress. Like friction and corrosion, the fatigue is also sensitive to surface condition of the component. The characteristics such as surface roughness, surface hardness, surface defects produced during primary manufacturing etc. have a great effect on the component's performance. There are many surface treatments procedures which can influence these surface conditions/imperfections positively and enhance the fatigue performance of the product. The shot peening (SP) and laser shock peening (LSP) are known to induce the compressive stress on the surface, enhance surface hardness and result in the closure of surface cracks and defects. Finishing and super-finishing improves the surface finish and reduce the size of roughness asperities. Additionally, in some cases the in-service conditions damage also leads to cracks which propagate and cause fatigue failure. In particular, the fan and compressor blades in aero-engines are exposed to impact of foreign particle present in air cause foreign object damage (FOD). The FOD in aircraft components such as wings and blade cause the high cycle fatigue failure (Nie et al., 2014)(Nie et al., 2013).

The shot peening is based on the impacting the spherical shots made of metal, glass or other substances on the surface on which the shot peening is done. The impact of the shot created localized plastic deformation which due to residual stresses increases the surface hardness. The LSP utilizes pulsed laser beam shot impinging on the surface of material. The region which LSP is done is (termed as opaque ablating layer) is opaque to the laser and the opaque layer is submerged/covered by a layer of materials which is transparent such as water or glass. The LSP is performed as a series of laser pulsed shot which are impinged over the surface with specified laser poser, shot pitch and pulse duration. The laser during LSP passes through the transparent layer and strikes the ablating layer. The opaque layer (i.e. the work surface) absorbs the highly focused very short duration laser pulse creating intensity of several GPa over a small impacting region. This vaporizes a small volume of surface materials creating plasma which expands rapidly and develops shock wave. The high pressure shock wave plastically deforms the region of impact and generates compressive residual stress and microstructure changes. Thus the shot peening and LSP result in increasing the surface hardness, closure of asperities and positively affect the fatigue performance (Nie et al., 2014). There are other techniques to impart the surface compressive stresses targeted to improve the fatigue performance, these include similar and other methods such as submerged water jet peening (SWJP), cladding by hard materials, surface hardening through heat treatment etc. (G. Wang et al., 2024)(Lone et al., 2024)

3.1.1 Roller Burnishing and Peening

The severity of stress is maximum on the surface. Consequently, the surface of components is most adversely affected under fatigue. Surface modifications via treatments such as peening (shot/laser shock/water/ultrasonic/roll peening etc.), SPD, and burnishing etc., are directed to improve the fatigue performance of materials (Wagner & Wollmann, 2008). The peening treatment induces plastic deformation on the surface and significantly enhances dislocation density. The cold work and surface compression are produced during the burnishing treatment. These changes have significant effects on fatigue performance. Cold work retards the crack nucleation significantly and, the crack propagation characteristic also gets retarded. Compressive residual stresses induced during burnishing significantly retard the rate of crack propagation (Leverant et al., 1979). The manner of how the burnishing is performed decides the outcome on the material's fatigue performance. While both the peening and burnishing improve the surface finish and minimize surface asperities, which otherwise act like existing micro/macro crack and provide favorite crack nucleation sites. As opposed to shot peening, burnishing results in very smooth surfaces.

Figure 6. (a) Deep rolling- principle of operation (Wonga et al., 2014) and (b) roller burnishing (Kundan & Prasad, 2015).

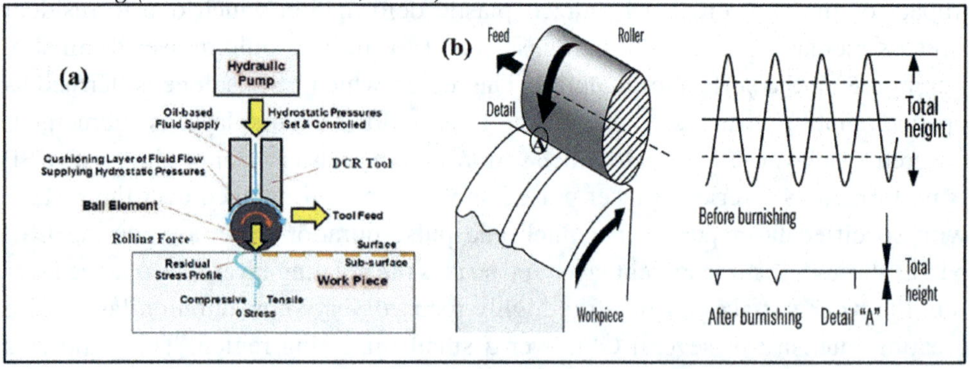

3.1.2 Deep Rolling

Deep rolling (schematically shown in Figure 6 (a) is the process of compressing the surface by a disc or ball shaped roller to create plastically deformed track along the path traced by the roller. Pressure exerted by the roller creates residual stress on and near the surface which decides whether the fatigue performance will enhance or deteriorate. The deep rolling process parameters which play a critical role on the surface characteristics include shape and profile of deep rolling tool, deep rolling force, number of passes and the feed rate. Too high pressure has a risk of surface crack which severely reduces the fatigue performance, whereas too low pressure may not produce significant effect on the fatigue strength (Delgado et al., 2016). Thus, careful control of deep rolling process parameters is essential during surface treatment. A treatment with optimum pressure can induce residual stress up to a certain depth high enough to not get relieved and be retained at temperatures as high as 450°C in case of Ti alloys. Studies have shown that in comparison to deep rolling, the performance of shot peening produces better results for low cycle fatigue. Whereas, for high cycle fatigue the deep rolling is found to be far more beneficial than shot peening (Majzoobi et al., 2009). The strategies such as multiple passes and the preheat temperature during deep rolling are also selected based on the desired results. The elevated temperature deep rolling, for example, significantly improves the fatigue strength in comparison to room temperature treatment(Delgado et al., 2016).

3.2 Severe Plastic Deformation

The SPD is a technique to produce ultra-fine-grained (UFG, grain size range 100 nm to 1 μm) and nano-crystalline (average grain size is below 100 nm) microstructure of metallic materials (Bagherpour et al., 2019). There are various methods through which SPD has been performed. These methods include ECAP, accumulated roll bonding (ARB), high pressure torsion (HPT) and multistep forging (MF) etc. (Liu et al., 2025; Z. Wang et al., 2025; Xin et al., 2025; Yadav et al., 2025). Recently, much attention has been directed towards the behavior of UFG and nano-structured materials under cyclically varying loads (Li et al., 2019). The effect of SPD on fatigue performance of materials is presented in this section.

ECAP is a bulk processing route in which the material is pressed to pass through a channel of uniform cross section with the angle between entry and exit (known as channel intersection angle) is generally 90°. The materials undergo simple shear as it passes through the bend. The multiple passes results in an increase in strength (Semenova et al., 2016).

The ECAP can be performed at room temperature or in hot condition below recrystallization temperature and it generates cumulative logarithmic strains of the order of 4–6 resulting in UFG structure. Unlike coarse grain, in case of UFG materials, the compensation of cyclic strains by dislocation arrangements during fatigue is not possible. Room temperature ECAP may, however, result in surface and internal cracks, and adversely affect fatigue performance. In case of age-hardened materials such as 2XXX and 7XXX series Al alloys, the effect of initial condition of the materials affects differently to the same ECAP treatment. During the processing of age-hardenable AA2011 by ECAP, it was demonstrated that a solution-treated state gives better strength in comparison to T6 state. The ECAP pressing in T6 found to provide better ductility and an inferior cyclic loading performance (Namdar & Jahromi, 2015). The UFG formed during ECAP is found to show different behavior during high cycle fatigue (HCF) and low cycle fatigue (LCF). In the former case, the ECAPed parts showed higher endurance. In the latter case a cross-over of the fatigue plots of coarse and UFG were observed.

3.3 Accumulated roll bonding (ARB)

The ARB can produce laminated structures or composite laminated structures and simultaneously produce UFGs. This aspect makes the ARB process a promising technology for light weight material designs (Haus"ol et al., 2010). Whereas the UFGs cannot accommodate large number of dislocations during cyclic loading, the coarse grain can. Interestingly, in case of UFGs produced in mono-metallic Cu via ARB, grain coarsening and an associated softening during cyclic deformation was observed. The coarsening and consequently softening too is found to be suppressed significantly with the presence of foreign atoms (e.g., alloying) even in very small amounts. The composite laminated structure (CLS) with interlayer in-between produced by ARB is reported to possess significantly higher fatigue life in comparison to mono laminated structure (MLS) (Wittenauer & Sherby, 1987). This is because the deformation of softer interlayer in the CLS delaminates blunts the crack propagation front. Thus, a simultaneous UFG formation coupled with suitably layered composite structure, produced by ARB, can provide good fatigue resistance. The design of layered structures can be articulated by the number of ARB passes, layer thickness and interface between layers.

It can be observed from the preceding literature that various processes that lead to UFG and nano-structured materials can enhance the material's yield strength via Hall-Petch effect but only at the cost of tensile ductility. Furthermore, grain-refinement alone may increase or reduce the fatigue performance which is influenced by the type of fatigue condition and the evolved microstructure of material.

3.4 Hydrostatic Extrusion

Hydrostatic extrusion (HE) process can apply very high isostatic stress and produces UFG with considerable microstructural changes. In addition to refinement, it also results in void closure and densification. It is reported that when commercially pure Ti (Ti-Cp) is subjected to a total strain of 3.21 by HE, produced average grain size of 72 nm (Garbacz et al., 2011). The authors reported an increase in strength by 120% and fatigue limit by 60%. The HE processing of commercially Ti-Cp is found to have produced complex microstructure comprising of super fine lamellae with elongated grains/subgrains oriented along extrusion direction. High anisotropy was found along and normal to the extrusion direction, with strengthening and work-hardening along and normal to extrusion direction respectively. This anisotropy was found to reduce fatigue performance along the extrusion direction (EC et al., 2019). The fatigue strength of materials can be improved by grain refinement by HE processing. It was found during unnotched or smooth test samples that the fatigue strength of Ti-Cp improved by 60% after 7 passes of HE. This was attributed to severe grain refinement from 12.4 μm to 72 nm (Garbacz et al., 2011). However, the notched specimen shows reduced fatigue life which was attributed to notch sensitivity of nano-structured material or the effect of residual stresses generated during HE process (Estrin & Vinogradov, 2010; Woo-Jin et al., 2006).

3.5 Friction Stir Processing (FSP)

The evolution of mechanical properties in the materials bears the consequence of its microstructure. Every fabrication technology produces a characteristic microstructure which depends on its process parameters. Like the processing given in the preceding sections, every process adapts the material properties differently. In the FSP, typical mechanical property evolution is obtained. It becomes essential to understand the influence of processing parameters of the processes on the microstructure(Z. Y. Ma, 2008).

The FSP is a processing technology which utilized in put material produced by other primary processes upstream (such as casting, rolling, extrusion, powder metallurgy and additive manufacturing etc.). The processed region comprises of a heat affected zone (HAZ), thermo- mechanically affected zone (TMAZ) and stir zone (SZ). The SZ is marked with dynamically recrystallized (DRX) equiaxed UFG structure. The TMAZ possesses partially recrystallized structure having elongated grains. It is possible to tailor the zones and their structure with suitable selection of FSP process parameters(F Khan et al., 2019). The pre-existing void, defects, internal microcrack etc. get annihilated and homogenized significantly. Consequently, considerable enhancement in properties is achieved(Jana et al., 2010). The reduction and near elimination of defects results in significant improvement in fatigue properties as and the defects (such as voids. Porosity, inclusion, dendrites etc., which act as crack nucleation sites and propagation facilitators) gets finely subdivided and dispersion during processing to become benign to fatigue damage (Couper et al., 1990). Typically, fatigue life enhancement of the order of five times (in case of hypoeutectic Al–Si–Mg alloy) cast alloy has been demonstrated because of repair of casting imperfections(Jana et al., 2010). The inferences were drawn in the light of SEM+EDS analyses showed that the crack initiation, in the as investment cast alloy, was associated with porosities. In case of FSP performed on casting, it was associated to Si particle-to-matrix interface. The authors confirmed the results by matching the size ratio of defects to the paris crack growth equation. They found one order higher magnitude of crack growth rate in the cast condition than in the FSPed condition. Similar results were also observed for NiAl bronze and AZ91 (Mg alloy) in which case the fatigue performance of the material showed a clear enhancement in case of FSPed samples viz-a-viz as cast samples (Cavaliere & De Marco, 2007; Fuller et al., 2007).

3.6 Additive Manufacturing of Aerospace Alloys

The additive manufacturing (AM) route is gaining popularity due to benefits such as ability for near net shape manufacturing resulting is waste reduction without the use of tooling, mould and dies. The traditional AM routes such as Laser Engineered Net Shaping (LENS), Selective Laser Melting (SLM) and Electron Beam Melting (EBM) are effective and has been well adopted for commercial industrial production mainly for the custom products. The ability of these methods to produce the cellular structures, materials with graded properties has made them popular too. But, these methods suffer from the low throughput rate, and consequently not suited for mass production or large sized bulk products. The recent entrant to the AM technology space is wire and arc additive manufacture (WAAM). The WAAM is an arc welding based layer-by-layer deposition process and has the ability to produce with several tens of times that of laser/electron beam based methods. The WAAM find accelerated applications in the aircraft/space, aerospace and rapid transport system applications. To be specific, manufacturing of parts of landing gears and aircraft engines parts are being explored (Y. Ma et al., 2010). The manufacturing and crack repairs of strategic materials such as Ti alloys by AM assumes critical importance as the size of components is large (L. Wang et al., 2023). The reactivity of Ti with atmosphere and its phase transformation during processing temperature are real challenge for AM. The poor conductivity and thermal expansion coefficient makes it difficult to machine. The AM defects such as voids, oxide inclusions, interface bonding, variation of build microstructure in the build direction offer additional challenge due to thermal characteristics of Ti alloys. The structural components used in aerospace are large and are subjected to fatigue loads. This makes the fatigue growth rate (FCG) of AMed Ti-alloy parts very important. The presence defects critically affects fatigue behavior of AMed Ti–6Al–4V parts. An experimental investigation reported that the crack initiation sight were present at the defects. The size and location of the defect greatly influenced the failure under fatigue. Post AM fabrication a hot-isostatic-pressing (HIP) treatments resulted in the closure/repair of defects and greatly enhanced the fatigue performance(Wycisk et al., 2015).

4. SUMMARY

The growing applications of light materials such as Mg, Al, Ti based alloys have shifted the onus to better understanding of these materials under vulnerable conditions. Fatigue is regarded as one of the most critical situations as the materials fail at considerably lower values of stress than their static strength without warning. The present structured review aimed to provide status related to materials, their applications, microstructure, and susceptibility under fatigue conditions. The processes that produce microstructure and properties to better resist conditions of fatigue are also given. Thus, the present review provides useful insights for all stake holders with interest in light materials and fatigue at a single platform. The summary of major facts outlined in this article are given below

4.1 Al-Li Based Alloys

The addition of Li in Al-Cu reduces density and produces δ' (Al_3Li) which has effect like T1 and θ'. Recently 2nd and 3rd generation Al-Li alloys and ultra-light Al-Be, Al-Be-Mg alloys etc. find critical applications in space and aerospace. The alloying of Al with Li makes it 7-10% lighter improves corrosion resistance, formability and low temperature toughness. Despite remarkable properties, however, these alloys also suffer from low ductility, issues with fracture toughness, poor short-transverse properties and significant fatigue crack extension rates especially. AA2020, AA2050, AA2055, AA2060, AA2065, AA2076, AA2090, AA2091, AA2096, AA2098, AA2099, AA8090, etc. are the Al-Li alloys widely used in aircraft applications.

The addition of Mg in the Al-Li system while reducing density also contributes to strength whereas Zn though heavier but contributes to solid solution strengthening and improves corrosion resistance. Addition of Mn controls grain size and microstructural texture that improves fracture toughness. The 2nd generation alloys (e.g. AA8090-T86) show significant anisotropy in comparison to 3rd generation alloys (e.g. AA2199). AA2199-T8E80 alloy is 8% lighter and better FCGR performance in comparison to AA2024 HDT-T351. Anisotropy leads to unpredictable fatigue, crack propagation and thus creates design difficulty. The anisotropy is attributed to the presence of grain boundary precipitates and clustering of precipitates.

4.2 Mg-Based Alloys

The apprehension of Mg for its inflammability restricted its but of late Mg has proven to be safe. The basal planes and less slip systems in the hcp make critical shear 40 times less than the prismatic slip. Addition of rare earth metal to Mg forms "WEXX and EVXX" series alloys. The AZXX (AZ61, AZ80, ZK30, ZMC711) series are high strength alloys whereas AZ31, ZM21 etc. are moderate strength alloys.

In Mg-alloys pre-crack initiation is associated with microstructural changes including dislocation generation. The precipitates in Mg alloys may be either coherent or incoherent. At or near room temperature the dislocation movement about the basal slip planes is greatly favored whereas at elevated temperatures the prismatic slip planes become favorable. Thus, structural integrity, alloys composition, age-hardening and strain-hardening treatment all affect the fatigue response in Mg alloys.

Mn (even in 0.4%) improves the strength of Mg-alloys and up to 0.58% the alloy's fatigue properties are also improved. Beyond 0.58% excess Mn forms large intermetallic which act as stress raiser and preferred crack initiation site, consequently, reduces fatigue performance.

The anisotropy of Mg makes room temperature (and warm forming below 250°C) formability difficult. The addition of rare-earth and elements such as Ca, Be favor weak texture and improve formability.

4.3 Ti-Based Alloys

The commercially pure (Cp)Ti-alloys changes from alpha(α) hcp structure to beta (β) bcc at 883°C. Ti-alloys are classified as α, near α or super α, α/β, β and γ alloys. Al is primary alloying element in α-alloys, on adding small amount of β stabilizing elements (Mo and V) the alloys respond well to thermo-mechanical treatments. They can be processed to the α/β and β phases to attain either equiaxed α, acicular transformed β microstructure, or duplex microstructures. Slow cooling from β gives lamellar structure containing no primary α. The rapid cooling from β results in martensitic acicular structure. Lamellar microstructure, bimodal microstructure, and acicular microstructures are generally found in rolled, forged, and welded regions, respectively.

The α-alloys offer superior resistance to creep and low-cycle fatigue, whereas, β-alloys possess superior fatigue strength, higher ductility, fracture toughness and crack initiation resistance. VT-20, Timetal-1100 and Ti6242 are common near-α alloys, Ti-6Al-4V, Ti-6Al- 6V-2Sn and Ti-6Al-2Sn-4Zr–6Mo are α/β alloys, TB6 is near-β alloy is. VT-20 alloy is used in load carrying bulk frame of supersonic fighter aircraft, Timetal-1100 and Ti6242 find application in high strength and high temperature (as high as 760°C) automobile engine valves, compressor disc and impellers. TB6 is used in landing gear parts, helicopter rotor hub system. Aircraft parts such as turbine engine blades and discs are made of γ-Ti Al alloys. These parts are subjected to low and high cycle fatigue. Fatigue resistance of these materials is a critical issue for such applications.

The fatigue crack initiation and FCGR are significantly affected by microstructure of Ti alloys. Lamellar structure offers crack closure effect in large crack, whereas, impeding effects for large crack do not occur in bimodal microstructure and bimodal structure shows higher FCGR. However, bimodal structure shows good small crack growth resistance than lamellar structure.

It is important to address all aspects to limit the harm due to fatigue and enhance the life and reliability of the components and system. A snippet of processing routes that alter materials structure which safeguard against fatigue are also summarized.

The effect of fatigue is most severely felt at the surface, consequently surface modifications peening (shot/laser shock/water/ultrasonic/roll peening etc.)/burnishing enhances the fatigue performance. The plastic deformation on the surface induced by peening enhances dislocation density and the burnishing produces the compressive cold work on surface. The peening retards the crack nucleation rate and burnishing retards the rate of crack propagation. The peening and burnishing both improve the surface finish and reduce surface asperities which otherwise act as preexisting crack.

The plastic deformation rendered on the surface and near surface by deep rolling affects fatigue performance. But the deep rolling pressure is a controlling factor, if it is too high it may generate surface or near surface crack which on the contrary hard the fatigue life. When the deep rolling is performed under optimum pressure the residual stress field may penetrate enough depth and is safe from getting relieved at as high as 450°C in Ti alloys. In comparison to deep rolling the shot peening produces better results for low cycle fatigue. Whereas, for high cycle fatigue the deep rolling is found to be far more beneficial. Elevated temperature deep rolling is found to significantly improve the fatigue strength in comparison to room temperature treatment.

The processing routes such as ECAP, ARB and FSP produce heavy microstructural changes including UFG through SPD, the multiple passes of ECAP significantly enhance strength but the elongation decreases. The ECAP can generates accumulated logarithmic strains of the order of 4–6 resulting in UFG. Room temperature ECAP may result in surface and internal cracks. ECAP, when performed on solution treated age-hardenable results in better strengthening than age-hardened condition. ECAP on age-hardened condition improves ductility and lower cyclic loading performance. ECAP processed materials show higher endurance limit for HCF whereas when subjected to LCF a cross-over of the fatigue plots of coarse and ultrafine grains are observed.

The processing via ARB produces UFG in the laminated morphology. When ARB is performed on dissimilar stacked materials a CLS can be produced. The CLS is found to possess significantly higher fatigue life in comparison to MLS. UFG formation coupled with suitably designed composite layered structure produced by ARB can provide good fatigue resistance.

The materials consolidation due to void closure, densification and refinement via HE derives excellent fatigue benefits. An increase in strength by120% and fatigue limit by 60% by HE on Ti alloys has been reported. A superfine lamella with elongated grains/sub- grains were found during HE of commercially pure Ti. The anisotropy produced because HE reduces fatigue performance along with the extrusion direction.

The FSP produces typical characteristic zones that comprise dynamically recrystallized UFGs in the central SZ and partially recrystallized elongated grains in the TMAZ. It is possible to tailor the zones and their structure with suitable selection FSP process parameters. The asperities, defects the existing microstructure gets refined significantly and homogenized; and consequently, considerable enhancement in properties is obtained. The reduction and near elimination of defects brings in significant improvement in fatigue properties as asperities and defects that get eliminated and fatigue damage is avoided. The crack growth rate was found to reduce one order of magnitude when FSP was performed on the investment casting.

DECLARATIONS

Availability of Data and Material

The original contributions presented in this study are included in the article. Further inquiries can be directed to the corresponding authors.

Funding

This work was supported by the Deanship of Scientific Research, Vice Presidency for Graduate Studies and Scientific Research, King Faisal University, Saudi Arabia. Grant No. KFU252099

Authors' Contributions

Conceptualization, N.G. and A.N.S.; methodology, A.S., and N.G.; validation, N.G., A.N.S., and A.S.; formal analysis, M.F.K., and S.Q.M.; investigation, A.S. and M.E.A.M; resources, M.F.K., S.Q.M., and M.E.A.M.; data curation, M.F.K., and S.Q.M.; writing—original draft preparation, N.G., A.N.S., and M.F.K.; writing—review and editing, N.G., M.E.A.M., M.F.K., and A.N.S., and M.E.A.M; visualization, S.Q.M., and A.S.; supervision, A.N.S and M.E.A.M; project administration, A.N.S and M.E.A.M.; funding acquisition, M.F.K., S.Q.M., and M.E.A.M.

Acknowledgements

The authors thank the Deanship of Scientific Research, Vice Presidency for Graduate Studies and Scientific Research, King Faisal University for supporting this work.

Conflict of Interest

The authors declare that they have no known competing financial interests or personal relationships that could have appeared to influence the work reported in this paper.

REFERENCES

Akram, J., Pal, D., & Stucker, B. (2019). Establishing flow stress and elongation relationships as a function of microstructural features of ti6al4v alloy processed using slm. *Designs*, *3*(2), 1–11. DOI: 10.3390/designs3020021

Bagherpour, E., Pardis, N., Reihanian, M., & Ebrahimi, R. (2019). An overview on severe plastic deformation: Research status, techniques classification, microstructure evolution, and applications. *International Journal of Advanced Manufacturing Technology*, *100*(5-8), 1647–1694. DOI: 10.1007/s00170-018-2652-z

Baragetti, S., Cavalleri, S., & Tordini, F. (2011). Fatigue behavior of notched Ti-6Al-4V in air and corrosive environment. *Procedia Engineering*, *10*, 2435–2440. DOI: 10.1016/j.proeng.2011.04.401

Benedetti, M., & Fontanari, V. (2004). The effect of bimodal and lamellar microstructures of Ti–6Al–4V on the behaviour of fatigue cracks emanating from edge-notches. *Fatigue & Fracture of Engineering Materials & Structures*, *27*(11), 1073–1089. DOI: 10.1111/j.1460-2695.2004.00825.x

Bhuiyan, M. S., Mutoh, Y., & McEvily, A. J. (2012). The influence of mechanical surface treatments on fatigue behavior of extruded AZ61 magnesium alloy. *Materials Science and Engineering A*, *549*, 69–75. DOI: 10.1016/j.msea.2012.04.007

Boyer, R. R. (1996). An overview on the use of titanium in the aerospace industry. *Materials Science and Engineering A*, *213*(1-2), 103–114. DOI: 10.1016/0921-5093(96)10233-1

Cavaliere, P., & De Marco, P. P. (2007). Superplastic behaviour of friction stir processed AZ91 magnesium alloy produced by high pressure die cast. *Journal of Materials Processing Technology*, *184*(1–3), 77–83. DOI: 10.1016/j.jmatprotec.2006.11.005

Chamos, A. N., Pantelakis, S. G., Haidemenopoulos, G. N., & Kamouts, E. (2008). Tensile and fatigue behaviour of wrought magnesium alloys AZ31 and AZ61. *Fatigue & Fracture of Engineering Materials & Structures*, *31*(9), 812–821. DOI: 10.1111/j.1460-2695.2008.01267.x

Collins, J. A., & Tovey, F. M. (1972). Fretting fatigue mechanisms and the effect of direction of fretting motion on fatigue strength. *Testing and Evaluation*, 420–424.

Couper, M. J., Neeson, A. E., & Griffiths, J. R. (1990). Casting defects and the fatigue behaviour of an aluminium casting alloy. *Fatigue & Fracture of Engineering Materials & Structures*, *13*(3), 213–227. DOI: 10.1111/j.1460-2695.1990.tb00594.x

Crupi, V., Epasto, G., Guglielmino, E., & Squillace, A. (2017). Influence of microstructure [alpha+beta and beta] on very high cycle fatigue behaviour of Ti-6Al-4V alloy. *International Journal of Fatigue*, *95*, 64–67. DOI: 10.1016/j.ijfatigue.2016.10.002

Cvijovic-Alagic, I., Gubeljak, N., Rakin, M., Cvijovic, Z., & Geric, K. (2014). Microstructural morphology effects on fracture resistance and crack tip strain distribution in Ti–6Al–4V alloy for orthopedic implants. *Materials & Design*, *53*, 870–880. DOI: 10.1016/j.matdes.2013.07.097

Delai, O. Y. (2011). *Research on microstructure evolution and dynamic recrystallization behavior of titanium alloy TB6 and TA15 during β forging*. Nanjing University of Aeronautics and Astronautics.

Delgado, P., Cuesta, I. I., Alegre, J. M., & D'az, A. (2016). State of the art of Deep Rolling. *Precision Engineering*, *46*, 1–10. DOI: 10.1016/j.precisioneng.2016.05.001

Deschamps, A., Decreus, B., Geuser, F. D., Dorin, T., & Weyland, M. (2013). The influence of precipitation on plastic deformation of Al–Cu–Li alloys. *Acta Materialia*, *61*(11), 4011–4021. DOI: 10.1016/j.actamat.2013.03.015

Dwivedi, P., Maheshwari, S., & Siddiquee, A. N. (2021). Issues and Requirements for Aluminum Alloys Used in Aircraft Components: State of the Art. *Russian Journal of Non-Ferrous Metals*, *62*(2), 212–225. DOI: 10.3103/S1067821221020048

Estrin, Y., & Vinogradov, A. (2010). Fatigue behaviour of light alloys with ultrafine grain structure produced by severe plastic deformation: An overview. *International Journal of Fatigue*, *32*(6), 898–907. DOI: 10.1016/j.ijfatigue.2009.06.022

Fadag, H. A., Mall, S., & Jain, V. K. (2008). A Finite Element Analysis of Fretting Fatigue Crack Growth Behavior in Ti–6Al–4V. *Engineering Fracture Mechanics*, *75*(6), 1384–1399. DOI: 10.1016/j.engfracmech.2007.07.003

Frunz verde, D., Muntean, S., M rginean, G., Campian, V., Marşavina, L., Terzi, R., & Şerban, V. (2010). Failure analysis of a Francis turbine runner. *IOP Conference Series. Earth and Environmental Science*, *12*(1), 12115. DOI: 10.1088/1755-1315/12/1/012115

Fuller, C. B., Mahoney, M. W., Bingel, W. H., Calabrese, M., & London, B. (2007). Tensile and fatigue properties of friction stir processed NiAl bronze. *Materials Science Forum*, *539*, 3751–3756. DOI: 10.4028/www.scientific.net/MSF.539-543.3751

Gangil, N., Siddiquee, A. N., & Maheshwari, S. (2020). *Composite Fabrication on Age-Hardened Alloy using Friction Stir Processing*. CRC Press. DOI: 10.1201/9781003003038

Gao, Z., Yang, S., Meng, X., Zhentao, W., & Peng, Z. (2021). Study on fatigue crack growth of electron beam selective melting of titanium alloy. *Materials Research Express*, *8*(9), 96521. DOI: 10.1088/2053-1591/ac2444

Garbacz, H., Motyka, M., Ziaja, W., Lewandowska, M., Sieniawski, J., & Topolski, K. (2011). High cycle fatigue strength of hydrostatically extruded nanocrystalline CP titanium. *Proceedings of the 12th World Conference on Titanium*, 1137–1140.

Ghonem, H. (2010). Microstructure and fatigue crack growth mechanisms in high temperature titanium alloys. *International Journal of Fatigue*, *32*(9), 1448–1460. DOI: 10.1016/j.ijfatigue.2010.02.001

Giummarra, C., Thomas, B., & Rioja, R. J. (2007). New aluminum lithium alloys for aerospace applications. *Proceedings of the Light Metals Technology Conference*, 313.

Golden, P. J., & Grandt, A. F.Jr. (2004). Fracture Mechanics Based Fretting Fatigue Life Predictions in Ti–6Al–4V. *Engineering Fracture Mechanics*, *71*(15), 2229–2243. DOI: 10.1016/j.engfracmech.2003.10.005

Guru, P. R. F., Khan, M. D., Panigrahi, S. K., & Ram, G. D. J. (2015). Enhancing strength, ductility and machinability of a Al-Si cast alloy by friction stir processing. *Journal of Manufacturing Processes*, *18*, 67–74. DOI: 10.1016/j.jmapro.2015.01.005

Haus"ol, T., H"oppel, H. W., & G"oken, M. (2010). Tailoring materials properties of UFG aluminium alloys by accumulative roll bonded sandwich-like sheets. *Journal of Materials Science, 45*, 4733–4738.

Huang, C., Zhao, Y., Xin, S., Zhou, W., Li, Q., & Zeng, W. (2017). Effect of microstructure on tensile properties of Ti–5Al–5Mo–5V–3Cr–1Zr alloy. *Journal of Alloys and Compounds*, *693*, 582–591. DOI: 10.1016/j.jallcom.2016.09.233

Jana, S., Mishra, R. S., Baumann, J. B., & Grant, G. (2010). Effect of friction stir processing on fatigue behavior of an investment cast Al–7Si–0.6 Mg alloy. *Acta Materialia, 58*(3), 989–1003. DOI: 10.1016/j.actamat.2009.10.015

Jaseem, J. I., Immanuel, R. J., Rao, P. N., Khan, F., Sahoo, B. N., Panigrahi, S. K., & Kamaraj, M. (2018). Synergetic effect of cryorolling and post-roll aging on simultaneous increase in wear resistance and mechanical properties of an Al-Cu alloy. *Journal of Tribology*, *140*(November), 1–11. DOI: 10.1115/1.4040162

Khan, F. (2018, July). M., Karthik, G. M., Panigrahi, S. K., & Ram, G. D. J. (2019). Friction stir processing of QE22 magnesium alloy to achieve ultra fi ne- grained microstructure with enhanced room temperature ductility and texture weakening. *Materials Characterization*, *147*, 365–378. DOI: 10.1016/j.matchar.2018.11.020

Khan, F., & Rokkala, U. (2023). Development of high strength and corrosion resistance Mg-Zn-Dy/HA-Ag composite for temporary implant applications. *Materials Letters*, *347*, 134604. Advance online publication. DOI: 10.1016/j.matlet.2023.134604

Khan, N. Z., Siddiquee, A. N., & Khan, Z. A. (2017). *Friction stir welding: dissimilar aluminum alloys.*

Kim, W. J., Hong, S. I., Kim, Y. S., Min, S. H., Jeong, H. T., & Lee, J. D. (2003). Texture development and its effect on mechanical properties of an AZ61 Mg alloy fabricated by equal channel angular pressing. *Acta Materialia*, *51*(11), 3293–3307. DOI: 10.1016/S1359-6454(03)00161-7

Kong, H., & Liu, C. (2018). Analysis of the effect of fretting wear on key structural components of TB6 titanium alloy with actual failure cases. *IOP Conference Series. Materials Science and Engineering*, *439*, 22011. DOI: 10.1088/1757-899X/439/2/022011

Kumar, A. F., Khan, M., Panigrahi, S. K., & Chaudhari, G. P. (2021). Microstructural evolution and corrosion behaviour of friction stir-processed QE22 magnesium alloy. *Corrosion Reviews*, *39*(4), 351–360. DOI: 10.1515/corrrev-2020-0083

Kundan, D. K., & Prasad, E. K. (2015). An experimental work on multi-roller burnishing process on aluminum alloy. [IJAPME]. *International Journal of Advances in Production and Mechanical Engineering*, *1*(4), 24–28.

Lanning, D. B., Haritos, G. K., Nicholas, T., & Maxwell, D. C. (2001). Low-cycle fatigue/high-cycle fatigue interactions in notched Ti–6Al–4V. *Fatigue & Fracture of Engineering Materials & Structures*, *24*(9), 565–577. DOI: 10.1046/j.1460-2695.2001.00411.x

Leverant, G. R., Langer, B. S., Yuen, A., & Hopkins, S. W. (1979). Surface residual stresses, surface topography and the fatigue behavior of Ti-6AI-4V. *Metallurgical Transactions. A, Physical Metallurgy and Materials Science*, *10*(2), 251–257. DOI: 10.1007/BF02817635

Li, J. I., Huang, J. I., Liu, D. Y., Chen, Y. I., Zhang, X. H., & Ma, P. C. (2019). Distribution and evolution of aging precipitates in Al−Cu−Li alloy with high Li concentration. *Transactions of Nonferrous Metals Society of China*, *29*(1), 15–24. DOI: 10.1016/S1003-6326(18)64910-6

Liang, R., Ji, Y., Wang, S., & Liu, S. (2016). Effect of microstructure on fracture toughness and fatigue crack growth behavior of Ti17 alloy. *Metals*, *6*(8), 186. Advance online publication. DOI: 10.3390/met6080186

Liu, R., Xu, S., Xu, C., Sun, K., Ju, X., Yang, X., Pan, Y., Li, J., Ren, G., & Wang, L. (2025). Study on the effect of ultrasonic rolling on the microstructure and fatigue properties of A03600 aluminum alloy. *Journal of Alloys and Compounds, 1010*, 178124. https://doi.org/https://doi.org/10.1016/j.jallcom.2024.178124. DOI: 10.1016/j.jallcom.2024.178124

Lone, N. F., Ali, N., Qazi, A. M., Bajaj, D., & Siddiquee, A. N. (2024). New Developments in Bend Forming of Welded Sheet Metals. In *Comprehensive Materials Processing (Second Edition)* (Vol. 12, pp. 27–40). DOI: 10.1016/B978-0-323-96020-5.00034-0

Lutjering, G. (1999). Property optimization through microstructural control in titanium and aluminum alloys. *Materials Science and Engineering A, 263*(2), 117–126. DOI: 10.1016/S0921-5093(98)01169-1

Ma, Y., Li, J., Lei, J., Tang, Z., Liu, Y., & Yang, R. (2010). Influences of Microstructure on Fatigue Crack Propagating Path and Crack Growth Rates in TC4ELI Alloy. *Chin Shu Hsueh Pao, 46*(9), 1086–1092. DOI: 10.3724/SP.J.1037.2010.00155

Ma, Z. Y. (2008). Friction Stir Processing Technology: A Review. *Metallurgical and Materials Transactions. A, Physical Metallurgy and Materials Science, 39*(3), 642–658. DOI: 10.1007/s11661-007-9459-0

Majzoobi, G. H., Azadikhah, K., & Nemati, J. (2009). The effects of deep rolling and shot peening on fretting fatigue resistance of Aluminum-7075-T6. *Materials Science and Engineering A, 516*(1–2), 235–247. DOI: 10.1016/j.msea.2009.03.020

Md, F. K., & Panigrahi, S. K.F Khan MD. (2018). Achieving excellent superplasticity in an ultrafine-grained QE22 alloy at both high strain rate and low-temperature regimes. *Journal of Alloys and Compounds, 747*, 71–82. DOI: 10.1016/j.jallcom.2018.02.294

Meenia, S., & Khan, F. (2016). Particle refinement and fine-grain formation leading to enhanced mechanical behaviour in a hypo-eutectic Al-Si alloy subjected to multi-pass friction stir processing. *Materials Characterization, 113*, 134–143. DOI: 10.1016/j.matchar.2016.01.011

Mishra, S. F., Khan, M. D., & Panigrahi, S. K. (2022). A crystal plasticity based approach to establish role of grain size and crystallographic texture in the Tension–Compression yield asymmetry and strain hardening behavior of a Magnesium–Silver–Rare Earth alloy. *Journal of Magnesium and Alloys, 10*(9), 2546–2562. DOI: 10.1016/j.jma.2021.08.021

Moshiri, A., Zarei-Hanzaki, A., Anousheh, A. S., Abedi, H. R., Sohn, S. S., Yang, J., & Berto, F. (2021). On the fatigue and dwell-fatigue behavior of a low-density steel and the correlated microstructure origin of damage mechanism. *Journal of Materials Research and Technology*, *15*, 6136–6154. DOI: 10.1016/j.jmrt.2021.10.135

Nakai, M., Niinomi, M., Komine, K., Liu, H., Morisada, Y., & Fujii, H. (2019). High-cycle fatigue properties of an easily hot-workable (α+ β)-type titanium alloy butt joint prepared by friction stir welding below β transus temperature. *Materials Science and Engineering A*, *742*, 553–563. DOI: 10.1016/j.msea.2018.10.103

Namdar, M., & Jahromi, S. A. J. (2015). Influence of ECAP on the fatigue behavior of age-hardnable 2xxx aluminum alloy. *International Journal of Minerals Metallurgy and Materials*, *3*(3), 285–291. DOI: 10.1007/s12613-015-1072-4

Neve, S. L. (2019). *Fire Behaviour of Magnesium Alloys.*

NIE, J. F. (2012). Precipitation and Hardening in Magnesium Alloys. *Metallurgical and Materials Transactions. A, Physical Metallurgy and Materials Science*, *43A*, 3891–3939.

Nie, X., He, W., Li, Q. P., Long, N., & Chai, Y. (2013). Experimental Investigation on Microstructure and Mechanical Properties of TC17 Titanium Alloy Treated by Laser Shock Peening with Different Laser Fluence. *Journal of Laser Applications*, *25*(4), 042001–1–042001–042006.

Nie, X., He, W., Zhou, L., Li, Q., & Wang, X. (2014). Experimental Investigation of Laser Shock Peening on TC6 Titanium Alloy to Improve High Cycle Fatigue Performance. *Materials Science and Engineering A*, *594*, 161–167. DOI: 10.1016/j.msea.2013.11.073

Polmear, I. J. (1989). *Light alloys: metallurgy of the light metals* (2nd ed.). Edward Arnold.

Potzies, C., & Kainer, K. U. (2004). Fatigue of Magnesium Alloys. *Advanced Engineering Materials*, *6*(5), 281–289. DOI: 10.1002/adem.200400021

Prasad, N. E., Srivatsan, T. S., Wanhill, R., Malakondaiah, G., & Kutumbarao, V. V. (2014). Fatigue Behavior of Aluminum–Lithium Alloys. In *Aluminum–Lithium Alloys* (pp. 341–349). Processing, Properties, and Applications. DOI: 10.1016/B978-0-12-401698-9.00011-2

Ravi Kumar, N. V., Blandin, J. J., Suéry, M., & Grosjean, E. (2003). Effect of alloying elements on the ignition resistance of magnesium alloys. *Scripta Materialia*, *49*(3), 225–230. DOI: 10.1016/S1359-6462(03)00263-X

Rioja, R. J., & Liu, J. (2012). The Evolution of Al-Li Base Products for Aerospace and Space Applications. *Metallurgical and Materials Transactions. A, Physical Metallurgy and Materials Science*, *43*(9), 3325–3337. DOI: 10.1007/s11661-012-1155-z

Runge, J. M. (2018). The metallurgy of anodizing aluminum: Connecting science to practice. In *The Metallurgy of Anodizing Aluminum: Connecting Science to Practice*. Springer International Publishing., DOI: 10.1007/978-3-319-72177-4

Sahoo, B. N., & Khan, F. (2018). Microstructural modification and its effect on strengthening mechanism and yield asymmetry of in-situ TiC-TiB2/ AZ91 magnesium matrix composite. *Materials Science and Engineering A*, *724*(March), 269–282. DOI: 10.1016/j.msea.2018.03.060

Semenova, I. P., Polyakov, A. V., Polyakova, V. V., Huang, Y., Valiev, R. Z., & Langdon, T. G. (2016). High-Cycle Fatigue Behavior of an Ultrafine-Grained Ti–6Al–4V Alloy Processed by ECAP and Extrusion. *Advanced Engineering Materials*, *18*(12), 2057–2062. DOI: 10.1002/adem.201500630

Song, W., Wang, P., Wan, D., Qian, G., Correia, J., & Berto, F. (2021). Fatigue crack growth behavior of Ni-Cr-Mo-V steel welded joints considering strength mismatch effect. *International Journal of Fatigue*, *151*, 106389. DOI: 10.1016/j.ijfatigue.2021.106389

Song, Y., Ma, Y., Chen, H., Chen, H., Zhang, T., & Gao, Z. (2021). The effects of tensile and compressive dwells on creep-fatigue behavior and fracture mechanism in welded joint of P92 steel. *Materials Science and Engineering A*, *813*, 141129. DOI: 10.1016/j.msea.2021.141129

Takeuchi, E., Furuya, Y., Nagashima, N., & Matsuoka, S. (2008). The effect of frequency on the giga-cycle fatigue properties of a Ti-6Al-4V alloy. *Fatigue & Fracture of Engineering Materials & Structures*, *31*(7), 599–605. DOI: 10.1111/j.1460-2695.2008.01257.x

Teng, Z., Wu, H., Huang, Z., & Starke, P. (2021). Effect of mean stress in very high cycle fretting fatigue of a bearing steel. *International Journal of Fatigue*, *149*, 106262. DOI: 10.1016/j.ijfatigue.2021.106262

Verdhan, N., Bhende, D. D., Kapoor, R., & Charavarty, J. K. (2015). Effect of microstructure on the fatigue crack growth behaviour of a near-α Ti alloy. *International Journal of Fatigue*, *74*, 46–54. DOI: 10.1016/j.ijfatigue.2014.12.013

Verma, R., Kumar, P., Jayaganthan, R., & Pathak, H. (2022). Extended Finite Element Simulation on Tensile, Fracture Toughness and Fatigue Crack Growth Behaviour of Additively Manufactured Ti–6Al–4V Alloy. *Theoretical and Applied Fracture Mechanics, 117*, 103163. DOI: 10.1016/j.tafmec.2021.103163

Wagner, L., & Wollmann, M. (2008). Shot peening of non-ferrous alloys to enhance fatigue performance. *ICSP-10, Tokyo, Japan.*

Wang, G., Yao, S., Chi, Y., Zhang, C., Wang, N., Chen, Y., Lu, R., Li, Z., & Zhang, X. (2024). Improvement of Titanium Alloy TA19 Fatigue Life by Submerged Abrasive Waterjet Peening: Correlation of Its Process Parameters with Surface Integrity and Fatigue Performance. *Chinese Journal of Aeronautics, 37*(1), 377–390. DOI: 10.1016/j.cja.2023.03.033

Wang, L., Li, Y., Zhou, L., Lou, Y., Liu, S., Zheng, D., & Yi, M. (2023). Progress in Additive Manufacturing, Additive Repair and Fatigue Evaluation of Aviation Titanium Alloy Blades. *Materials Research Letters, 11*(12), 973–1012. DOI: 10.1080/21663831.2023.2275599

Wang, S. C., & Starink, M. J. (2005). Precipitates and intermetallic phases in precipitation hardening Al–Cu–Mg–(Li) based alloys. *International Materials Reviews, 50*(4), 193–215. DOI: 10.1179/174328005X14357

Wang, X., Shi, Q., Wang, X., & Zhang, Z. (2010). The influences of precrack orientations in welded joint of Ti–6Al–4V on fatigue crack growth. *Materials Science and Engineering A, 527*(4-5), 1008–1015. DOI: 10.1016/j.msea.2009.09.031

Wang, Y., & Zhao, G. (2020). Hot Extrusion Processing of Al–Li Alloy Profiles and Related Issues: A Review. In *Chinese Journal of Mechanical Engineering (English Edition)* (Vol. 33, Issue 1). Springer. DOI: 10.1186/s10033-020-00479-7

Wang, Z., Wu, S., Kang, G., Li, H., Wu, Z., Fu, Y., & Withers, P. J. (2021). In-situ synchrotron X-ray tomography investigation of damage mechanism of an extruded magnesium alloy in uniaxial low-cycle fatigue with ratchetting. *Acta Materialia, 211*, 116881. DOI: 10.1016/j.actamat.2021.116881

Wang, Z., Xing, Z., Lei, Y., Hu, B., Hu, J., Li, H., Feng, X., & Kang, G. (2025). A review on cyclic plasticity, damage, and fatigue failure of magnesium alloys. *Journal of Materials Science & Technology*. https://doi.org/https://doi.org/10.1016/j.jmst.2025.02.038

Weiss, I., & Semiatin, S. L. (1999). Thermomechanical processing of alpha titanium alloys – an overview. *Materials Science and Engineering A, 263*(2), 243–256. DOI: 10.1016/S0921-5093(98)01155-1

Westengen, H., & Rashed, H. (2016). Magnesium Alloys: Alloy and Temper Designation System. *Reference Module in Materials Science and Materials Engineering*, 1–3.

Wittenauer, J., & Sherby, O. D. (1987). *Factors influencing the fatigue behavior of ferrous laminates*.

Wonga, C. C., Hartawana, A., & Teo, W. K. (2014). Deep Cold Rolling of Features on Aero-Engine Components. *Procedia CIRP*, *13*, 350–354. DOI: 10.1016/j.procir.2014.04.059

Woo-Jin, K., Chang-Young, H., & Ho-Kyung, K. (2006). Fatigue strength of ultrafine-grained pure Ti after severe plastic deformation. *Scripta Materialia*, *54*(10), 1745–1750. DOI: 10.1016/j.scriptamat.2006.01.042

Wu, L., Li, X., & Wang, H. (2012). The effect of major constituents on microstructure and mechanical properties of cast Al-Li-Cu-Zr alloy. *Materials Characterization*, *171*, 110800. DOI: 10.1016/j.matchar.2020.110800

Wu, Y., He, W., Ma, H., Nie, X., Liang, X., Pan, J., Wang, S., Shang, M., & Cheng, L. (2024). Titanium Alloy Materials with Very High Cycle Fatigue: A Review. In *Materials* (Vol. 17, Issue 12). Multidisciplinary Digital Publishing Institute (MDPI). DOI: 10.3390/ma17122987

Wycisk, E., Siddique, S., Herzog, D., Walther, F., & Emmelmann, C. (2015). Fatigue Performance of Laser Additive Manufactured Ti–6Al–4V in Very High Cycle Fatigue Regime up to 10^9 Cycles. *Frontiers in Materials*, *2*, 72. DOI: 10.3389/fmats.2015.00072

Xiao-Bing, S., Xin-Yi, L., & Ting, Z. (2010). Research progress in fretting-resistance technology of titanium alloys. *Heat Treatment of Metals*, *35*(3), 83–86.

Xin, W., Wang, S., & Yan, Z. (2025). Study on the high temperature fatigue performance of AZ31B magnesium alloy. *Materials Today. Communications*, *43*, 111584. https://doi.org/https://doi.org/10.1016/j.mtcomm.2025.111584. DOI: 10.1016/j.mtcomm.2025.111584

Xu, Z., Tang, G., Tian, S., Ding, F., & Tian, H. (2007). Research of electroplastic rolling of AZ31 Mg alloy strip. *Journal of Materials Processing Technology*, *182*(1-3), 128–133. DOI: 10.1016/j.jmatprotec.2006.07.019

Yadav, J., Gangil, N., Siddiquee, A. N., Ali Mohsin, M. E., Alnamasi, K., Moinuddin, S. Q., Khan, M. F., & Maheshwari, S. (2025). Effect of Tool Pin Profiles on Defects Formation during Friction Stir Welding of AA 6082-T6 Straight-Pipes-Butt-Weld Joints. *Journal of Materials Research and Technology*, *36*, 1645–1657. Advance online publication. DOI: 10.1016/j.jmrt.2025.03.185

Yan, M., & Yu, P. (2015). An Overview of Densification, Microstructure and Mechanical Property of Additively Manufactured Ti-6Al-4V — Comparison among Selective Laser Melting, Electron Beam Melting, Laser Metal Deposition and Selective Laser Sintering, and with Conventional Powder. In *Sintering Techniques of Materials*. InTech. DOI: 10.5772/59275

Yang, Y., Li, M., & Li, K. R. (2014). Comparison and analysis of main effect elements of machining distortion for aluminum alloy and titanium alloy aircraft monolithic component. *International Journal of Advanced Manufacturing Technology, 70*(9-12), 1803–1811. DOI: 10.1007/s00170-013-5431-x

Yoshimura, R., Konno, T. J., Abe, E., & Hiraga, E. (2003a). Transmission electron microscopy study of the early stage of precipitates in aged Al–Li–Cu alloys. *Acta Materialia, 51*(10), 2891–2903. DOI: 10.1016/S1359-6454(03)00104-6

Yoshimura, R., Konno, T. J., Abe, E., & Hiraga, K. (2003b). Transmission electron microscopy study of the evolution of precipitates in aged Al–Li–Cu alloys: The θ′ and T1 phases. *Acta Materialia, 51*(14), 4251–4266. DOI: 10.1016/S1359-6454(03)00253-2

Zhu, J., Jie, Z., Chen, C., Zheng, H., & Wang, W. (2023). Fatigue Crack Propagation of Corroded High-Strength Steel Wires Using the XFEM and the EIFS. *Materials (Basel), 16*(13), 4738. Advance online publication. DOI: 10.3390/ma16134738

Zuo, J. H., Wang, Z. G., & Han, E. H. (2008). Effect of microstructure on ultra-high cycle fatigue behavior of Ti–6Al–4V. *Materials Science and Engineering A, 473*(1-2), 147–152. DOI: 10.1016/j.msea.2007.04.062

Chapter 2

Optimizing Coating Thickness and Tool Geometry in Turning Nano Hybrid Aluminium Composite:

A Numerical Simulation Approach

C. Dhanesh
https://orcid.org/0000-0001-6053-7940
S.A. Engineering College, India

A. Thanikasalam
https://orcid.org/0000-0003-3876-5036
Academy of Maritime Education and Training, India

Palanivendhan Murugadoss
https://orcid.org/0000-0003-0388-1547
SRM Institute of Science and Technology, India

C. K. Dhinakarraj
https://orcid.org/0000-0003-4405-0598
Adhiparasakthi Engineering College, India

Muhammad Azhar Ali Khan
College of Engineering, Prince Mohammad Bin Fahd University, Saudi Arabia

Abdul Aziz Afzal
https://orcid.org/0000-0001-9983-7713
College of Engineering, Prince Mohammad Bin Fahd University, Saudi Arabia

ABSTRACT

This study employs a numerical model of the turning procedure and optimized for ideal coating thickness and tool geometry during the machining of aluminium composites. Tungsten carbide (WC) is coated with aluminium oxide (Al_2O_3) ceramic and

DOI: 10.4018/979-8-3373-9454-1.ch002

considered for machining aluminium alloy 6061 reinforced with silicon carbide and multiwalled carbon nanotube (AA6061+15%SiC+3%MWCNT) reinforced composite by wt.%. Three coating thicknesses (5, 10, and 15 μm) were considered with relief angle (2, 4, and 6°), and nose edge radius (0.4, 0.6, and 0.8 mm). An orthogonal array developed by Taguchi (L_9) served as the basis for the experiments, and the output responses, temperature at interface (IT), and resultant cutting forces (RCF) were optimized using signal-to-noise ratio. Observation shows that the RCF and IF tend to lower with lower coating thickness. In contrast, a moderate 4° relief angle and lower (0.4 mm) nose radius deliver lower RCF and lesser IT

1. INTRODUCTION

Enhancements in product quality and lower component costs are the main factors of current manufacturing technology advancements. These days, producers aim to provide goods with reduced surface roughness, good form, accurate calibration, and affordable prices. When dealing with materials that are challenging to process, it is preferable to achieve higher material removal rate (MRR), longer tool life, and reduced machining costs refrain from sacrificing surface quality (Chen et al., 2020; Sahoo et al., 2024). Composites are much more challenging to cure than metals due to their anisotropy, lack of uniformity, and quite abrasive reinforcing with hard ceramic particles. Despite their numerous beneficial qualities, metal matrix composites' (MMCs') machinability remains a significant barrier to their broad use. When MMCs with robust reinforcements are machined, cutting tools wear out rapidly. The reinforcements may be substantially more destructive than the materials utilized in the cutting tool due to their toughness. These abrasive particles damage surface integrity, reduce tool life, and significantly reduce MMC machinability. Due to their greater strength than the matrix, these enhancements alter the plastic deformation characteristics of the material (Dhinakarraj et al., 2025; Pramanik, 2021). Particulate-reinforced MMCs may be cut using a variety of techniques that might provide cutting forces (CFs). Among them encompass the potential for reinforcing particles to shear and plough the metallic matrix, break or detach from the matrix, and come into proximity to rounded cutting edges (Kishawy & Hosseini, 2018).

Because of their unusual structure, great hardness, brittleness, and diverse composition, nanocomposites are exceptionally challenging to manufacture. Machining such materials often produces delamination, low surface quality, and fast tool wear (Nayak et al., 2022). Because the mechanical behavior varies between the matrix and reinforcement phases, inconsistent carbon fiber distribution and thermal stresses might cause surface defects. Using cutting tools with significant wear resistance, like coated carbide or diamond tools (Babu & Kumar, 2021), would help one to address these problems. Reduced cutting speeds (CSs) and regulated feed rates (FRs) under optimal machining settings help decrease heat damage and enhance surface quality. Two other techniques that may reduce tool wear and heat generation thus improving performance are cryogenic cooling and ultrasonic-aided machining (Gupta & Davim, 2020). Using adaptive control systems, continuous monitoring, and strategic techniques, one may further minimize these challenges by means of exact machining, prolonged tool life, and preserving of the nanocomposite structure (Shyha & Huo, 2021).

Simulating machining techniques helps one to develop useful abilities applicable in the real world. One can address the challenges resulting from machining processes. Initially, the strain rates are very high even at moderate cutting speeds which needs to be studied elaborately (Kyratsis et al., 2023). Choosing the proper boundary conditions is also challenging, as plastic deformation occurs in small areas, such as the cutting edge, the main deformation region, and the secondary deformation region, among other similar locations. The modelling of tool-chip interface friction and the workpiece are two crucial components of FEM cutting models that significantly impact the final analysis results (Markopoulos, 2012). Simulations were conducted on an Al-Si-Mg alloy using an HSS cutting tool with MWCNTs as the lubricating oil. All of the process factors had an impact on tool wear, but cutting depth was the most important. Cutting speed (CS) is inversely proportional to feed rate (FR) in terms of tool wear. Chips were created in a continuous pattern at greater cutting speeds and with a discontinuous pattern at lower and medium rates (Imhade P. Okokpujie et al., 2022).

The experimental and simulation findings showed an average error of 9.94% when machining Ti6Al4V with TiAlN-coated inserts. The experiments evaluated CS and FR with a constant DoC of 0.25 mm. Additionally, they investigated the impact of tool geometry, including relief and rake angle, as well as nose radius (NR), on the temperature at the interface and CFs (Satyanarayana et al., 2013). Using a ceramic insert, a DEFORM-3D simulation was carried out to investigate the turning properties of AISI H13 and found that CFs have risen in line with rises in DoC, CS, and FR. Because the sliding mechanism generates more heat at the chip, the rake face in the primary cutting region approaches the highest temperature (Parihar et al., 2017a). Using an L_9 orthogonal array, researchers examined the relationships between process variables, tool wear, and turning time. They discovered that internal machining variables and machining time were crucial in finding a solution to the turning AA6061 tool wear issue. At maximum DoC and FR, the tool wear was at its greatest; FR has the highest influence on the cutting time, and vibration occurs at high feed rates (I.P. Okokpujie et al., 2023). A finite element model (FEM) was employed to simulate the forces generated during machining of an Al-Si7Mg alloy using a coated insert. The model found that DoC and FR have a significant influence on the CFs. The workpiece's thermal behaviour causes the CS to increase while chip formation (CF) is somewhat reduced. Cutting at a slower pace results in faster material removal; however, increasing the CS doesn't influence the surface roughness (SR)(Sridhar et al., 2023).

Cutting tool coating, feed rate, and speed, as well as other cutting variables, were studied to determine their impacts on cutting forces (CFs) and torque while drilling AISI 1050. DEFORM-3D allows for indirect analysis of cutting tool stresses. Drill ends and the helical-shaped flute take the brunt of the loads, while the flank faces and, to a lesser extent, the outside corner also feels the pressures (Yıldız et al., 2020). Models for turning using the DEFORM-3D FEA software, the Cockcroft-Latham ductile fracture criteria, and the Johnson-Cook material constitutive model were developed for AISI 1045 steel with a complex grooved insert. In comparison to the flat rake insert, the intricately grooved one is more effective at promoting chip breakage, and the findings demonstrate that the chip-breaking groove can drastically alter the geometry and angle of chip flow (L. Wu et al., 2020). A statistical and experimental study on the machinability of the AZ91 alloy, cutting temperature and surface roughness (SR) were examined using Deform-3D. According to the results of this study, SR and CF components are best described by FR. On the other hand, for the temperature, the CS influence is higher. In terms of CFs, the average discrepancy among experimental and computational findings is 4.5% (Akgün, 2023).

The objective of this research is to investigate the influence of coating thickness (CT), cutting insert relief angle (RA), and NR on the turning performance of a nano aluminium composite material composed of aluminium alloy 6061 (AA6061) reinforced with SiC and multi-walled carbon nanotubes (MWCNTs). The Taguchi technique is used in conjunction with numerical simulations to investigate the impact of specific machining factors through experimental design. The signal-to-noise (S/N) ratio is used to analyze key response variables, specifically the interface temperature (IT) and the resultant cutting force (RCF), which are crucial to tool performance and workpiece integrity. Based on the optimal parameter combination identified through this analysis, a confirmation experiment is carried out to validate the simulation results.

Despite extensive research on the machining of aluminium-based composites, a significant literature gap remains in understanding the combined effects of CT, RA, and NR during the turning of hybrid nano-reinforced composites, particularly AA6061 integrated with SiC and MWCNTs. The novelty of this study lies in its comprehensive approach, which incorporates advanced finite element simulations with a Taguchi-based experimental design to optimize machining parameters for this specific class of nanocomposites. The machinability of MMCs has shown a notable improvement compared to conventional experimental techniques through FEM-based optimization using Taguchi's method. By integrating the predictive power of the Finite

Element Method with the robustness of Taguchi's design of experiments, researchers can efficiently analyze the influence of multiple machining parameters with fewer simulation runs while accurately modelling tool-workpiece interactions. Considering their abrasive reinforcements, this integrated method helps to identify ideal cutting conditions that decrease tool wear, lower cutting forces (CFs), and improve surface finish; critical issues in machining metal matrix composites (MMCs). Less time and resources are needed for the FEM-Taguchi hybrid approach than for comprehensive factorial or conventional trial-and-error studies; it offers deeper understanding of parameter sensitivity and interaction effects. This strategy increases machinability of MMCs in a more scientific, economical, and exact way than earlier approaches of machining optimization. Moreover, the work especially addresses assessing IT and CFs, which are important markers of tool wear and thermal stability under different geometric tool circumstances, thus offering important information for the high-performance machining of hybrid materials of future generation.

2. MATERIALS AND METHODS

2.1 Matrix and Reinforcements

The primary alloying constituents of AA6061 are silicon and magnesium, which cause the alloy to precipitation-harden. AA6061 has a nominal composition of Al-1wt.%Mg-0.6wt.%Si, with 0.25wt.%Cu added to enhance mechanical strength and 0.2%Cr added to mitigate the copper's effect on corrosion resistance Yuvaraj et al., 2023). Railroad coaches, truck frames, ships, bridges, military crossings, pylons, towers, rivets, motorboats, and aerospace parts (such as rotor skins for helicopters) are among the many applications of AA6061. In addition to its 154 W/m-K thermal conductivity, 68.9 GPa modulus of elasticity, and 145 MPa yield strength, AA6061 has a density of 2.7 g/cm^3 (Bhat Kakandikar, 2019). As the most fundamental substance composed of silicon (Si) and carbon (C) atoms bound together in a covalent connection, silicon carbide (SiC) has exceptional chemical and mechanical stability. SiC particulates of 75 μm are considered for reinforcements in the AA6061 matrix. SiC has a more refraction index than diamond (Siddharthan et al., 2020). SiC provides, low coefficient of thermal expansion, excellent heat conduction, extraordinary toughness, and resistance to corrosion (Wankhade et al., 2021). MWCNTs have extraordinary physical and mechanical characteristics, as well as great elasticity and resilience, which let them resist large deformations without lasting damage because to their special structure. Moreover, MWCNTs are perfect for complex electronic applications and thermal management systems because of their outstanding thermal conductivity that of diamonds and great electrical conductivity. These qualities, along with their low density, stable chemical properties, and notable aspect ratio, provide several possible applications including in aerospace, nanocomposites, and energy storage systems (Ramachandran et al., 2022).

2.2 Fabrication of Nanocomposite

The ultrasonic-assisted stir casting technique is one of advanced techniques for generating AA6061-SiC-MWCNT composites. This approach creates a vortex by melting aluminum 6061 at 750°C and thereafter physically agitating the slurry for 10 minutes at 500 rpm, therefore enabling the dispersion of reinforcing phases to be accomplished uniformly (Al-Mhanna et al., 2023). SiC particles and MWCNTs are heated to 350°C then added to the molten aluminium to eliminate any moisture. The melt is then exposed to ultrasonic vibrations, which produce cavitation and strong acoustic streaming, thus enhancing the wettability within the matrix and reinforcements, breaking down agglomerates, and so promoting particle spreading (Raj et al., 2023). SiC's and MWCNTs' complimentary strengthening actions produce a composite with extraordinary tensile strength, wear resistance, and hardness. This method guarantees the consistent distribution of reinforcements, therefore reducing flaws and improving the general composite performance. Demanding usage in the structural, transportation, and aerospace industries (Venkatesh et al., 2023; S. Wu et al., 2018) fit this material well. The ultrasonic-assisted stir-casting system employed in this study is shown in Figure 1.

Figure 1. Schematic and photographic view of Ultrasonic aided stir casting technique

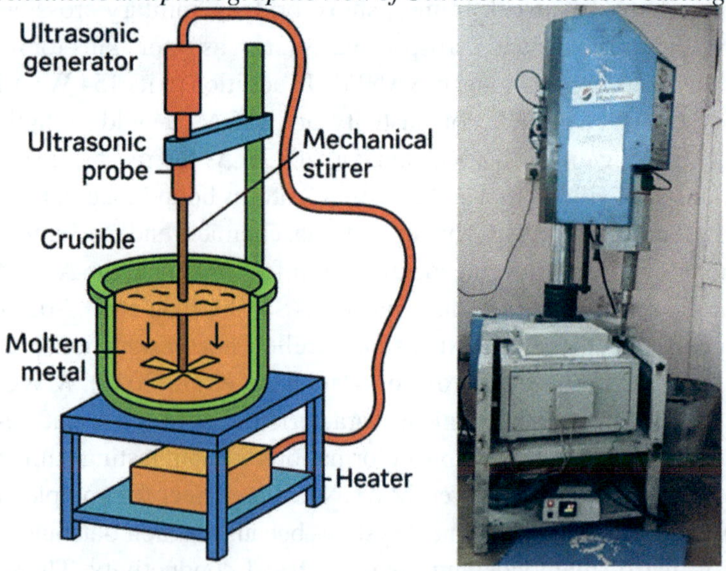

2.3 Characterization Methods of Nanocomposites

The cast specimens were tested for mechanical characteristics, including ductility, stress-strain association, and tensile strength, using the procedures outlined in ASTM E8 standards (Perumal et al., 2025; Srinivasan et al., 2024). The vulnerability of a metal to plastic distortion, as measured by indentation, is known as its hardness. The micro-hardness testing machine developed by Vickers is used to measure micro-hardness following ASTM E384 (Kociubczyk et al., 2023).

2.4 Taguchi's Methodology

Statisticians use Taguchi's Design of Experiments (DoE) to conduct cost-effective, multi-aspect explorations. (Reddy & Esanakula, 2023; Roy, 2010). The optimal mix of components may be identified by investigating the impact of each element on the outcomes. Taguchi's method examines quality features with a minimum number of tests using a unique design of Orthogonal Array (OA) (Pai et al., 2023; Shetty et al., 2009). To analyze the machinability performance of the cast composite, a L_9 orthogonal array (OA) design was employed to conduct experiments with different parameter configurations. The experimental matrix was created using the statistical package Minitab-a8 (Bhowmick et al., 2024). Table 1 lists the parameters examined along with their corresponding values. CT influences heat dissipation and tool wear resistance; the RA affects cutting-edge strength and chip flow, while the NR plays a role in surface finish and tool durability. The specific levels for each parameter are selected based on literature, industrial standards, and simulation constraints, ensuring relevance, practicality, and a balanced experimental design that yields accurate and meaningful results. As in Eq. (1), the outputs are analyzed based on the lower-the-better signal-to-noise (S/N) ratio.

$$\frac{S}{N} = -10log\left(\frac{1}{n}\sum_{i=1}^{n} y_i^2\right) \quad (1)$$

Table 1. Data used in a simulation analysis

Parameter	Unit	-1	0	+1
Coating Thickness (CT)	μm	5	10	15
Relief Angle (RA)	°	2	4	6
Nose Radius (NR)	mm	0.4	0.6	0.8

In turning of MMCs, geometric parameters such as CT, RA, and NR play a critical role in determining machining performance and interface behaviour due to the abrasive nature and heterogeneous structure of MMCs. CT influences tool durability and thermal resistance; a thicker coating can provide better protection against abrasive particles but may also alter heat dissipation and cutting-edge sharpness, affecting surface finish and tool–chip interaction. The RA determines the clearance between the tool and the workpiece; an optimal RA reduces rubbing and friction at the tool-workpiece interface, thereby minimizing tool wear and improving dimensional accuracy. However, a too-small RA increases the contact area and heat generation, thereby accelerating wear. NR affects chip flow, surface finish, and stress distribution at the cutting edge; a larger NR generally improves surface quality and distributes CFs more evenly, but it can also increase CFs and the risk of built-up edge formation. Together, these geometric parameters significantly influence output responses such as CFs, temperature, tool wear, and surface integrity, making their selection and optimization essential for efficient and high-quality machining of MMCs. In every trial, the following parameters are kept constant: CS (300 m/min), DoC (0.5 mm), and FR (0.25 mm/rev).

2.5 Finite Element Method

FEM analysis of machining processes plays a vital role in both research and industry due to its ability to simulate and predict complex physical phenomena with high accuracy, eliminating the need for costly and time-consuming experiments. Machining includes thermal, mechanical, and material deformation elements in nonlinear, transient, highly dynamic interactions among the cutting tool and workpiece. Among the interactions that could be better comprehended using FEM than with experimental techniques include stress distribution, temperature rise, chip building, and tool wear. Unlike conventional experimental methods, FEM allows virtual testing under many conditions with fine-grained regulation of parameters, therefore removing the need for expensive and time-consuming arrangement and trials. FEM is a great alternative to or complement for testing as it promotes industrial innovation by means of its predictive powers. The FEM uses a multifarious method to replicate real-time manufacturing operations. Plastic deformation, frictional contact, thermo-mechanical connection, and processes for chip production are among the many physical events engaged in the modeling of the turning process (Tamizharasan & Senthil Kumar, 2012).

The 2D flow of complex metal forming operations, such as machining, welding, and joining, can be evaluated using DEFORM-2D with either an axisymmetric or planar strain assumption. Based on FEM, it is an advanced system for simulating processes (Winiczenko et al., 2024). Figure 2 illustrates the meshing process of a modelled tool-workpiece system, typically used in finite element simulations for material deformation or machining analysis. The mesh-sized tool and workpiece are also shown, along with the location of the workpiece and tool during machining. In Figure 2(a), a meshed tool with delicate and structured elements is shown in contact with the workpiece, indicating regions of critical interaction. Figure 2(b) presents an overall view of the workpiece, featuring a structured yellow mesh and a noticeable refinement near the contact zone with the tool, indicating the need for higher resolution in this high-stress region. Figure 2(c) shows the tool-workpiece interface, clearly showing mesh distortion and refinement at the contact location, which is essential for capturing accurate stress and strain gradients. Figure 2(d) provides a detailed close-up of the most refined meshing at the contact point, highlighting individual mesh nodes and elements used for precise simulation results. Adaptive meshing is considered during simulation. To ensure accurate modelling of the physical interaction between the tool and the workpiece.

Figure 2. Modeled tool-workpiece in meshed condition (a) tool (b) workpiece and tool position (c) meshed tool (d) meshing of the workpiece at tool contact

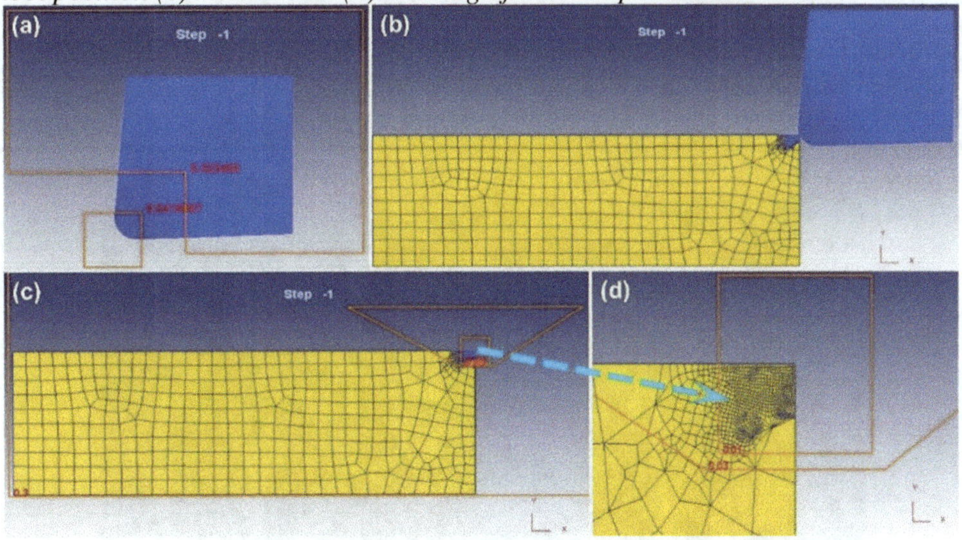

3. RESULTS AND DISCUSSIONS

3.1 Characterization of Hybrid Composite

After fabricating the hybrid nanocomposite, its microstructure is characterized to identify the distribution of reinforcements. The SEM micrograph (Figure 3) shown at 1000× magnification reveals the microstructural morphology of a hybrid MMC (AA6061+15%SiC+3%MWCNT). The spherical and agglomerated structures observed in the image are characteristic of the presence of MWCNTs. At the same time, the smaller, more angular particles likely represent the SiC reinforcements embedded within the aluminium matrix. The interface between the matrix and reinforcements appears irregular, with some degree of porosity and microcracking, which may be caused by differential thermal expansion and mechanical stresses during processing. The MWCNTs are embedded in the matrix, indicating possible near-uniform dispersion, which may influence the composite's mechanical behaviour, particularly under cutting or load conditions. Despite the presence of some voids and pull-out regions, the reinforcements are generally well-integrated within the AA6061 matrix, suggesting effective composite formation. This hybrid structure will enhance hardness, thermal stability, and wear resistance; however, the microstructural defects may act as stress concentrators during machining or mechanical loading.

Figure 3. SEM micrograph of AA6061+15%SiC+3%MWCNT

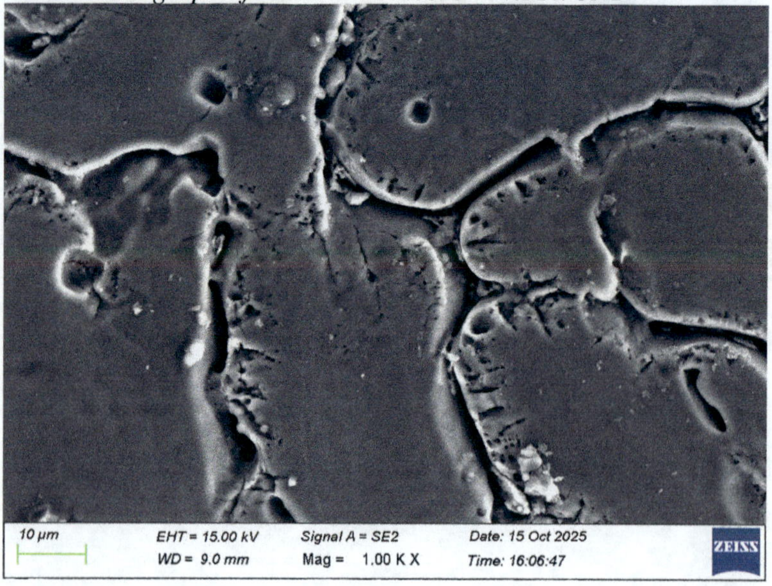

The mechanical characterization of the AA6061+15%SiC+3%MWCNT hybrid composite reveals a tensile strength of 362 MPa, an elongation of 2.78%, and a hardness of 153 HV, reflecting the synergistic reinforcement effect of both ceramic and nano-scale constituents, governed by multiple strengthening mechanisms. The high tensile strength arises primarily from load transfer, dislocation strengthening, and grain refinement. The rigid SiC particles act as barriers to dislocation movement, thereby enhancing strength through the Orowan mechanism. Meanwhile, MWCNTs, with their exceptionally high aspect ratio and stiffness, facilitate efficient load transfer from the ductile aluminium matrix to the reinforcement phase (Singhal et al., 2025). Additionally, the mismatch in the coefficient of thermal expansion (CTE) between the matrix and reinforcements induces a high dislocation density around the particles during cooling, further improving strength through thermal mismatch strengthening. The relatively low elongation (2.78%) indicates reduced ductility, typical of particle-reinforced composites, due to restricted plastic flow and the presence of stress concentration sites, as seen in the SEM image, such as voids, agglomerations, or debonded particles. These microstructural discontinuities can initiate early crack propagation under tensile loading. The elevated hardness of 153 HV is primarily attributed to the hard SiC particles and the reinforcing MWCNTs, which resist localized plastic deformation (Divakar et al., 2025).

The incorporation of 15% SiC and 3% MWCNT into the AA6061 matrix results in a significant improvement in mechanical performance compared to both monolithic AA6061 and its single-reinforced counterpart, AA6061+15%SiC. The tensile strength increases from approximately 270 MPa in monolithic AA6061 to 310 MPa with 15% SiC (a 14.8% increase) and further to 362 MPa in AA6061+15%SiC+3%MWCNT hybrid composite, marking a 34.1% enhancement over the base alloy. This strength gain is attributed to combined strengthening effects from both SiC and MWCNTs, including load transfer, dislocation pinning, and thermal mismatch. However, this improvement comes at the expense of ductility, where elongation drops drastically from 14% in AA6061 to 6% in the AA6061+15%SiC composite (a 57% reduction) and further down to 2.78% in AA6061+15%SiC+3%MWCNT, reflecting an 80% reduction. The loss in ductility is due to the rigid and brittle nature of the reinforcements, which introduce stress concentration sites and inhibit plastic deformation. Similarly, hardness increases from 95 HV in AA6061 to 125 HV in AA6061+15%SiC (31.6% rise), and up to 153 HV in AA6061+15%SiC+3%MWCNT, resulting in a 61.1% increase. The synergistic effect of SiC's abrasive resistance and MWCNTs' high stiffness contributes to this substantial enhancement.

3.2 FEM-based Simulation Outcomes

For the turning simulation, the tool is meshed using adaptive meshing, which allows its shape to change during the simulation. The mesh size at the contact position will be smaller, whereas, at extreme locations, the mesh size will be larger, which is not subjected to the shearing procedure (Friderikos et al., 2020; Tanarro et al., 2020). Table 2 presents the combination of input parameters used to design and execute the experiments. The output responses of IT and RCFs are determined from the simulation studies and recorded. For analysis, the minimization of both responses is considered simultaneously, and the corresponding S/N ratio is calculated, which is also recorded in Table 2. Observation shows that with higher CT, the IT rises as the coating retains the heat and restricts it from dissipating it, as depicted in Figure 4. With a higher RA, the IT decreases due to reduced contact between the workpiece and theool surface (Dabees et al., 2022). A higher NR increases the IT, as a larger cutting tool portion will contact the workpiece, resulting inore friction at the interfaces (Bhushan, 2020). The RFCs increase with increasing CT and RA due to the higher pressure at the interface, as seen in Figure 5 (Li et al., 2021). CFs are expected to rise when the NR is changed from 0.4 to 0.6 mm but decrease with a change in NR to 0.8 mm. The CF and NR are directly proportional due to the higher flow stress, but theymprove the tool life (Dutta & Narala, 2021).

Table 2. Input factors, output responses and S/N ratio for minimization

Exp. No	CT (μm)	RA (°)	NR (mm)	IT (°C)	RCF (N)	Combined S/N ratio
1	5	2	0.4	362	21.1	-48.179
2	5	4	0.6	372	42.6	-48.457
3	5	6	0.8	381	53.1	-48.692
4	10	2	0.6	374	36.1	-48.487
5	10	4	0.8	381	48.3	-48.677
6	10	6	0.4	361	35.1	-48.181
7	15	2	0.8	393	61.5	-48.983
8	15	4	0.4	363	20.7	-48.202
9	15	6	0.6	374	102	-48.759

The contour plots (Figure 4) illustrate the variation of IT during the turning of MMCs as a function of three key tool geometric parameters: CT, RA, and NR. The temperature ranges from approximately 361°C to 393°C, indicating a total variation of 32°C across the studied parameter space. The IT increases with decreasing RA and increasing CT, with a maximum variation of around 9% from the lowest to highest temperature observed in that plot. A similar trend is seen where temperature increases with both higher CT and larger NR, suggesting a variation of up to 8.9% across this combination. Larger NR and smaller RA result in the highest IT, contributing to 9% variation. These trends suggest that excessive NR, thicker coatings and lower RA tend to increase thermal buildup at the tool-workpiece interface, likely due to the greater contact area and reduced heat dissipation.

Figure 4. Influence of Input Variables on IT

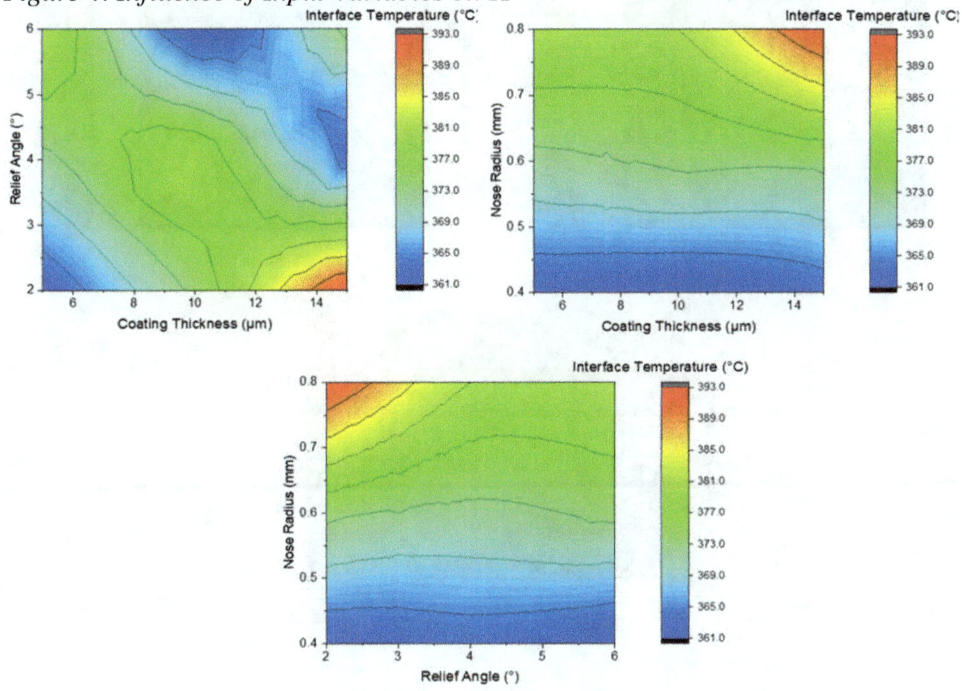

The contour plots in Figure 5 represent the variation of RCF during turning for varying CT, NR, and RA. The RCF ranges from approximately 20.6 N to 102 N, reflecting a significant variation of 81.4 N, which corresponds to a 395% increase from minimum to maximum. The RCF tends to increase sharply with increasing CT beyond 12 µm and with decreasing RA, indicating a strong interaction where excessive coating and low clearance cause greater cutting resistance. Increasing CT and NR contribute to higher RCF, with a pronounced rise observed at higher values of both, suggesting an increase of 300–350% in RCF. The lower RA, combined with a higher NR, causes a steep rise in RCF, again highlighting the role of increased contact area and mechanical load. Overall, it is observed that some increases in geometric parameters can enhance tool strength and durability; excessive values significantly raise RCFs, potentially leading to tool deflection, vibration, and premature wear, which is also reported by other researchers (Liu et al., 2019). An optimal combination of these geometrical parameters is thus essential for efficient and low-resistance machining of MMCs.

Figure 5. Influence of input variables on RCF

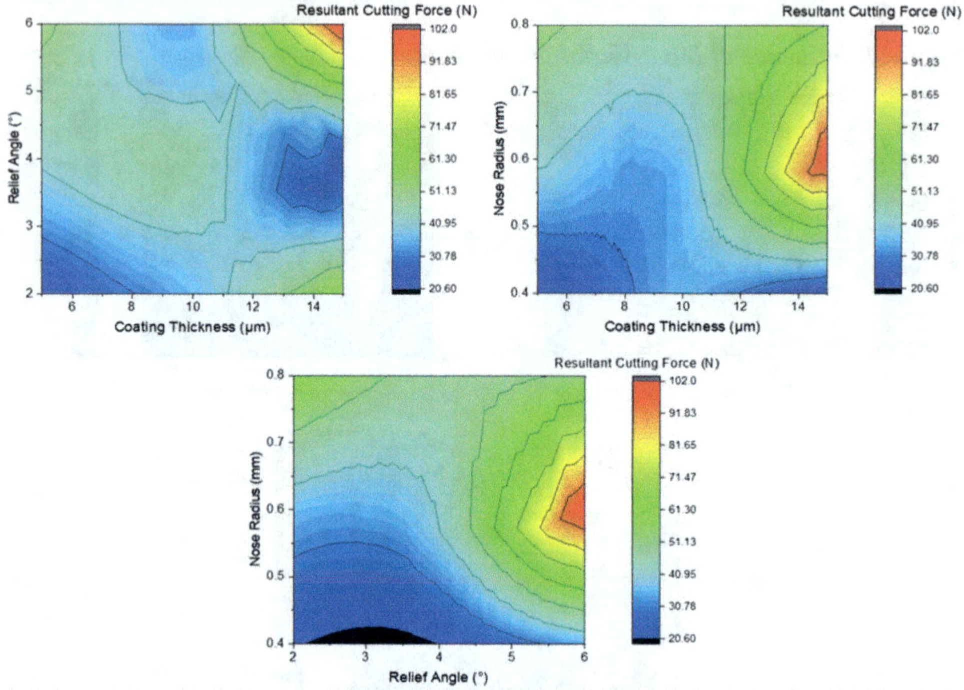

To analyze the output responses while minimizing IT and RCFs simultaneously, Eq. (1) is employed. To identify the optimal condition, a response table is formulated (Table 3), as shown in Figure 6, by calculating the average S/N ratio for each input value. The main effects plot for S/N ratios illustrates the influence of CT, RA, and NR on the response variable, intending to optimize the response based on the "Smaller is Better" criterion. For CT, the S/N ratio decreases as the thickness increases from 5 to 15 μm, indicating that a lower CT (5 μm) yields a better performance. In the case of RA, the middle level (4°) provides the highest S/N ratio, suggesting that it is the most effective among the tested values. For NR, a significant drop in S/N ratio is observed with increasing radius, with 0.4 mm showing the highest S/N ratio and, therefore, being the optimal level. Based on this analysis, the optimized conditions for minimizing the response are a CT of 5 μm, RA of 4°, and NR of 0.4 mm, as these levels contribute to higher S/N ratios and better performance consistency.

Table 3. Response table for S/N ratio

Level	CT	RA	NR
1	-48.44	-48.55	-48.19
2	-48.45	-48.45	-48.57
3	-48.65	-48.54	-48.78
Delta	0.21	0.10	0.60
Rank	2	3	1

Figure 6. Main effects plot for combined S/N ratio

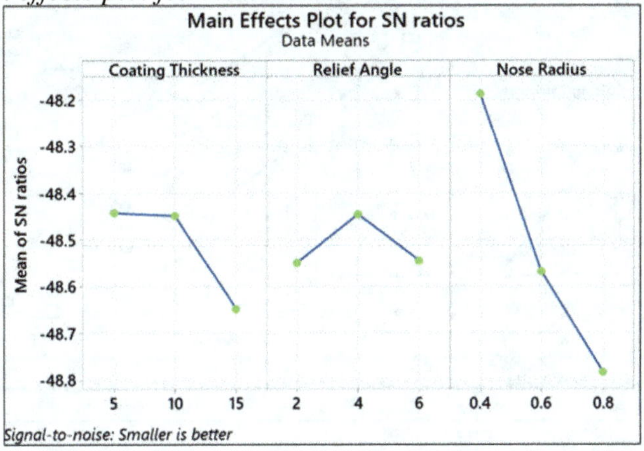

The relationship between the inputs and the determined S/N ratio of IT and RCFs is examined (Ahmad et al., 2022). Among CT and RA, a significant relationship exists that significantly affects the outputs, as seen in Figure 7. However, no considerable relationship is observed between CT and NR. But, among NR and RA, a moderate relationship exists, as represented by non-parallel lines (Thirumalvalavan et al., 2025). To identify the most influential parameter that impacts the outputs the most, the analysis of variance (ANOVA) tool is considered. Table 4 presents the ANOVA outputs for the determined signal-to-noise (S/N) ratio. The R^2 value obtained is 98.12%, with an adjusted R2 value of 92.49%, which are relatively close to each other. Hence, a significant model is developed with a 95% confidence interval (Bhardwaj et al., 2025). It is revealed that NRs have the most dominant role, contributing 82.68% to the overall variation. A larger NR increases the contact area between the tool and the workpiece, leading to greater frictional heat generation and higher IT and resulting in higher RCFs due to increased material deformation resistance. Following this, CT contributes 12.36%, where a thicker coating improves thermal insulation and wear resistance, moderately affecting heat dissipation and force transmission. Although its impact is less than NR, it plays a critical role in prolonging tool life under elevated thermal loads. The RA shows the least influence at 3.09%, mainly because it has a more subtle effect on the contact mechanics. A larger RA reduces the tool-workpiece contact length, slightly lowering friction and heat, but its contribution to overall temperature and force generation is comparatively minimal (Ranjan & Hiremath, 2023).

Figure 7. Interaction plot for inputs and S/N ratio

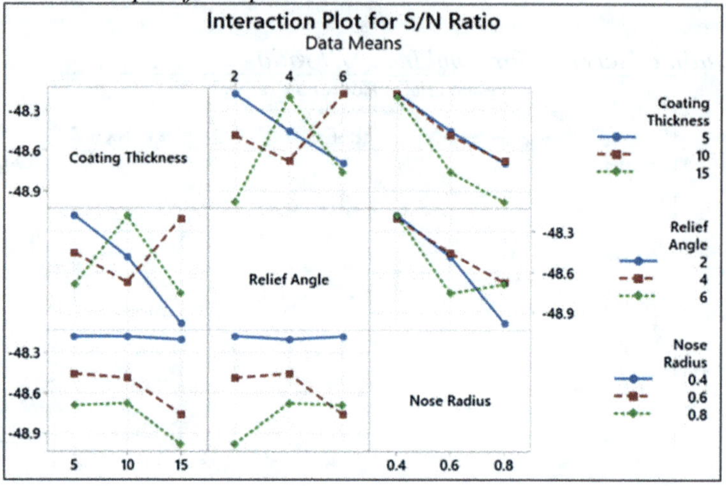

Table 4. ANOVA for SN ratios

Source	DoF	Seq SS	Adj MS	F	P	% Contribution
Coating Thickness	2	0.0819	0.0409	6.58	0.132	12.36
RA	2	0.0205	0.0103	1.65	0.378	3.09
NR	2	0.54789	0.2739	44.05	0.022	82.68
Residual Error	2	0.01244	0.0062			1.88
Total	8	0.66269				

Figure 8 presents the results from the simulation for each trial with variations in IT of a tool-workpiece interaction, accompanied by a comparative temperature-time graph. Each plot (Trial 1 to Trial 9) shows the temperature gradient at the tool-workpiece interface, with colour scales ranging from 30°C to nearly 400°C. Notably, Trial 7 exhibits the highest peak IT at approximately 393°C, whereas Trial 6 records the lowest at about 361°C. The line graph confirms this trend, showing that while all trials follow a similar heating curve, the peak temperatures vary slightly. The percentage difference between the highest and lowest peak IT indicates a modest 8.14% variation in maximum IT across the trials. Changes in cutting conditions, such as tool geometry or the thermal conductivity of the coating, are likely to influence the variation. Most trials stabilize within the 370–390°C range, indicating a generally consistent thermal performance with minor fluctuations, which may affect material deformation and tool wear behaviour.

Figure 8. Simulation output of each trial with variation in IT

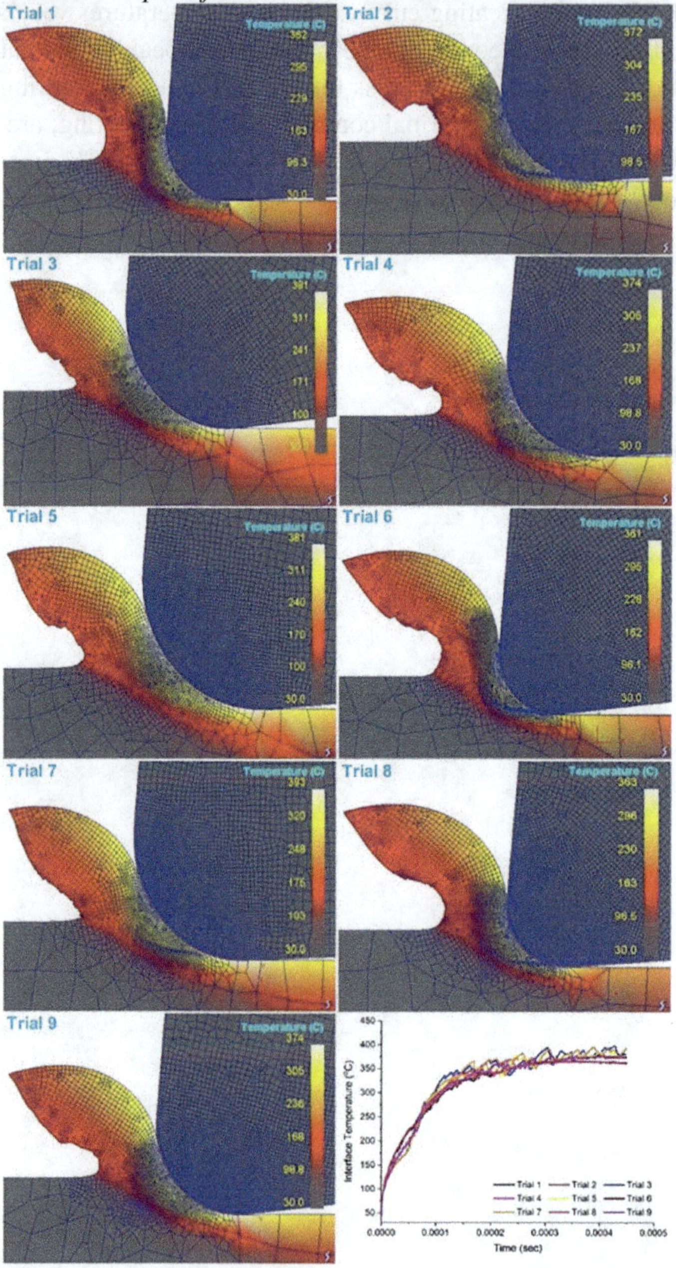

With the optimal condition obtained, it is necessary to conduct a confirmatory experiment under the same numerical conditions, which achieved an IT of 361°C and an RCF of 38.1 N. The obtained values are lower by 8.14% and 3.33% when compared with the maximum and average values of IT, respectively. Additionally, they are 62.65% and 18.45% lower than the RCF. Figure 9 presents the results of setting the optimum condition under steady-state transient conditions. In Figure 9(a), the temperature distribution during the cutting process is illustrated, with a maximum temperature of approximately 361 °C. The highest temperature is concentrated near the tool-chip interface, where intense plastic deformation and friction occur. This localized heating facilitates chip formation but also necessitates effective thermal management to maintain tool life as concluded by other researchers (Parihar et al., 2017b). Figure 9(b) depicts the RCF distribution, which peaks at around 38.1 N. The force is highly concentrated near the cutting edge, consistent with the primary deformation zone. Figure 9(c) presents the finite element mesh, which demonstrates a refined mesh near the cutting zone to ensure an accurate simulation of thermal and mechanical gradients. Figure 9(d) shows the evolution of the IT, revealing a steep increase in the initial phase of contact, with the temperature stabilizing at approximately 360°C within 0.0005 seconds. These results indicate that the chosen tool geometry and CT offer an optimal balance, minimizing CFs while effectively managing thermal loads and enhancing chip formation characteristics (Senthilkumar, 2024).

Figure 9. Optimum condition performance evaluation (a) CT (b) RCF (c) Chip after steady-state machining (d) IT variation at optimal condition

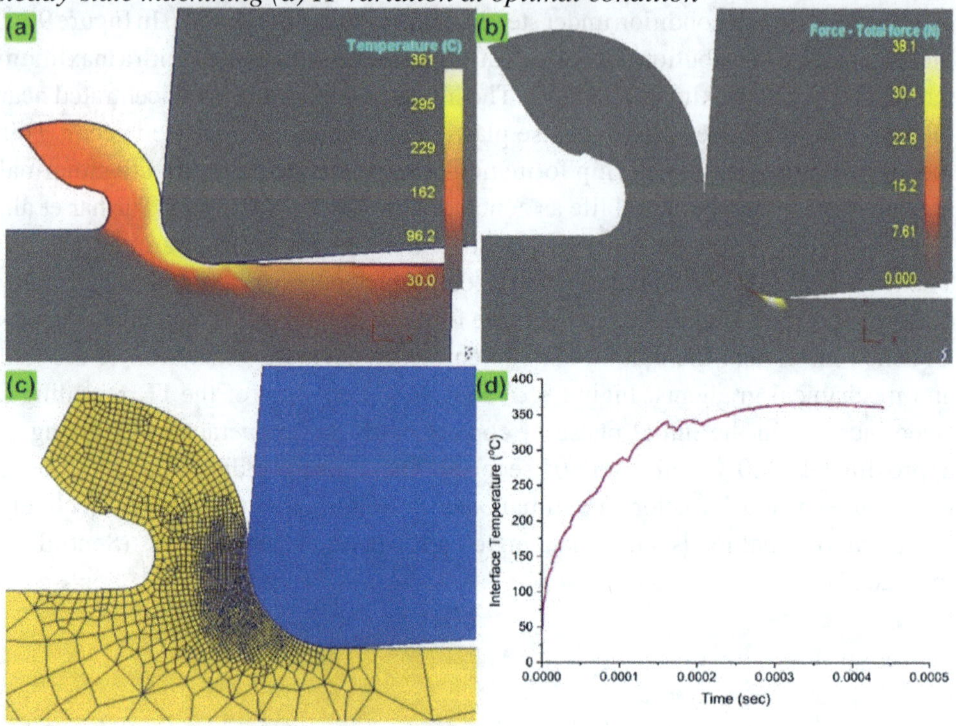

The findings from FEM-based optimization of machining MMCs have profound implications for industrial scalability and cost-effectiveness. This method reduces the time and effort required for physical testing by enabling the precise prediction of optimal machining settings and tool shapes. Due to the hard reinforcing particles, MMCs are extremely difficult and costly to process; this tool helps producers simulate various scenarios to determine the optimal conditions for machining MMCs. This improves product quality and uniformity while significantly reducing material waste and tool wear. It is also easier to increase operations without spending excessive expenditures, owing to the capacity to digitally test and refine methods, which accelerates the transfer from prototype to manufacturing. FEM-based optimization makes industrial usage more realistic and sustainable by ensuring a more simplified, productive, and cheap way of machining MMCs.

The outcomes of both future research employing nano-reinforced composites and present-day industrial machining techniques stand to be much appreciated. Reduced CFs, improved surface smoothness, and ideal machining parameters gained from FEM-based Taguchi analysis provide the path for increased industrial effectiveness. Direct influence of these elements on manufacturing efficiency, tool cost, and component quality. These developments may be readily applied to CNC machines for milling challenging-to-cut materials such as MMCs, thereby enabling manufacturers to achieve greater throughput with less interruption and more dimensional precision. Because of their temperature sensitivity, brittleness, and ultra-hardness of the nano-fillers, which cause even more machining difficulties, the methodology and results of the research provide a fantastic basis for investigations on nano-reinforced composites. By changing ideal parameter trends and using a simulation-driven method, engineers may create more exact machining techniques for newly created composite materials. This will let these materials be used more broadly in high-performance applications like aviation, health care implants, and micro-electromechanical systems (MEMS).

4. CONCLUSION

The investigational outcomes from the machining simulation of AA6061+15%SiC+3%MWCNT composite using carbide inserts coated with Al_2O_3 are as follows.

1. Higher CT raises the IT as the coating limits its dissipation and stores heat. With a higher RA, the IT decreases due to the lower contact surface, and a higher NR increases the IT due to increased friction at the interfaces.
2. RCFs increase with increasing CT and RA due to the higher pressure at the interface. RCFs rise with increasing NR; after the optimal condition, they tend to lower as NR and RCFs are directly proportional due to higher flow stress.
3. The optimal conditions achieved are a CT of 5 μm, a relief of 4 °, and a thickness of 0.4 mm. CT and RA are significantly related, but no considerable relationship is seen between CT and NR. A moderate interaction is observed between RA and NR.
4. ANOVA results show that the influence of NR is higher at 82.68%, CT at 12.36%, and RA at 3.09%, with an R^2 value of 98.12%.
5. The confirmation experiment achieves an IT of 361°C and an RCF of 38.1 N, which are lower by 8.14% and 3.33% when compared with the maximum and average values of IT and 62.65% and 18.45% when compared with RCF.

The limitations of this study include reliance on simulated data through the FEM model, which, although highly informative, may not fully capture all the real-world complexities, such as dynamic tool wear, material inhomogeneity, and machine vibrations, that occur during actual machining. Additionally, the study assumes idealised boundary conditions and material behaviour, which may differ in industrial settings where environmental and operational variabilities exist. Moreover, the study focuses primarily on orthogonal cutting and does not extend to multi-axis or complex machining operations commonly used in industry.

REFERENCES

Ahmad, M. N., Ishak, M. R., Mohammad Taha, M., Mustapha, F., Leman, Z., Anak Lukista, D. D., Irianto, , & Ghazali, I. (2022). Application of Taguchi Method to Optimize the Parameter of Fused Deposition Modeling (FDM) Using Oil Palm Fiber Reinforced Thermoplastic Composites. *Polymers*, *14*(11), 2140. DOI: 10.3390/polym14112140

Akgün, M. (2023). Measurement and optimization of cutting forces, surface roughness and temperature in turning of AZ91 Mg alloy. *Sadhana*, *48*(2), 60. DOI: 10.1007/s12046-023-02113-3

Al-Mhanna, M. A.-M. H., Al-Ethari, H., & Jasim, T. A. (2023). Influence of Warm Squeezing on Some Properties of AA7075 Alloy Prepared by Stir Casting. *2023 14th International Conference on Mechanical and Aerospace Engineering (ICMAE)*, 60–63. DOI: 10.1109/ICMAE59650.2023.10424449

Babu, B. S., & Kumar, K. (2021). *Nanomaterials and Nanocomposites: Characterization, Processing, and Applications*. CRC Press. Retrieved from https://books.google.co.in/books?id=RxwfEAAAQBAJ

Bhardwaj, A. R., Meshram, P. D., Chakule, R. R., & Gupta, P. (2025). Processing of aluminum metal matrix composite with titanium carbide reinforcement. *Materials and Manufacturing Processes*, *40*(9), 1–15. DOI: 10.1080/10426914.2025.2491413

Bhat, A., & Kakandikar, G. (2019). Manufacture of silicon carbide reinforced aluminium 6061 metal matrix composites for enhanced sliding wear properties. *Manufacturing Review*, *6*, 24. DOI: 10.1051/mfreview/2019021

Bhowmick, S., Prajapati, G., Biswas, N., & Majumdar, G. (2024). Fuzzy Modelling and Grey Relation Analysis of Electric Discharge Machining Performed on Inconel 718 Superalloy. *2024 4th International Conference on Computer, Communication, Control & Information Technology (C3IT)*, 1–6. DOI: 10.1109/C3IT60531.2024.10829413

Bhushan, R. K. (2020). Impact of nose radius and machining parameters on surface roughness, tool wear and tool life during turning of AA7075/SiC composites for green manufacturing. *Mechanics of Advanced Materials and Modern Processes*, *6*(1), 1. DOI: 10.1186/s40759-020-00045-7

Chen, J.-P., Gu, L., & He, G.-J. (2020). A review on conventional and nonconventional machining of SiC particle-reinforced aluminium matrix composites. *Advances in Manufacturing*, *8*(3), 279–315. DOI: 10.1007/s40436-020-00313-2

Dabees, S., Mirzaei, S., Kaspar, P., Holcman, V., & Sobola, D. (2022). Characterization and Evaluation of Engineered Coating Techniques for Different Cutting Tools—Review. *Materials (Basel)*, *15*(16), 5633. DOI: 10.3390/ma15165633

Dhinakarraj, C. K., Senthilkumar, N., Palanikumar, K., & Deepanraj, B. (2025). Machinability evaluation of magnesium composite using response surface methodology and nature-inspired metaheuristic algorithms. [IJIDeM]. *International Journal on Interactive Design and Manufacturing*, *19*(4), 2607–2627. DOI: 10.1007/s12008-024-01866-3

Divakar, S., Sardar, S., Sah, S., & Das, D. (2025). A state-of-the-art review on SiC and MWCNTs reinforced hybrid metal matrix composites: Processing, properties, and applications. *Hybrid Advances*, *10*, 100454. DOI: 10.1016/j.hybadv.2025.100454

Dutta, S., & Narala, S. K. R. (2021). Effect of tool nose radius in turning of novel Mg alloy. *Materials Today: Proceedings*, *38*, 2675–2679. DOI: 10.1016/j.matpr.2020.08.334

Friderikos, O., Sagris, D., David, C. N., & Korlos, A. (2020). Simulation of Adiabatic Shear Bands in Orthogonal Machining of Ti6Al4V Using a Rigid-Viscoplastic Finite Element Analysis. *Metals*, *10*(3), 338. DOI: 10.3390/met10030338

Gupta, K., & Davim, J. P. (2020). *High-Speed Machining*. Academic Press. Retrieved from https://books.google.co.in/books?id=fZ_NDwAAQBAJ

Kishawy, H. A., & Hosseini, A. (2018). *Machining Difficult-to-Cut Materials: Basic Principles and Challenges*. Springer International Publishing. Retrieved from https://books.google.co.in/books?id=vYtoDwAAQBAJ

Kociubczyk, A. I., Grau, J. E., Sarutti, J. L., Méndez, C. M., Gregorutti, R. W., & Ares, A. E. (2023). Solidification Analysis, Microhardness, and Corrosion Resistance of a Cast Hip Prosthesis in ASTM F745 Austenitic Stainless Steel. *Journal of Materials Engineering and Performance*, *32*(3), 1065–1075. DOI: 10.1007/s11665-022-07164-x

Kyratsis, P., Tzotzis, A., & Davim, J. P. (2023). *3D FEA Simulations in Machining*. Springer International Publishing. Retrieved from https://books.google.co.in/books?id=p9mzEAAAQBAJ

Li, S., Lin, H., Zhang, T., Sui, J., & Wang, C. (2021). High-Speed Machining of Malleable Cast Iron by Various Cutting Tools Coated by Physical Vapor Deposition. *Chinese Journal of Mechanical Engineering*, *34*(1), 46. DOI: 10.1186/s10033-021-00561-8

Liu, G., Huang, C., Su, R., Özel, T., Liu, Y., & Xu, L. (2019). 3D FEM simulation of the turning process of stainless steel 17-4PH with differently texturized cutting tools. *International Journal of Mechanical Sciences*, *155*, 417–429. DOI: 10.1016/j.ijmecsci.2019.03.016

Markopoulos, A. P. (2012). *Finite Element Method in Machining Processes*. Springer London. Retrieved from https://books.google.co.in/books?id=FM8_dRux4sgC

Nayak, R. K., Pradhan, M. K., & Sahoo, A. K. (2022). *Machining of Nanocomposites*. CRC Press LLC. Retrieved from https://books.google.co.in/books?id=ZEW9zgEACAAJ

Okokpujie, I. P., Chima, P. C., & Tartibu, L. K. (2022). Experimental and 3D-Deform Finite Element Analysis on Tool Wear during Turning of Al-Si-Mg Alloy. *Lubricants (Basel, Switzerland)*, *10*(12), 341. DOI: 10.3390/lubricants10120341

Okokpujie, I. P., Tartibu, L. K., Chima, P. C., & Ogundipe, A. T. (2023). A finite element based investigation of tool wear rate via machining of Al6061 alloys using deform-3D. *Materials Today: Proceedings*. Advance online publication. DOI: 10.1016/j.matpr.2023.05.566

Pai, A., Shenoy Baloor, S., & Kini, C. R. (2023). Taguchi Analysis of the Deformation Characteristics of Single-Point Cutting Tool with Micro-Tool Coatings During Orthogonal Machining. *Journal of The Institution of Engineers (India): Series D*, *104*(2), 811–820. DOI: 10.1007/s40033-022-00437-8

Parihar, R. S., Sahu, R. K., & Srinivasu, G. (2017a). Finite Element Analysis of Cutting Forces Generated in Turning Process using Deform 3D Software. *Materials Today: Proceedings*, *4*(8), 8432–8438. DOI: 10.1016/j.matpr.2017.07.188

Parihar, R. S., Sahu, R. K., & Srinivasu, G. (2017b). Finite Element Analysis of Cutting Forces Generated in Turning Process using Deform 3D Software. *Materials Today: Proceedings*, *4*(8), 8432–8438. DOI: 10.1016/j.matpr.2017.07.188

Perumal, G., Senthilkumar, N., Satheshkumar, D., & Anandhakumar, S. (2025). Comparative analysis of solution treatment with short and long-term aging on mechanical, tribological, and corrosion potential of Ti-6Al-4V alloy. *Multiscale and Multidisciplinary Modeling. Experiments and Design*, *8*(3), 159. DOI: 10.1007/s41939-025-00759-6

Pramanik, A. (2021). *Machining and Tribology: Processes, Surfaces, Coolants, and Modeling*. Elsevier. Retrieved from https://books.google.co.in/books?id=RksiEAAAQBAJ

Raj, P., Biju, P. L., Deepanraj, B., & Senthilkumar, N. (2023). Optimizing the machining conditions in turning hybrid aluminium nanocomposites adopting teaching–learning based optimization and MOORA technique. [IJIDeM]. *International Journal on Interactive Design and Manufacturing*. Advance online publication. DOI: 10.1007/s12008-023-01450-1

Ramachandran, K., Boopalan, V., Bear, J. C., & Subramani, R. (2022). Multi-walled carbon nanotubes (MWCNTs)-reinforced ceramic nanocomposites for aerospace applications: A review. *Journal of Materials Science*, *57*(6), 3923–3953. DOI: 10.1007/s10853-021-06760-x

Ranjan, P., & Hiremath, S. S. (2023). Finite element modeling and experimental validation of turning process using bio-mimicked structured tool for outcomes relevant to industry. *Machining Science and Technology*, *27*(3), 209–246. DOI: 10.1080/10910344.2023.2215304

Reddy, G. K., & Esanakula, J. R. (2023). Optimization of Operating Parameters of Wire EDM Using Design of Experiments Criteria. *2023 International Conference on Self Sustainable Artificial Intelligence Systems (ICSSAS)*, 1504–1509. DOI: 10.1109/ICSSAS57918.2023.10331679

Roy, R. K. (2010). *A Primer on the Taguchi Method* (2nd ed.). Society of Manufacturing Engineers., Retrieved from https://books.google.co.in/books?id=k5VBsRZfzQsC

Sahoo, A. K., Kumar, R., & Panda, A. (2024). *Sustainable Hard Machining: Implementation and Assessment*. CRC Press. Retrieved from https://books.google.co.in/books?id=OaALEQAAQBAJ

Satyanarayana, K., Venu Gopal, A., & Bangaru Babu, P. (2013, November 15). Finite Element Simulation of Cutting Forces in Turning Ti6Al4V Using DEFORM 3D. *Volume 2A. Advances in Manufacturing*. Advance online publication. DOI: 10.1115/IMECE2013-62868

Senthilkumar, N. (2024). Machining Studies on AISI 4140 Steel as Automotive Axle and Drive Shafts in Automobile via Numerical Simulation. *SAE Technical Paper*, 2024-01-5232. DOI: 10.4271/2024-01-5232

Shetty, R., Pai, R. B., Rao, S. S., & Nayak, R. (2009). Taguchi's technique in machining of metal matrix composites. *Journal of the Brazilian Society of Mechanical Sciences and Engineering*, *31*(1). Advance online publication. DOI: 10.1590/S1678-58782009000100003

Shyha, I., & Huo, D. (2021). *Advances in Machining of Composite Materials: Conventional and Non-conventional Processes.* Springer International Publishing. Retrieved from https://books.google.co.in/books?id=aVw0EAAAQBAJ

Siddharthan, B., Rajiev, R., Saravanan, S., & Naveen, T. K. (2020). Effect of Silicon Carbide in Mechanical Properties of Aluminium Alloy Based Metal Matrix Composites. *IOP Conference Series. Materials Science and Engineering*, *764*(1), 012040. DOI: 10.1088/1757-899X/764/1/012040

Singhal, V., Shelly, D., Saxena, A., Gupta, R., Verma, V. K., & Jain, A. (2025). Study of the Influence of Nanoparticle Reinforcement on the Mechanical and Tribological Performance of Aluminum Matrix Composites—A Review. *Lubricants (Basel, Switzerland)*, *13*(2), 93. DOI: 10.3390/lubricants13020093

Sridhar, N., Aezhisai Vallavi, M. S., & Mugilan, T. (2023). An integrated approach of FEM analysis using DEFORM 3D and experimental investigation of forces developed in Al-Si7Mg. *Materials Today: Proceedings*, *80*, 888–895. DOI: 10.1016/j.matpr.2022.11.324

Srinivasan, D., Senthilkumar, N., Ganesh, M., & Perumal, G. (2024). *Characterization studies on nano aluminium composite reinforced with montmorillonite nanoclay and titanium carbide.* Multiscale and Multidisciplinary Modeling, Experiments and Design., DOI: 10.1007/s41939-024-00374-x

Tamizharasan, T., & Senthil Kumar, N. (2012). Optimization of cutting inserts geometry using DEFORM-3D: Numerical simulation and experimental validation. *International Journal of Simulation Modelling*, *11*(2), 65–76. DOI: 10.2507/IJSIMM11(2)1.200

Tanarro, Á., Mallor, F., Offermans, N., Peplinski, A., Vinuesa, R., & Schlatter, P. (2020). Enabling Adaptive Mesh Refinement for Spectral-Element Simulations of Turbulence Around Wing Sections. *Flow, Turbulence and Combustion*, *105*(2), 415–436. DOI: 10.1007/s10494-020-00152-y

Thirumalvalavan, S., Perumal, G., Senthilkumar, N., & Selvarasu, S. (2025). Enhancing Tribological Characteristics of Titanium Grade-5 Alloy through HVOF Thermal-Sprayed WC-Co Nano Coatings by TOPSIS and Golden Jack Optimization Algorithm. *Recent Patents on Nanotechnology*, *19*(4), 544–567. DOI: 10.2174/0118722105306841240808092616

Venkatesh, V. S. S., Rao, G. P., Patnaik, L., Gupta, N., Kumar, S., Saxena, K. K., Sunil, B. D. Y., Eldin, S. M., & Kutham Al- kafaji, F. H. (2023). Processing and evaluation of nano SiC reinforced aluminium composite synthesized through ultrasonically assisted stir casting process. *Journal of Materials Research and Technology*, *24*, 7394–7408. DOI: 10.1016/j.jmrt.2023.05.030

Wankhade, L. N., Rathod, D., Nukulwar, M. R., Agrawal, E. S., & Chavhan, G. R. (2021). Characterization of aluminium-silicon carbide metal matrix composites. *Materials Today: Proceedings*, *44*, 2740–2747. DOI: 10.1016/j.matpr.2020.12.699

Winiczenko, R., Skibicki, A., & Skoczylas, P. (2024). The Experimental and FEM Studies of Friction Welding Process of Tungsten Heavy Alloy with Aluminium Alloy. *Applied Sciences (Basel, Switzerland)*, *14*(5), 2038. DOI: 10.3390/app14052038

Wu, L., Liu, J., Ren, Y., Yang, Z., Wang, G., Sun, J., Song, J., Xu, W., & Liu, X. (2020). 3D FEM simulation of chip breakage in turning AISI1045 with complicate-grooved insert. *International Journal of Advanced Manufacturing Technology*, *108*(5–6), 1331–1341. DOI: 10.1007/s00170-020-05460-1

Wu, S., Yuan, D., Lü, S., Hu, K., & An, P. (2018). Nano-SiCP particles distribution and mechanical properties of Al-matrix composites prepared by stir casting and ultrasonic treatment. *China Foundry*, *15*(3), 203–209. DOI: 10.1007/s41230-018-8009-2

Yıldız, A., Kurt, A., & Yağmur, S. (2020). Finite element simulation of drilling operation and theoretical analysis of drill stresses with the deform-3D. *Simulation Modelling Practice and Theory*, *104*, 102153. DOI: 10.1016/j.simpat.2020.102153

Yuvaraj, N., Koli, Y., Vedabouriswaran, G., & Issar, P. (2023). Mechanical and Tribological Properties of AA6061/SiC/Aloe Vera Powder Hybrid Al Composites Fabricated by Stir Casting. *Silicon*, *15*(5), 2451–2465. DOI: 10.1007/s12633-022-02168-7

Chapter 3
Strategies to Optimize the Performance of Aluminum-Lithium Alloy-Based Parts for High-Speed Aircraft

Rafael Vargas-Bernal
https://orcid.org/0000-0003-4865-4575
Polytechnical University of Guanajuato, Mexico

Barbara Bermudez-Reyes
https://orcid.org/0000-0002-0417-6172
National Autonomous University of Mexico, Mexico

Emmanuel Reyes-Gordillo
https://orcid.org/0009-0001-8073-7445
Secretary of Science, Humanities, Technology, and Innovation, Mexico

Francisco Reyes-Calderón
https://orcid.org/0000-0002-2745-4833
National Technological Institute of Mexico, Morelia's Technological Institute, Mexico

ABSTRACT

This chapter examines strategies to optimize aluminum-lithium (Al-Li) alloys for high-speed aeronautics, where lightweight, high-strength materials must endure extreme aerodynamic and thermal stresses. Al-Li alloys, with lower density and increased stiffness, are crucial for supersonic and hypersonic aircraft. The discussion

DOI: 10.4018/979-8-3373-9454-1.ch003

covers various generations of Al-Li alloys, their mechanical and corrosion-resistant properties, and key manufacturing challenges for aerospace components. Emerging methods like rapid solidification, powder metallurgy, and additive manufacturing create refined microstructures and improve mechanical performance. Advanced joining techniques such as laser beam and friction stir welding enhance damage tolerance and manufacturability. Future trends emphasize digitalization, sustainable production, multifunctional hybrid materials, and tailored alloy compositions to meet next-generation aerospace demands. Overall, Al-Li alloys offer vital weight savings, durability, and cost-effectiveness for high-speed aircraft structures.

INTRODUCTION

The high-speed aircraft sector is evolving, as is shown by historical developments, current technologies, and recent prospects (Pollock & Wild, 2024a). The recent advancements and upcoming aircraft designs within industry, driven by novel technological developments and emerging scientific research. There is a financial viability of civil high-speed airlines for achieving a Mach 4 level as the economic threshold for profitable operations. Historical data suggests flights beneath Mach 2.12 carry relatively low risks, while commercial hypersonic aircraft development is projected for later this century. Accident investigations so far have primarily focused on supersonic and hypersonic aircraft (Pollock & Wild, 2024b). These developments bring forward challenges in safety and reliability, affecting engine performance, structural soundness, and economics, particularly in military aerospace sectors. A key topic is the vital role of continuous monitoring systems, which both support autonomous operation and enable early detection of potential failures. The risk of high-speed aviation related to extreme operational stresses and sensitive technological vulnerabilities requires specialized approaches to safety that differ significantly from those used for subsonic aircraft. This risk can be mitigated by integrating human factors, simulation, advanced training programs, and policy strategies to effectively incorporate high-speed aircraft into existing airspace frameworks. Besides, high-speed safety and reliability cannot be managed using traditional practices and calls for ongoing research and innovation – including enhanced health monitoring and adaptive policy development – to address these evolving challenges and improve future accident prevention.

Aluminum-lithium (Al-Li) alloys are pivotal in high-speed aeronautics primarily owing to their lightweight high strength, and stiffness (Xiao et al., 2024). Compared to traditional aluminum, these alloys reduce density by 7-10% and increase elastic modulus by 10-15%, yielding valuable weight savings critical for fuel efficiency (Stoliar et al., 2021). Their exceptional strength-to-weight ratio and rigidity make them well-suited for critical aerospace components subject to cyclic loading and requiring high fatigue resistance, such as wings, fuselage sections, leading and trailing edges, control surfaces, and seal tracks. These alloys also exhibit good corrosion resistance and maintain performance regardless of low temperatures, tailoring to demanding aeronautical environments and including space applications like cryogenic tanks in booster systems (Chen et al.,, 2024b). Additionally, aluminum-lithium alloys contribute to cost savings by reducing manufacturing and maintenance expenses by up to 30% compared to planes built with composite materials (Zhou et al., 2021). They enable lighter, more rigid, stiffer, and more damage-tolerant structures, which are essential for high-speed and high-performance aircraft. The alloys are used in military aircraft, rockets, satellites, and commercial airplanes such as the Airbus A350 and Gulfstream G650. Therefore, scientific research must continue helping aluminum-lithium alloys improve the sustainability and efficiency goals of modern and future high-speed aeronautics through weight reduction, enhanced mechanical performance, and economic benefits.

Scientific research on strategies to optimize the performance of aluminum-lithium (Al-Li) alloy-based parts for high-speed aircraft must be developed because these alloys offer critical advantages for aerospace applications. Al-Li alloys have a lower density than conventional aluminum alloys, enabling significant weight reduction in aircraft components – often 8-20% lighter depending on lithium content (Li H. et al., 2023). This weight savings translates directly into improved fuel efficiency, increased range, higher speed capabilities, and lower operational costs for high-speed aircraft where such performance improvements are crucial. Additionally, Al-Li alloys provide superior stiffness, high specific strength, excellent fatigue resistance, corrosion resistance, and enhanced structural performance, which are essential for the safety and durability of aircraft operating under high stress and dynamic conditions typical of high Mach flight (Wang et al., 2023). Research aimed in this chapter has the aim for enhancing these properties through composition optimization, advanced manufacturing techniques, and targeted heat treatments to tailor the microstructure and performance, while overcoming challenges like anisotropy and cracking during processing. Optimizing these alloys supports the development of lighter, stronger, and more efficient aircraft, which meets the growing demands for fuel economy, environmental impact reduction, and high-speed aerodynamic performance in modern aerospace engineering.

This chapter provides a comprehensive overview of strategies required to optimize the performance of aluminum-lithium (Al-Li) alloy-based parts for high-speed aircraft focusing on presenting the past, present, and future perspectives approaches. This has been divided into five sections as follows: In the section titled "*Background*", the fundamentals associated with the status of the high-speed aeronautics have been explained. Then, the section titled "*The Role of the Aluminum-Lithium in Aircraft*", reviews existing aluminum-lithium (Al-Li) alloys applied to the high-speed aircraft are discussed. Next, in the section titled "*Next-Generation Aluminum-Lithium Alloys*", the innovations that are being introduced to optimize the performance of these materials in the parts used in high-speed aeronautics. Subsequently, the section titled "*Future Research Directions*", explores forthcoming technologies and materials to optimize the Al-Li parts under conditions of high-speed aeronautics. Finally, the section titled "Conclusions" offers a summary of the contributions that aluminum-lithium alloys will promote for developing parts for high-speed aircraft around the world in this century through technological development and scientific research around the world.

BACKGROUND

In this section, the fundamentals of high-speed aeronautics are discussed, highlighting the challenges encountered in this area of technological research and development, particularly in the application of the different types of materials used in various aircraft components. The principles under which high-speed aviation operates are discussed in the next subsection.

Fundamentals of High-Speed Aeronautics

High-speed aeronautics is a specialized field within aerospace engineering focused on designing, developing, and operating aircraft capable of flying substantially faster than the speed of sound, typically classified as supersonic (Mach 1 to Mach 5) and hypersonic (above Mach 5) flight. This topic presents enormous challenges, including aerodynamics, structural integrity, propulsion, thermal management, and materials science, which must be addressed to achieve reliable, efficient, and safe flight for high-speed applications. Furthermore, under this regime, the compressibility of air becomes important, unlike at low speeds, where air is incompressible. This leads to distinct aerodynamic behaviors, such as changes in pressure, density, and temperature of the airflow around the aircraft.

The Mach number, a ratio of aircraft speed to the speed of sound in air, defines flight regimes including subsonic (below Mach 1), transonic (near Mach 1), supersonic (above Mach 1), and hypersonic (above Mach 5). It influences aircraft design, performance, stability, and control during high-speed flights. The term "Mach" is named in honor of Austrian physicist Ernst Mach, who conducted research on shock waves and high-speed phenomena.

Key aspects for the development of high-speed aeronautics include (Pollock & Wild, 2024a, Pollock & Wild, 2024b):

- *Aerodynamics*: Managing shock waves, minimizing drag, and optimizing lift at supersonic and hypersonic speeds are central challenges. Sophisticated computational fluid dynamics (CFD) and practical wind tunnel experiments are extensively utilized.
- *Propulsion Systems*: Conventional jet engines are replaced or supplemented with supersonic and hypersonic propulsion systems such as turbojets, ramjets, scramjets, and combined-cycle engines, designed to perform efficiently at high speeds.
- *Thermal Protection*: High-speed flight generates extreme aerodynamic heating. Materials with high-temperature resistance and active cooling methods are essential to protect structural components and maintain aircraft integrity.
- *Materials and Structures*: Lightweight, heat-resistant composites and alloys are crucial for durability and safety under dynamic and thermal stresses encountered during high-speed flight.
- *Flight Control and Stability*: Advanced guidance and control systems are necessary to maintain stability and maneuverability as aerodynamic forces become complex at high velocities.

These required aspects to develop high-speed aeronautics are illustrated in Figure 1.

Figure 1. Critical technical aspects for the development of high-speed aeronautics

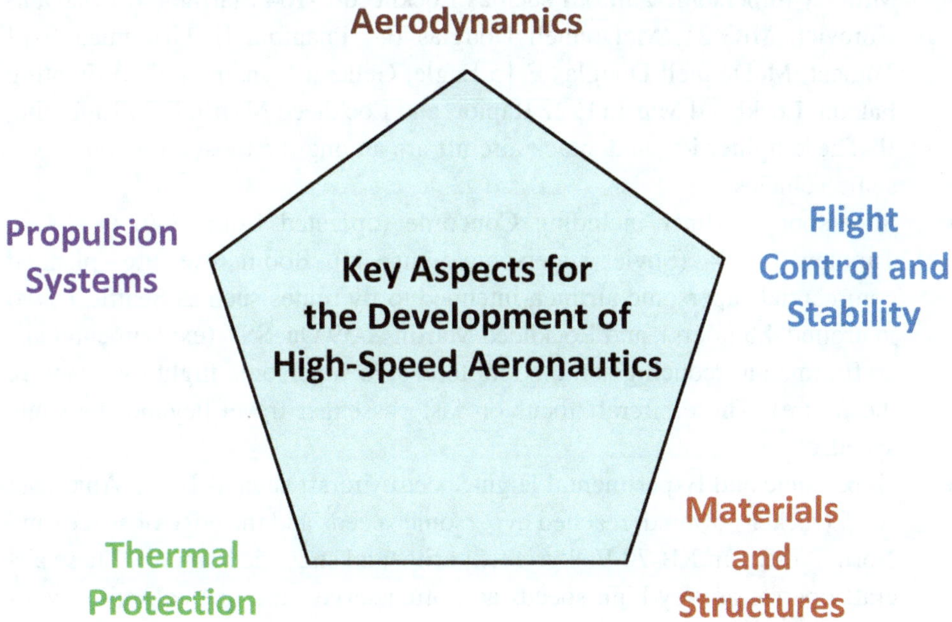

Characteristics of high-speed aeronautics include: 1) Airflow becomes compressible, meaning changes in pressure cause significant changes in air density, 2) The onset of shock waves and transonic flow regimes, where parts of the airflow over the aircraft can become supersonic even if the aircraft itself is subsonic, 3) Increased aerodynamic drag, known as wave, drag, and associated stability and control challenges, 4) Use of specialized aircraft designs such as swept wings to delay critical Mach effects and improve high-speed performance, and 5) Speed regimes typically studied include transonic (around Mach 0.8 to 1.2), supersonic (above Mach 1), and hypersonic speeds (above Mach 5).

The field of high-speed aeronautics is crucial for comprehending how aircraft perform, are controlled, and are structurally designed when flying at or close to the speed of sound. This includes military fighter jets, supersonic commercial planes, and spacecraft traveling through the atmosphere.

Once the fundamentals of high-speed aeronautics have been analyzed, aircraft operating at high speed are now discussed in the next subsection.

Aircraft Operating at High Speed

Examples of aircraft that operate within the realm of high-speed aeronautics, particularly in supersonic and fast flight categories, include the following:

- Military Supersonic Aircraft such as Lockheed F-104 Starfighter, Mikoyan-Gurevich MiG-21, McDonnell Douglas F-4 Phantom II, Grumman F-14 Tomcat, McDonnell Douglas F-15 Eagle, General Dynamics F-16 Fighting Falcon, Lockheed Martin F-22 Raptor, and Lockheed Martin F-35 Lightning II. These fighter jets and attack aircraft are among the most common supersonic vehicles.
- Supersonic Airliner including Concorde (operated from 1976 to 2003), Tupolev Tu-144 (Soviet supersonic transport), Boom Overture (planned commercial supersonic airliner, intended to fly routes such as Seattle-Tokyo in around 4.5 hours), and Lockheed Martin X-59 QueSST (experimental aircraft aimed at reducing sonic boom to enable supersonic flight over land in the future). These aircraft focus on fast passenger travel beyond the sonic speed.
- Hypersonic and Experimental High-Speed Aircraft such as North American X-15 (rocket-powered, reached hypersonic speeds and the edge of space) and North American XB-70 Valkyrie (experimental strategic bomber). These aircraft operate at very high speeds and often serve experimental, military, or specialized missions.

- Notable High-Speed Surveillance Aircraft as Lockheed Sr-71 Blackbird (reached speeds above Mach 3 for reconnaissance missions). This aircraft is iconic for sustaining high supersonic speeds for surveillance.

These airships represent a range of high-speed aeronautics applications, from military fighters and reconnaissance jets to commercial supersonic transport and experimental aircraft pioneering the future of fast flight.

After reviewing aircraft operating at high speed, the current status of materials applied to high-speed aeronautics is discussed in the following subsection.

Current Status of the Materials Applied to High-Speed Aeronautics

Materials have always played a crucial role in advancing aviation design, production, and safety. As the demand for quicker and more fuel-efficient planes rises, researchers and engineers are turning more into innovative materials to meet these challenges. Modern aircraft benefit significantly from composites, strong alloys, ceramics, and nanomaterials, which improve their performance, efficiency, and environmental sustainability (Lin, 2024).

The weight of an aircraft is essential in its design and functionality, directly impacting fuel efficiency, travel distance, velocity, and total operational expenses. Decreasing the aircraft's weight leads to lower fuel consumption, which benefits economic savings and reduces environmental harm. For example, trimming just 1% of weight can cut fuel usage by about 0.75%. Therefore, the aviation industry is always striving for ways to cut weight while preserving safety, strength, and longevity.

Cutting-edge materials are transforming aircraft design by increasing durability, reducing weight, and improving fuel economy. Developments like titanium alloys, aluminum-lithium alloys, composite materials, and other novel substances are essential to the aviation sector. The impact of these materials is outlined as follows (Lin, 2024; Soni et al., 2024):

- *Metal Matrix Composites (MMCs)* combine metals with ceramics or other materials to create composites with improved characteristics. MMCs offer outstanding thermal stability and strength while also being lightweight. They are particularly useful for high-temperature components such as engine parts (Soni et al., 2024).
- *Carbon Fiber Reinforced Polymers (CFRPs)* rank among the most used advanced materials in modern aircraft. Comprising carbon fibers embedded in a polymer matrix, they deliver an outstanding strength-to-weight ratio. CFRPs

are approximately five times lighter than aluminum and offer greater strength and stiffness (Simöes, 2024).

- *Smart materials,* which respond to external factors such as electrical fields, pressure, or temperature, are emerging as transformative components in aviation. These materials can change their shape, stiffness, or other characteristics in response to environmental variations. The use of smart materials allows for the creation of adaptive wing designs that improve aerodynamic efficiency, resulting in better fuel savings and enhanced flight performance (Rodinò et al., 2024).
- *Titanium alloys* are highly regarded for possessing outstanding strength along with high resistance to corrosion and fatigue, making them ideal for aircraft applications (Srivastava et al., 2024). While titanium is heavier than composite materials, its superior strength allows for the creation of thinner components, which contributes to lowering the overall weight. These alloys are particularly important for vital components such as engines, landing gear, and the airframe. The Airbus A350 XWB extensively employs titanium, providing durability while minimizing weight.
- *Aluminum-Lithium (Al-Li) alloys* are a rising advanced material within aeronautics (Xiao et al., 2024). Adding lithium to aluminum decreases the alloy's density, leading to weight reduction without sacrificing strength. These alloys are becoming increasingly favored in military and commercial aircraft due to their excellent fatigue resistance and improved performance. For example, the Boeing 777 incorporates Al-Li alloys, resulting in lighter airframes and enhanced overall performance.

The different engineering materials used in high-speed aeronautics that need further investigation are shown in Figure 2.

Figure 2. Cutting-edge materials for high-speed aircraft design

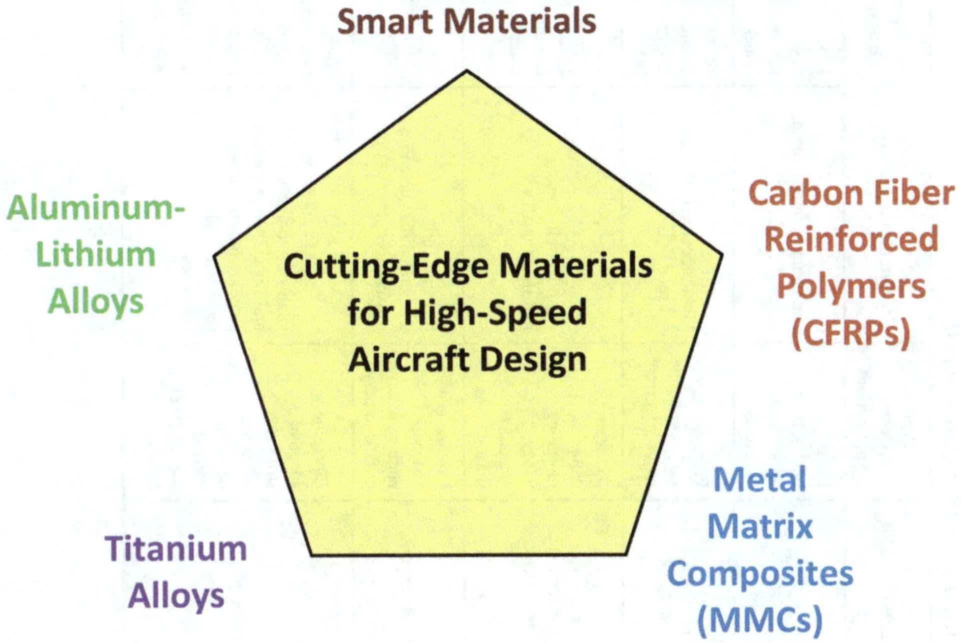

Table 1 presents a comparative summary of the performance characteristics of metal matrix composites, carbon fiber reinforced polymers, smart materials, titanium alloys, and aluminum-lithium alloys in high-speed aeronautics. This comparison focuses on key attributes such as strength-to-weight ratio, stiffness, temperature resistance, corrosion resistance, fatigue resistance, manufacturability, and typical applications in aircraft structures and components (Peters, 2024 et al.; Kessel et al., 2025; Roy, 2025).

Table 1. Comparison of the performance characteristics of materials used in high-speed aircraft

Material Type	Strength-to-Weight Ratio	Stiffness	Temperature Resistance	Corrosion Resistance	Fatigue Resistance	Manufacturability	Typical Aircraft Applications	Notable advantages	Main Limitations
Metal Matrix Composites (MMCs)	Very high	High	Excellent (esp. SiC/Ti)	Good	High	Complex, costlier	Engine blades, high-temperature structures	High temperature strength, reduced weight	Processing cost, fabrication issues
Carbon Fiber Reinforced Polymers (CFRP)	Very High	Very High	Moderate (thermoset:limited)	Excellent	Excellent	Widely adopted scalable	Fuselage, wings, surfaces, engine covers, interiors	Lightest, corrosion/fatigue resistant	Brittle failure, delamination risk
Smart Materials	Medium-High (varies)	Medium-High	Good (some up to 150°C)	Good	Varies	Application-dependent	Vibration damping, morphing wings, actuators, sensors	Adaptive/self-sensing, multifunctionality	Limited load bearing, complexity
Titanium Alloys	High	High	Excellent (up to ~600°C)	Excellent	Excellent	Difficult, expensive	Landing gear, fasteners, engine/airframe, critical joinery	Strength, temperature/fatigue/corrosion resistance	High cost, density
Metamaterials	High	Very high	Moderate to High	Variable	Variable	Complex	Emerging: adaptive structures, vibration damping, thermal management layers	Highly customizable multifunctional properties; potential for integrating sensing and adaptive behaviors.	High fabrication complexity and cost: limited operational heritage; scale-up challenges

continued on following page

Table 1. Continued

Material Type	Strength-to-Weight Ratio	Stiffness	Temperature Resistance	Corrosion Resistance	Fatigue Resistance	Manufacturability	Typical Aircraft Applications	Notable advantages	Main Limitations
Ceramic Materials	High	Very high	Very High (<2000 °C)	Excellent	Moderate	Challenging	Thermal protection systems, nose cones, and engine components exposed to extreme heat	Exceptional thermal, chemical, and oxidation resistance: lightweight and stiff	Brittleness; low fracture toughness; difficult to machine and repair.
Aluminum-Lithium Alloys	High	High (~10%↑)	Good (~150-200°C)	Excellent	High	Well established, improving	Fuselage, stringers, ribs, skin panels, empennage	8-16% lighter, good damage/fatigue tolerance	Anisotropy, crack sensitivity

The integration of these cutting-edge materials in aircraft design not only reduces weight but also greatly improves performance. The following are keyways where these materials enhance the capabilities of aircraft (Taleghani & Torabi, 2024):

1. *Enhanced fuel consumption*: Decreasing an aircraft's weight is directly linked to better fuel economy. A lighter plane needs less thrust, leading to reduced fuel use. This improvement is especially crucial since fuel expenses make up a large part of an airline's operating costs. Utilizing advanced materials enables manufacturers to build aircraft that consume less fuel, thereby reducing expenses and decreasing carbon emissions.
2. *Improved structural optimization*: Advanced materials have outstanding mechanical characteristics, allowing engineers to design aircraft structures that can withstand greater loads and stress. This enhanced durability can result in extended service life and lower maintenance expenses.
3. *Enhanced aerodynamics*: The use of advanced materials enables the creation of more streamlined designs. Their lightweight enables the creation of intricate shapes and structures that reduce drag and enhance lift. The design versatility offered by composites allows engineers to craft more efficient wing designs, leading to better performance during takeoff, cruising, and landing.
4. *Reduction of noise*: Advanced materials play a role in decreasing noise levels during aircraft operation. For example, materials with strong acoustic insulation abilities must be incorporated into cabin designs to enhance the degree of comfort. Additionally, lighter and more streamlined structures help achieve peaceful flights, improving the journey experience.

The different strategies used to improve high-speed aviation capabilities are represented in Figure 3.

Figure 3. Key technical strategies to enhance the capabilities of high-speed aircraft

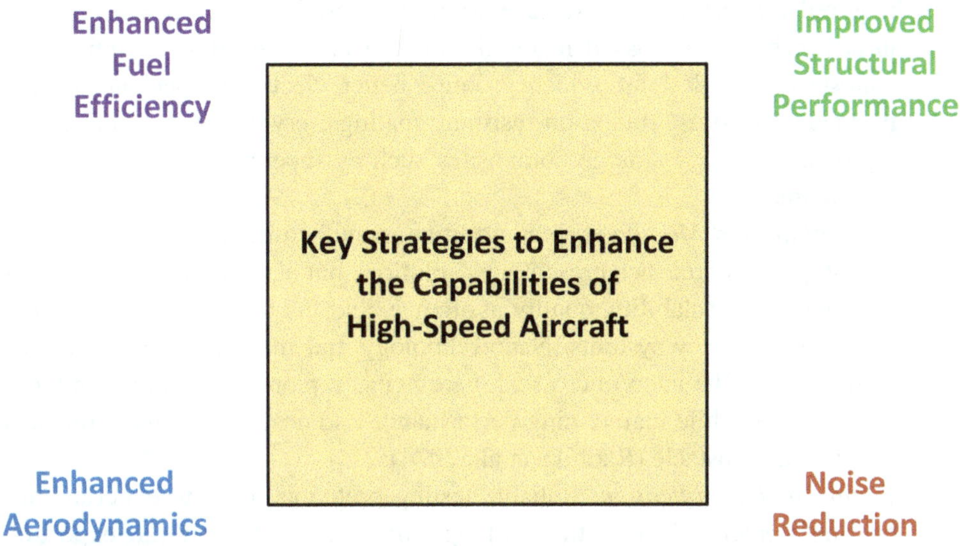

The incorporation of innovative materials like smart materials, MMCs, Al-Li Alloys, titanium alloys, and CFRPs is revolutionizing flight engineering and capabilities. These materials contribute to weight reduction while enhancing fuel efficiency, structural integrity, and aerodynamic performance.

High-speed aeronautics faces significant challenges in both materials and parts, as highlighted in recent high-impact journal articles. The main difficulties in high-speed aeronautics are described below. Having reviewed the current state of materials applied in high-speed aeronautics, the key technical challenges in aviation materials are discussed in the following subsection.

Key Technical Challenges in Aviation Materials

In the area of materials, the main challenges are shown in Figure 4 and described below:

- *Extreme Environment Performance*: Materials must withstand high temperatures, stresses, and rapid temperature changes during flight. They need to be strong, lightweight, non-corrosive, and resistant to UV radiation and impact. For space vehicles, materials must also endure low temperatures, low pressures, and high levels of radiation, all factors that increase the risk of material deterioration and catastrophic failure from impacts (e.g., space debris, meteoroids).

- *Advanced Composite Development*: Traditional alloys may not offer the required properties for next-generation vehicles. There is growing interest in advanced carbon-based materials, multilayer composites, ceramics, and nanostructured designs, which promise better electrical, mechanical, and thermal properties. Radiation-resistant coatings, cryogenic fuel tanks, and lightweight energy storage composites such as supercapacitors are key research areas.
- *Integration and Manufacturing*: Additive manufacturing, digital twins, and Industry 5.0 approaches present new solutions but also introduce challenges in production scalability, quality assurance, and the integration of new materials with legacy systems. Nanotechnology and metamaterials (materials engineered at the nano/micro-scale) are being explored yet scaling up these technologies while maintaining performance and cost-effectiveness remains a significant obstacle (Radhika et al., 2024).
- *Testing and Certification*: Reliable testing protocols for new materials, including accelerated-life testing for long-duration, high-temperature exposure, are essential. Understanding the long-term metallurgical changes in structural materials due to the sustained extreme environment of high-speed flight is critical to ensuring safety and longevity (Ghimire & Raji, 2024).
- *Radiation and Heat Management*: Materials for high-speed aeronautics not only need to be strong and light but also excel in dissipating heat and safeguarding against space radiation. Graphene, for example, demonstrates impressive heat dissipation and radiation shielding capabilities but faces hurdles in mass production and integration within composite aerospace structures (Kanti et al., 2025).

Figure 4. Key challenges in materials for optimizing the high-speed aircraft design

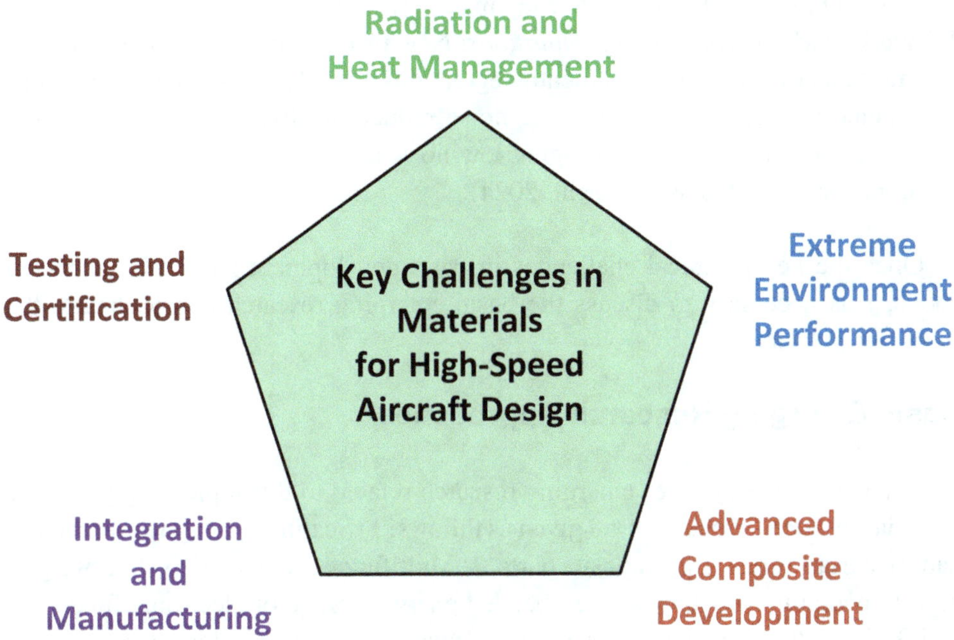

After describing the key technical challenges in aviation materials, the following subsection analyzes the key technical challenges in the development of aviation parts.

Key Technical Challenges in the Development of Aviation Parts

In the area of parts, the main challenges are described below:

1. *Propulsion System Components*: Engine components (such as turbine disks and nozzle assemblies) must endure far longer durations at high stress and temperature compared to subsonic or military aircraft. This places high demands on part durability, manufacturing precision, and material selection (Yu et al., 2025).
2. *Complex Geometries and Integration*: The push for mixed-functionality parts—such as morphing, deployable, or adaptive surfaces—demands materials compatible with advanced manufacturing (e.g., additive, digitally-driven processes) and capable of integrating sensors, thermal controls, and electromagnetic shielding directly into the part design (Farsadi et al., 2024).
3. *Structural Integrity and Weight Optimization*: Aircraft parts must maintain structural integrity at supersonic/hypersonic speeds while remaining ultralight to maximize fuel efficiency and payload capacity. This requires continual ma-

terial innovation and precise engineering, particularly in the assembly of large and complex propulsion systems (Lin, 2024).
4. *Noise and Environmental Regulations*: New propulsion materials and designs must meet stringent environmental regulations related to atmospheric emissions (including ozone depletion), and noise reduction, driving ongoing research into novel combustor technologies, low-noise nozzle systems, and eco-friendly composites (Giladi & Menachi, 2024).

Once the key technical challenges in parts development have been analyzed above, it is necessary to discuss the basic emerging research approaches in the following subsection.

Basic Emerging Research Approaches

Some critical aspects of emerging research related to this topic to improve the materials used in aeronautics are given as follows: 1) Scaling up graphene and other nanomaterials for use in composite parts, 2) Multifunctional materials: lightweight, high-temperature resistant with embedded energy storage or electronic functions, and 3) Data-driven materials design and smart manufacturing for quality control (Moenck et al., 2025).

In summary, solving these materials and parts challenges is foundational to advancing high-speed aeronautics, requiring interdisciplinary collaboration between materials science, mechanical engineering, manufacturing technologies, and aerospace operations.

After discussing the basic approaches of emerging research, possible methods for optimizing parts used in high-speed aeronautics are required to be discussed in the next section.

Possible Methods to Optimize the Parts Used in High-Speed Aeronautics

Recent advances in optimizing parts for high-speed aeronautics, as reported in high-impact journals, focus on sophisticated methods that balance computational efficiency, design accuracy, and multi-variable optimization across aerodynamics, structural integrity, and propulsion systems. Key optimization approaches include:

- *Phased collaborative design strategies*: Utilizing platforms such as OpenVSP (Open Vehicle Sketch Pad) combined with surrogate modeling (e.g., Kriging models) and advanced algorithms (differential evolution) enable rapid parametric modeling and iterative optimization (Du et al., 2025). This helps optimize critical geometric parameters of wing-body-propeller configurations to improve wing loading and propulsion efficiency while managing computational costs.

- *High-Fidelity Simulation Coupled with Low-Fidelity Models*: Integration of RANS (Reynolds-Averaged Navier-Stokes) for detailed aerodynamic simulations with lower fidelity methods like Vortex Lattice Method (VLM) and Actuator Disk Theory (ADT) achieves high-accuracy optimization results with reduced computational loads, enabling the exploration of numerous design variables such as wing twist, airfoil shape, and propeller geometry (Nikkhoo & Esmaeili, 2023).
- *Metaheuristic and Multi-Objective Evolutionary Algorithms*: Techniques such as Non-dominated Sorting Genetic Algorithm II (NSGA-II) and NSGA-III are effectively applied to optimize complex components like angular contact ball bearings designed for high-speed applications used in aircraft gearboxes (Kim et al., 2024). Objectives include minimizing power loss, maximizing lifespan, ensuring safety factors, and optimizing the lubricating film thickness under high-speed, high-load conditions. Robust algorithms modeled after natural evolutionary processes to address intricate design challenges. Some examples are Genetic Algorithms, Differential Evolution, and Quantum-Inspired Evolutionary Optimization (QIEO).
- *Multi-Fidelity Data Mining and Machine Learning Frameworks*: These frameworks leverage large datasets, combining inexpensive estimates with high-fidelity data, to accelerate aircraft design optimization and improve prediction accuracy of design performance in multi-disciplinary contexts. Tools such as MIT's TASOPT, TU Braunschweig's PrADO, and specialized frameworks for conceptual and preliminary design, incorporating multi-objective genetic algorithms and Monte Carlo methods, support these optimizations (Leng et al., 2024).
- *Adaptive and Local Optimization Techniques*: For control and guidance systems in interceptors or highly dynamic high-speed aircraft, local optimization methods minimize the performance functional for real-time trajectory control, simplifying implementation without solving complex boundary problems (Zhang et al., 2024). These imply combining aerodynamic shape optimization with active control technology to enhance high-speed vehicle performance.
- *Advanced Machining and Manufacturing Methods*: Approaches such as full 5-axis machining, high-torque/high-speed milling, and gantry profiling contribute to precision manufacturing of optimized parts with complex geometries, improving surface finish and structural properties essential for high-speed aeronautics (Zhao et al., 2024).

- *Structural Optimization*: Minimizing weight while maximizing the strength and durability of components like wings and fuselages. Techniques include topology optimization and material configuration optimization. It involves designing lightweight yet sufficiently strong parts (e.g., aerospace components) with sophisticated structural designs, often combined with additive manufacturing to produce optimized geometries (Hanush & Manjaiah, 2022).
- *Aerodynamic Shape Optimization*: Refining aircraft part shapes to improve aerodynamic performance, decreasing drag, and increasing fuel efficiency. Standard methods include genetic algorithms, hierarchical genetic algorithms (HGAs), and lift-constrained drag minimization algorithms (Serani et al., 2024).
- *Multidisciplinary Design Optimization (MDO)*: Integration of various engineering disciplines (aerodynamics, structures, propulsion) to optimize complex aerospace systems (Liu et al., 2024). Coupled aerodynamic-structural optimization and impact dynamics optimization for crashworthiness and performance. Some example methods are the Multidisciplinary Feasible (MDF) and Collaborative Optimization (CO).
- *Trajectory Optimization*: Optimizing flight paths and trajectories to reduce fuel consumption and increase mission efficiency. Relevant for spacecraft and high-speed aircraft missions, involving orbital mechanics and flight path algorithms (Gao et al., 2024).
- *Surrogate and Reduced Order Modeling*: Using surrogate models to approximate complex simulations, reducing computational costs while retaining critical accuracy. Reduced Order Models (ROM) capture essential physics for faster computational performance (Pang et al., 2024).
- *High-Performance Computing Techniques*: Leveraging GPU (Graphics Processing Unit) acceleration, parallel processing, and specialized hardware to speed up optimization simulations. So far, significant efficiency improvements have been achieved using Message Passing Interface (MPI) and GPU-accelerated computational fluid dynamics (CFD) solvers (Liu et al., 2024).

From the above, it can be concluded that optimizing high-speed aeronautical components requires the extensive application of evolutionary computational algorithms, comprehensive computational software, and multi-robust simulations, combined with flexible manufacturing systems. These tools make it possible to improve operational performance, mechanical reliability, and aerodynamic efficiency under the restrictive conditions of high-speed flight. Furthermore, they lead to weight and fuel savings, thereby improving structural integrity and optimizing overall system performance in high-speed aeronautics. The specific choice of one of these methods, or a combination of them, depends on the various multidisciplinary factors required by the specific component or system. The role that aluminum-lithium alloys play in the design of aircraft parts is described below, with an emphasis on traditional alloys, explaining their evolution, as well as their advantages and disadvantages.

Once the optimization methods of high-speed aviation parts have been analyzed, it is required that the practical restrictions for the comprehensive use of high-speed Al-Li alloys are discussed in the following subsection.

Practical Constraints for the Exhaustive Use of Aluminum-Lithium Alloys

The use of aluminum-lithium (Al-Li) alloys in high-speed aeronautics entails several key considerations (Dongming, 2024):

1. *Certification*: Aluminum-lithium alloys require full certification under existing aerospace regulations, but due to lithium's unique properties (flammability and toxicity), special certification conditions may apply. Research and testing are needed to address safety aspects such as fire toxicity and structural performance under crash scenarios.
2. *Cost and Manufacturability*: These alloys are more expensive than traditional aluminum alloys because of specialized production processes and stringent quality control. However, advancements in manufacturing, including laser beam welding and friction stir welding, improve manufacturability at scale and offer weight savings for aircraft components. The market for aerospace Al-Li alloys is growing steadily, driven by demand for lightweight, fuel-efficient aircraft.
3. *Sustainability and Lifecycle*: Aluminum-lithium alloys support sustainability due to their recyclability without degradation of mechanical properties. Closed-loop recycling systems enable the reuse of production scrap and end-of-life aircraft materials, reducing environmental impact and supporting circular economy principles. Recycling reduces energy consumption and greenhouse gas emissions compared to primary production.

In summary, Al-Li alloys in high-speed aeronautics balance advanced certification requirements and higher costs with benefits in manufacturability, weight reduction, and sustainability throughout the material lifecycle.

After describing the possible restrictions for the comprehensive practical use of Al-Li alloys, it is necessary to review the challenges of certification and qualified testing requirements in the following subsection.

Certification Challenges and Qualified Testing Requirements for Al-Li Alloys

The certification challenges and qualified testing requirements for aluminum-lithium (Al-Li) alloys used in high-speed aeronautics are notable due to the advanced properties and complexities of these materials. Al-Li alloys, especially third-generation ones, offer significant benefits such as lower density, higher specific stiffness and strength, and improved damage tolerance compared to traditional aluminum alloys (Bouchaala, 2023). However, these alloys exhibit anisotropic mechanical properties and require tailored thermo-mechanical processing to ensure microstructural stability, fracture toughness, corrosion resistance, and weldability. Certification challenges include ensuring consistent material quality, managing anisotropy, and validating performance under complex, multidirectional stresses typical in aerospace structures. Qualified testing demands a thorough evaluation of mechanical properties, fracture toughness, crack growth resistance, thermal stability, and corrosion behavior. Joining processes such as laser beam welding and friction stir welding require qualification due to their impact on material performance. The certification process reflects the need for extensive mechanical testing under operational conditions, ensuring reliability and performance for critical aircraft components like fuselage panels and wing structures.

The following section analyzes the role of aluminum-lithium alloys in aircraft design to understand the status and why additional research is needed for emerging generations.

THE ROLE OF THE ALUMINUM-LITHIUM ALLOYS IN AIRCRAFT

Aluminum-lithium (Al-Li) alloys are sophisticated, lightweight materials widely utilized in aeronautics because they possess a distinctive blend of properties that satisfy the stringent demands of aerospace applications.

Key technical characteristics to optimize the aluminum-lithium alloys, depicted in Figure 5, include (Wanhill, 2014; Rambabu et al., 2017):

- *Low Density and Weight Savings*: Adding lithium to aluminum reduces the alloy's density by approximately 7-10%, leading to significant weight reduction in aircraft structures. This weight savings directly contributes to enhanced fuel economy and increased load-carrying capacity.
- *High Strength and Stiffness*: Al-Li alloys have 10-15% higher modulus of elasticity, meaning they are stiffer than conventional aluminum alloys. They

also offer increased specific strength, making them suitable for structural components that must endure high stress while minimizing weight.

- *Improved Fatigue and Corrosion Resistance*: They exhibit excellent fatigue resistance and better corrosion resistance, which enhances the durability and lifespan of aircraft parts.
- *Thermal Stability*: New generation Al-Li alloys demonstrate good thermal stability, maintaining mechanical properties at a range of operating temperatures, including extremely low temperatures, which makes them ideal for aerospace environments.
- *Applications in Aerospace Components*: These alloys are extensively utilized in essential aircraft components including leading and trailing edges, fuselage panels, seat tracks, access covers, and control surfaces. They also play vital roles in space applications, such as cryogenic fuel tanks for rockets, due to their superior performance at low temperatures.
- *Challenges*: Although aluminum-lithium alloys offer many benefits, they may exhibit lower ductility and fracture toughness compared to certain traditional aluminum alloys. Their tendency to oxidize at elevated temperatures requires careful handling during manufacturing, especially in processes involving molten metal. Welding and joining methods remain areas of ongoing development to optimize the use of Al-Li alloys.

Figure 5. Key technical characteristics to optimize the aluminum-lithium alloys

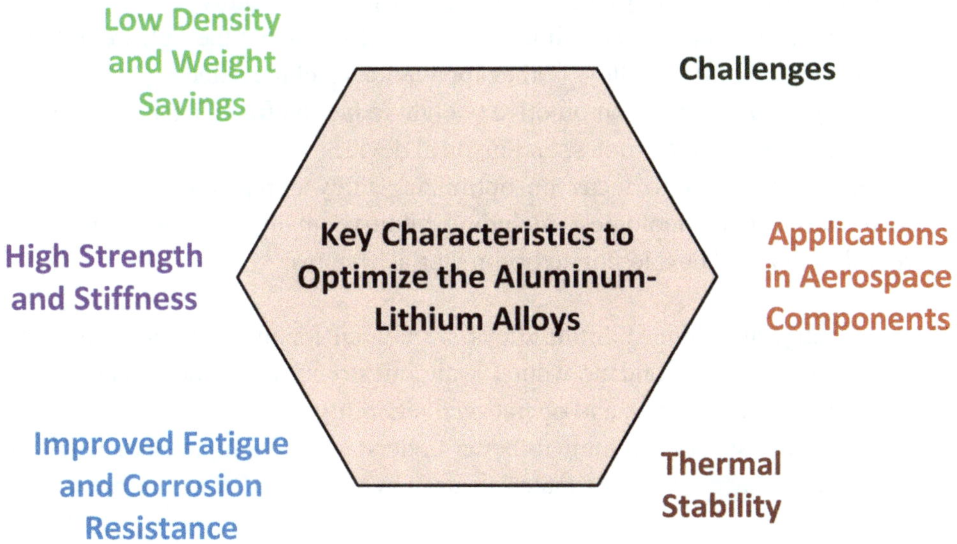

Manufacturing techniques for Al-Li alloys have evolved to include advanced processes like rapid solidification and powder metallurgy to refine their microstructure and optimize mechanical properties for aerospace needs. The impact of manufacturing methods on the characteristics of aluminum-lithium (Al-Li) alloys is listed as follows (Mishra, 2025):

1. The manufacturing process involves multiple essential stages: raw material selection, alloy design, melting and casting, alloy modification, degassing to remove impurities, heat treatment, and quality control.
2. Raw materials such as high-purity aluminum, lithium, and other alloying elements (zirconium, magnesium, copper) are precisely selected and melted under controlled conditions to ensure a homogeneous molten metal.
3. Innovative manufacturing methods like rapid solidification and powder metallurgy have been increasingly adopted (Li S et al., 2023). Rapid solidification leads to refined microstructures, improving mechanical strength and fatigue resistance. Powder metallurgy involves blending fine powders under heat and pressure for uniform lithium distribution and enhanced properties.
4. Casting methods vary depending on product complexity and volume, including sand casting, die casting, and investment casting.
5. After casting, heat treatment methods like solution heat transfer and precipitation hardening are used to further improve mechanical properties including strength and toughness.
6. Quality control techniques, including non-destructive testing and chemical analysis, ensure the alloys meet required aerospace standards.
7. The aerospace sector is the primary user, benefiting from the lightweight and durable nature of Al-Li alloys. Other applications include defense (military aircraft, armored vehicles), automotive (weight reduction for efficiency), marine, sports equipment, electronics, and medical devices.
8. Research continues to focus on optimizing alloy compositions, improving manufacturing technologies, including additive manufacturing, enhancing recyclability, and developing hybrid material systems.

In summary, aluminum-lithium alloys are crucial for modern aeronautics by enabling lighter, stronger, and more durable aircraft structures that deliver enhanced performance, fuel efficiency, and operational reliability.

The following subsection aims to better understand the unique characteristics that Al-Li alloys have over other materials used in high-speed aeronautics.

Why Choose Al-Li Alloys on Others Aeronautical Materials for High-Speed Aeronautics?

Aluminum-lithium (Al-Li) alloys are often chosen over other aeronautical materials for high-speed aeronautics because they provide an excellent combination of lightweight and high performance, crucial for this demanding application:

- They have 7-10% lower density compared to conventional aluminum alloys, significantly reducing structural weight, which is critical for fuel efficiency and speed in aircraft.
- They offer 10-15% higher modulus of elasticity (stiffness) and up to a 26% increase in specific modulus compared to common alloys like 7075-T651, providing enhanced structural rigidity without weight penalty.
- These alloys achieve considerable weight savings while maintaining or improving strength and damage tolerance, which improves overall aircraft performance, maneuverability, and payload capacity.
- Al-Li alloys exhibit good fatigue resistance and damage tolerance that is vital in high-cycle, high-stress environments typical of high-speed flight.
- They perform well at both room and cryogenic temperatures, making them suitable for aerospace systems exposed to extreme thermal conditions, such as liquid oxygen and hydrogen fuel tanks.
- Their relatively good corrosion resistance compared to older aluminum alloys helps ensure the durability and longevity of aircraft structures.
- Additionally, Al-Li alloys facilitate the manufacture of complex and thin structural sections due to their combination of mechanical properties and formability.

In summary, Al-Li alloys stand out because they balance low weight, high stiffness, fatigue resistance, and thermal performance, all critical to the structural demands and weight sensitivity of high-speed aerospace vehicles. This balance makes them superior to many traditional aluminum alloys and certain composites or other metals used in aeronautics.

Next, the generations of aluminum-lithium alloys that have been designed so far will be reviewed.

Generations of Aluminum-Lithium (Al-Li) Alloys

The following provides an overview of the different generations of aluminum-lithium (Al-Li) alloys used in aeronautic applications (Wu, 2022; Xiao et al., 2024):

1. First Generation (1920s-1960s)
    - Early alloys had lower lithium content (~0.1% to ~2.1%).
    - Examples: Alloys 1230, 1420, and 2020.
    - Used in early military aircraft like Tu-144, MiG-29, and the A-5 Vigilante.
    - Intended as replacements for conventional aluminum, they had limitations such as lower strength or toughness.
2. Second Generation (1970s-1980s)
    - Higher lithium content (around 2% or more).
    - Examples: Alloys 1430, 1440, 1450, 2090, 2091, and 8090.
    - Provided significant density reduction (about 8-10%), leading to considerable weight savings.
    - Used in larger and more advanced aircraft including Airbus A330/A340, C-17 Globemaster, Fokker aircraft, and the EH-101 helicopter.
    - However, higher lithium caused issues like reduced fracture toughness and anisotropic mechanical properties, limiting widespread adoption.
3. Third Generation (1990s-Present)
    - Lithium content reduced to approximately 0.75-1.8%, balancing weight savings and mechanical performance.
    - Examples: Alloys 2050, 2096, 2196, 2198 (Chen et al., 2022), and 2297.
    - Widely accepted and used in commercial and military aerospace programs.
    - Exhibits improved fracture toughness, corrosion resistance, and thermal stability compared to earlier generations.
    - These alloys are employed in various aircraft including Boeing 787, Airbus A380 and A350, Falcon 9 rockets, as well as military aircraft like the F-16.

Overall, as generations progressed, lithium content was optimized to reduce density and improve stiffness while addressing deficiencies in toughness and resistance to corrosion. Third-generation aluminum-lithium alloys currently offer a good combination of lightweight properties, strength, and durability for modern high-performance aerospace applications. In Table 2, a summary of the evolutive process of aluminum-lithium alloys is presented.

Table 2. Evolution of emerging requirements for aluminum-lithium alloys

Historical Period	Physical Properties
1903-1930s	Strength under static load.
1930s-1960s	Strength under static load, anti-corrosive properties, and fatigue strength.
1960s-1970s	Strength under static load, anti-corrosive properties, thermal stability, durability, resistance to harm, and superior behavior in thick structural parts.
1980's	Significant weight reduction.
1990's-today	Strength under static load, anti-corrosive properties, durability, resistance to harm, superior behavior in thick structural parts, symmetrically balanced attributes, lower expenses for manufacturing and assembly, affordable servicing costs, significant reduction in weight, and an advanced level of security.

Now that the generations of Al-Li alloys have been analyzed, the role of traditional alloys used in aircraft design is discussed in the next subsection.

Traditional Al-Li Alloys Used in Aircraft

Traditional aluminum-lithium (Al-Li) alloys used in aeronautics are specialized lightweight materials designed mainly to decrease the weight of aircraft and aerospace structures while maintaining or enhancing strength and performance. These alloys incorporate lithium into aluminum, which significantly lowers the alloy's density and increases the specific strength – meaning high strength relative to weight. The addition of lithium also improves fatigue and corrosion resistance with respect to conventional aluminum alloys (Sun et al., 2022; Jia et al., 2025).

Key characteristics and uses of traditional aluminum-lithium alloys in aeronautics include:

1. *Weight reduction*: Because lithium has a low density, these alloys contribute to reducing the weight of aircraft, thereby enhancing fuel efficiency and allowing for greater payload capacity.
2. *High load-bearing capacity*: This blend enables the production of components that are both lighter and structurally robust, including wings, fuselage panels, and landing gear elements.
3. *Improved fatigue resistance*: Aluminum-lithium alloys endure cyclic loading better, which is important for long-life aircraft.
4. Satisfactory corrosion resistance: Under harsh environments, this increases the durability of the parts.

5. Typical alloy types: Traditionally used alloys include materials developed between the 1970s and mid-1980s, including CP276, Weldalite 049, 2091, 2090, and 8090.
6. Manufacturing process: Manufacturing and processing techniques such as casting, alloying, and various heat treatments must be scrupulously controlled to optimize the operational performance of these alloys.

It can be summarized, then, that thanks to their ability to offer significant weight savings without compromising mechanical strength or durability, traditional aluminum-lithium alloys are an essential material in aerospace engineering, as they have facilitated more advanced and efficient aircraft designs.

The key characteristics of the traditional aluminum-lithium alloys 8090, 2090, 2091, Weldalite 049, and CP276 used in aeronautics have been compiled in Table 3 to present a comparative summary of their properties.

Table 3. Key features of traditional aluminum-lithium alloys used in aeronautical parts

Alloy	Lithium Content (wt%)	Copper Content (wt%)	Other Major Elements	Key Features	Typical Applications
8090	~2.45	~1.3	Mg (~0.95), Zr (~0.12)	Damage tolerant, medium - strength, 10% lower density than 2024, higher modulus, good corrosion resistance.	Sheets, plates, extrusions, forgings, and welded structures.
2090	~2.2	~2.7	Zr (~0.12)	High strength, ~8% lower density than 7075-T6, improved fatigue resistance, and toughness.	High-strength aircraft structure components.
2091	~2.0	~2.1	Zr (~0.10)	Like 2090 but with variable fatigue performance, substitution for 2024 aluminum.	Aircraft structural parts.
Weldalite 049	~1.3	~5.4	Mg (~0.4), Ag (~0.4), Zr (~0.14)	Very high strength (tensile up to 700 MPa), excellent natural aging, good weldability, and improved toughness.	Plates and weldable aerospace components.
CP276	~2.2	~2.7	Mg (~0.5), Zr (~0.12)	Advanced strength and fracture toughness with balanced mechanical properties.	Aerospace structural applications.

It is also important to analyze some specific characteristics:

- Considerable weight reduction is achieved because all these alloys have a lower density.
- The combination of adequate mechanical and corrosion resistance in the 8090 alloy favors its excellent damage tolerance.
- High strength with varying fatigue life and toughness can be achieved by using the 2090 and 2091 alloys (Abdullatef et al., 2023).
- The addition of silver and magnesium in the Weldalite 049 alloy contributes to high strength and weldability (Kridli et al., 1998).
- A balance between strength, toughness, and ease of fabrication can be achieved by using the CP276 alloy (Meyer et al., 1987).

The advantage of using these alloys lies in the ability to achieve the appropriate balance between damage tolerance and corrosion resistance, as well as the ideal manufacturing methods required by the various aerospace structural components. Thanks to the evolution of materials, it is possible to further optimize their operational performance, as described in the following section, which reviews recent and innovative technical advances in aluminum-lithium alloys for these applications.

After analyzing traditional Al-Li alloys used in aviation, the effect of human factors and integration with design on the safety and reliability of aluminum-lithium parts is now discussed in the next subsection.

The Effect of Human Factors and Integration With Design in Safety and Reliability of Aluminum-Lithium Parts

In the design of aluminum-alloy parts for high-speed aeronautics, it is essential to integrate human factors, maintainability, operational considerations, inspection, repair in real operational contexts, and ease of repair to ensure safety, reliability, and cost-effectiveness (Pollock et al., 2024b).

Human factors require that parts be designed for ease of access and handling by maintenance personnel during inspections and repairs, minimizing complexity and reducing the risk of human error. Maintainability involves designing components for straightforward removal, replacement, or servicing, which reduces aircraft downtime and operational costs. Operational considerations include ensuring parts can withstand the environmental and mechanical stresses experienced during high-speed flight, such as fatigue, corrosion, and thermal effects, while also enabling efficient inspection techniques.

Inspection and repair in real operational contexts must allow for reliable detection of damage or wear (e.g., cracks or corrosion) using mechanisms compatible with typical maintenance facilities. The ease of repair mandates that the alloy parts are amenable to standardized repair processes, such as mechanical fastening, welding, or patching, without compromising material properties. Designing for non-destructive evaluation compatibility and accessible fasteners or weld zones facilitates timely and effective maintenance.

Overall, incorporating these factors ensures that aluminum-alloy components in high-speed aeronautics not only meet stringent performance criteria but also promote operational efficiency, safety, and cost-effectiveness throughout the aircraft lifespan.

The safety and reliability issues of aluminum-lithium-based parts are relevant for choosing which characteristics should be optimized in alloys dedicated to high-speed aviation, as studied in the following section.

NEXT-GENERATION ALUMINUM-LITHIUM ALLOYS

A summary of novel and recent aluminum-lithium (Al-Li) alloys used in aeronautics (Rao & Padal, 2022) is given here:

1. The third generation of Al-Li alloys features considerably reduced lithium content (below 2%) compared to earlier versions. This reduction improves thermal stability, fracture toughness, and overall mechanical performance.
2. Recent alloys incorporate additional elements like scandium (Sc) and magnesium (Mg), which enhance strength, weldability, and damage tolerance, while maintaining low density and high stiffness. The multi-element combinations are customized for aerospace uses.
3. These novel Al-Li alloys show excellent fatigue and corrosion resistance, as well as fracture toughness, rendering them appropriate for essential structural components such as fuselage skins, wings, and internal load-bearing parts.
4. In the pursuit of improved microstructures and superior mechanical performance, the application of emerging manufacturing techniques such as rapid solidification, powder metallurgy (Li S, et al. 2023), friction stir welding (FSW) (Chen J et al., 2024a), and laser beam welding (LBW) (Vivas et al., 2024) is possible.
5. The competitive advantage among alloys and composites to enable cost-effective, lightweight, and durable metallic structures has allowed companies such as Airbus and Alcoa to integrate these new alloys into aircraft such as the A350, A400M, and the Boeing 787.
6. The feasibility of increasing the life cycle of parts through high recyclability, longer material lifespans, and reduced maintenance has led to the development of highly desirable "worry-free" structures in aeronautical parts.
7. New Al-Li alloys show potential for emerging aerospace markets because they offer a balance of toughness, strength, weight savings, and manufacturability,

addressing evolving design challenges in supersonic and hypersonic aircraft as well.

In summary, these recent Al-Li alloys represent a significant evolution from traditional versions, emphasizing lower lithium levels, novel alloying elements, and advanced processing methods to satisfy the rigorous demands of future aerospace applications.

Key comparative technical features of novel and recent aluminum-lithium alloys, primarily representing the last generation Al-Li alloys used in aeronautics, are summarized in Table 4.

Table 4. A technical comparison of novel and recent aluminum-lithium alloys used in aeronautics

Alloy	Lithium Content (wt%)	Copper Content (wt%)	Other Key Alloying Elements	Key Features and Advances	Typical Applications
2050	~1.0-1.5	High (~2.5)	Mg, Zr	Improved thermal stability and fracture toughness; strong T1 (Al_2CuLi) phase strengthening; better fatigue and corrosion resistance.	High-performance aerospace structural parts.
2096	~1-5-2.0	High (~3.0)	Mg, Zr, Mn	Balanced strength and toughness with improved manufacturability, and good stress corrosion cracking resistance.	Aircraft structural applications.
2195	~1.0-1.2	Moderate (~1.7)	Cu, Ag, Mg, Zr	Used in Boeing 787 fuselage; moderate lithium for excellent weldability and damage tolerance.	Fuselage panels and aerospace structural parts.
2198	~1.2-1.4	Elevated (~2.0)	Mg, Zr	Enhanced fracture toughness and corrosion resistance; engineered precipitate microstructure.	Advanced aerospace structural components.
2297	~1.5	High (~2.5)	Mg, Zr	Optimized for high strength, fracture toughness, and fatigue resistance; advanced weldability.	Wing and fuselage structure components.

Some specific characteristics must also be discussed:

- Third-generation Al-Li alloys generally contain lower lithium percentage (<2%) than earlier generations to improve thermal stability and toughness.
- The Cu/Li ratio is enhanced, which favors the formation of the T1 phase (Al_2CuLi) that strongly strengthens the alloy (Lv et al., 2019; Chen et al., 2024).
- These alloys exhibit higher fracture toughness, improved fatigue crack growth resistance, and better corrosion resistance with respect to older Al-Li alloys.

- New production processes tailor the microstructure for optimal mechanical properties, including additive manufacturing advancements (Li S. et al., 2023).
- Widely applied in modern commercial aircraft fuselage and wing components, these alloys balance structural efficiency with manufacturability and durability.

This summary reflects the evolution of aluminum-lithium alloys in aerospace towards alloys with more balanced compositions, enhanced mechanical properties, and improved lifecycle performance.

Having studied the basic properties of aluminum-lithium alloys, the contributions of potential alloying agents to enhance high reliability performance for high-speed aeronautical applications are reviewed in the next subsection.

Emerging Alloying Elements for Al-Li Alloys

In aluminum-lithium (Al-Li) alloys used in high-speed aeronautics, magnesium (Mg), zirconium (Zr), and scandium (Sc) play critical roles in enhancing mechanical performance, manufacturability, and structural efficiency (Gancarz et al., 2021), as is described below:

Role of Magnesium (Mg):

- It acts as a solid solution strengthener in Al-Li alloys by dissolving into the aluminum matrix, that enhances the alloy’s strength and hardness.
- It modifies mechanical properties during the precipitation process by reducing the solubility of lithium in the alloy.
- It improves the damage tolerance and fatigue resistance of aerospace structures under cyclic loading.
- It increases the alloy's mechanical strength through the introduction of the T1 reinforcing phase (Al_2CuLi).
- It optimizes performance and increases structural longevity by increasing the corrosion resistance and weldability of parts made with these alloys during high-speed flights.

Role of Scandium (Sc):

- It improves grain refinement by inhibiting recrystallization through the formation of coherent Al_3Sc precipitates in the alloy's microstructure.
- It increases mechanical strength through the hardness introduced by Al_3Sc precipitates, which anchor dislocations and grain boundaries.

- It maintains the high-temperature strength typical of high-speed aeronautics by improving thermal and microstructural stability.
- It optimizes traditional aerospace aluminum alloys by increasing fatigue and damage tolerance, as is the case with Scalmalloy®.
- It enables the manufacture of complex aerospace components by providing improved weldability and corrosion resistance.
- It reduces weight and increases the performance of critical structural components such as wing spars and fuselage elements, although it increases the final cost of the part.

Role of Zirconium (Zr):

- It improves corrosion resistance, structural stability, and mechanical properties of aeronautical parts.
- It optimizes strength and toughness through grain refinement and recrystallization control due to the presence of coherent Al_3Zr precipitates during heat treatment, which inhibit grain growth and recrystallization.
- It increases alloy strength through precipitation hardening caused by dislocations introduced by Al_3Zr particles.
- It maintains mechanical properties during high-speed flights that produce significant thermal loads, thanks to Zr-containing precipitates, which are stable at high temperatures.
- It reduces susceptibility to corrosion by refining the grain structure due to the presence of Zr, which increases the structural hardness of aeronautical parts.
- It offers improved weldability and workability, required for the manufacture of complex aerospace parts requiring high mechanical performance and fatigue resistance.

The possible alloying elements for Al-Li alloys in high-speed aircraft are illustrated in Figure 6.

Figure 6. Emerging alloying elements for Al-Li alloys in high-speed aeronautics

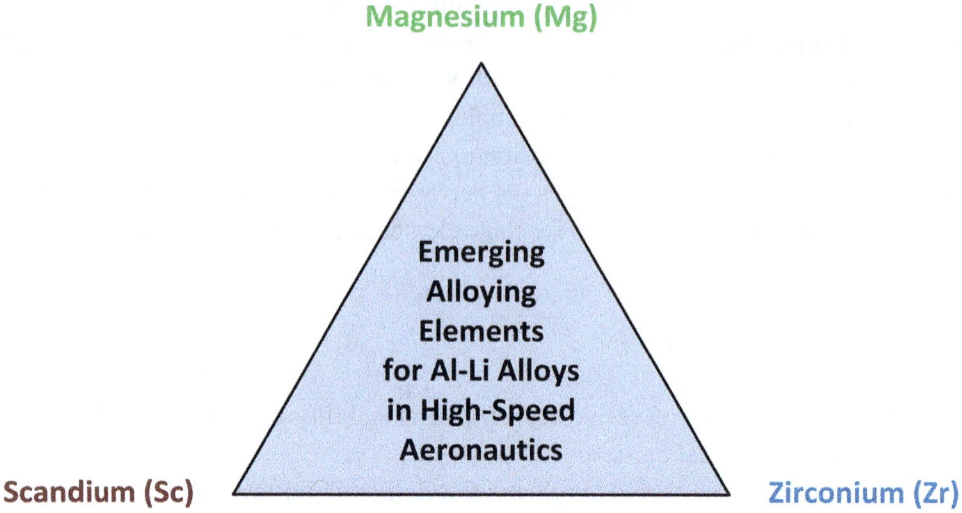

The addition of Mg and Sc to Al-Li alloys results in durable, strong, and lightweight materials necessary for optimizing high-speed applications where increased service life, improved safety, and reduced fuel consumption are required. Zirconium alloying is vital for applications where high-speed aircraft structures subjected to mechanical and thermal stresses require optimization of corrosion resistance, microstructural stability, and strength using emerging flexible manufacturing methods.

Undoubtedly, emerging alloying elements have reinforced the interest in new alloys that can be achieved, but it is now necessary to discuss the advances that can be achieved with respect to manufacturing processes and performance to optimize their operation at high speeds of parts based on aluminum-lithium alloys in the following subsection.

Recent Technical Advances in Aluminum-Lithium Alloys

To meet the stringent demands of aerospace applications such as sustainability, optimization of mechanical properties and their manufacturing techniques, this subsection describes the recent technical advances in aluminium-lithium (Al-Li) alloys used in high-speed aeronautics that can be exploited for the design of aeronautical parts (Zhang, 2022; Guo et al., 2025; Mishra, 2025; Zheng, 2025).

Advances in Alloy Composition and Microstructure

- New generations of Al-Li alloys are being improved with optimized alloying elements, micro-alloying, and nano-structuring to enhance strength, ductility, fatigue resistance, and corrosion resistance.
- Lithium additions continue to reduce alloy density while increasing stiffness and specific strength, critical for lightweight aerospace structures.
- Refined microstructures are achieved through rapid solidification processing, resulting in smaller grain sizes that improve mechanical performance (Li S. et al., 2023).

Advanced Manufacturing Processes

- Powder metallurgy and rapid solidification techniques allow for more uniform lithium distribution and better control of microstructure, producing alloys with superior mechanical properties (Li S. et al., 2023).
- Additive manufacturing (3D printing) technologies have begun to be applied, enabling complex geometries with tailored material properties suited for aerospace components.
- These advanced manufacturing methods contribute to weight reduction while maintaining or improving structural strength and durability.

Sustainability and Recycling

- Emerging manufacturing and processing processes must be environmentally friendly by promoting the reduction of energy consumption, waste, and greenhouse gas emissions when alloys are produced, disposed of, and recycled.
- The possibility of recovering lithium and other alloying materials from aluminum alloys should be a novel strategy to promote sustainable manufacturing processes.

Hybrid Material Systems and Multifunctionality

- The combined use of composite materials and aluminum-lithium alloys is being investigated to produce lightweight, high-strength hybrid structures for high-speed aeronautics.
- The potential to exploit properties such as improved sensing and thermal management, as well as self-healing, may be possible with the use of multifunctional materials of multiple natures and physical sizes.

Digitalization and Industry Impact

- Optimizing manufacturing processes and improving the predictability and quality of materials can be achieved using data analytics, artificial intelligence, and digital technologies.
- In addition to the space sector, sectors such as renewable energy, automotive, and others are seeking viable solutions based on lightweight, high-performance aluminum-lithium alloys.

From the above, it can be concluded that the optimization of lightweight and durable aluminum-lithium alloys for high-speed aeronautical applications must incorporate numerous current and emerging technical advances to improve alloy properties, novel design techniques, and manufacturing and processing processes that offer multifunctional performance while ensuring low environmental impact and high sustainability.

The recent technical advances described above were highlighted in the required depth, but multidisciplinary optimization also needs to leverage emerging manufacturing and processing techniques, as discussed in the next subsection.

Emerging Techniques in Manufacturing and Processing of Aluminum-Lithium Alloys

To complete the above discussion, it is necessary to analyze recent technical advances in the manufacturing and processing of aluminum-lithium (Al-Li) alloys used in high-speed aeronautics. These focus on improving material quality, manufacturing efficiency, mechanical performance, and microstructure control (Hajjioui et al., 2023; Mishra, 2025). A list of the main technical advances is provided below:

Raw Material Preparation and Alloy Design

- The precise control of the chemical composition of alloy elements such as copper, magnesium, zirconium and scandium in the aluminum-lithium alloy must be optimized according to the specific application of the part to manufacture.
- Properties such as corrosion resistance, lightness, and strength can be optimized from alloy design using computer -advanced modeling and simulation.

Melting and Casting Techniques

- The manufacture of aeronautical parts requires reaching a homogeneous molten alloy, which can be performed through controlled processes based on gas crucible or induction ovens.

- The definitive elimination of hydrogen impurities in the alloy, which prevents porosity and provides mechanical integrity, is achieved using degassing processes using vacuum treatment and degassing agents.
- The possibility of combining different manufacturing processes such as those based on investment, matrix, and the sand casting favors the possibility of producing complex parts and different production volumes.

Rapid Solidification and Powder Metallurgy

- The reduction of grain size, microstructure refinement, as well as the improvement of mechanical fatigue and strength are possible through rapid solidification processes.
- A uniform distribution of the alloy elements in an alloy provides improved mechanical properties through powder metallurgy, where the particles are consolidated to high pressures and temperatures.
- These methods enable tailoring microstructures to withstand the demanding conditions of high-speed aeronautics better.

Heat Treatment and Quality Control

- Heat treatments processes like solution heat treatment and precipitation hardening enhance the alloy strength and toughness of the alloy.
- Extensive quality control, involving mechanical testing, chemical analysis, and non-destructive inspection (X-ray, ultrasonic), ensures rigid compliance with aerospace standards.

Emerging Strategies

- Additive manufacturing (3D Printing) is gaining attention for its capability to create complex, lightweight aerospace components with aluminum-lithium alloys (Li S. et al., 2023).
- Hybrid manufacturing approaches and integration with composites are being explored to optimize performance further.
- Industry efforts also focus on improving recycling and sustainable production processes while maintaining alloy quality.

In summary, recent advances in manufacturing aluminum-lithium alloys for high-speed aeronautics involve refined raw material preparation, novel rapid solidification and powder metallurgy techniques, precision casting and melting, rigorous heat treatment, and enhanced quality control. The extreme conditions found in the aerospace area imply the need to develop fatigue resistant aluminum alloys, mechanically resistant, and lighter exploiting different manufacturing and processing processes such as those mentioned above. Next, possible research lines for technological development and future research are described.

In Table 5 a balance of the key advantages, limitations, and future opportunities of Al-Li alloys for high-speed aeronautics is summarized.

Table 5. Key advantages, limitations, and future technical opportunities of Al-Li alloys for high-speed aeronautics

Aspect	Advantages	Limitations	Future Opportunities
Density & Weight	Lower density (7–10% reduction) enables significant aircraft weight savings.	Reduced ductility and lower fracture toughness compared to some alloys.	Further density reduction with new alloy generations and improved metallurgy.
Modulus & Strength	Higher modulus of elasticity (10–15% increase), improved specific stiffness.	Anisotropy in grain orientation can affect mechanical property uniformity.	Tailored microstructure via AI/ML and metamaterials for specific components.
Corrosion Resistance	Good resistance (especially with Sc, Zr microalloying) and lower fatigue sensitivity.	Susceptibility to overaging (PFZs at grain boundaries reducing corrosion resistance).	Advanced corrosion protection, coatings, and real-time health monitoring.
Manufacturability	Compatible with modern joining methods (FSW, laser welding), adaptable to existing processes.	Sensitivity during joining (welding hazards, oxidation at high temps, moisture risk).	Innovative joining/ machining techniques, additive manufacturing integration.
Costs & Markets	Fuel savings and extended component life; competitive with CFRP for structure.	Higher costs, supply chain volatility, and certification process challenges.	New applications in electric aviation, UAVs, and space structures.
Environmental Impact	Lower fuel use, lighter aircraft, lower emissions.	Recyclability and environmental effects of lithium handling are not fully resolved.	Sustainable manufacturing and recycling technologies.

Also, it can be highlighted that:

- Lower density and higher stiffness directly enable lighter, faster, and more efficient aircraft.
- Manufacturability and corrosion resistance are evolving with new processes and microalloying.
- Future opportunities are tied to emerging alloy generations, advanced manufacturing, and integration in next-gen aeronautics.

After describing the emerging techniques in manufacturing and processing of Al-Li alloys, some case studies of the adoption of Al-Li alloy parts in commercial aircraft are presented in the following subsection.

Case Studies of the Adoption of Al-Li Alloys Parts in Commercial Aircraft

Two case studies of the adoption of Al-Li alloys in Airbus A350 and Boeing 787 aircraft are described below:

1. **Airbus A350.** The Airbus A350 uses the advanced Al-Li alloy 2198 extensively for its fuselage skin sheets. This alloy was chosen for its high static strength, excellent damage tolerance (capable of withstanding large damage like two-frame bay cracks), and high thermal stability which is critical for fuselage applications. The 2198 alloy in T8X temper also supports advanced joining methods such as friction stir welding and laser beam welding, essential for manufacturing large aerospace structures with complex shapes. Airbus has signed multi-year agreements with Alcoa for the supply of aluminum lithium products used not only on the A350's fuselage but also in wing skins and structural components, demonstrating wide adoption of Al-Li alloys across Airbus commercial programs. The use of Al-Li alloys in the A350 contributes to significant weight savings compared to conventional aluminum alloys or other metals, improving fuel efficiency and payload capabilities.
2. **Boeing 787.** The Boeing 787 relies more heavily on carbon fiber composites for its primary structures, but it still uses aluminum-lithium alloys in certain secondary structures and structural reinforcements where metal properties are advantageous alongside composites. Al-Li alloys in the 787 provide a lightweight yet strong alternative to conventional aluminum, especially where damage tolerance and fatigue resistance are needed. Its adoption in Boeing 787 structures highlights a hybrid material strategy integrating composites with advanced aluminum alloys to optimize performance and manufacturability.

In summary, Al-Li alloys like 2198 are favored in high-speed commercial aircraft for their excellent balance of high strength, low weight, thermal stability, fatigue and damage tolerance, and compatibility with advanced welding techniques. Airbus A350 showcases one of the most ambitious uses of Al-Li alloys in fuselage and wing components, while Boeing 787 uses Al-Li alloys strategically in conjunction with composites for enhanced structural performance.

If desired, a more detailed breakdown of specific component applications and mechanical performance improvements can be provided.

Once some case studies have been discussed, actual performance data and industry adoption timelines are now reviewed in the next subsection.

Real-World Performance Data and Industry Adoption Timelines of the Emerging Al-Li Alloys

A summary of real-world performance data and industry adoption timelines for emerging Aluminum-Lithium (Al-Li) alloys in high-speed aeronautics based on recent studies and reports are provided below:

- *Real-World Performance Data*. The latest generation of Al-Li alloys, such as 2198-T8 and 2196-T8511, have been extensively characterized by mechanical properties including static strength, fatigue behavior and fatigue crack growth resistance. These alloys generally exceed the strength and stiffness values of traditional aerospace aluminum alloys like 2024-T3 and 7075-T6, validating their suitability for structural aerospace applications. Fatigue resistance of 2198-T8, a common Al-Li alloy used in aerospace skins, performs similarly or better than 2024-T3 under threshold and stable crack growth conditions, with improved crack growth resistance near instability, especially in certain grain orientations. Al-Li alloys show some anisotropy in mechanical properties dependent on grain orientation but remain well within design allowable based on MMPDS standards. Supplemental tests demonstrate comparable corrosion resistance and mechanical behavior under harsh environmental conditions, such as 5% NaCl aqueous solutions and high temperature flames (~1800°F), indicating robustness in operational environments.
- *Industry Adoption Timelines*. Al-Li alloys began to see aerospace adoption primarily in the 1990s and early 2000s with early generations being used in military and commercial aircraft; however, earlier limitations related to manufacturing and material behavior slowed widespread use. The third-generation Al-Li alloys (since mid-2010s), including 2198 and others, have improved manufacturability, damage tolerance, and property consistency, leading to broader adoption in modern aircraft like the Airbus A350 and parts of the Boeing 787. Industry adoption accelerated over the late 2010s and early 2020s as manufacturers adopted advanced joining methods (friction stir welding, laser welding) and digital quality controls supporting these alloys' use in fuselage skins, wing skins, and other critical structures. Regulatory qualification and certification programs have progressed considerably, with comprehensive FAA-backed characterization programs providing needed data for design allowable and safety validation. Looking forward, continued R&D in alloy development supported by AI-driven materials discovery and

green aerospace initiatives is expected to further optimize and expand Al-Li alloy use through the 2020s to 2030s in next-generation hypersonic and supersonic vehicles.

Overall, emerging Al-Li alloys combine superior mechanical properties, validated fatigue and corrosion resistance, and improved manufacturability, making them well-suited for the structural demands of high-speed aeronautics. Their adoption, initially slow with earlier generations, has gained strong momentum with advanced third-generation alloys and industry-wide certification efforts. This positions Al-Li alloys as key materials for many current and future high-speed aerospace programs.

After reviewing the real-world performance data and industry adoption timelines for Al-Li alloys, it is necessary to review the policy, regulation, sustainability, and eco-friendly manufacturing of the alloys in the next subsection.

Policy, Regulation, Sustainability and Eco-Friendly Manufacturing of Al-Li Alloys

An overview of the policy, regulatory, and sustainability discussions surrounding the certification and eco-friendly manufacturing of Aluminum-Lithium (Al-Li) alloys for high-speed aeronautics is given below:

- *Policy and Regulatory Context.* Due to the unique chemical properties of lithium, especially its flammability and toxicity, regulatory bodies like the FAA have imposed special certification conditions for Al-Li alloys used in aircraft construction. Although Al-Li alloys are not considered "novel" materials by some stakeholders, the FAA emphasizes that existing regulations do not fully address the hazards related to lithium content, particularly regarding post-crash fire safety. Special regulatory conditions remain in place until standards are updated to explicitly cover the behavior of Al-Li alloys under extreme scenarios such as crashes and fires. This includes requirements for thorough testing of flammability, toxicity, and crashworthiness. Certification processes require comprehensive material characterization data on mechanical properties, crack growth, corrosion resistance, and manufacturing-induced variability to validate safety and durability for aerospace applications. Industry efforts continue to collect robust datasets and advance materials testing protocols to support regulatory approvals for wider adoption of advanced Al-Li alloys.
- *Sustainability and Eco-friendly Manufacturing.* Lightweight Al-Li alloys contribute directly to the sustainability goals in aerospace by reducing aircraft weight and consequently decreasing fuel consumption and emissions

throughout the aircraft lifecycle. There is increasing emphasis on greener manufacturing processes for Al-Li alloys, including reductions in energy use during melting, casting, forming, and heat treatment operations. Research into recycling and reclaiming lithium and aluminum from aerospace components aims to improve circularity and reduce environmental impact. Manufacturing innovations alongside Industry 5.0 integration and digital twins also help to minimize waste, improve quality control, and reduce scrap rates in production lines processing Al-Li alloys. Materials suppliers and aerospace OEMs are aligning with global frameworks for carbon neutrality and emissions reduction, ensuring Al-Li alloy use supports broader green aerospace initiatives.

In summary, the certification of Al-Li alloys for high-speed aeronautics involves navigating special regulatory conditions arising from lithium's hazardous properties, requiring rigorous testing focused on fire safety and crashworthiness. Concurrently, eco-friendly manufacturing trends strive to reduce the material's environmental footprint via energy-efficient processes, recycling, and lean manufacturing aligned with global sustainability goals. These frameworks collectively enable safe, sustainable adoption of advanced Al-Li alloys in next-generation aerospace vehicles.

The advances described so far are extremely relevant, however, there are still areas that have not een satisfied, developed or investigated as presented in the following section.

FUTURE RESEARCH DIRECTIONS

Key areas aimed at enhancing performance, sustainability, and expanding applications for aluminum-lithium (Al-Li) alloys in high-speed aeronautics must have the next future research directions:

1. *Design of advanced alloy compositions*: The composition of Al-Li alloys must be optimized through research and technological development that exploits new alloying elements, as well as microalloying techniques and nanomaterials, with the goal of improving damage tolerance, ductility, strength, and corrosion resistance.
2. *Emerging and advanced manufacturing techniques*: Exploiting refined microstructures, superior mechanical properties, and complex geometries that enhance performance and customize Al-Li alloy-based designs may be possible through novel rapid solidification methods and the use of emerging additive manufacturing.

3. *Environmentally friendly production and recycling*: The use of manufacturing processes that minimize waste, greenhouse gas emissions, and energy consumption, and increase the cost-effectiveness of recycling, while effectively recovering alloying elements from scrap and end-of-life products, is highly desirable.
4. *Hybrid Material Designs*: The combined use of alloys and other materials can enhance the design and application of parts with properties such as damage resistance, strength, and lightness.
5. *Digitization and analysis of material and process data*: Optimization of process control, manufacturing, and quality, as well as the development of predictive maintenance, can be achieved through big data analysis, machine learning, and artificial intelligence, with the goal of better understanding alloy behavior until a product is produced.
6. *Exploitation of alloys in various technological sectors*: Al-Li alloys can be used in applications where high performance and significant weight reduction are required, such as electronics, defense, renewable energy, and electric vehicles.
7. *Multi-functional Materials*: Exploration of new functionalities within Al-Li alloys, including self-healing capabilities, integrated sensing, and thermal management, to create materials that offer enhanced operational performance and safety.

These future trends reflect the aerospace industry's ongoing need for lighter, stronger, and more sustainable materials. The new research must be aimed at pushing the limits of Al-Li alloys to meet increasingly stringent performance and environmental requirements while broadening their industrial utility.

In the previous discussion, it has been possible to know the key areas that need to be developed to improve both the performance, sustainability and expansion to extensively use aluminum-lithium alloys for high-speed applications, but now it is necessary to know the perspectives of emerging and future techniques to optimize the alloys in the next subsection.

Emerging and Future Techniques to Optimize the Aluminum-Lithium Alloys

Emerging and future techniques to optimize aluminum-lithium (Al-Li) alloys for high-speed aeronautics focus on improving their mechanical properties, weight reduction, damage tolerance, manufacturability, and joining methods. Key advancements include:

1. Alloy Composition and Metallurgy Innovations

- New tailored alloys in the Al-Li family, such as Al-Li-X and Al-Mg-Sc, including third-generation low-density Al-Li alloys, are developed to enhance stiffness, strength, corrosion resistance, and damage tolerance.
- Lithium additions decrease density by about 3% per 1 wt.-% Li as well as enhance the elastic modulus by 6%, improving weight-to-stiffness ratio crucial for aeronautics.
- Addition of elements like scandium boosts strength significantly through precipitation strengthening.
- Control of microstructure through grain refinement techniques includes severe plastic deformation, grain refinement additions, rapid solidification, vibration, and stirring during solidification to optimize mechanical properties and ductility.
- Formation of strengthening precipitates, for instance the δ' phase (Al_3Li), enhances hardness and strength.

2. Advanced Manufacturing Techniques
    - Rapid solidification techniques produce refined microstructures with smaller grains, leading to improved strength and fatigue resistance.
    - Powder metallurgy methods allow precise blending and compacting of aluminum and lithium powders, resulting in alloys with uniform lithium distribution and superior properties.
    - Novel heat treatments improve the formability of challenging Al-Li alloys, enabling complex shapes through methods like spin forming and stretch forming while reducing cracking risks.
3. Innovative Joining Technologies
    - Laser Beam Welding (LBW) and Friction Stir Welding (FSW) are advanced joining processes being integrated to overcome fusion welding drawbacks, improving damage tolerance, durability, and service life of structural components.
    - New joining methods are critical to provide wider application of Al-Li alloys in aeronautics and related high-speed aircraft components with enhanced performance.
4. Design and Application-Specific Alloy Development
    - Alloys are being optimized for specific aerospace parts like fuselage structures, access covers, and seat tracks where weight reduction and stiffness are key design goals.
    - Continuous development aims to reduce environmental impacts, improve product life cycle, and achieve competitive cost-performance balance compared to carbon fiber composites and other materials.
5. Future Trends

- Emphasis on third-generation Al-Li alloys to improve part quality and performance in additive manufacturing processes.
- Continued focus on integrating new microalloying, nano-structuring, and forming technologies to push performance boundaries while ensuring manufacturability.
- Development of alloys combining low density, high modulus, and high damage tolerance tailored to next-generation commercial and military aircraft requirements.

Undoubtedly, the stringent demands of high-speed aeronautics can be met through the design of parts based on aluminum-lithium alloys to offer features such as high rigidity, improved mechanical properties and corrosion resistance, and light weight through cost-effective manufacturing and assembly.

After analyzing emerging and future techniques for optimizing aluminum-lithium alloys, the role that artificial intelligence and machine learning research areas will have on the emerging design of aluminum-lithium alloys is analyzed in the next subsection.

Future Research Gaps of Artificial Intelligence and Machine Learning Integration in the Design of Emerging Al-Li Alloys

Artificial Intelligence (AI) and Machine Learning (ML) will play a transformative role in the design, predictive maintenance, and optimization of Aluminum-Lithium (Al-Li) alloys (Jiang et al., 2023).

1. *Alloy Design*: AI and ML will enable accelerated and rational design of Al-Li alloys by analyzing vast data sets to predict the optimal compositions and microstructural properties. These techniques streamline the alloy development process by automating tasks such as data analysis, composition optimization, and property prediction. Through AI-driven modeling, it is possible to tailor the alloy for enhanced strength, corrosion resistance, lightweight, and other targeted performance criteria. This reduces costly trial-and-error experimentation traditionally required in materials development and enhances sustainability by optimizing the use of alloying elements.
2. *Predictive Maintenance*: In aerospace applications, AI algorithms will analyze sensor data embedded in Al-Li alloy components to detect wear patterns, anomalies, and potential failures before they occur. Machine learning models will forecast maintenance needs proactively, enabling scheduled interventions that minimize downtime and reduce overall maintenance costs while ensuring

optimal component performance. This predictive maintenance approach integrates real-time monitoring and historical data to enhance safety and reliability.

3. *Optimization*: AI and ML will optimize the manufacturing and processing parameters of Al-Li alloys, fine-tuning strength, ductility, corrosion resistance, thermal, and electrical properties. Optimization methods leverage multiscale simulations combined with AI models to understand precipitation phenomena and microstructural evolution at different length scales. This supports improved mechanical properties and functional performance of alloys tailored for high-performance aerospace environments. Furthermore, digitalization supports the integration of digital twin technologies for process optimization and quality control during alloy production.

In summary, digitalization combined with AI and ML revolutionizes Al-Li alloy design by enabling data-driven, efficient development; supports predictive maintenance by anticipating material failures; and enhances optimization through multiscale modeling and smart manufacturing processes. These technologies together contribute to improved performance, cost savings, and sustainability in aerospace alloy applications.

Among the prospects for future aluminum-lithium alloys for high-speed applications is the use of metamaterials, which are described in the next subsection.

Expected Role of Metamaterials in Al-Li Alloys for Emerging High-Speed Aeronautics

Metamaterials contribute significantly to Aluminum-Lithium (Al-Li) alloys for emerging high-speed aeronautics by enabling unique material properties and enhancements in aircraft performance that traditional materials cannot provide such as (Li, 2025):

- *Lightweight and Structural Integrity:* Metamaterials are engineered structures designed to have tailored physical properties beyond those naturally found. In Al-Li alloys, which already offer low density and high stiffness due to lithium additions, metamaterials can further enhance structural integrity without compromising weight. This is critical in high-speed aeronautics where weight reduction combined with high strength and stiffness improves fuel efficiency and overall aircraft performance.
- *Aerodynamics and Fuel Efficiency*: Metamaterials will impact aerodynamics by creating surfaces that control electromagnetic waves and physical interactions with airflow. Such engineered surfaces can reduce drag and improve lift, optimizing fuel consumption and maneuverability for high-speed aircraft. By

integrating metamaterial concepts with Al-Li alloys used in aircraft skins and structural components, designers can achieve superior aerodynamic performance, leading to energy savings and lower emissions.

- *Mechanical Metamaterials for Energy Reduction*: Mechanical Metamaterials integrated with Al-Li alloys will help to control turbulent flow dynamics and fluid-structure interactions in flight conditions. Research programs sponsored by aerospace and defense organizations aim to tailor Al-Li alloy structures at micro and macro scales to enhance maneuverability and reduce energy requirements during high-speed flight. This enables achieving better flight envelopes while maintaining material performance under stress.

Therefore, the integration of metamaterials in Al-Li alloys advances high-speed aeronautics by providing:

- Enhanced strength-to-weight ratios beyond conventional materials,
- Improved aerodynamic surfaces for drag reduction and better lift,
- Control over mechanical and fluid dynamic interactions to reduce energy use in high-speed flight.

Thus, metamaterials extend the already favorable properties of Al-Li alloys, making them pivotal in next-generation high-speed aerospace applications.

Innovation for Al-Li Alloys Based on Emerging and Novel Technological Trends

Trends in the design and application of Aluminum-Lithium (Al-Li) alloys in high-speed aeronautics are strongly influenced by three key modern drivers: AI-driven materials discovery, Industry 5.0 integration, and green aerospace initiatives. Here's a description of each:

1. *AI-Driven Materials Discovery*. Artificial Intelligence (AI) and machine learning are accelerating the discovery and optimization of Al-Li alloys with improved properties such as strength, corrosion resistance, and temperature tolerance. AI models can predict key material properties based on composition and processing parameters, reducing reliance on slow, costly trial-and-error experimentation. Autonomous platforms, combining robotics, machine learning, and high-throughput experimentation (e.g., the APEX platform), are shortening alloy development times from year to month by automatically designing, producing, and testing alloy samples. AI helps identify novel alloying strategies, including optimal lithium content and microalloying elements, specifically targeted for

aerospace performance requirements under extreme temperature and mechanical loading.

2. *Industry 5.0 Integration.* Industry 5.0 emphasizes human-centric automation, smart manufacturing, and resilience in aerospace production. In Al-Li alloy fabrication, this means integrating AI, robotics, and advanced sensing into manufacturing processes like friction stir welding, additive manufacturing, and precision forming to improve quality, reduce defects, and enhance reproducibility. Digital twins of Al-Li components allow real-time monitoring and predictive maintenance through sensor integration, optimizing lifecycle performance and reducing operational costs. The integration enhances customization and rapid prototyping of alloy parts adapted for specific aerostructure needs in high-speed aircraft.
3. *Green Aerospace Initiatives.* Lightweight Al-Li alloys contribute to lower aircraft weight, translating directly into reduced fuel consumption and CO_2 emissions, a core goal of green aerospace evolution. Alloy development increasingly targets materials and processes with reduced environmental footprints, including eco-friendlier alloying and recycling approaches and minimized energy inputs in manufacturing. Sustainable alloy design balances high performance with circular economic principles, aiding compliance with evolving regulatory and societal pressures for carbon neutrality in aviation.

In summary, it can be concluded that in high-speed aeronautics, the fusion of AI-driven alloy discovery, Industry 5.0 smart manufacturing, and green aerospace imperatives is transforming Al-Li alloy development. These trends enable the creation of next-generation materials with superior performance, manufacturing efficiency, and environmental sustainability tailored for the extreme demands of hypersonic and supersonic flight.

After having reviewed the importance of optimizing aluminum-lithium alloys for high-speed aeronautical applications, it is required to conclude the chapter in the next section.

CONCLUSION

High-speed aeronautical parts require characteristics such as light weight, stiffness, high strength, and low density, which can be achieved using aluminum-lithium alloys to optimize aircraft fuel consumption. The chapter highlighted the evolution of Al-Li alloys through three generations developed so far, where properties such as corrosion and fatigue resistance, fracture toughness, and thermal stability required for parts in supersonic and hypersonic environments were optimized. Parts with complex aerospace geometries exploit Al-Li alloys produced through novel techniques such as powder metallurgy, rapid solidification, and/or additive manufacturing. The use of alloying elements such as zirconium, scandium, and magnesium with optimized composition can increase the performance, structural stability, weldability, and overall performance of Al-Li alloys. Furthermore, emerging manufacturing process methods such as friction stir welding and laser beam welding were analyzed, highlighting their potential to improve component longevity and tolerance under extreme conditions such as those required in high-speed aeronautics. Future research prospects, such as the incorporation of hybrid and multifunctional materials, sustainability initiatives, and advanced digital modeling, offer exciting prospects for the expanded use of aluminum-lithium alloys in high-speed aeronautics, with a focus on improved mechanical and chemical properties, cost-effectiveness, and environmental responsibility. Continued interdisciplinary research and innovation will be pivotal in realizing aluminum-lithium alloys' full potential, facilitating safer, lighter, and more efficient aircraft capable of meeting the rigorous demands of high-speed aeronautics. Their superior mechanical properties and damage tolerance make Al-Li alloys ideal for critical aircraft components such as fuselage panels and wing skins, enabling safer and more efficient high-speed flights. Continued research advances manufacturing and joining techniques, ensuring these alloys meet the stringent demands of modern aerospace structures while reducing costs and environmental impact. Thus, Al-Li alloy research is essential to developing next-generation high-speed aircraft with improved performance and sustainability.

ACKNOWLEDGMENT

The first author is glad for the help of the Polytechnical University of Guanajuato (UPG, in Spanish) in developing this scientific research. Also, the authors are grateful for their valuable financial support to the Secretariat of Science, Humanities, Technology, and Innovation (SECIHTI) Project CBF2023-2024-333.

REFERENCES

Abdullatef, M. S., Alzubaidi, F. N., Al-Tamimi, A., & Mahmood, Y. A. (2023). Fatigue Life Estimation of High Strength 2090-T83 Aluminum Alloy under Pure Torsion Loading using Various Machine Learning Techniques. *Fluid Dyn. Mater. Process.*, *19*(8), 2083–2107. DOI: 10.32604/fdmp.2023.027266

Bouchaala, K. (2023). Review of Aluminium Lithium Alloy Manufacturing Procedures, Mechanical Characteristics and Precipitations for Usage in Aerospace Applications. *Journal of Astrophysics & Aerospace Technology*, *11*, 1. DOI: 10.37421/2329-6542.2023.11.249

Chen, J., Chen, R., Liao, H., He, Y., Chen, P., Tian, Y., Zhang, B., & Zhang, Z. (2024a). Improving Joint Performance of Friction Stir Welded 2195-O Al-Li Alloy by Post-Weld Treatment and Rolling Deformation. *Journal of Materials Research and Technology*, *29*, 5048–5059. DOI: 10.1016/j.jmrt.2024.03.009

Chen, Y., Pan, X., Peng, J., & Wang, F. (2022). Effect of Coupling Damage on Fatigue Behavior of 2198-T8 Aluminum-Lithium Alloy Thin Sheet for Fuselage. *Fatigue & Fracture of Engineering Materials & Structures*, *45*(3), 754–769. DOI: 10.1111/ffe.13631

Chen, Y., Yu, J., Ge, X., Sun, Y., Sun, L., Zhou, W., & Zhao, G. (2024b). Cryogenic Deformation Behavior, Constitutive Modeling and Microstructure Evolution of Solution-Treated 2195 Al-Li Alloy at High Strain Rates. *Journal of Materials Research and Technology*, *31*, 2915–2929. DOI: 10.1016/j.jmrt.2024.07.013

Dongming, G. (2024). High-Performance Manufacturing. *International Journal of Extreme Manufacturing*, *6*(6), 060201. DOI: 10.1088/2631-7990/ad7426

Du, Y., Liu, Z., Shen, M., Wu, J., Zhang, K., Jiang, C., & Zhing, J. (2025). An Efficient Parametric Modeling, Evaluation and Optimization Strategy for Aerodynamic Configuration Design of EVTOL Aircraft. *Aerosp. Res. Commun.*, *3*, 14986. DOI: 10.3389/arc.2025.14986

Farsadi, T., Ahmadi, M., Sahin, M., Khodaparast, H. H., Kayran, A., & Friswell, M. I. (2024). High Aspect Ratio Composite Wings: Geometrically Nonlinear Aeroelasticity, Multi-Disciplinary Design Optimization, Manufacturing, and Experimental Testing. *Aerospace (Basel, Switzerland), 11*(3), 193. DOI: 10.3390/aerospace11030193

Gancarz, T., Dobosz, A., Bogno, A.-A., Cempura, G., Schell, N., Chulist, R., & Henein, H. (2021). Characterization of Rapidly Solidified Al-Mg-Sc Alloys with Li Addition. *Materials Characterization, 178*, 111290. DOI: 10.1016/j.matchar.2021.111290

Gao, W., Bi, Y., Li, X., Dong, A., Wang, J., & Yang, X. (2024). Variational Method-based Trajectory Optimization for Hybrid Airships. *Aerospace (Basel, Switzerland), 11*(4), 250. DOI: 10.3390/aerospace11040250

Ghimire, R., & Raji, A. (2024). Use of Artificial Intelligence in Design, Development, Additive Manufacturing, and Certification of Multifunctional Composites for Aircraft, Drones, and Spacecraft. *Applied Sciences (Basel, Switzerland), 14*(3), 1187. DOI: 10.3390/app14031187

Giladi, R., & Menachi, E. (2024). Validating Aircraft Noise Models: Aviation Environmental Design Tool at Heathrow. *Journal of Air Transport Management, 116*, 102557. DOI: 10.1016/j.jairtraman.2024.102557

Guo, F., Duan, S., Matsuda, K., Liu, X., & Zou, Y. (2025). The Microstructure and Corrosion Behavior Analysis of Different Aging Al-Zn-Mg-Cu Alloys Manufactured by Additive Friction Stir Deposition. *Materials Characterization, 225*, 115187. DOI: 10.1016/j.matchar.2025.115187

Hajjioui, E. A., Bouchaâla, K., Faqir, M., & Essadiqi, E. (2023). A Review of Manufacturing Processes, Mechanical Properties and Precipitations for Aluminum Lithium Alloys used in Aeronautic Applications. *Heliyon, 9*(3), e12565. DOI: 10.1016/j.heliyon.2022.e12565

Hanush, S. S., & Manjaiah, M. (2022). Topology Optimization of Aerospace Part to Enhance the Performance by Additive Manufacturing Process. *Materials Today: Proceedings, 62*, 7373–7378. DOI: 10.1016/j.matpr.2022.02.074

Jia, Z., Dou, H., Liu, Y., Wang, Z., Zhang, H., Yue, T., & Mu, L. (2025). Study on Ductile Fracture Model considering Pre-Strain and its Application during Incremental Forming for 2091 Aluminum *Alloy. Engineering Fracture Mechanics, 322*, 111177. DOI: 10.1016/j.engfracmech.2025.111177

Jiang, L., Zhang, Z., Hu, H., He, X., Fu, H., & Xie, J. (2023). A Rapid and Effective Method for Alloy Materials Design via Sample Data Transfer Machine Learning. *npj Computational Materials, 9*(1), 26. DOI: 10.1038/s41524-023-00979-9

Kanti, P. K., Kumar, H. G. P., Wanatasanappan, V. V., Kumar, A., & Regasa, M. B. (2025). Graphene's Frontier in Aerospace: Current Applications, Challenges, and Future Directions for Space Engineering. *Nanoscale Advances, 7*(12), 3603–3618. DOI: 10.1039/D4NA00934G

Kessel, F., Friess, M., Rauh, C., & Wagner, A. (2025). Tailored Directional Porosity in Ceramic Matrix Composites (CMCs) for Hypersonic Applications. *CEAS Space Journal, 17*(4), 623–633. DOI: 10.1007/s12567-024-00566-8

Kim, J.-H., Kim, B.-S., Im, D., Park, J. H., Moon, D.-J., & Park, Y.-J. (2024). Optimization of High-Speed Angular Contact Ball Bearing for Aircraft Gearbox utilizing and Evolutionary Multi.Objective Algorithm, NSGA-III. *Advances in Mechanical Engineering, 16*(9), 1–14. DOI: 10.1177/16878132241272197

Kridli, G. T., El-Gizawy, A. S., & Lederich, R. (1998). Development of Process Maps for Superplastic Forming of Weldalite™ 049. *Materials Science and Engineering A, 244*(2), 224–232. DOI: 10.1016/S0921-5093(97)00545-5

Leng, J.-X., Shen, Y., Huang, W., An, K., Zhou, C.-C., & Wang, Z.-G. (2024). Multi-Fidelity Data Mining-based Design Optimization Framework for Scramjet-based Aircraft. *Energy, 311*, 133448. DOI: 10.1016/j.energy.2024.133448

Li, H., Li, X., Li, Y., Gao, G., Wen, K., Li, Z., Zhang, Y., & Xiong, B. (2023). Exploration of Alloying Elements of High Specific Modulus Al-Li Alloy based on Machine Learning. *Materials (Basel), 17*(1), 92. DOI: 10.3390/ma17010092

Li, S.-S., Yue, X., Li, Q.-Y., Peng, H.-L., Dong, B.-X., Liu, T.-S., Yang, H.-Y., Fan, J., Shu, S.-L., Qiu, F., & Jiang, Q.-C. (2023). Development and Applications of Aluminum Alloys for Aerospace Industry. *Journal of Materials Research and Technology, 27*, 944–983. DOI: 10.1016/j.jmrt.2023.09.274

Li, X., Wang, L., & Lu, Y. (2025). Three-Dimensional Disordered Alloy Metamaterials: A New Platform of Structure-Function Integration. *Materials Futures, 4*(1), 012001. DOI: 10.1088/2752-5724/ada172

Lin, S. (2024). Advanced Materials and Their Impact on Aircraft Performance. *J. Aeronaut. Aerospace Eng., 13*(3), 355.

Liu, J., & Kulak, M. (2000). A New Paradigm in the Design of Aluminum Alloys for Aerospace Applications. *Mater. Sci. Forum, 331-337*, 127-142. doi: DOI: 10.4028/www.scientific.net/MSF.331.337.127

Liu, Y., Zhang, S., Zhang, J., Yao, K., Luo, M., & Liu, Y. (2024). Towards the Design of Future Aircraft: A Critical Review on the Tools and Methodologies. *Aerosp. Res. Commun., 2*, 13096. DOI: 10.3389/arc.2024.13096

Lv, K., Zhu, C., Zheng, J., Wang, X., & Chen, B. (2019). Precipitation of T_1 phase in 2198 Al-Li Alloy Studied by Atomic-Resolution HAADF-STEM. *Journal of Materials Research, 34*(20), 3535–3544. DOI: 10.1557/jmr.2019.136

Meyer, P., Cans, Y., Ferton, D., & Reboul, M. (1987). The Metallurgy of Industrial Al-Li Alloys. *Journal de Physique. Colloque, 48*(C3), 131–138. DOI: 10.1051/jphyscol:1987316

Mishra, D. M. (2025). A Review of Manufacturing Processes for Aluminum-Lithium Alloys and Their Applications. *Premier Journal of Engineering, 3*, 100007. DOI: 10.70389/PJE.100007

Moenck, K., Koch, J., Rath, J.-E., Büsch, L., Gierecker, J., Kähler, F., Kalscheuer, F., Masuhr, C., Kipping, J., Prünte, P., Schoepflin, D., Eschen, H., Wulff, L.A., Rodeck, R., Wende, G., Gomse, M., & Schüppstuhl, T. (2025). Industry 5.0 in Aircraft Production and MRO: Challenges and Opportunities. *CEAS Aeronaut. J.* doi: DOI: 10.1007/s13272-025-00832-3

Nikkhoo, A. & Esmaeili, A. (2023). Effect of Different Aero-Structural Optimization in the Commercial Airplane. J. Comput. Appl. Mech., 54(2), 268-284. doi: 10.22059/jcamesh.2023.355499.809

Pang, B., Zhang, Y., Li, J., Wang, X., Chang, M., & Bai, J. (2024). Data-Driven Surrogate Model for Aerodynamic Design using Separable Shape Tensor Method. *Chinese Journal of Aeronautics, 37*(9), 41–58. DOI: 10.1016/j.cja.2024.03.014

Peters, A. B., Zhang, D., Chen, S., Ott, C., Oses, C., Curtarolo, S., McCue, I., Pollock, T. M., & Prameela, S. E. (2024). Materials Design for Hypersonics. *Nature Communications, 15*(1), 3328. DOI: 10.1038/s41467-024-46753-3

Pollock, L., & Wild, G. (2024a). An Examination of High-Speed Aircraft - Part 1: Past, Present, and Future. *Transp. Eng., 18*, 100290. DOI: 10.1016/j.treng.2024.100290

Pollock, L., & Wild, G. (2024b). An Examination of High-Speed Aircraft – Part 2: Safety and Reliability. *Transp. Eng., 18*, 100269. DOI: 10.1016/j.treng.2024.100269

Radhika, C., Shanmugam, R., Ramoni, M., & Gnanavel, B. K. (2024). A Review on Additive Manufacturing for Aerospace Application. *Materials Research Express, 11*(2), 022001. DOI: 10.1088/2053-1591/ad21ad

Rambabu, P., Prasad, N. E., Kutumbarao, V. V., & Wanhill, R. J. H. (2017). Aluminium Alloys for Aerospace Applications. In Eswara Prasad, N., & Wanhill, R. J. H. (Eds.), *Aerospace Materials and Material Technologies* (Vol. 1, pp. 29–52). Springer., DOI: 10.1007/978-981-10-2134-3_2

Rao, M. S., & Padal, K. T. B. (2022). New Developments of Aluminum Alloys for Future Generation Applications – A Review. *AIP Conference Proceedings, 2648*, 030011. DOI: 10.1063/5.0114421

Rodinò, S., & Maletta, C. (2024). Design Considerations and Applications of Shape Memory Alloy-based Actuation in Morphing Structures: A Systematic Review. *Prog. Eng. Sci., 1*(4), 100021. DOI: 10.1016/j.pes.2024.100021

Roy, M. (2025). Degradation of C_f-SiC Composites under Hypersonic Conditions. *Wear, 576-577*, 206141. DOI: 10.1016/j.wear.2025.206141

Serani, A., Diez, M., & Quagliarella, D. (2024). Aerodynamic Shape Optimization in Transonic Conditions through Parametric Model Embedding. *Aerospace Science and Technology, 155*, 109611. DOI: 10.1016/j.ast.2024.109611

Simöes, S. (2024). High-Performance Advanced Composites in Multifunctional Material Design: State of the Art, Challenges, and Future Directions. *Materials (Basel), 17*(23), 5997. DOI: 10.3390/ma17235997

Soni, R., Verma, R., Garg, R. K., & Sharma, V. (2024). A Critical Review of Recent Advances in the Aerospace Materials. *Materials Today: Proceedings, 113*, 180–184. DOI: 10.1016/j.matpr.2023.08.108

Srivastava, M., Jayakumar, V., Udayan, Y., Sathishkumar, M., Muthu, S. M., Gautam, P., & Nag, A. (2024). Additive Manufacturing of Titanium alloy for Aerospace Applications: Insights into the Process, Microstructure, and Mechanical Properties. *Applied Materials Today, 41*, 102481. DOI: 10.1016/j.apmt.2024.102481

Stoliar, I. A., Shepelevich, V. G., Wendler, E., & Tashlykova-Bushkevich, I. I. (2021). Effect of Lithium on the Structural-Phase State of Rapidly Solidified Al-Mg-Li Alloy during Heat Treatment. *Journal of Surface Investigation. X-ray, Synchrotron and Neutron Techniques, 15*(4), 752–758. DOI: 10.1134/S1027451021040194

Sun, Y. J., Yang, R., Xie, L., Li, Y. B., Wang, S. L., Li, H. X., Wang, W. R., & Zhang, J. S. (2022). Interfacial Bonding and Corrosion Behaviors of HVOF-Sprayed Fe-based Amorphous Coating on 8090 Al-Li Alloy. *Surface and Coatings Technology, 436*, 128316. DOI: 10.1016/j.surfcoat.2022.128316

Taleghani, A. S., & Torabi, F. (2024). Recent Developments in Aerodynamics. *Frontiers of Mechanical Engineering, 10*, 1537383. DOI: 10.3389/fmech.2024.1537383

Vivas, J., Aldanondo, E., Dufour, P., Delgado, C., & Fernández-Calvo, A. I. (2024). Laser Beam Welding of T Joints for Aluminum-Copper-Lithium Aircraft Panels: Effect of Filler Wire on Recyclability and Weldability. *Journal of Materials Research and Technology, 33*, 5809–5820. DOI: 10.1016/j.jmrt.2024.10.223

Wang, H., Duan, Z., Guo, Q., Zhang, Y., & Zhao, Y. (2023). Machine Learning Design of Aluminum-Lithium Alloys with High Strength. *Computers, Materials & Continua*, *77*(2), 1393–1409. DOI: 10.32604/cmc.2023.045871

Wanhill, R. J. H. (2014). Aerospace Applications of Aluminum-Lithium Alloys. In Prasad, N. E., Gokhale, A. A., & Wanhill, R. J. H. (Eds.), *Aluminum-Lithium Alloys: Processing, Properties, and Applications* (pp. 503–535). Elsevier., DOI: 10.1016/B978-0-12-401698-9.00015-X

Wu, Y. (2022). Application of Aluminum Alloy in Aircraft. *Journal of Physics: Conference Series*, *2228*(1), 012024. DOI: 10.1088/1742-6596/2228/1/012024

Xiao, G., An, R., Yang, J., & Ma, J. (2024). Research Progress of New Generation of Aluminum-Lithium Alloys Alloying. *Advanced Engineering Materials*, *26*(12), 2301639. DOI: 10.1002/adem.202301639

Yu, Y., Yu, T., Mao, Y., Yang, Y., & Liang, S. (2025). Key Factors affecting Over-expanded Flow Separation in Design of Large Expansion Ratio Single Expansion Ramp Nozzle. *Aeronautical Journal*, *129*(1333), 717–736. DOI: 10.1017/aer.2024.70

Zhang, J., Wang, B., & Tian, Y. (2022). Study of Pre-Strain before Artificial Aging on Mechanical Property and Stress Corrosion Behavior of 2297 Al-Cu-Li Alloy. *Materials Science and Engineering A*, *831*, 142174. DOI: 10.1016/j.msea.2021.142174

Zhang, Z., Ao, Y., Li, S., & Gu, G. X. (2024). An Adaptive Machine Learning-based Optimization Method in the Aerodynamic Analysis of a Finite Wing under Various Cruise Conditions. *Theoretical & Applied Mechanics Letters*, *14*(1), 100489. DOI: 10.1016/j.taml.2023.100489

Zhao, B., Wang, Y., Peng, J., Wang, X., Ding, W., Lei, X., Wu, B., Zhang, M., Xu, J., Zhang, L., & Das, R. (2024). Overcoming Challenges: Advancements in Cutting Techniques for High Strength-Toughness Alloys in Aero-Engines. *International Journal of Extreme Manufacturing*, *6*(6), 062012. DOI: 10.1088/2631-7990/ad8117

Zheng, W., Cui, M., Song, Q., & Yuan, W. (2025). Effect of Low-Temperature Deformation and Deep-Cryogenic Treatment on the Mechanical and Corrosion Resistance of 2050 Aluminum-Lithium Alloy. *Journal of Alloys and Compounds*, *1034*, 181347. DOI: 10.1016/j.jallcom.2025.181347

Zhou, B., Liu, B., & Zhang, S. (2021). The Advancement of 7XXX Series Aluminum Alloys for Aircraft Structures: A Review. *Metals*, *11*(5), 718. DOI: 10.3390/met11050718

ADDITIONAL READING

Ahmed, B., & Wu, S. (2014). Aluminum Lithium Alloys (Al-Li-Cu-X)-New Generation Material for Aerospace Applications. *Applied Mechanics and Materials, 440*, 104–111. . DOI: 10.4028/www.scientific.net/AMM.440.104

Bell, L. (2023). Properties, and Precipitations of Aluminium Lithium Alloys in Aeronautics. *J. Aeronaut. Aerospace Eng., 12*(3), 1000320.

Boyer, R. R., Cotton, J. D., Mohaghegh, M., & Schafrik, R. E. (2015). Materials Considerations for Aerospace Applications. *MRS Bulletin, 40*(12), 1055–1066. DOI: 10.1557/mrs.2015.278

Charette, R. O., Leonard, B. G., Bozich, W. F., & Deamer, D. A. (1998). Russian Aluminum-Lithium Alloys for Advanced Reusable Spacecraft. *AIP Conference Proceedings, 420*, 926–936. DOI: 10.1063/1.54896

Gloria, A., Montanari, R., Richetta, M., & Varone, A. (2019). Alloys for Aeronautic Applications: State of the Art and Perspectives. *Metals, 9*(6), 662. DOI: 10.3390/met9060662

Ning, J., Li, Q., Zou, Z., Liu, T., & Chen, L. (2021). Hot Tensile Deformation Behavior and Microstructural Evolution of 2195 Al-Li Alloy. *Vacuum, 188*, 110176. doi: . vacuum.2021.110176DOI: 10.1016/j

Omiyale, B. O., Olugbade, T. O., Abioye, T. E., & Farayibi, P. K. (2022). Wire Arc Additive Manufacturing of Aluminium Alloys for Aerospace and Automotive Applications: A Review. *Materials Science and Technology, 38*(7), 391–408. DOI: 10.1080/02670836.2022.2045549

Radan, L., Songmene, V., Samuel, A. M., & Samuel, F. H. (2025). Effects of Alloy Composition, Hardness, and Milling Parameters on the Cutting Forces of Al-Li-based Alloys. *Materials (Basel), 18*(12), 2683. DOI: 10.3390/ma18122683

KEY TERMS AND DEFINITIONS

Aluminum-Lithium Alloys: A group of alloys composed primarily of aluminum and lithium, often with additional elements such as copper, zirconium, and magnesium, and characterized by their significantly lower density compared to conventional aluminum alloys due to the lightweight nature of lithium.

Casting: A manufacturing process in which molten metal is poured into a mold containing a hollow cavity of the desired shape, where the metal is cooled and solidified in the mold, then removed to produce a solid metal part that replicates the cavity's shape.

Friction Stir Welding (FSW): A solid-state joining process that uses a non-consumable rotating tool to join two facing workpieces without melting them.

High-Speed Aeronautics: A branch of aeronautics that studies and deals with aircraft and flight conditions where air speed approaches or exceeds significant fractions of the speed of sound.

Laser Beam Welding (LBW): A manufacturing technique in which two or more pieces of material are joined together by using a concentrated heat source in the form of a laser beam.

Mach number: A dimensionless quantity that represents the ratio of the speed of an object (or flow velocity) to the speed of sound in the surrounding medium.

Powder Metallurgy: A manufacturing process in which metal powders are used as raw materials to create parts or components using four steps: powder production, mixing and blending of powders, compacting the powder mixture into a desired shape, and sintering the compact shape at a temperature below the melting point of the melting point of the metal to bond particles together.

Precipitation: The process in which small particles known as precipitates are formed within a supersaturated solid solution, usually during heat treatment or aging of an alloy, in a phase transformation, causing the excess solute atoms to be clustered together and giving finely dispersed particles within the metal matrix, which provide hardness and strength.

Rapid Solidification: A process of cooling molten metal at an extremely high rate (>10,000 K/s) to prevent the formation of equilibrium microstructures and results in unique material properties, such as the creation of metastable phases, amorphous structures, and nanocrystalline materials to enhance strength, corrosion resistance, and other functional properties of metals.

Chapter 4
Abrasive Jet Machining of Aluminum and Aluminum Alloys

Lauren iu Slătineanu
https://orcid.org/0009-0006-5246-1281
"Gheorghe Asachi" Technical University of Ia i, Romania

Irina Besliu-Bancescu
https://orcid.org/0000-0003-3132-7818
University Stefan cel Mare From Suceava, Romania

Oana Dodun
"Gheorghe Asachi" Technical University of Ia i, Romania

Gheorghe Nagî
"Gheorghe Asachi" Technical University of Ia i, Romania

Adelina Hri uc
"Gheorghe Asachi" Technical University of Ia i, Romania

ABSTRACT

It is known that the aluminum and aluminum alloys are materials frequently preferred for the aerospatial equipment due to their properties to ensure a certain strength and a low weight of the equipment components. For such reasons, the aluminum and aluminum alloys were utilized just from the first use of metallic materials in the aerospatial equipment. The abrasive jet machining is a machining method based on the effects generated on the workpiece surface by the abrasive particles placed in a high-speed gas stream. Essentially, these effects are materialized by a material removal process and/or a change of the surface aspect. The abrasive jet machining

DOI: 10.4018/979-8-3373-9454-1.ch004

is applied as a subtractive machining method. The objective of the proposed chapter was to offer a general image concerning the theoretical principles and applications of using the abrasive jet machining in the manufacturing of aluminum and aluminum alloys parts included in the aerospatial equipment. A synthesis concerning the main aspects specific to the abrasive jet machining of the aluminum and aluminum alloys was achieved.

INTRODUCTION

Abrasive jet machining (AJM) uses the effects produced by the contact of abrasive particles in a high velocity gas stream with the workpiece material. Usually, the abrasive stream is created by machining equipment with a nozzle for directing the particles to the workpiece.

This machining method is generally classified within the broader category of *nonconventional machining methods*, which encompasses electro-physical and chemical processes. If the main subgroups of nonconventional machining methods are considered, the abrasive jet machining could be placed in the subgroup of nonconventional machining methods based on mechanical effects or in the group of machining methods that use moving fluids (Slătineanu et al., 2004).

The significance of abrasive jet machining and the research conducted in this domain can be accentuated by the multitude of global analyses that have been meticulously developed over the course of years by researchers(Chen et al., 2017; Madhu and Balasubramanian, 2015; Melentiev and Fang, 2018; Mistry et al., 2018; Ramachandran & Ramakrishnan, 1993).

The abrasive machining process has certain similar aspects to the so-called *sandblasting process,* and other process names like "pencil blasting" or "microabrasive blasting" could be met. As *main differences* between the abrasive jet machining and sandblasting, the following could be mentioned: abrasive particles are much finer, diameters of the nozzle orifice are much smaller, better control of the process parameters, and, as a result, a higher machining accuracy. Another difference corresponds to the fact that for the abrasive jet machining, the surfaces that must not be affected by the material removal process can be shielded using *an erosion-resistant mask* (Jain, 2013). As a method of distinguishing abrasive air jet machining from sandblasting, Melentiev and Fang (2018) referred to it as abrasive air jet machining (AAJM). In their opinion, the names micro-AAJM, microblasting, and powder blasting could be used for nozzles with an exit diameter and feature smaller than 1 mm.

There is a large set of solid materials that could be machined using abrasive jets. Thus, various metallic materials (pure metals and alloys) and nonmetallic materials (plastics, ceramics, semiconductors, composites, wood, etc.) could be machined using the abrasive jet.

Abrasive jet machining is usually implemented for machining workpieces made of hard and brittle materials as glass, ceramics, germanium, silicon, tungsten, etc. Some such materials could be used in the aerospace industry), but also to machine workpieces made of ductile materials, such as aluminum and aluminum alloys.

The main manufacturing processes that could be achieved by abrasive jet machining are the deburring applied inclusively in zones less accessible, drilling and cutting of thin workpieces, marking, cleaning, removing of deposited layer or corroded layers or contaminated layers, preparing of surfaces to be bonded, engraving, hand-held polishing, etching, radiusing, obtaining of micro-mechanical systems (MEMS) (Todd et al., 1994, Jain, 2013).

Among the metallic materials that are used in the aerospace industry, aluminum and aluminum alloys are materials whose characteristics ensure good machinability when the abrasive jet machining is applied to obtain various parts. The aluminum and aluminum alloys were used in the aerospace equipment since 1903 (Danylenko, 2018), when the Wright brothers built a cylinder block and other engine parts made of such materials for the aircraft structural components since about 1930 (Starke and Staley, 2011). In Table 1 are presented the main proprieties of the aluminum and aluminum alloys most commonly used in the aerospace industry.

In comparison with other metallic materials, due to their convenient ratio of weight to strength, the aluminum alloys are frequently preferred in the so-called *critical weight applications*. Aluminum alloys, for instance, are utilized in airplanes to create fuselages, wings, and support systems unique to military cargo and commercial airlines (Yashpal et al. 2015).

Table 1. Properties of aluminum and some aluminum alloys used in the aerospace industry (adapted from Guest, 2018; Heinz et al., 2000; Yashpal et al.)

Material		Density, ρ, g/cm^3	Yield strength, R_p, MPa	Ultimate strenght, R_m, MPa	Major alloying elements	Used in the construction of:
Aluminum		**2.70**	**350**	**460**	**-**	
Aluminum alloys	2024 T3 HP L	2.78	359	460	Copper	Wings and fuselages
	Al-Mg-Li-B L	2.71	363	486	Magnesium, lithium	Floor sections, seat structure, fuselage structure
	Al-Mg-Li-SF L	2.71	359	485	Magnesium, lithium	Floor sections, seat structure, fuselage structure
	AA 2014	2.80	230-435	370-480	Copper, manganese, iron	Internal portions of aircraft
	AA 7075	2,81	140	280	Zinc, magnesium, copper	Wings and fuselages
	AA 7068	2,85	683	710	Magnesium, copper, zirconium	Valve body, connecting rod (one of the strongest alloys utilized today)

Sometimes, the application of abrasive jet machining in the cases of soft materials is not considered effective enough, and increasing the performance and the optimization of the abrasive jet machining process could be necessary (Gherissi and Einasri, 2019).

HISTORICAL EVOLUTION OF THE SCIENTIFIC AND TECHNICAL INFORMATION CONCERNING THE ABRASIVE JET MACHINING

The first official document that contained information concerning equipment used to generate an abrasive jet machining process was the patent application granted to Tilghman in 1870. As carrier fluids, Tilghman was proposing the vapors, the water, and the air.

The airless centrifugal wheel was created by the American Foundry Equipment Company as mechanical equipment that uses centrifugal force to propel abrasive particles to the surface of the workpiece. The equipment allowed cleaning of the industrial surfaces. In 1932, the company changed its name to „Wheelabrator Corporation", taking into consideration the extended use of the above-mentioned equipment (Connecticut Resources Recovery Authority, 2013).

It is mentioned that the AJM was promoted in 1968 by Franz to cut laminated paper tubes. The first equipment for AJM was introduced in 1983. In the years `90, John Olsen promoted the use of abrasive jet cutting in traditional machine shops (Mistry et al., 2018). The abrasive jet was used not only to materialize machining processes but also for the restoration of buildings and art objects, in paleontology, etc.

GENERATION OF THE ABRASIVE STREAM

Generation of the Abrasive Stream Using a Gun

A first way of obtaining the abrasive stream is based on the use of high velocity gas stream, in which the abrasive particles are subsequently introduced. In such a case, some of the main components of the equipment used for abrasive jet machining could be seen in Figure 1. A compressor adsorbs the air by a filtering subsystem and sends the compressed air to the air desiccator. The filtering subsystem is necessary to avoid the possible contamination of the abrasive jet with substances that could diminish the efficiency of the abrasive jet machining process. The adsorption and the introduction of the abrasive particles in the high-velocity stream could be materialized using a certain decrease of the pressure, eventually in association with a fluidizing action of the powder mass.

A fluidization of the abrasive powder could be obtained using the vibration of the recipient that contains the abrasive powder with a frequency of 50-60 Hz.

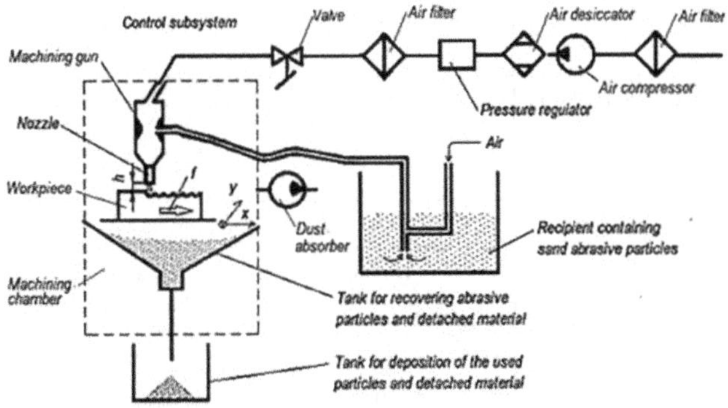

As carrier gases, compressed air, carbon dioxide, and nitrogen could be used. The carrier gas must be cheap, easily available, and non-toxic. Its density is about 1.3 kg/m^3, while the processing pressure is 2-10 bar (0.196 – 0.96 MPa). If the in-carrier gas initial speed is 500-700 m/s, the abrasive jet is of 100-300 m/s.

The amount of abrasive material directed to the workpiece divided by the time is known as the abrasive flow rate, and it can vary from 2-4 g/min for fine machining to 10-20 g/min for abrasive machining. In these conditions, the consumption of carrier gas is about 0.008 m^3/min at a pressure of 5 bar (0.49 MPa).

The abrasive stream thus constituted is sent to the workpiece through a nozzle in which there is a hole with a small dimensions of the cross-section. Bedcauthe intense abrasive effect exerted by the abrasive jet, the surface of the nozzle orifice can be affected by a wear process. To diminish this effect, the nozzle is made of materials characterized by a high wear resistance, like sapphire, tungsten carbide, boron carbide, or other various ceramic materials (Deng et al., 2006).

Holes with cylindrical cross-section and diameters of 0.13-1.25 mm or rectangular cross-section with the dimensions of 0.075x0.5 – 0.15x3.8 mm^2 are achieved in the nozzle. The service life of the nozzle made of tungsten is 8-15 hours if AJM is carriedout with silicon carbide abrasive particles, or 20-35 hours if the abrasive particles made of aluminum oxide are preferred. If the nozzle is made of sapphire, the orifice diameter is 0.2-0.66 mm, and the service life is about 300 hours.

Some succinct information concerning the abrasive particle materials was included in Table 2.

Table 2. AJM abrasive particles characteristics (adapted from Jain, 2013; Metals Handbook, 1989; Shi et al., 2017; Gupta, 2024)

Material	Particle size	Application
Aluminum oxide (Al_2O_3)	10, 20, 50 μm	Cutting, cleaning, deburring, and fine machining of hard and less hard materials
Silicon carbide (SiC)	25, 40 μm	Cleaning, cutting, and deburring of hard material
Glass beads	0.635-1.27 μm	Matte finishing, cleaning, deburring, polishing
Magnesium carbonate		Fine cleaning
Dolomite	200 mesh	Etching, polishing
Sodium bicarbonate	27 μm	Cleaning, cutting, and deburring of soft materials Light finishing below 50 ^{O}
Diamond	13 μm	Texturing
Garnet	80 mesh-200 mesh	More often used in AWJ; if used in AJM, it provides robust removal of harder materials.

Usually, the abrasive particles are not reused. After their use in the abrasive jet machining process, the abrasive particles from the abrasive jet and the particles detached from the workpiece material are recuperated in a first tank. Subsequently, they are stored in a tank for deposition of the used particles and detached material.

A controlled relative movement between the abrasive jet and the workpiece could be obtained using a table whose movements are coordinated utilizing a computer numerical control subsystem. If the table can be moved in an *XY* rectangular coordinate system, a guide could ensure the movement of the gun along a third axis of the Cartesian coordinate system.

More simple equipment for the abrasive jet machining could be based on the manual movement of the gun, when the abrasive jet could be directed inclusively in a zone less accessible but found within the operator's attention. Such a last solution is used, for example, by using cabinets similar to those used as blasting cabinets.

It may be noticed that the small particles detached from an aluminum workpiece could have high explosive properties. For this reason, a wet dust collector could be included in the blasting equipment (Holzknecht, 2005).

As above-mentioned, the absorption of the abrasive particles occurs due to the decrease of the pressure in the zone of the gun where the cross-sectional area of the compressed air stream diminishes.

Following Bernoulli's law, if the fluid is considered incompressible, in this zone an increase in the fluid speed occurs. A drop in either the potential fluid energy or the pressure is linked to the rise in fluid speed. A general equation based on Bernoulli's law could thus be written:

$$\frac{v^2}{2} + g_a z + \frac{p}{\rho} = C_1 \qquad (1)$$

where v is the fluid speed in a point found in the fluid stream (m/s), g_a – the acceleration due to the gravity (m/s^2), z – the elevation of the point above the reference plane along the positive direction of z-axis, so in an opposite direction to the gravity acceleration (m), p – the static pressure in the considered point (Pa), ρ – the density of the fluid in all points of the stream (kg/m^3) and C_1 – a constant.

All the terms of the above-written equation could be multiplied by the fluid density ρ, and the following equation could thus be written:

$$\frac{\rho v^2}{2} + \rho g_a z + p = C_2 \qquad (2)$$

where C_2 is a constant that corresponds to the fluid pressure where v=0 and z=0.

According to Isupov and Lukin (2008), the velocity of the abrasive particles in the nozzle that are thought of as a solid phase could be calculated as follows:

$$v_p = kv_g\sqrt{\frac{3c_x\rho_g l}{2\rho_p D}} \tag{3}$$

where k is a coefficient that takes into consideration the asymmetric shape of the abrasive particle and its rotation in various directions as it moves, v_g – the gas velocity, c_x is the drag coefficient, ρ_g – the gas density, l – the nozzle length, ρ_p – the density of the abrasive particle material, and D – the diameter of the abrasive particle.

Generation of the Abrasive Stream Based on the Centrifugal Force

In the field of abrasive jet blasting, equipment based on the effects of the centrifugal force is used (Patel, 2011). Such equipment could also work when the sand is not used as a material for the particles directed to the workpiece surface, but other materials.

The equipment (Figure 2) includes a wheel found in a fan and whose rotating speed is about 3000 m/min. The particles used as tools could be made of cast-off material, and they are introduced along the rotation axis of the shaft with blades.

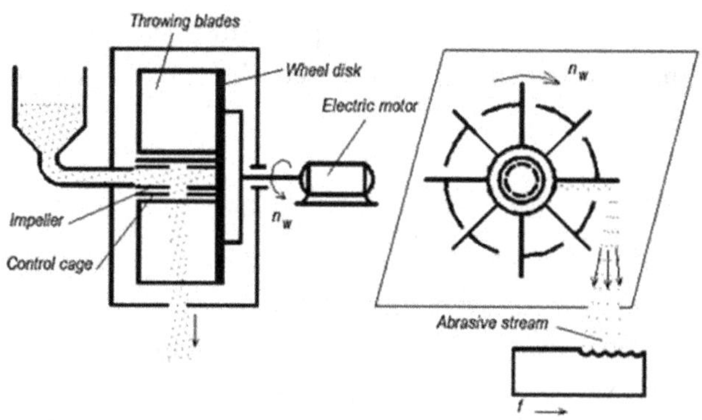

The particles move along the throwing blades as a consequence of the centrifugal force and are directed out of the impeller house along a direction tangential to the wheel periphery with a speed of about 80 m/s. As advantages of this abrasive machining method, the low energy and the uniformity and homogeneity of a large machined area are mentioned (The Masters, 2012).

As can be seen in Figure 2, the equipment for wheel blasting includes a wheel disk to which the blades are attached. The wheel disk has a high-speed rotation. The abrasive particles enter the impeller through the rotation shaft. A control cage facilitates the adjustment of the quantity of the abrasive particles, which are afterward directed to the workpiece using the throwing blades attached to the wheel disk.

Abrasive Particle Movement

As mentioned above, the abrasive jet machining supposes the controlled sending of a high-velocity gas stream that includes abrasive particles to the workpiece surface. Due to the relative movements between the abrasive jet and the workpiece, small quantities of the workpiece material and other substances found on the workpiece surfaces are removed. The abrasive particle action can also contribute to the plastic deformation of the metallic crystals found in the workpiece surface layer.

The abrasive jet has an approximately cylindrical shape along a short distance after leaving the nozzle (Figure 3). Subsequently, the abrasive jet takes an approximately conical shape, due to the decrease of the pressure from a high value in the jet axis to low values corresponding to the atmospheric pressure. Therefore, a divergent shape of the abrasive jet could be observed.

Along their trajectories, the abrasive particle speed diminishes because of the resistance generated by the air molecules. Thus, when the distance from the nozzle and jet axis is higher, the abrasive particle speed is lower. For such a reason, Gauss's normal distribution could be taken into consideration to characterize the speed distribution (Figure 3). Ghobeity et al. (2012) appreciated that a Weibull distribution of particle velocities along a radial distance from the center of the impacting jet could be more adequate.

For cleaning, the standoff distance between the nozzle and the workpiece surface is between 0.5 and 15 mm, and for fine finishing, it is between 13 and 75 mm. When using the cylindrical shape of the abrasive jet, lower standoff distance values are desirable.The impingement angle of 60 – 90 $^{\circ}$.

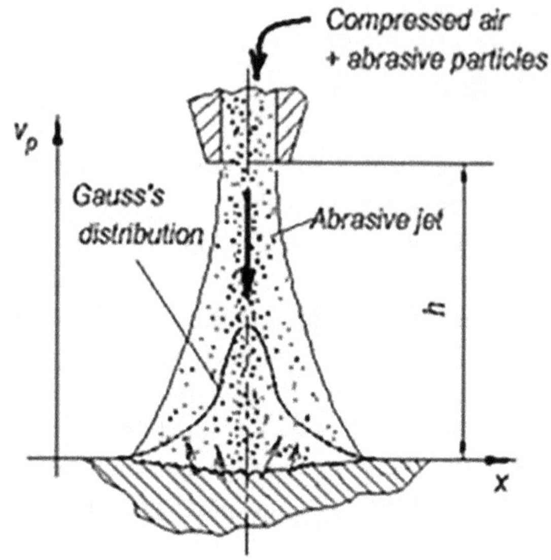

EFFECTS EXERTED BY THE ABRASIVE PARTICLES ON THE SURFACE LAYER

At the contact of the abrasive particles with the workpiece surface, a part of the abrasive particles' kinetic energy E_k is transmitted to the workpiece surface layer, where some specific processes could develop. From the process efficiency that corresponds to the abrasive jet machining, the following effects are interesting:

1. Development of the microcracks;
2. Developing the micro-cutting processes;
3. Generation of the plastic deformation of the surface layer.

Another effect is the elastic deformation of the workpiece surface layer, but since this effect does not affect the material removal process and the aspect of the machined surface, it is not of interest.

It is expected that when the abrasive particles leave the nozzle, they have distinct speeds that are higher near the jet axis and lower near the nozzle wall, due to the friction phenomenon. Taking into consideration the maximum speed v_p (m/s) of the abrasive sand particles found in the jet axis, the particle kinetic energy E_k (J) is given by equation (4).

$$E_k = \frac{m_p v_p^{\ 2}}{2}$$

[J] (4)

where m_p (kg) is the mass of the abrasive particle and v_p (m/s) is the maximum speed of the abrasive particles.

Essentially, the effects exerted by the abrasive particles on the workpiece surface layer depend on the abrasive particle shape and energy, and on workpiece material properties.

If the abrasive particles have rounded surfaces or they hit the workpiece surface with their rounded surfaces, and if the abrasive kinetic energy E_k is low, there is no significant effect on the workpiece surface (Figure 4, a).

When the kinetic energy E_k is higher, a plastic deformation could be generated (Figure 4, b), with a certain material lifting around the impact area. If the workpiece material is fragile, cracks could develop in the zone of impact, but a plastic deformed zone could also be observed (Figure 4, c). The generation and propagation of the cracks could determine the removal of small particles from the workpiece material.

If the particle kinetic energy E_k is low, no effect might be seen if the abrasive particle has sharp edges and is directed perpendicular to the workpiece surface (Figure 4, d). The shape of the particle's active surface may be partially replicated by a tiny cavity and plastic deformation if the kinetic energy E_k is larger (Figure 4, e). A certain material lifting around the impact zone could also be observed. If the workpiece material is fragile, it is possible that small chips to be detached from the workpiece material, as a consequence of the propagation and joining of small cracks (Figure 4, *f* and *g*).

When the abrasive particle is directed along a direction inclined to the workpiece surface, there is no effect if its kinetic energy E_k is low (Figure 4, *h*). If the kinetic energy E_k is higher, the effects of plastic deformation accompanied by a material lifting around the impact area could develop (Figure 4, *i*), or just a microcutting process could determine a real detaching of small chips from the workpiece material (Figure 4, *j*).

After the impact, due to the elasticity of the workpiece material and abrasive particle material, the particle could be rejected and found in the recuperation tank, where there are also the particles detached from the workpiece material. Situations when the abrasive particles penetrate the workpiece surface layer and remain there could also be met. It is also possible that the abrasive particle remains stuck in the workpiece material (Figures 4, *k*), especially when the workpiece material is affected by plastic deformation and adhesion phenomena develop. This is sometimes the case for aluminum and aluminum alloys. When plastic deformation phenomena develop, a thin layer with deformed crystals could be generated around the impact zone.

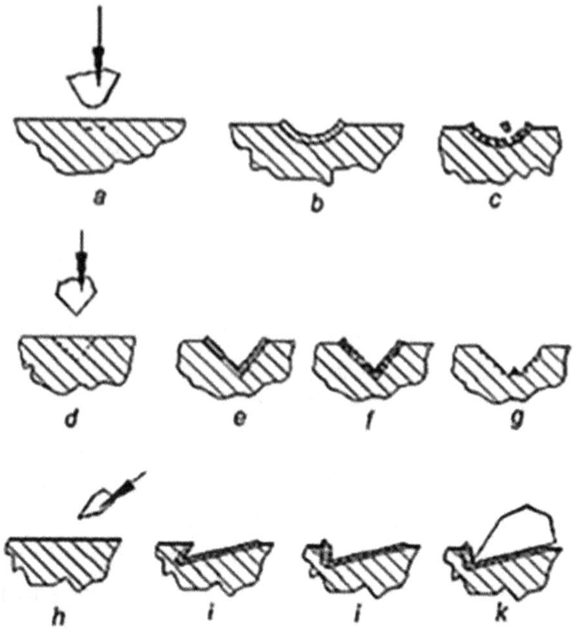

SYSTEMIC APPROACH OF ABRASIVE JET MACHINING PROCESS

The systemic analysis is an analysis method applied to better understand a process, phenomenon, solid ensemble, etc., by considering it as a system, that is, as a set of independent elements that form an organized whole, and in which there are certain relationships among its components.

The systemic approach could take into consideration the system components as subsystems able to interact with each other and place the system in a certain environment. The system has a certain dynamic of the phenomena and processes that develop in the system and in its environment.

When approaching the abrasive machining process as a system, the input factors, disturbing factors, and output parameters could be highlighted (Figure 5).

As *the main input factors of the abrasive jet machining*, the following could be mentioned:

- *Chemical composition and physical–mechanical properties (hardness, tensile strength, impact resistance, brittleness, plasticity, etc.) of the workpiece material.* It is known that brittle materials are considered better machinable when applying the abrasive jet machining;

- *Chemical composition and physical properties (hardness, tensile and compression strength, etc.) of the abrasive particles' materials.* It is desirable to use abrasive particle materials characterized by high hardness and simultaneously good toughness;
- *Chemical composition and physical–mechanical properties (mechanical resistance, adherence to the workpiece material) of the materials to be removed from the workpiece surface,* when the abrasive jet machining process is used as a cleaning process. The properties of the material to be removed from the workpiece exert an essential influence on the process output parameters. They must be considered when designing the abrasive jet machining technology;
- *The general shape and geometrical characteristics of the abrasive particles* (shape, dimensions, sharp edges, rounding radius, etc.). If the abrasive particles have sharp edges, there is a higher probability of obtaining more convenient values for the material removal rate. At the same time, smaller dimensions of the abrasive particles could generate diminished values of the surface roughness parameters that characterize the asperities' heights;
- *Characteristics of the transport gas circulation* (chemical composition, pressure, speed). The industrial use of the abrasive jet machining proved that high values of the pressure and speed of the transport gas can significantly influence the values of the process output parameters. At the same time, the transport gas must not interact with the workpiece material or with the abrasive particle material;
- *Geometrical characteristics of the nozzle orifice (shape, dimensions).* When establishing the machining conditions, the shape and the dimensions of the cross-section through the nozzle orifice must be considered;
- *Distance between nozzle and surfaces to be machined (standoff).* Since along the abrasive particles' trajectories in the air, their movement speed diminishes and the jet takes a conical shape, it is expected that the standoff has a low value or not to exceed certain values;
- *The angle of incidence between the abrasive jet axis and the surface to be machined.* In general, the abrasive jet must be inclined toward the workpiece surface in order for the abrasive particles to have a cutting impact on the workpiece material;;
- *Mixing ratio* could be defined as the ratio of the volume flow rate of abrasive particles to the volume flow rate of carrier gas (Sankar, 2017). Essentially, the increase of the mixing ratio determines an increase of the material removal rate up to a value when a decrease of the jet velocity occurs, and may just block the abrasive particles circulation through the nozzle;

- - Characteristics of relative movements between abrasive jet and workpiece. When the speed of the relative movement between the abrasive jet and workpiece is low, it is expected that the effect exerted on the workpiece material is stronger. For example, when grooves are generated using the abrasive jet machining, a higher depth of the groove could be obtained when using lower values of the speed that correspond to the relative movement between the workpiece and the abrasive jet.

When designing the abrasive jet machining process, the input factors could be set at convenient values from the desired results (material removal rate, surface roughness, etc.). On the other hand, the factors able to exert a disturbing effect (*disturbing factors)* could at least be taken into consideration, even if their values could not be strictly controlled. Such *disturbing factors* could be the following:

- Inhomogeneity of the materials that correspond to the abrasive particles and workpiece to be machined;
- Not controlled variation of some of the processing parameters (pressure and speed of transport gas, the incidence angle between the jet axis and the surface to be machined, adherence between the depot materials and the workpiece surface, etc.).

The main *output parameters of the abrasive jet machining are the following*:

- *Macrogeometry of the machined surface*, which means the shape and the accuracy of the machined surface. An accuracy of ±0,05…±0,13 mm could be obtained when using the abrasive jet machining;
- *Surface roughness parameters;* nowadays, there are multiple possibilities and various parameters that could be used to assess the roughness of the machined surface. The value of the surface roughness parameter *Ra* could be 0.15-1.5 μm;
- *The thickness of the workpiece surface layer affected by the abrasive jet machining process* (hardening, presence of residual stresses, changes in the dimensions of the crystals, etc.);
- *Rate of removing deposited materials on the workpiece surface* and eventually *removing a part of the workpiece surface layer*. For example, the material removal rate could reach values of 16 mm^3/min in the case of glass, 1.6-4.1 mm^3/min in the case of metallic materials, and ~50 mm^3/min for aluminum alloys (Saravanan et al., 2022);
- *Force exerted by the abrasive jet on the workpiece surface and the devices* used to clamp the workpiece and the gun;

- Energy consumption for the abrasive jet machining process;
- *Possible heating of the workpiece material;* sometimes, due to the processes that develop in the workpiece surface layer, a certain increase in the material temperature could be highlighted.

It is expected that each process output parameter is affected in a certain way by the values of certain process input factors.

A graphical representation corresponding to the above-mentioned systemic approach of the abrasive jet machining process can be observed in Figure 5. To highlight the fact that other categories of process input factors, disturbing factors, and process output parameters could be identified in the future, three rectangles including the question sign were introduced in the graphical representation in Figure 5.

A succinct highlighting of the factors able to exert influence on the material removal rate and surface roughness using the Ishikawa diagram was presented by Chen et al. (2017).

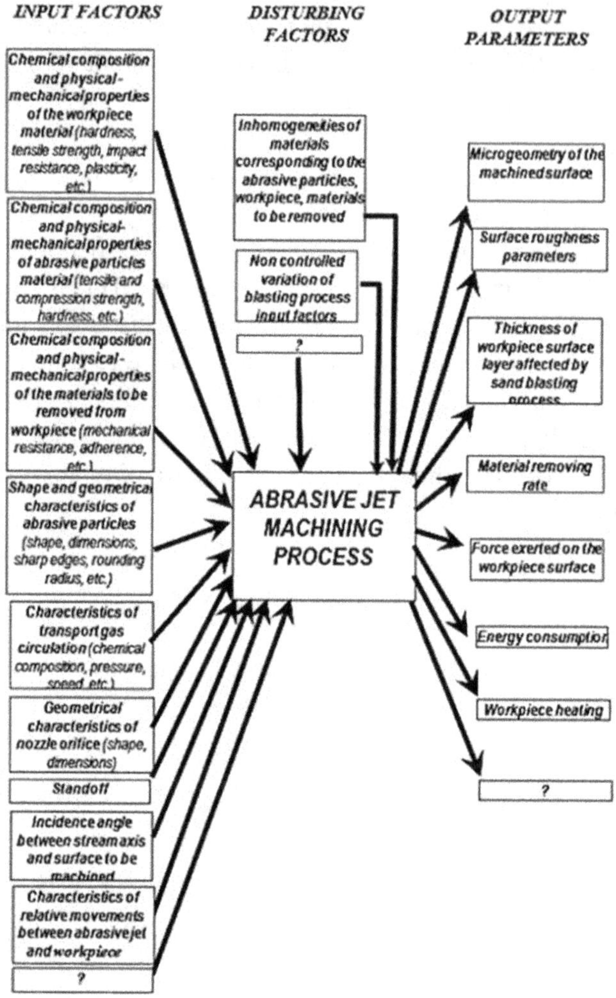

EFFECTS EXERTED BY DISTINCT FACTORS ON THE RESULTS OF ABRASIVE JET MACHINING PROCESS

As shown above, there are many factors that can exercit influence on the performance of the abrasive jet machining. Over the years, researchers have been interested in and developed many investigations that focused on the possibilities of optimizing the abrasive jet machining process, when the influence exerted by the various process input factors on the machining results is known.

The main output parameters of the abrasive jet machining process have been mentioned previously. Table 3 provides a fictitious representation of the impact that the process input elements have on a few of the process output characteristics. The information included in the table was found in the accessed published works. In some of the locations in Table 3, where adequate information was not identified, question marks have been placed, with the possibility that such information may be identified in the future.

Next, only a few particular issues regarding the influence of different factors on the results of the use of abrasive jet machining will be addressed.

Thus, in Table 4, some information concerning the machinability ratings of certain materials when using various materials can be observed. It can be noticed that in comparison with the other materials (used inclusively in the aerospace industry), the aluminum machinability is considered fair or just poor.

In Figure 6, the influence exerted by the value of the standoff on the shape of the groove generated by the abrasive particles can be seen. If for small values of the standoff there is a rectangular shape of the cross-section through the groove, as the value of the standoff increases, the shape of the cross-section through the groove changes.

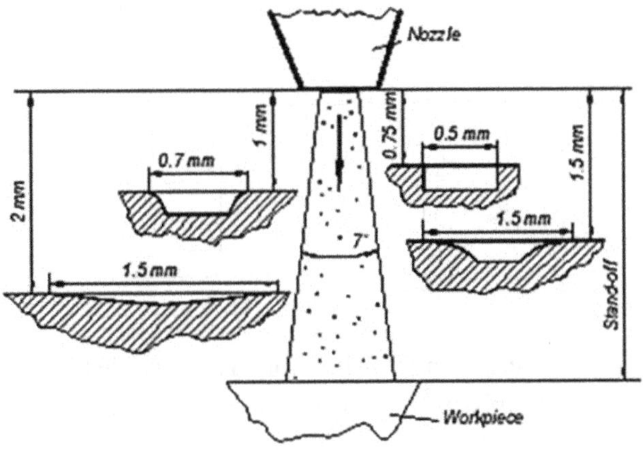

Table 3. Effects exerted by difrent input factors on the values of some interest parameters in the case of the abrasive jet machining (adopted from Ally et al., 2012; Chen et al., 2017; Gherissi & Einasri, 2019; Jain, 2013; Kumar & Neeraj, 2015;

Melentiev and Fang, 2018; Sahu and Mohapatra, 2015; Shi et al., 2017; Sinha et al., 2016; Slătineanu et al., 2018; Tsai et al., 2009; Verma & Lal, 1984, Kwon & Lee, 2022)

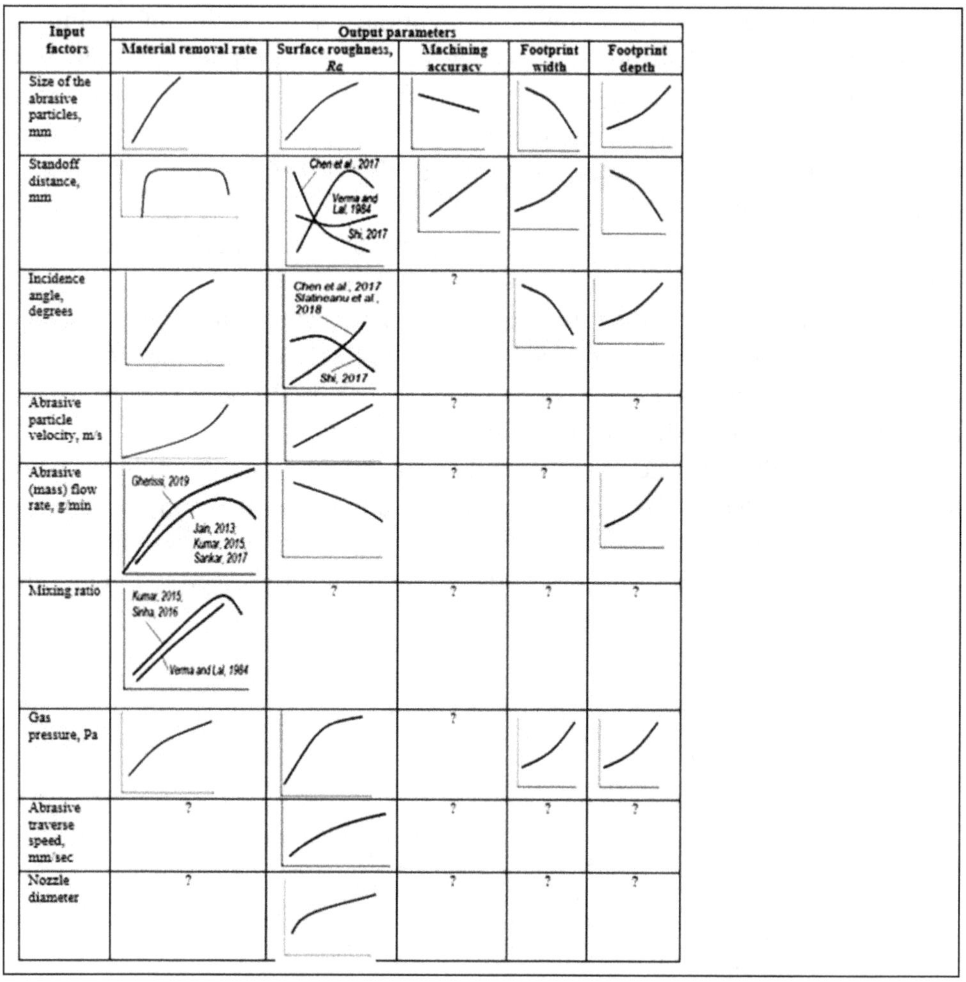

The increase in the average size of the abrasive particles leads to an increase in the value that corresponds to the surface roughness parameter *Ra*. An image concerning the intensity of the influence exerted by the average size of the abrasive particles on the value of the surface roughness parameter *Ra* is given by the empirical mathematical model:

$$Ra = 3.095\, g^{0.531}\, h^{-0.068}\, \alpha^{0.0922} \tag{5}$$

where the h is the standoff and α is the angle between the test plan surface and the abrasive jet axis (Slătineanu, 2013).

The increase of the content of the abrasive particles in the abrasive jet determines a diminishing of the exit velocity of the carrier gas (fig. 7). To increase the exit velocity of the carrier gas, an increase of the internal gas pressure up to nearly twice the pressure at the nozzle exit when the abrasive particle density is zero could be used. In such a case, it is considered that there is a critical value of the exit velocity of the gas (Kumar, 2019).

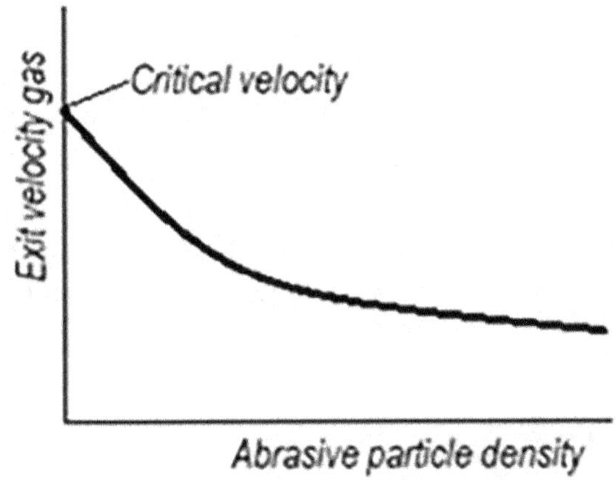

Table 4. Machinability evaluation of various materials when applying the abrasive jet machining (Todd et al., 1994)

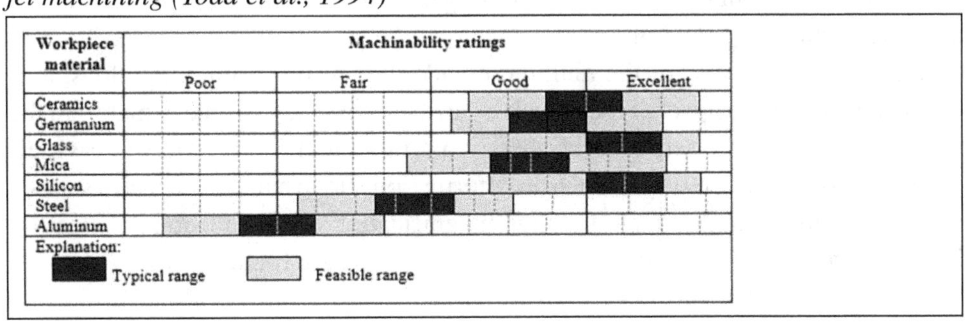

Explanation:
Typical range Feasible range

For a certain process duration, the mass of the material removed from the workpiece could correspond (Jain, 2013) to the following analytic relation:

$$\Delta m = K\frac{\rho}{H} \cdot \frac{1}{2}m_p v^2 \tag{6}$$

where K is a coefficient (K>2), ρ – density of the workpiece material, H – hardness of the workpiece material, m_p – the abrasive particle mass, and v – abrasive particle velocity.

The erosion rate E_r could be assessed by means of the quotient of mass loss and the amount of abrasive. The abrasive particle velocity v affects the erosion rate in concordance with the relation (Jain, 2013):

$$E_r \approx v^k \tag{7}$$

where k=2.3-3 for metallic workpieces, k=2 4 for glass and k=1.8 – 3.2 for elastomers.

The value of the surface roughness parameter Ra, and the erosion rate E_r are given (in Jafar et al., 2013, pp. 138-145), conform (Slikkerveer et al., 1998) by the analytic relations:

$$Ra \approx \frac{b}{4} = \frac{0.63}{4}\left(\frac{3}{2\pi}\right)^{\frac{1}{3}}\left(\frac{E}{H}\right)^{\frac{1}{2}}\frac{U^{\frac{1}{3}}}{H^{\frac{1}{3}}} \tag{8}$$

$$E_r = \frac{m_t}{m_p} = \frac{\pi}{2}\frac{\rho_t b c_l^2}{m_p} \tag{9}$$

where b is the size of the plastic zone radius, E – elastic modulus, of the workpiece material, H - the workpiece (target) hardness (Pa), U – particle kinetic energy, c_l – lateral crack length, ρ_t – workpiece (target) material density, m_p – mass of the abrasive particle, m_t- workpiece (target) mass.

Due to the plasticity of the aluminum and some of the aluminum alloys, as a consequence of the impact of the abrasive particles with the workpiece material, the surface affected by the action of the abrasive particles includes a variety of cavities, prominent zones affected by the plastic deformation, etc. An image of such a surface can be seen in Figure 8; due to the high magnification obtained using the electron beam scanning, many details of the machined surface are highlighted.

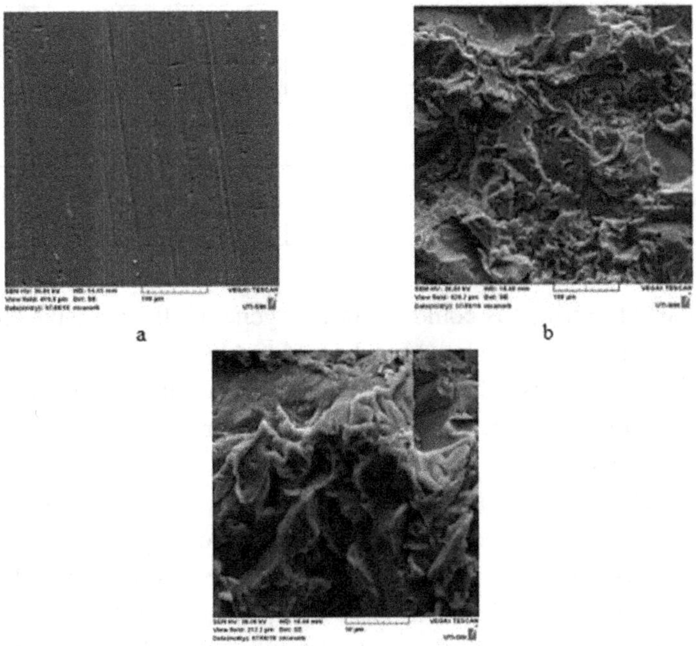

a b c

ESTABLISHING THE ABRASIVE JET MACHINING CONDITIONS

Taking into consideration the type of workpiece material and the processing method, the type of abrasive particles' material could be established. If the workpiece material is aluminum or an aluminum alloy, this material could be sodium bicarbonate, but also the other materials recommended in Table 2 could be selected.

The next step could be the selection of the abrasive particle grit; the information included in Table 2 could be taken into consideration. In correspondence with the abrasive particles' dimensions, the diameter of the nozzle orifice could be established (Table 5).

Table 5. Establishing the nozzle orifice diameter (adapted from Todd et al., 1994)

Abrasive material	Grit size (mm)	Nozzle orifice diameter (mm)
Aluminum oxide	0.000254-0.00127	0.127-0.457
Silicon carbide	0.000635-0.00127	0.2032-0.457
Dolomite	0.0635	0.6604-1.27
Glass beads	0.0635-1.27	0.6604-1.27
Diamond	0.013	-

Some types of carrier gas could be used to generate the abrasive jet. The compressed air is the cheapest version, but when the workpiece material could be affected by an oxidation process, the other carrier gases could be preferred. If a greater material removal rate is necessary, higher values for the carrier gas pressure and speed could be imposed.

The devices for clamping the workpiece could be selected considering the mass of the workpiece to be machined and the forces and torques generated by the abrasive jet machining process. Even if the forces and the torques are not high, they could move the small mass workpieces, if these workpieces are not clamped in adequate devices or they are not manually moved relatively to the abrasive jet position.

ADVANTAGES, DISADVANTAGES, AND LIMITATIONS OF THE ABRASIVE JET MACHINING

The main benefits that have been reported for the abrasive jet machining are the following (Ally et al., 2012; Jain, 2013; Kumar, 2019; Kumar and Neeraj, 2015; Melentiev and Fang, 2018; Sankar, 2017; Hu, 2025):

- Low capital and operating costs;
- Ease of operation of the machining equipment;
- Thin surface layer that is impacted by the machining process;
- A certain hardening process could increase the wear resistance of the machined surfaces;
- Very low increase of the surface layer temperature and therefore the possibility of machining temperature-sensitive materials (such as glass, ceramics, etc.), without significantly affecting the surface layer physical properties and the eventual crystalline structure;
- Low values of the forces and torques generated by the machining process, and therefore, there are no significant plastic deformations of the workpiece;
- Lack of vibrations of significant size;

- Low values of the asperities' heights of the machined surfaces; the surface roughness is essentially dependent on the abrasive particle size;
- Possibility of processing very thin workpieces made of hard and brittle materials like silicon, germanium, glass, ceramics, etc.;
- The machining results are not greatly impacted by the workpiece material's electrical and thermal characteristics.;
- Workpieces made of conductive and nonconductive materials could be machined;
- Good capacity to cut intricately shaped holes in fragile and hard materials;
- When cutting is used, the width of the slid could be very small and, consequently, there is a low loss of material;
- Narrow slots could be cut (0.12-0.25 mm);
- The process facilitates the obtaining of angular cutting of complex contours;
- Different values for the material removal rate and surface roughness might be obtained by adjusting the abrasive processing settings.;
- The abrasive jet machining could be applied to obtain surface microreliefs convenient for tribological properties.

As for *the disadvantages* or *limitations* of the abrasive jet machining, the following can be mentioned:

- Quite low material removal rate;
- Risk of remaining embedded abrasive particles in the surface layer obtained by abrasive jet machining, especially when materials characterized by high plasticity are machined;
- Possible impairment of machining accuracy by tapering of the entry edges of the holes by the particles found at the periphery of the abrasive jet;
- For certain operating conditions, the machined surface could present striations;
- The necessity of a dust collection subsystem to avoid or diminish the atmospheric pollution and the generation of health hazards;
- The noise level could sometimes be appreciated as dangerous;
- Quite a low service life of the nozzle;
- Usually, the abrasive particles are not reused, since their sharp edges are rounded and small particles could generate the nozzle clogging;
- When low values of the standoff distance are used, there is the risk of nozzle damage;
- In the cases of machining deep holes and cutting, a certain inclination of the machined surfaces is possible;

- When the machining space cannot be tightly sealed, there is a risk of environmental pollution.

ABRASIVE JET MACHINING PROCESSES

Surface finishing or polishing. Because abrasive jet machining could be used to lower the height of surface asperities, specific finishing techniques relying solely on the abrasive jet's action were developed (Teja and Reddy, 2017; Tsai et al., 2009). For example, an abrasive jet polishing process was used to finish the surfaces of parts made of mold steel that were previously obtained by electrical discharge machining.

The processing conditions were characterized by an abrasive material to the additive ratio of 1:2, an impact angle of 30 °, gas pressure p=0.4 MPa, the nozzle-to-workpiece height of 10 mm, platform rotational velocity of 200 rpm, and a platform travel speed of 150 mm/s. Over the course of 20 minutes, the surface roughness decreased from an initial value of Ra=1.7 μm to a final value of Ra=0.27 μm (Tsai et al., 2009).

Surface texturing is a processing method used to change the geometrical characteristics of the workpiece surface so that an improvement in some tribological and sealing performances (wear resistance, friction resistance, microlubrication processes) is possible. When using micro-abrasive jet machining for surface texturing, micro-dimples or micro-grooves (with dimensions from 50 μm to millimeters in diameter or length and 5-20 μm in depth) are generated on the workpiece surface.

The standoff distance, jet pressure, abrasive flow rate, process duration, and other factors all affect the micro-abrasive jet machining results. The machining process makes it easier to recycle the abrasive particles, which reduces environmental pollution. Usually, masks are used to protect the surfaces that must not be affected by the texturing process (Su et al., 2016; Shi et al., 2017).

Micro-AJM blasting is a processing method in which fine abrasive particles are accelerated by a high-pressure carrier gas and directed through a nozzle onto the part surface. The particle impacts remove material at a micro-scale, resulting in surface modification.

Depending on the process parameters (injection pressure, nozzle diameter, abrasive type, and size), the outcome can range from simple surface cleaning to creating a general matte/rough surface. In μ-AJM of aluminum alloys, careful control of input parameters—such as abrasive particle type, nozzle diameter, operating pressure, standoff distance, and injection time—is essential to achieve the desired surface roughness. Surface roughness values between 0.7 and 1.5 μm were achieved at 25 kPa. The roughness increased progressively with higher injection pressure and larger nozzle diameters (Kwon et al, 2022).

Usually, Laval line-type nozzles are used for micro AJM blasting and texturing. These nozzles have a specially designed convergent-divergent nozzle that accelerates abrasive particles entrained in a carrier gas to supersonic speeds (Gadge et al., 2022; Ge et al., 2024; Hu, 2025). Unlike conventional circular nozzles, its outlet has an elongated, slot-like shape, which generates a jet with a linear profile. This geometry allows a more uniform distribution of particles over a larger impact area and improves the process where controlled erosion along a line is advantageous.

Cutting of grooves of low depth and that are placed in complex configuration could be applied in the case of hard and brittle materials. The dimensions of the groove are defined as it can be seen in Figure 10.

A so-called air-abrasive cutting apparatus (Whisper-jet, KCP-1000, American Dental Technology Denix, U.S.A.) was used in the field of pediatric dentistry to prepare the cavity of dentin using alumina or chitin-chitosan particles (Nakamura et al., 2006).

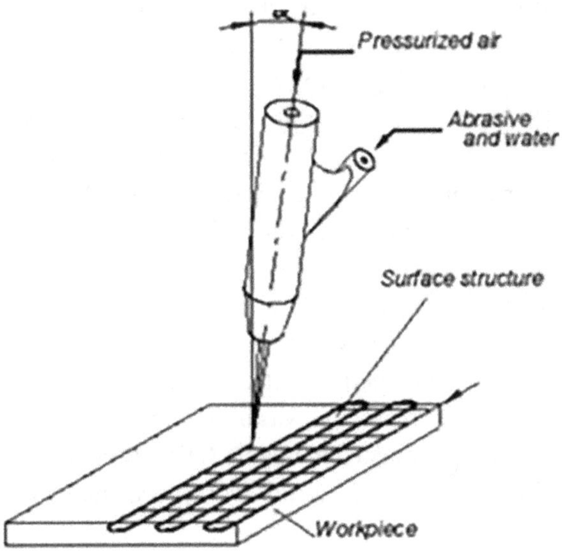

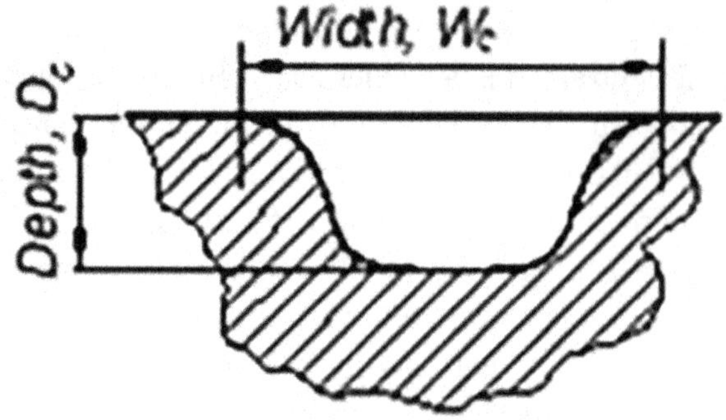

Cuts could be achieved with minimal effect on the margins in gallium arsenide, silicon, or ceramic crystal wafer. Cutting processes using the abrasive jet facilitate preparing the cavities in the dental molds for crowns (High precision, 2012).

To obtain a certain contour of the machined surfaces, sometimes masks are applied to the surfaces of the workpieces. The mask material must be wear-resistant. The mask could be applied using an operations succession (Jain, 2013; Park et al., 2004) of type 1. Preheating the substrate; 2. Applying the mask film by laminating; 3. Exposure; 4. Development; 5. Water rinse; 6. Acid neutralization. A 5-axis CNC-controlled micro-abrasive air jet turning system and a spring mask were used to generate helical grooves on bovine bone using white alumina abrasive particles (Li et al., 2025).

Drilling and *cutting* of low-thickness workpieces could be applied in the case of workpieces made of glass, ceramics, and hard metallic materials. Holes with diameters of about 1 mm were obtained in a steel plate of thickness 0.6 mm (Vanmore and Dabade, 2018). A drilling process was used to obtain a hole with a diameter of 0.203 mm in the side of a hypodermic needle (High Precision, 2012).

Engraving could be used in the case of workpieces made of metallic materials and plastics.

Preparing cutting-edge by pressure abrasive air wet jet machining supposes the application of a processing method based on the use of pressured abrasive jet machining (PAWAJM). The method was applied to increase the working capacities of the cutting edges sharp cutting tool (Wang et al., 2020). The researchers confirmed the possibility of obtaining an average cutting edge rounding from 14 to 56 μm, higher surface hardness, and compressive residual stresses.

Essentially, the pressurized air wet abrasive jet machining uses accelerated unbound abrasive particles found in water through a jet nozzle. The introduction of the abrasive particle in the air jet is realized either by suction or by compression.

The diameter of the nozzle orifice is higher than in the case of other proper abrasive jet machining processes (for example, d_{no}=8 mm, following Freiburg et al., 2019).

Deburring is a machining process aiming at removing the burrs resulted in the case of drilling holes or milling. It could be applied inclusively in the case of small-dimension parts.

Removing oxides or other substances that exist on metallic and ceramic materials could be made using the abrasive jet.

Removal of stain or of the layers deposited over time on documents, on museum objects, or just on the building walls could be a particular application of the abrasive jet machining.

The possibility of removing oxidation appeared on masterpieces without affecting the painted surface could be taken into consideration (High precision, 2012).

Calibration, tapering, and cleaning could be applied in the case of electronic components.

A cleaning process using the abrasive jet is sometimes applied to clean and dress the CBN (cubic boron nitride) and diamond wheels (High precision, 2012).

The abrasive jet micromachining ensures the control of the machined surfaces using masks. This machining process can be applied inclusively in the case of necessity to obtain channels in the anisotropic patterns and suspending structures (Ghobeity et al., 2012).

COMPARATIVE ANALYSIS OF THE ABRASIVE MACHINING PROCESSES WITH OTHER CONVENTIONAL OR NONCONVENTIONAL MACHINING PROCESSES

To develop a comparison of the abrasive jet machining process with other industrial machining processes (Ravishankar, 2017; Saran et al., 2025), the water jet machining (WJM) and the abrasive water jet machining (AWJM) were selected.

The water jet machining consists of using a water jet, which acts as a tool in the form of a water saw. The water jet has a high velocity and pressure, and it can slice workpieces made of various materials. In the case of abrasive water jet machining, the abrasive particles ensure an increase in the machining process efficiency.

Both the abrasive jet and water jet are greatly accelerated and further focused on the zone to be affected by the machining process. The use of water as a carrier fluid contributes to the development of specific processes in the machining zone. The brittle and hard materials are more suitable for the WJM than the AJM. As most of the metal cutting techniques generate so-called heat-affected zones (HAZ), it is important to mention that the AJM, WJM, and AWJM can be considered as practically free from introducing such an effect.

Everything from soft metallic and nonmetallic materials like rubber and foam to hard, brittle ceramics and glass can be machined using the AWJ cutting method.

Examining the unique benefits of water jet machining and abrasive water jet machining, it should be noted that alterations in material properties may result from the increased heat produced during the milling process. Since heat is effectively eliminated by the water circulation, there is no thermal increase in the machining zone during abrasive water jet machining or water jet machining. Such machining techniques are therefore highly helpful when it comes to the machining of temperature-sensitive materials such as glass fiber laminate, aluminum, cutting tool steels, etc.

One significant benefit of the water jet cutting technique is that, unlike some cutting and grinding operations, it doesn't produce any dust or other particles that could be dangerous to breathe in.

Materials that break down rapidly in the presence of moisture cannot be cut using water jet machining or abrasive water jet machining.

TRENDS IN THE USE AND INVESTIGATION OF THE ABRASIVE JET MACHINING

The new or improved manufacturing processes that have been derived from the abrasive jet machining were a result of the intense research work that was carried out in this field. Figure 11 presents the evolution of the number of papers published on two of the main database platforms (Science Direct and Clarivate Analytics) in the last decade in the field of abrasive jet machining. The analysis took into consideration the abrasive water jet machining, since some of the researchers appreciated that this last machining method (abrasive water jet machining) could be included in the larger field of abrasive jet machining. The tendency lines illustrated in Figure 11 show an exponential growth for the abrasive jet machining processes-related scientific and technical work.

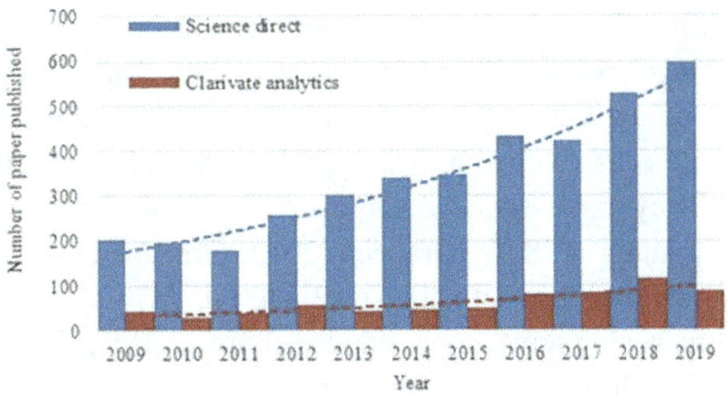

The main development directions of the abrasive jet machining refer to:

-Type of abrasives: ice-assisted jet machining (Melentiev & Fang, 2018), magnetic abrasive (Ke et al., 2012; Melentiev & Fang, 2018);

- Types of energy carrier fluids: abrasive air-jet machining, abrasive water jet machining (Melentiev & Fang, 2018), magnetorheological jet finishing (Melentiev & Fang, 2018; Chen et al., 2017; Kordonski & Shorey, 2007; Kordonski et al., 2006);
- Supplementary machining conditions (cryogenic-assisted abrasive jet machining, dry-ice blasting, hybrid machining including the electrical discharge machining and abrasive jet machining, intermittent jet machining, etc.);
- Type of machining operations (turning, milling, drilling, polishing, etc.).

The flexible abrasive jet machining could be defined as an abrasive jet machining method that uses flexible or elastic abrasive particles. There is a version of the flexible abrasive jet machining in which a certain restraining of the abrasive jet ensures a high material removal rate and improved abrasive path control. Such a restraining could be obtained by placing a permanent magnet in an opposite position to the common abrasive gun. The solution was applied in the case of the so-called *flexible magnetic abrasive jet machining*, including the use of self-made abrasive (Ke et al., 2012).

The micro-abrasive intermittent jet machining is distinguished by the use of a short period of time when the abrasive particle's access is prevented. The method was used to facilitate the machining of small holes. The gas stream without abrasive particles removes the material accumulated in the hole, ensuring better conditions for continuing the microabrasive jet drilling process (Zhang et al., 2005).

Cryogenic-assisted abrasive jet machining was applied in the case of quite soft materials and alloys. Their low hardness and ductility do not facilitate material removal by cracking processes. The use of a cryogenic liquid can significantly diminish this disadvantage. An example of using the cryogenic abrasive jet machining in the case of workpieces made of polydimethylsiloxane highlighted the increase of the process efficiency expressed as a volume of material removed per unit kinetic energy at approximately -178 °C (Gradeen, 2012; Gradeen et al., 2012). Alumina particles of 25 μm were used as an abrasive material.

Due to the remarkable advantages these techniques offer, they can be applied in the processing of aluminum alloy parts used in the aerospace and automotive industries, where precision and surface quality requirements are high.

Ultrasonic vibration-assisted abrasive jet machining. Ultrasonic vibrations have been found to influence particle trajectories and the impact process in the stagnation zone. They significantly contribute to improved performance in micro-hole making. The positive effects are reflected in both increased material removal rates and improved morphology at the bottom of the hole. In addition, the periodic vibration of the part modifies the fluid flow lines in the stagnation zone. Thus, they allow a greater number of particles to participate in the removal process and cause the particles, after impact, to roll and smooth the surface, resulting in a better finish (Qi et al., 2021; Ge et al., 2024).

Dry-ice blasting uses a carbon dioxide solid form introduced in a pressurized air stream. The method was applied to clean surfaces without leaving chemical residues (Dry-ice blasting, 2018).

The hybrid process of abrasive jet machining and electrical discharge machining was developed to increase the efficiency of the material removal and to improve the quality of the machined surface. A high-speed gas jet including abrasive particles was added to the electrical discharge process to remove the recast layer. The experiments proved the increase in the efficiency of the material removal process and the diminishing of the surface asperities' heights (Lin et al., 2012; Lin et al., 2018; Wankhade et al., 2016).

Hot abrasive jet machining (hAJM) — or Abrasive Hot Air Jet Machining / Hot Air Abrasive Jet Machining — is a variant of AJM that consists of heating the carrier air and/or particles before acceleration. In some experimental installations, particles are heated directly in a preheating zone (either by contact with heated walls or by passing through a jet of hot air before acceleration). Heating increases the jet energy and locally embrittles brittle materials (glass, ceramics), resulting in a higher material removal rate (MRR) and sometimes better roughness compared to conventional AJM (Pradhan & Dhupal, 2022; 2023; Prasad et al., 2025).

Plastic abrasive jet machining (PAJM) is another development direction for the AJM. PAJM is a variant of AJM in which particles of plastic materials (melamine, urea, polymethacrylate, etc.) are accelerated by an air stream. They are projected towards the target surface to remove paint, coatings, or contaminants, or to clean/pretreat sensitive surfaces. Softer /more elastic particles than aluminum oxide or SiC are used as abrasives to remove layers with relatively low working pressures (0.2–0.6 MPa) without damaging the substrate. These processes proved to be useful for CFRP and aeronautical structural aluminum, sensitive parts (Zhu et al., 2022; Zhao et al., 2024).

Abrasive Jet Machining (AJM) is the unconventional machining process in which a jet of fine abrasive particles, propelled by compressed air or inert gas at high pressure, erodes the part material through repeated impact. The technology is used in particular for fragile materials (glass, ceramics, carbides), precision micro-machining, surface cleaning, and deburring.

Its applicability for aluminum alloys is based on several premises:

- Ductile materials can be processed by optimizing the parameters (particle size, abrasive flow rate, gas pressure), which allows control of the erosion rate and the final roughness.
- The lack of direct mechanical contact eliminates cutting forces, reducing the risk of warping thin or complex-shaped parts.

The possibility of localized machining and microstructuring makes AJM attractive for finishing applications, selective cleaning, or making microchannels in aluminum components used in electronics, aerospace, and automotive.

CONCLUSIONS

Abrasive jet machining is classified as a machining method that utilizes the kinetic energy of abrasive particles to remove thin layers of material from the workpiece. Machining processes based on this method have been promoted to solve specific categories of problems in the manufacture of machines and devices usable in various fields. The initial applications of the method were documented in the United States during the late 19th century. The principle underlying the phenomenon under consideration is that material removal from the workpiece occurs as a result of the development of microcracks and through microchipping effects. The machining process also results in the formation of a thin layer of plastically deformed material. The material removal rate, the values of certain roughness parameters, the machining accuracy, the footprint width, and the footprint depth are utilized as output parameters of the process. A multitude of input factors influence the machining process, with the values of these factors potentially impacting the output parameters of the abrasive jet machining process. A critical evaluation of the merits and drawbacks inherent to abrasive jet machining has enabled the discernment of scenarios in which this machining method is deemed efficacious. In recent decades, a variety of application processes have been developed for the aforementioned method, including micro-AJM blasting, ultrasonic vibration-assisted abrasive jet machining, dry-ice blasting, hot abrasive jet machining, and plastic abrasive jet machining, among others. It is hypothesized that in the future, the diversification of constructive solutions for parts and equipment, as well as the emergence and promotion of new materials, will result in the identification of new applications for the abrasive jet machining.

REFERENCES

Ally, S., Spelt, J. K., & Papini, M. (2012). Prediction of machined surface evolution in the abrasive jet micro-machining of metals. *Wear*, *292–293*, 89–99. DOI: 10.1016/j.wear.2012.05.029

Chen, F., Miao, X., Tang, Y., & Yin, S. (2017). A review on recent advances in machining methods based on abrasive jet polishing (AJP). *International Journal of Advanced Manufacturing Technology*, *90*(1-4), 785–799. DOI: 10.1007/s00170-016-9405-7

Connecticut Resources Recovery Authority. Projects. Bridgeport Project/Southwest Division History (2013), Retrieved June 8, 2016, from http://www.crra.org/pages/proj_bridgeport_history.htm

Danylenko, Y. A. (2018). Characteristics and classification of innovation and innovation process. Science and innovation, 14(3), 14-26.

Deng, J., Zhang, X., Niu, P., Liu, L., & Wang, J. (2006). Wear of ceramic nozzles by dry sand blasting. *Tribology International*, *39*(3), 274–280. DOI: 10.1016/j.triboint.2004.07.026

Dry-ice blasting (2018). Retrieved November 12, 2019, from http://yourmoveinfo.blogspot.com/2018/06/dry-ice-blasting.html

Freiburg, D., Aßmuth, R., Garcia Carballo, R., Biermann, D., Henneberg, J., & Merklein, M. (2019). Adaption of tool surface for sheet-bulk metal forming by means of pressurized air wet abrasive jet machining. *Production Engineering*, *13*(1), 71–77. DOI: 10.1007/s11740-018-0858-4

Gadge, M., Lohar, G., & Chinchanikar, S. (2022). A review on micro-blasting as surface treatment technique for improved cutting tool performance. *Materials Today: Proceedings*, *64*, 725–730. DOI: 10.1016/j.matpr.2022.05.196

Ge, H., Zhang, X., & Liu, Y. (2024). Numerical and Experimental Study on the Jet Behavior of Ultrafine Dry Powder Based on a Supersonic Nozzle. *ACS Omega*, *9*(46), 46574–46587. DOI: 10.1021/acsomega.4c08542

Ge, J., Lin, Y., Qi, H., Li, Y., Li, X., Li, C., Li, Z., & Xu, K. (2024). The impact of ultrasonic-induced jet morphology on polishing efficiency. *International Journal of Mechanical Sciences*, *284*, 109764. DOI: 10.1016/j.ijmecsci.2024.109764

Gherissi, A., & Einasri, I. (2019). Experimental and numerical study on jet abrasive machining of aluminum 2024 – T3. *Acta Technica Napocensis, Series: Applied Mathematics. Mechanical Engineering (New York, N.Y.), 62*(1), 149–156.

Ghobeity, A., Krajac, T., Burzynski, T., Papini, M., & Spelt, J.K. Surface evolution models in abrasive jet micromachining. *Wear, 264*(3-4), 185–198. https://doi.org/ DOI: 10.1016/j.wear.2007.02.020

Gradeen, A. G. (2012). *Cryogenic abrasive jet machining of polydimethylsiloxane and polytetrafluoroethylene at different temperatures*. Ryerson University, Toronto., DOI: 10.32920/ryerson.14652027

Gradeen, A. G., Spelt, J. K., & Papini, M. (2012). Cryogenic abrasive jet machining of polydimtylsiloxane at different temperatures. *Wear, 274-275*, 335–344. DOI: 10.1016/j.wear.2011.09.013

Guest, D. E. (2018). Aluminum alloys in the aerospace industry. Retrieved October 28, 2019, from https://www.thomasnet.com/insights/aluminum-alloys-in-the-aerospace-industry/

Gupta, K. (2024). Developments in Abrasive Water Jet Machining of Nonmetals. *Jurnal Poli-Teknologi, 23*(2), 73–82. DOI: 10.32722/pt.v23i2.6910

Handbook, M. (1989). *Machining* (9th ed., Vol. 16). ASM International.

Heinz, A., Haszler, A., Keidel, C., Moldenhauer, S., Benedictus, R., & Miller, W. S. (2000). Recent development in aluminium alloys for aerospace applications. *Materials Science and Engineering A, 280*(1), 102–107. DOI: 10.1016/S0921-5093(99)00674-7

Holzknecht, E. (2005). Shot blasting goes high tech. Case studies document how new systems reduced manufacturing costs for end users. *Metal Finishing, 103*(7-8), 25–27. DOI: 10.1016/S0026-0576(05)80460-X

Hu, Y. (2025). Characteristics of abrasive jet erosion manufacturing technology: A comparative review. *International Journal of Advanced Manufacturing Technology, 138*(9-10), 1–51. DOI: 10.1007/s00170-025-15698-2

Hu, Y. (2025). Characteristics of abrasive jet erosion manufacturing technology: A comparative review. *International Journal of Advanced Manufacturing Technology, 138*(9-10), 1–51. DOI: 10.1007/s00170-025-15698-2

Isupov, M. G., & Lukin, L. L. (2008). Characteristics of gas–abrasive jets. *Russian Engineering Research, 28*(6), 566–570. DOI: 10.3103/S1068798X08060117

Jafar, R. J. M., Spelt, J. K., & Papini, M. (2013). Surface roughness and erosion rate of abrasive jet micro-machined channels: Experiments and analytical model. *Wear, 303*(1-2), 138–145. DOI: 10.1016/j.wear.2013.03.005

Jain, V. K. (2013). *Abrasive jet micromachining*. Retrieved November 1, 2019, from https://home.iitk.ac.in/~vkjain/L-4-AJMM_ME698D-21-03-13.pdf

Ke, J.-H., Tsai, F.-C., Hung, J.-C., & Yan, B.-H. (2012). Characteristics study of flexible magnetic abrasive in abrasive jet machining. *Procedia CIRP*, *1*, 679–680. DOI: 10.1016/j.procir.2012.05.025

Kordonski, W., & Shorey, A. (2007). Magnetorheological (MR) jet finishing technology. *Journal of Intelligent Material Systems and Structures*, *18*(12), 1127–1130. DOI: 10.1177/1045389X07083139

Kordonski, W. I., Shorey, A. B., & Tricard, M. (2006). Magnetorheological jet (MR JET™) finishing technology. *Transactions of the ASME, 128,* 20-26. DOI: 10.1115/1.2140802

Kumar, R., & Neeraj. (2015). Minimization of material removal rate in abrasive jet machining of tempered glass using Taguchi method. *International Journal of Recent Research in Civil and Mechanical Engineering*, *2*(1), 253–260.

Kumar, S. (2019). Course Name: Unconventional Machining Processes. Retrieved November, 8, from http://docplayer.net/137171374-Course-name-unconventional-machining-processes-semester-v-course-code-ame507.html

Kwon, D.-K., & Lee, J.-H. (2022). Performance Improvement of Micro-Abrasive Jet Blasting Process for Al 6061. *Processes (Basel, Switzerland)*, *10*(11), 2247. DOI: 10.3390/pr10112247

Li, J., Li, Q., Deng, Y., Zhang, W., & Yin, H. (2025). Micro-Abrasive Air Jet Machining Technology for Fabrication of Helical Grooves on Bovine Bone. *Micromachines*, *16*(2), 149. DOI: 10.3390/mi16020149

Lin, Y.-C., Chen, Y.-F., Wang, A.-C., & Sei, W.-E. (2012). Machining performance on hybrid process of abrasive jet machining and electrical discharge machining. *Transactions of Nonferrous Metals Society of China*, *22*, 775–780. DOI: 10.1016/S1003-6326(12)61803-2

Lin, Y.-C., Hung, J.-C., Lee, H.-M., Wang, A.-C., & Fan, S.-F. (2018). Machining performances of electrical discharge machining combined with abrasive jet machining. *Procedia CIRP*, *68*, 162–167. DOI: 10.1016/j.procir.2017.12.040

Madhu, S., & Balasubramanian, M. (2015). A review on abrasive jet machining process parameters. *Applied Mechanics and Materials, 766-767*, 629–634. . DOI: 10.4028/www.scientific.net/AMM.766-767.629

The Master of Blasters (2012). Retrieved November 3, 2019, from http://www.raga .at/download/en.pdf.

Melentiev, R., & Fang, F. (2018). Review. Recent advances and challenges of abrasive jet machining. *CIRP Journal of Manufacturing Science and Technology, 22*, 1–20. DOI: 10.1016/j.cirpj.2018.06.001

Mistry, M., Palkhade, S., Patel, V., Shah, D., Vyas, K., & Shah, S. (2018). A review on abrasive jet machining. *International Journal of Scientific Research in Science, Engineering and Technology, 4*(1), 571–574. DOI: 10.1016/0924-0136(93)90005-Q

Nakamura, H., Nakamura, M., Osuga, N., & Miyazawa, H. (2006). Application of an air-abrasive cutting apparatusin the pediatric dental field: Cutting using chitin-chitosan grains. *Pediatric Dental Journal: International Journal of Japanese Society of Pediatric Dentistry, 16*(1), 57–66. DOI: 10.1016/S0917-2394(06)70066-0

Park, D.-S., Cho, M.-W., Lee, H., & Cho, W.-S. (2004). Micro-grooving of glass using micro-abrasive jet machining. *Journal of Materials Processing Technology, 146*(2), 234–240. DOI: 10.1016/j.jmatprotec.2003.11.013

Patel, P. A. (2011). Effective use and maintenance of shot-blasting machines. Retrieved November 3, 2019, from https://www.pshotblast.com/articles/effective-use -maintenance-page-1.pdf

Pradhan, S., & Dhupal, D. (2022). An integrated approach of simulation, modeling and computer-aided design of hot abrasive jet machining setup. *Journal of Advanced Manufacturing Systems, 21*(03), 427–472. DOI: 10.1142/S0219686722500123

Pradhan, S., & Dhupal, D. (2023). Machining of aluminium nitride ceramic with a hAJM process using SiC abrasives: Experiments and simulation. *Journal of Manufacturing Processes, 95*, 68–90. DOI: 10.1016/j.jmapro.2023.04.011

Prasad, S. R., Rishendra, D., Krishna, N. V., Dinesh, S., & Hamraj, S. (2025). Machinability characteristics of hot air abrasive jet machining. *Journal of Nonlinear Analysis and Optimization, 16*(1).

Precision, H. Abrasive Jet Machining (2012). Retrieved October 31, 2019, from-https://web.archive.org/web/20120304025100/http://www.texasairsonics.com/ Applications.html

Qi, H., Qin, S., Cheng, Z., Teng, Q., Hong, T., & Xie, Y. (2021). Towards understanding performance enhancing mechanism of micro-holes on K9 glasses using ultrasonic vibration-assisted abrasive slurry jet. *Journal of Manufacturing Processes*, *64*, 585–593. DOI: 10.1016/j.jmapro.2021.01.048

Ramachandran, N., & Ramakrishnan, N. (1993). A review of abrasive jet machining. *Journal of Materials Processing Technology*, *39*(1-2), 21–31. DOI: 10.1016/0924-0136(93)90005-Q

Sahu, S. B. B. J., & Mohapatra, S. R. (2015). Design of AJM, analysis of process parameter and their effect on MRR. *International Journal of Analytical. Experimental and Finite Element Analysis*, *2*(1), 24–28.

Sankar, M. R. (2017). Introduction to abrasive machining and finishing processes. Retrieved November, 15, from https://nptel.ac.in/content/storage2/nptel_data3/html/mhrd/ict/text/112103250/lec14.pdf

Saran, R. N., Mali, H. S., & Malik, A. (2025). Surface texturing of axisymmetric thin-walled cylindrical components: A review. *Surface Engineering*, *41*(5), 532–563. DOI: 10.1177/02670844251351765

Saravanan, K., Francis, J. X., Sudeshkumar, M. P., Maridurai, T., Suyamburajan, V., & Jayaseelan, V. (2022). Optimization of SiC abrasive parameters on machining of Ti-6Al-4V alloy in AJM using Taguchi-Grey relational method. *Silicon*, *14*, 997–1004. DOI: 10.1007/s12633-020-00918-z

Shi, L., Fang, Y., Dai, Q., Huang, W., & Wang, X. (2017). Surface texturing on SiC by multiphase jet machining with microdiamond abrasives. *Materials and Manufacturing Processes*, *33*(13), 1–7. DOI: 10.1007/s00170-015-8272-y

Sinha, M. K., Setti, D., Ghosh, S., & Rao, P. V. (2016). An alternate method for optimisation of minimum quantity lubrication parameters in surface grinding. *International Journal of Machining and Machinability of Materials*, *18*(5-6), 586–605. DOI: 10.1504/IJMMM.2016.078983

Slătineanu, L. (2013). Abrasive jet finishing. In M. Korzynski (Ed.), *Nonconventional finishing technologies* (pp. 61-82). Warsaw: Polish Scientific Publishers PWN. ISBN-10 : 8301172649

Slătineanu, L., Nagîţ, G., Dodun, O., Coteaţă, M., Chinesta, F., Gonçalves-Coelho, A., Pamies Teixeira J., San Juan, M., Santo, L., & Santos, F. (2004). *Non-traditional manufacturing processes*. Chişinău, Republic of Moldova: Tehnica Info Publishing House. ISBN (Print)9975-63-164-9

Slikkerveer, P. J., Bouten, P. C. P., Veld, F. H., & Scholten, H. (1998). Erosion and damage by sharp particles. *Wear, 217*(2), 237–250. DOI: 10.1016/S0043-1648(98)00187-2

Starke, E. A., & Staley, J. T. (2011). 24 - Application of modern aluminium alloys to aircraft. In Lumley, R. (Ed.), *Fundamentals of Aluminium Metallurgy: Production, Processing and Applications* (pp. 747–783). Woodhead Publishing., DOI: 10.1533/9780857090256.3.747

Su, X., Shi, L., Huang, W., & Wang, X. (2016). A multi-phase micro-abrasive jet machining technique for the surface texturing of mechanical seals. *International Journal of Advanced Manufacturing Technology, 86*(5-8), 2047–2054. DOI: 10.1007/s00170-015-8272-y

Teja, N. S., & Reddy, G. V. (2017). Experimental investigation of MRR and surface finish using abrasive jet machining. *International Journal of Scientific and Engineering Research, 8*(7), 622–629. https://www.ijtimes.com/index.php/ijtimes/article/view/2256

Todd, R. H., Allen, D. K., & Alting, L. (1994). *Manufacturing processes. Reference guide*. Industrial Press Inc.

Tsai, F.-C., Yan, B.-H., Kuan, C.-Y., Hsu, R.-T., & Hung, J.-C. (2009). An investigation into superficial embedment in mirror-like machining using abrasive jet polishing. *International Journal of Advanced Manufacturing Technology, 43*(5-6), 500–512. DOI: 10.1007/s00170-008-1734-8

Vanmore, V. V., & Dabade, U. A. (2018). Experimental and parametric optimization of micro holes on ANSI 304 using micro abrasive jet machining (MAJM). *Procedia Computer Science, 133*, 620–626. DOI: 10.1016/j.procs.2018.07.096

Verma, A. P., & Lal, G. K. (1984). An experimental study of abrasive jet machining. *International Journal of Machine Tool Research, 24*(1), 19–29. DOI: 10.1016/0020-7357(84)90043-X

Wang, W., Biermann, D., Aßmuth, R., Arif, A. F. M., & Veldhuis, S. C. (2020). Effects on tool performance of cutting edge prepared by pressurized air wet abrasive jet machining (PAWAJM). *Journal of Materials Processing Technology, 277*, 116456. DOI: 10.1016/j.jmatprotec.2019.116456

Wankhade, S., Dabade, B., & Khond, M. (2016). Abrasive assisted electrical discharge machining: An innovative hybrid approach. *Proceedings of the International Conference on Communication and Signal Processing* (pp. 290-297). Lonere, Raigad (MS), India: Atlantis Press. DOI: 10.2991/iccasp-16.2017.46

Yashpal, K., Jawwalkar, C. S., & Kant, S. (2015). A review on use of aluminium alloys in aircraft components, *i-manager's. Journal of Materials Science*, *3*(3), 33–38. DOI: 10.26634/jms.3.3.3673

Zhang, L., Kuriyagawa, T., Yasutomi, Y., & Zhao, J. (2005). Investigation into micro abrasive intermittent jet machining. *International Journal of Machine Tools & Manufacture*, *45*(7-8), 873–879. DOI: 10.1016/j.ijmachtools.2004.11.003

Zhao, Y., Zhao, X., Chen, X., Song, J., Cui, Y., Niu, F., Du, X., Lu, W., & Zuo, D. (2024). Modeling and analysis of plastic abrasive jet flow characteristics. *Journal of Manufacturing Processes*, *123*, 96–111. DOI: 10.1016/j.jmapro.2024.05.064

Zhu, Y. S., Wu, J., Lu, W. Z., Zuo, D. W., Xiao, H. P., Cao, D. W., & Ko, T. J. (2022). Surface formation mechanics and its microstructural characteristics of AAJP of aluminum alloy by using amino thermosetting plastic abrasive. *International Journal of Precision Engineering and Manufacturing-Green Technology*, *9*(1), 59–72. DOI: 10.1007/s40684-020-00284-6

Chapter 5
Advances in Metal Additive Manufacturing for Smart Materials: Design to Deployment

Jagadeesh Baskaran
https://orcid.org/0000-0003-3081-7030
PSG Institute of Technology and Applied Research, India

ABSTRACT

Smart materials that can sense and respond to environmental stimuli such as temperature, stress, magnetic fields, and electric currents are transforming the design and functionality of engineering systems across diverse sectors. From aerospace actuators to biomedical implants, these materials offer tailored performance and adaptability, often mimicking biological responsiveness. Concurrently, metal additive manufacturing, or metal 3D printing, is redefining possibilities in fabrication, offering geometric complexity, customization, and material efficiency that traditional methods cannot match. The intersection of metal Additive Manufacturing and smart materials presents a powerful opportunity: manufacturing functional, responsive structures with tailored properties, spatially varying functionalities, and integrated sensors or actuators from the ground up. This chapter proposes to explore the current landscape, challenges, and prospects of fabricating smart materials using Additive Manufacturing.

DOI: 10.4018/979-8-3373-9454-1.ch005

INTRODUCTION

Materials have traditionally been viewed as passive components that are designed to resist, support, or transport energy. However, the emergence of smart materials has redefined this role, positioning materials as active participants in system behavior. Smart materials can sense, respond, and adapt to external stimuli in a predictable and often reversible manner, making them integral to modern high-performance systems in fields like aerospace, biomedical engineering, robotics, and civil infrastructure.

This chapter presents a comprehensive overview of smart materials, detailing their definitions, classifications, processing (3D Printing in particular) and the functional requirements of smart systems they enable.

1. DEFINITIONS OF SMART MATERIALS

1.1 Basic Definition

A **smart material** is a material that can **sense environmental changes and respond to them** in a useful, predictable, and reversible way.

This response may manifest as:

- A change in shape, stiffness, or volume.
- A change in physical properties (e.g., conductivity, opacity, magnetization)
- Signal generation (e.g., electrical output from mechanical strain)

1.2 Broader Definition

In more integrated applications, smart materials are part of intelligent systems where sensing, actuation, and control are combined. In this context, a smart system refers to a material or structure capable of:

- Sensing its environment (input)
- Processing the input (signal interpretation)
- Acting upon the environment (output)

This makes smart materials the foundation for adaptive, responsive, and multi-functional systems.

2. CLASSIFICATIONS OF SMART MATERIALS

Smart materials can be classified based on the type of **stimulus** as shown in Table 1.

Table 1. Classification of Smart Materials based on stimulus

Stimulus Type	**Material Response**	**Types**
Thermal	Shape change, expansion	Shape Memory Alloys (NiTi), Hydrogels
Mechanical stress	Electric charge, shape change	Piezoelectric ceramics, Magnetostrictive alloys
Electric field	Strain or actuation	Electroactive polymers, Electrostrictive ceramics
Magnetic field	Deformation, heating	Magnetostrictive materials, Ferrofluids
Light (Photonic)	Bending, opacity change	Photomechanical polymers, Photochromic glasses
Chemical/pH	Swelling, conductivity change	pH-sensitive hydrogels, Conductive polymers

Based on Functional Mechanism, Smart Materials can be classified as follows:

a. Shape Memory Materials
   - Return to a pre-defined shape after deformation upon heating.
   - **Examples**: NiTi (metallic), Polyurethane (polymeric)

b. Piezoelectric Materials
   - Generate electric charge under mechanical load, or vice versa.
   - **Examples**: PZT ceramics, $BaTiO_3$

c. Magnetostrictive Materials
   - Change dimensions in response to magnetic fields.
   - **Examples**: Terfenol-D, Fe-Ga alloys

d. Electroactive Polymers (EAPs)
   - Undergo large strains under electric field.
   - **Examples**: Ionic polymer–metal composites (IPMCs)

e. Photostrictive and Chromogenic Materials
   - Change shape or optical properties under light exposure.
   - **Examples**: Liquid crystal elastomers, photochromic lenses

f. Self-Healing Materials
   - Capable of autonomously repairing internal or surface damage.
   - **Examples**: Microcapsule-embedded polymers, shape memory metals

g. Hydrogels and Responsive Gels
   - Change volume in response to humidity, pH, or temperature.
   - **Applications**: Biomedical scaffolds, microfluidics

3. FUNCTIONAL REQUIREMENTS OF SMART SYSTEMS

To implement smart materials effectively in engineering systems, several functional requirements must be considered. These relate to material properties, system integration, reliability, and responsiveness.

3.1 Sensing Capability

The system must accurately detect external stimuli such as stress, temperature, magnetic fields, or chemical changes. Sensing may be intrinsic (material changes detectable directly) or extrinsic (integrated sensors).

3.2 Actuation Function

The material should be capable of generating a measurable physical response, such as force, displacement, or signal. Actuation should be sufficient in magnitude and repeatable over multiple cycles.

3.3 Signal Processing and Feedback

Smart systems must include data interpretation or feedback control, which can be integrated (e.g., in responsive polymers) or coupled via external electronics. For real-time applications, low-latency response is essential.

3.4 Reversibility and Adaptivity

Smart responses must be reversible to allow reuse, especially in cyclic environments. Materials should adapt to changing external conditions or damage.

3.5 Multi-functionality

The ability to combine multiple functions (e.g., sensing + actuation + healing) is highly desirable. This can reduce system complexity and weight in aerospace or biomedical applications.

3.6 Material Compatibility

Integration into systems requires compatibility with existing substrates, fabrication methods, and operational environments (e.g., temperature, corrosion resistance).

For hybrid systems, mechanical and thermal compatibility between smart and host materials is critical.

3.7 Scalability and Manufacturability

The smart material system should be scalable to different sizes and manufacturable using reliable processes like Additive Manufacturing (AM) or molding. Performance should be uniform across scale, especially for micro- or nano-scale devices.

3.8 Durability and Fatigue Resistance

Long-term performance under **repeated stimuli**, environmental exposure, or mechanical loading is essential. Fatigue resistance is particularly important in SMAs and piezoelectric devices used in aerospace or automotive applications.

4 ADDITIVE MANUFACTURING OF SMART MATERIALS

4.1. General concept of Additive manufacturing

Commonly referred to as 3D printing, AM has revolutionized the landscape of materials engineering and manufacturing by enabling the layer-by-layer fabrication of complex geometries directly from digital models(Gibson, Rosen, & Stucker, 2015). Among the various AM methods, metal 3D printing has emerged as a pivotal technology for fabricating high-performance components, especially in aerospace, biomedical, and energy sectors. A recent frontier in this field is the fabrication of smart materials—materials that respond adaptively to external stimuli such as temperature, magnetic or electric fields, stress, or moisture.

4.2. Metal 3D Printing

Metal 3D printing enables microstructural control, gradient composition, and complex topology optimization, all of which are crucial for fabricating smart materials with enhanced functionality. The types of AM used in SMA fabrication is provided in figure 1.

4.2.1. Principles of metal 3D printing

Figure 1. AM processes used for SMA fabrication

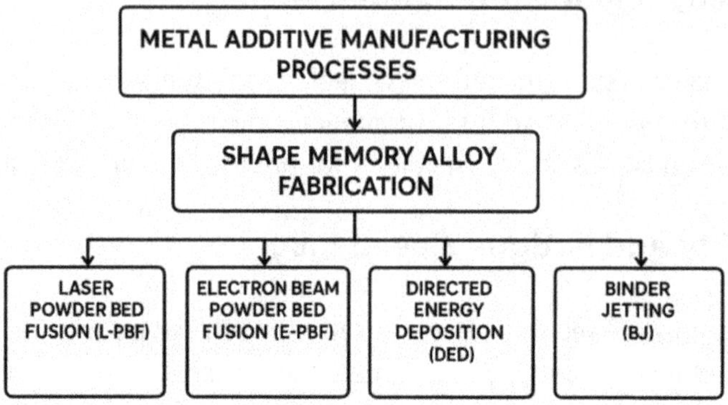

In Powder Bed Fusion (PBF), *a* thin layer of metal powder is spread on a build platform. A high-energy laser or electron beam selectively melts regions of the powder bed based on a CAD model(Kannan, Sivakumar, Jagadeesh, & Balaji, 2024; Sames, List, Pannala, Dehoff, & Babu, 2016). Materials built with this process: Ti-6Al-4V, NiTi (Nitinol), stainless steel, Inconel. Precise thermal gradients in SLM facilitate functional phase transformation in SMAs. EBM enables processing of reactive materials (e.g., titanium-based SMAs) with lower residual stress due to its vacuum environment and preheating capability.

In Directed Energy Deposition (DED)process, a focused energy source (laser, electron beam, or plasma arc) melts metal powder or wire as it is being deposited through a nozzle(Varatharajulu, Duraiselvam, Pradeep, & Jagadeesh, 2025). It is suitable for functionally graded materials (FGMs) and allows multi-material deposition, essential for smart material systems. This process also enables creation of metal-matrix composites and gradient SMAs by varying feedstock composition during printing.

In the Binder Jetting (BJ) process, a binder selectively joins powder particles in a bed; the green part is later sintered in a furnace(Trenfield, Madla, Basit, & Gaisford, 2018). The limitations of the process include lower density and mechanical strength; limited for structural smart applications. It is mainly suitable for prototyping porous smart structures or integrating metallic and ceramic particles.

Cold Spray Additive Manufacturing (CSAM) is a relatively new technique where in metal powders are accelerated through a high-pressure gas stream and adhere to a substrate via solid-state impact. The main advantage of the process is that minimal thermal input preserves intrinsic smart properties, such as magnetostrictive and piezoelectric behavior. It has a great potential for fabricating magnetically responsive coatings or components.

The fusion of **AM** and **smart materials** marks a transformative step in materials engineering and manufacturing science. While AM offers unprecedented design freedom and on-demand customization, smart materials provide dynamic responses to environmental stimuli such as heat, stress, electric fields, or magnetic fields. The convergence of these domains enables the fabrication of adaptive, responsive, and multifunctional structures—paving the way for next-generation systems across aerospace, biomedical, robotics, and energy sectors. AM is a layer-by-layer fabrication method that allows for:

- Complex geometries not possible via traditional subtractive methods.
- Material and structural customization at multiple length scales.
- Integration of multiple materials and functions into a single build.

Smart materials exhibit reversible and programmable changes in properties in response to specific stimuli. They include:

- **Shape Memory Alloys (SMAs)** – such as NiTi, which respond to temperature changes.
- **Piezoelectric Materials** – convert mechanical stress into electric charge and vice versa.
- **Magnetostrictive Materials** – deform under magnetic fields (e.g., Terfenol-D).
- **Electroactive Polymers** – change shape or size when electrically activated.
- **Thermo-responsive Hydrogels** – swell or contract with temperature or pH changes.

AM allows the precise placement of smart material elements within complex geometries, enabling localized functional zones such as sensors, actuators, or thermal switches. In 4D Printing, the fourth dimension refers to time-dependent behavior. Structures fabricated via AM using smart materials can change shape, behavior, or performance in response to external stimuli after printing. A 4D-printed biomedical stent made from shape memory polymer that self-expands at body temperature. AM allows embedding of smart functional layers into load-bearing components, creating self-monitoring or self-regulating systems. Printed piezoelectric layers within a turbine blade for real-time structural health monitoring. Personalized smart prosthetics, medical implants, and wearable devices are now feasible by combining patient-specific geometry (via 3D scanning) and responsive materials.

Applications Enabled by the Synergy of AM and smart materials are provided in Table 1.

Table 2. Applications of Smart materials enabled by AM

Field	Application Example
Biomedical	Thermally responsive scaffolds, self-expanding implants
Aerospace	Morphing wings using SMA actuators
Robotics	Soft robotic grippers using electroactive polymers
Civil Structures	Self-sensing and self-healing concrete reinforcement
Energy	Vibration energy harvesters with piezoelectric metals

4.3 Early Research Examples

a. 4D Printing with Shape Memory Polymers (SMPs)
   - Researchers have used fused deposition modeling (FDM) to print thermally activated SMPs that return to a memorized shape after deformation.

b. Direct Ink Writing (DIW) of Responsive Hydrogels
   - 3D-printed pH-sensitive hydrogel valves have been developed for drug delivery systems, responding to local acidity in the body.

c. Multi-material Metal AM with Smart Alloys
   - Laser powder bed fusion (LPBF) has been used to fabricate components with **graded NiTi SMAs**, allowing actuation upon heating within structural frameworks.

d. Piezoelectric Inkjet Printing
   - Piezoelectric inks (e.g., $BaTiO_3$-based) have been printed directly onto flexible substrates to form sensors and energy harvesters.

4.4 Challenges in Combining AM and Smart Materials

High-temperature AM processes (e.g., LPBF, DED) can degrade smart material properties like piezoelectricity or shape memory effects. Reactive or multi-phase smart materials may segregate or lose functionality during printing. Simultaneously printing dissimilar smart materials (e.g., metal + polymer or ceramic + metal) remains complex due to mismatched thermal and mechanical properties. Smart materials often expand, contract, or deform significantly. Designing for predictable movement in printed parts requires multi-physics modeling. Unlike traditional materials, databases and standards for the mechanical and functional behavior of printed smart materials are sparse.

4.5 Topology Optimization Process for SMA Lattice Structures

Topology optimization (TO) is a computational design methodology used to find the most efficient material distribution within a given design space, subject to specific performance objectives and constraints. When applied to lattice structures, TO enables the generation of complex architectures with tailored mechanical and functional properties.

In the context of SMAs, TO must consider not only mechanical performance but also the functional behaviors such as the shape memory effect and superelasticity. This adds a layer of complexity beyond traditional structural materials.

The process starts with defining the **design domain**, **boundary conditions**, and **objective functions**.

- **Design domain:** The allowable region where material can exist.
- **Objective functions:**
    - Minimize compliance (maximize stiffness).
    - Maximize energy absorption.
    - Control deformation under stimuli (thermal, mechanical).
- **Constraints:**
    - Volume fraction (limit material usage).
    - Manufacturability constraints (e.g., minimum feature size for 3D printing).
    - SMA-specific constraints (e.g., phase transformation thresholds, temperature ranges).

Conventional TO assumes linear elastic materials, but SMAs require nonlinear, thermo-mechanically coupled constitutive models. These models must account for:

- Martensitic phase transformation.
- Hysteresis effects.
- Transformation strains.
- Thermal actuation.

These are typically represented by constitutive models(Machado & Lagoudas, 2008) such as:

- Boyd–Lagoudas model (thermomechanical).
- Auricchio model (for finite-strain SMA behavior).
- Phenomenological constitutive models tailored for finite element analysis (FEA).

A generalized workflow of the TO process is provided in Figure 2 .

Figure 2. Workflow of Topology Optimization

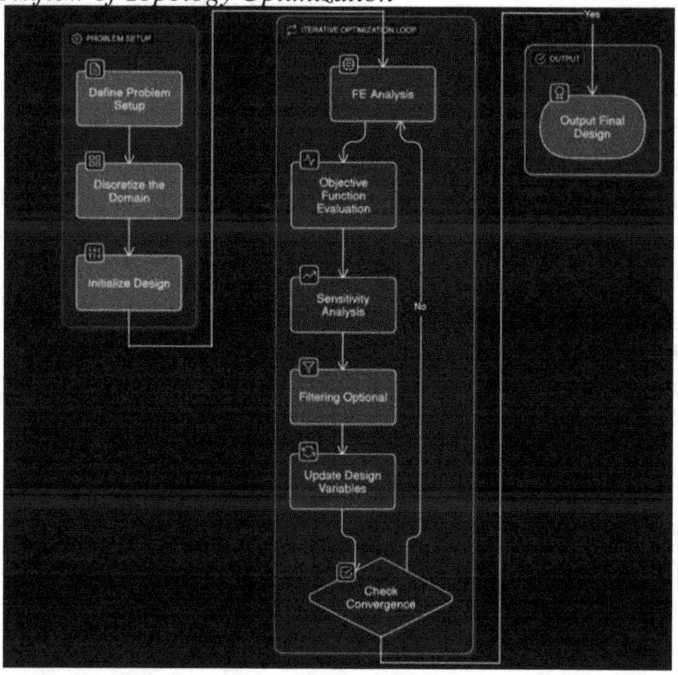

The design domain is discretized into finite elements (typically hexahedral or tetrahedral). For lattice structures, this step also includes defining:

- Periodic unit cell topology (e.g., BCC, FCC, octet truss).
- Graded or hierarchical lattice arrangements(Jagadeesh, Duraiselvam, & Prashanth, 2024).

The finite element model integrates both the mechanical equilibrium equations and thermal or phase transformation responses, making it multiphysics in nature. Several optimization algorithms can be used depending on the complexity and required fidelity:

- **SIMP (Solid Isotropic Material with Penalization):** Most common method, material distribution is optimized between 0 (void) and 1 (solid) resulting in an optimization of stiffness and weight reduction(Rozvany, 2000)
- **Level-set method:** Captures crisp boundaries and is well-suited for SMA morphing structures.
- **Density-based methods with filter techniques:** Used to avoid checkerboarding and mesh-dependency.
- **Gradient-based algorithms (e.g., MMA – Method of Moving Asymptotes):** Efficient for large-scale problems.
- **Evolutionary algorithms or genetic algorithms:** Useful when dealing with discrete transformations or multiple objectives.

When designing SMA-based lattices, it's common to integrate multiple objectives, such as:

- Maximizing stiffness and energy recovery.
- Minimizing thermal lag during actuation.
- Ensuring that transformation strains are within recoverable limits.

This requires solving coupled equations that combine mechanical equilibrium with SMA-specific constitutive equations and, often, thermal conduction models.

TO can also enable graded stiffness or phase transformation zones, where different parts of the lattice activate sequentially. This is particularly valuable in applications like self-deploying implants or soft robotic actuators. In SMA lattices, the geometry isn't just a structural decision—it governs how the lattice will deform under stimuli. TO algorithms can optimize the following to tune the thermomechanical response:

- Lattice cell orientation.
- Beam diameter variation.
- Porosity distribution.

For TO designs to be successfully manufactured:

- A minimum feature size must be enforced.
- Overhang angles must be optimized to reduce support structures.
- Thermal simulations must ensure homogenous material properties post-printing.

SLM-printed NiTi, for example, requires post-processing (e.g., heat treatment at ~500–600°C) to restore phase transformation behavior that may be disrupted during laser melting.

AI-guided design tools are being developed to predict performance and optimize geometry–function relationships in 4D-printed structures. Nanoscale AM techniques like two-photon polymerization can fabricate ultra-fine smart structures, expanding applications in micro-robotics and MEMS. Researchers are exploring hierarchical architectures (e.g., plant or insect-inspired) to combine passive structural strength with active smart responses(Cheng, Shen, & Zhang, 2024). Combining AM with self-healing polymers/metals and stimuli-responsive gels could lead to truly autonomous systems that adapt and recover from damage.

4.6 Post-Processing

The output of TO is typically a grayscale material density distribution. This needs to be:

- Interpreted into a printable geometry using thresholding and smoothing.
- Converted into a CAD model for 3D printing.
- Validated using structural simulations that include the actual SMA properties.

Topology must ensure that **localized stress/temperature states** do not exceed the recoverable transformation limits of the SMA. Optimization routines must enforce:

- Transformation strain bounds.
- Martensite-austenite phase compatibility.
- Actuation energy constraints.

5 KEY CONSIDERATIONS FOR PRINTING SMART METALS

5.1 Microstructural Control

Thermal cycling in SLM/EBM significantly affects grain size, texture, and phase transformation behavior—critical for smart functionalities.

Post-processing (e.g., heat treatment) is often required to activate or enhance responsiveness.

5.2 Multi-Material Integration

Techniques like DED allow real-time variation of material composition, crucial for gradient smart materials (e.g., SMA-ceramic composites). Interfacial bonding and thermal compatibility are critical design parameters.

5.3 Porosity and Lattice Architectures

AM enables fabrication of designed porosity, influencing actuation behavior, electrical conductivity, and surface reactivity(Ashby, 2006; Baskaran, Muthukannan, Shukla, & Gokuldoss, 2024; Jagadeesh, Duraiselvam, & Prashanth, 2023; Jagadeesh et al., 2024; Yan, Hao, Hussein, & Raymont, 2012). A SEM image showing a unit cell of AM-built primitive-type TPMS lattice is provided in Figure 3. The inset image shows the satellite powder particles sticking to the walls of the lattice before heat treatment.

Figure 3. Schwarz Primitive lattice structure showing the unmelted powder particles sticking to walls

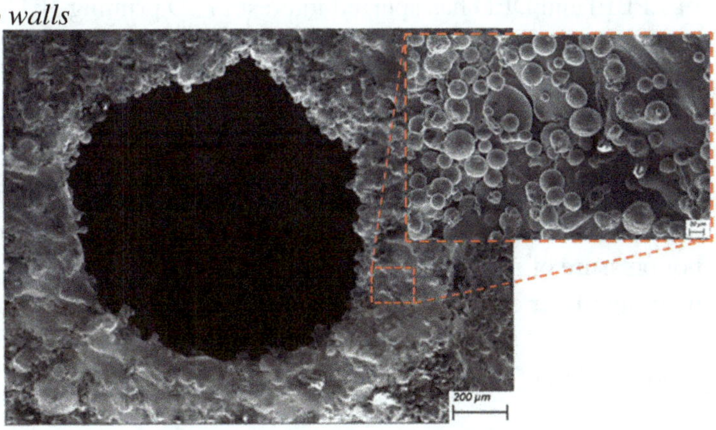

Lattice structures can be engineered for mechanically responsive systems (e.g., auxetic SMAs). Some of the applications of such smart materials are provided in Table 3.

Table 3. Application areas and examples of Porous Smart Materials

Application Area	Smart Material Functionality	Example
Biomedical Implants	Shape memory behavior	NiTi implants that deploy in vivo(Baskaran, 2025)
Aerospace Actuators	Thermo-mechanical response	SMA actuators for morphing wings
Robotics	Electrical-mechanical coupling	Piezoelectric metal sensors
Wearable Electronics	Magnetic/electrical response	Printed conductive fabrics with embedded metal sensors

6 SMART MATERIALS COMPATIBLE WITH METAL AM

6.1 Challenges in 3D Printing of NiTi Alloys

The rise of LPBF) and DED has spurred interest in 3D printing NiTi parts with customized geometries. Yet, NiTi presents unique difficulties:

a. Narrow Processing Window

NiTi has a very tight composition-performance relationship; even a slight deviation in Ni/Ti ratio (±0.1 at. %) can dramatically alter transformation temperatures. The high vapor pressure of **nickel** leads to preferential evaporation during melting, especially under high laser energy density, skewing the local composition.

b. Porosity and Cracking

LPBF tends to create porosity due to balling, keyholing, and lack of fusion, which severely affects fatigue performance and transformation behavior. Thermal stresses and repeated melting cause microcracking, particularly because NiTi has low thermal conductivity and high thermal expansion.

c. Residual Stresses and Texture

High thermal gradients in AM induce residual stresses, which can lead to premature failure or distorted parts. AM-processed NiTi often exhibits strong texture and anisotropy(M. Li et al., 2025), which affects mechanical and functional behavior (e.g., actuation directionality).

d. Post-processing Complexity

Heat treatment is essential to recover the shape memory properties and relieve residual stress, but finding the optimal parameters requires tuning both phase stability and microstructural homogeneity.

The functional behavior of NiTi alloys arises from a solid-state, diffusionless martensitic transformation between different crystallographic phases:

a. Phases Involved
   - **Austenite (B2):** High-temperature, cubic structure; stronger and stiffer.
   - **Martensite (B19′):** Low-temperature, monoclinic structure; easily deformable.
   - **R-phase:** An intermediate, rhombohedral phase that may appear during cooling before martensite forms, especially in Ni-rich compositions or heavily cold-worked material.

b. Transformation Temperatures
   - **Ms, Mf, As, Af:** Martensite start/finish and Austenite start/finish temperatures define the thermal window for SME. 0.1 at. % increase in Ni can reduce Af by ~10 °C.
   - Transformation hysteresis and width are influenced by grain size, defects, and impurities.

c. AM-induced Effects
   - **Segregation:** Rapid solidification during AM can trap solute atoms, locally affecting transformation behavior.
   - **Precipitates:** Ti-rich intermetallic phases (e.g., Ti_2Ni, Ti_4Ni_2O) can form during melting or solid-state transformation, acting as nucleation sites or impeding transformation.

6.2 Thermomechanical Processing of NiTi

To ensure desirable SME and mechanical properties, **post-AM thermomechanical treatments** are essential.

a. Heat Treatment
   - Performed typically between **400–600 °C** for aging and **800–950 °C** for solution annealing.
   - Purpose:
     - Eliminate residual stress
     - Dissolve undesirable precipitates
     - Modify grain size and texture

    - Adjust transformation temperatures
- Heat treatment must be tightly controlled to avoid the precipitation of Ni_4Ti_3 or Ti_2Ni, which can pin martensite variants or reduce ductility.

b. Hot and Cold Working
- **Cold working** (e.g., rolling or drawing) increases dislocation density and can refine grains, but also raises transformation hysteresis and suppresses full SME unless followed by annealing.
- **Hot working** (above recrystallization temperature) enables larger deformation without cracking and promotes isotropic microstructure.

c. Training and Thermo-mechanical Cycling
- Repeated thermo-mechanical cycling "trains" the material to favor specific martensitic variants, improving SME reliability(Benafan et al., 2014).
- Critical for actuator applications where cyclic stability and fatigue resistance are needed.

6.3 Additional considerations for post-processing of Ni-Ti

Solutionizing & Aging: Post-processing usually involves a high-temperature solution anneal (e.g., ~950 °C for several hours) to homogenize microstructure, reduce residual stresses, and sharpen phase transformation peaks as measured by DSC.(Tareq, Rahman, Poudel, Chung, & Kwon, 2024)

Aging Temperature & Time Trade-off: Aging in the 300–600 °C range induces precipitation of Ni-rich phases (Ni_4Ti_3, Ni_3Ti_2, Ni_3Ti), which deplete Ni in the matrix and raise transformation temperatures. For instance, aging at 350 °C for extended hours (e.g., up to 18 h) increased A (austenite finish) and achieved up to 4.2% strain recovery at 60 °C (Tareq et al., 2024)

SLM NiTi Specific Insights: Selective Laser Melting (SLM) printed NiTi, when aged at 350 °C for 18 h, shows peak transformation strain and minimal permanent deformation—whereas aging at 450 °C decreased performance due to reduced precipitate volume.(Khoo et al., 2018)

6.4 Impacts of Annealing on Microstructure & Properties

Precipitation-Controlled Behavior: Aging generates Ni-rich precipitates, which refine the martensitic transformation behavior. Fine Ni_4Ti_3 precipitates enhance superelasticity, while coarse precipitates can limit failure strain.

Transformation Temperature Tuning: Higher aging temperatures tend to lower A , while longer time at moderate temps (e.g., 350 °C) increases A significantly

Microstructural Evolution: Heat treatments contribute to grain boundary migration and slight grain growth in SLM NiTi parts—but not full recrystallization if prior deformation is minimal. Smaller grains improve fatigue resistance and transformation strain, but grain growth can soften the material and reduce mechanical strength(Khoo et al., 2018)

6.5 Surface Finishing Strategies to Preserve Smart Effects

Electrochemical Polishing (ECP): ECP delivers micro-level smoothness and corrosion resistance while minimizing mechanical damage. Acidic electrolytes (e.g., methanolic H_2SO_4) effectively reduce nickel leaching and enhance surface integrity of NiTi SMA.(Dutta et al., 2025)

Chemical Etching & Passivation: Techniques like nitric-acid etching and chemical polishing remove surface defects, oxide layers, and nickel-rich zones—enhancing corrosion resistance and maintaining transformation behavior(Chmielewska et al., 2022).

Shot Peening / Sand Blasting: Introduces surface compressive stress and removes superficial defects, potentially improving fatigue life. However, care is needed to ensure these mechanical treatments do not adversely affect small-scale structures or interfere with the SMA's microstructure.

Surface Coating / Cladding: Methods like laser cladding, arc or plasma spraying can deposit hard, corrosion-resistant layers—but these processes introduce heat-affected zones and diffusion from the coating, which may alter the underlying NiTi behavior

7. MAGNETOSTRICTIVE AND MAGNETIC ALLOYS

Magnetostrictive and magnetic alloys are critical materials in advanced electromechanical systems where the conversion between magnetic energy and mechanical strain is required. Applications range from precision actuators and sensors to sonar and energy harvesting systems. Among these, **Terfenol-D** ($Tb_{0.3}Dy_{0.7}Fe_2$) and **Fe-Ga** (Galfenol) have emerged as the most prominent magnetostrictive materials owing to their high magnetostriction, robustness, and adaptability to various operational environments.

7.1 Fundamentals of Magnetostriction

Magnetostriction is a property of ferromagnetic materials that causes them to change shape or dimensions during magnetization. Conversely, magnetostrictive materials can generate magnetic fields when subjected to mechanical stress—an effect exploited in magnetostrictive sensors and transducers. The magnetostriction arises due to the realignment of magnetic domains within the material, resulting in strain(Paul, Nguyen, & Shen, 2024).

Key performance metrics of magnetostrictive materials include:

- **Saturation magnetostriction (λs):** Maximum strain achieved under magnetic saturation.
- **Magnetic permeability and coercivity:** Influence magnetic energy requirements.
- **Anisotropy constants:** Govern directional behavior of magnetostriction.
- **Mechanical strength and machinability:** Essential for real-world applications.

7.2 Terfenol-D

7.2.1 Overview

Discovered at the Naval Ordnance Laboratory in the 1970s, Terfenol-D is a giant magnetostrictive material composed primarily of terbium (Tb), dysprosium (Dy), and iron (Fe). It exhibits exceptionally high magnetostriction (up to 2000 ppm), making it suitable for high-precision actuators and sonar systems.

7.2.2 Fabrication

Despite its high performance, Terfenol-D poses significant fabrication challenges:

1. **Brittleness:** The intermetallic structure of Terfenol-D makes it inherently brittle and difficult to machine or process via conventional mechanical methods.
2. **Rare-earth elements:** Tb and Dy are expensive and limited in availability, raising concerns about material sustainability and cost.
3. **Crystal growth complexity:** Single crystals or highly textured polycrystals are preferred for enhanced performance, but growing large single crystals via Bridgman or Czochralski methods is time-consuming and sensitive to processing conditions.
4. **Oxidation sensitivity:** Rare-earth elements are highly reactive with oxygen, requiring inert or vacuum environments during melting and casting.

7.2.3 Anisotropy

Terfenol-D exhibits **strong magnetocrystalline anisotropy**, primarily along the ⟨112⟩ or ⟨100⟩ crystallographic axes. This anisotropy is critical for performance, as magnetostriction varies significantly with orientation. Consequently, alignment of grains or use of single crystals is necessary for optimal magnetostrictive response.

Performance

- **Saturation magnetostriction:** 1500–2000 ppm
- **Operating frequency:** Up to tens of kHz
- **Magnetic field requirement:** ~0.1–0.5 T for saturation
- **Drawback:** Low tensile strength (~20–30 MPa)

7.3 Fe-Ga Alloys (Galfenol)

Fe-Ga alloys, commonly known as **Galfenol**, were developed by the Naval Surface Warfare Center in the late 1990s as a ductile alternative to Terfenol-D. Typical compositions range from 17 to 27 at.% Ga in iron. Although Fe-Ga offers lower magnetostriction than Terfenol-D (up to ~400 ppm), its mechanical robustness and ductility make it highly attractive for industrial and structural applications.

7.3.1 Fabrication

Compared to Terfenol-D, Fe-Ga is more forgiving in terms of processing, but still presents some challenges:

1. **Segregation during solidification:** Ga has a tendency to segregate, which can lead to inhomogeneous microstructures if not carefully controlled.
2. **Phase formation:** The formation of ordered phases such as DO_3 and B2, depending on Ga content and thermal history, can influence magnetostrictive behavior.
3. **Texture control:** Grain orientation plays a critical role in maximizing magnetostriction; therefore, texture development through rolling and annealing is essential.
4. **Crack formation:** Though ductile, Fe-Ga can form cracks due to residual stress and microsegregation in laser-based processes.

7.3.2 Anisotropy

Fe-Ga exhibits **weaker anisotropy** than Terfenol-D but still benefits from orientation control. The ⟨100⟩ direction typically provides the highest magnetostrictive response. Textured polycrystals or directional solidification can be employed to align grains.

Performance

- **Saturation magnetostriction:** Up to 400 ppm (bulk), 500+ ppm (textured/single crystal)
- **Operating frequency:** Can exceed 100 kHz in thin-film form
- **Magnetic field requirement:** ~0.05–0.2 T
- **Mechanical strength:** Significantly higher than Terfenol-D (≥ 300 MPa)
- **Ductility:** Fe-Ga can be rolled, welded, and machined using standard methods.

Table 4 summarizes the properties of Terfenol – D and Fe- Ga.

Table 4. Comparative Summary: Terfenol-D vs. Fe-Ga

Property	**Terfenol-D**	**Fe-Ga (Galfenol)**
Max Magnetostriction (ppm)	1500–2000	200–400
Magnetic Field for Saturation	0.1–0.5 T	0.05–0.2 T
Mechanical Strength	Low (~30 MPa)	High (≥ 300 MPa)
Fabrication Complexity	High	Moderate
Cost	High (rare earths)	Moderate (Fe-based)
Anisotropy	Strong (⟨112⟩, ⟨100⟩)	Moderate (⟨100⟩)
Ductility	Brittle	Ductile
Applications	Sonar, precision actuators	Structural sensors, energy harvesters

8 PIEZOELECTRIC AND ELECTROACTIVE METALS

Piezoelectric materials are capable of converting mechanical stress into electric charge and vice versa, while **electroactive materials** exhibit significant mechanical responses to electrical stimuli. Traditionally, these phenomena are associated with **ceramics** (e.g., PZT) or **polymers** (e.g., PVDF). However, there is growing interest in electroactive metals and metallic systems that can exhibit piezoelectric-like behavior through specific mechanisms—primarily due to their mechanical robustness, conductivity, and environmental resilience compared to brittle ceramics and soft polymers. The idea of combining functional electromechanical behavior with the ductility and processability of metals, especially through **AM**, is highly attractive for next-generation sensors, actuators, and adaptive structures.

8.1 Electroactive and Piezoelectric Behavior in Metals

Conventional metals do **not** exhibit intrinsic piezoelectricity because they possess **centrosymmetric crystal structures**, which preclude the emergence of net polarization under stress. However, metallic systems can show **pseudo-piezoelectric** or **electroactive-like behavior** via:

1. **Engineered non-centrosymmetric microstructures**: Through texturing, alloying, or surface modification.
2. **Composite systems**: Embedding piezoelectric ceramics in a metallic matrix.
3. **Surface piezoelectricity**: Observed at the nanoscale where surface symmetry is broken.
4. **Electrochemical actuation**: As seen in ionic metals or metallic hydrides (e.g., NiH), where ion transport causes volumetric change.

Examples include:

- **Epsilon-phase iron gallium (Fe-Ga)** alloys showing magnetoelectric coupling.
- **Engineered nanolattices of metallic elements** exhibiting flexoelectric-like behavior.
- **Electrochemical actuation in conducting metal-polymer hybrids**.

8.2 Additive Manufacturing Challenges

Absence of Intrinsic Piezoelectricity in Metals

- The fundamental barrier is **centrosymmetry**, which limits bulk piezoelectric behavior.
- In AM, achieving **asymmetric grain structures** or **multiphase interfaces** that could break inversion symmetry is extremely difficult, especially without precise control of microstructure at the nano- or atomic scale.

Compositional Homogeneity and Phase Control

- Piezoelectric performance in metals—if present—is highly sensitive to **stoichiometry** and **local atomic order**.
- Common AM processes like LPBF or DED can lead to:
    - o **Segregation of alloying elements**
    - o **Unwanted phase formation**
    - o **Disruption of texturing necessary for anisotropic response**

High Thermal Gradients and Residual Stresses

- These can induce:
    - o Cracking in functionally active regions
    - o Degradation of phase boundary integrity in composites
    - o Interfacial delamination in metal-ceramic hybrid structures

Surface and Interface Engineering

- Much of the pseudo-piezoelectric response in metals is **surface-driven**.
- AM techniques currently lack resolution to reliably tailor surface anisotropy at the **nanoscale**, limiting repeatability of electroactive behavior.

8.3 Early-Stage Research and Developments

Despite the limitations, a few promising research directions are emerging at the intersection of electroactive metals and AM:

a. Metal-Ceramic Piezoelectric Composites via AM
- Researchers are using AM to co-fabricate structures combining metallic matrices (e.g., stainless steel, Ti6Al4V) with embedded piezoelectric phases (e.g., $BaTiO_3$, ZnO).(Chen et al., 2020)
- Challenges include:
    - o Thermal mismatch between metal and ceramic
    - o Chemical incompatibility at interfaces
    - o Achieving connectivity of the piezoelectric phase within the metallic lattice

Fused filament fabrication (FFF) and binder jetting have been explored for low-temperature co-sintering of metal-ceramic blends(Y. Li et al., 2025).

b. Grain Orientation Engineering in Printed Metals

- Laser-based AM techniques can induce **columnar grain structures**, which may be exploited to engineer **non-centrosymmetric textures** under specific conditions.
- Example: Laser scanning strategies to promote ⟨111⟩-textured grains in certain alloys like Fe-Ga or Ni-Mn-Ga, which may support magnetostrictive and piezoelectric-like behavior.

c. Additive Manufacturing of Electrochemical Actuators
  - Early work focuses on printing metallic foams or porous electrodes (e.g., Ni, Pt, Ti) for use in electrochemical transducers.
  - When combined with ionic liquids or electrolytes, these can behave like soft actuators or artificial muscles.

Example: Inkjet or aerosol-jet printing of platinum microgrids for bending-type microactuators.

d. Nanoscale Architected Metals
  - Recent advances in two-photon lithography and nanoscale AM have enabled creation of metallic lattices with engineered asymmetry.
  - These architected metals may show flexoelectricity (polarization due to strain gradients), offering a metallic route to electromechanical response.

The concept of **piezoelectric or electroactive metals** remains in its infancy, but AM is a powerful tool for advancing this field. Emerging directions include:

- **Topology-optimized metallic lattices** with embedded smart behavior.
- **Hybrid AM techniques** combining direct ink writing (DIW) of functional inks with laser-based melting of structural metals.
- **Multi-material AM** enabling in-situ joining of active and passive phases.
- **Machine learning-guided microstructure design** to explore symmetry breaking in metals.

9 SELF-HEALING METAL SYSTEMS

The idea of self-healing materials—inspired by biological systems—aims to enable structures to autonomously repair damage without external intervention, thereby extending service life and improving safety. While self-healing behavior has been widely demonstrated in polymers, composites, and some ceramics, the extension of this concept to metals presents a unique challenge due to their high strength, crystalline structure, and lack of inherent mobility for damage remediation.

Despite these hurdles, the concept of self-healing metals is gaining momentum, driven by advances in materials design, microstructural engineering, and AM. The vision is to develop metallic systems capable of autonomously or externally stimulated healing of microcracks, voids, fatigue damage, or corrosion-induced degradation.

9.1 Conceptual Overview of Self-Healing in Metals

Unlike polymers, where reversible bonding or capsule-based healing is common, metals require **more thermodynamically robust mechanisms** due to their crystalline structure and relatively immobile atoms at service temperatures. Key conceptual strategies include:

Metals, particularly at elevated temperatures, exhibit atomic diffusion that can lead to the closure of microcracks or voids. At high homologous temperatures, boundary motion can effectively ‘sweep out’ or heal defects. Certain alloys may enable dislocation structures to rearrange in a damage-compensating manner under thermal or mechanical cycling.

Similar to polymers, capsules or channels filled with reactive agents (e.g., liquid metal or corrosion inhibitors) can be embedded in a metallic matrix. Upon crack formation, the healing agent is released and reacts with the surrounding metal to fill the void. Alloys capable of solid-state phase transformations (e.g., NiTi SMAs or Cu-based systems) may realign their structure during thermal cycling to ‘heal’ fatigue-induced microcracks. In high-temperature alloys (e.g., steels, Ni-based superalloys), selective oxidation or precipitation can locally fill cracks or pores by forming **oxide bridges or nano-precipitates** under service conditions.

9.2 Early Research

While still in nascent stages, several pioneering studies have demonstrated that **self-healing in metals is physically feasible** under specific conditions. Some key proof-of-concept approaches include:

a. Microencapsulated Healing Agents in Metal Matrices

Researchers have embedded **microcapsules filled with low-melting-point metal alloys** (e.g., indium-based eutectics) into structural metals such as aluminum or steel. (Blaiszik, Jones, Sottos, & White, 2014). Upon crack formation and local heating, these capsules rupture and the molten metal flows into the crack. On solidification, the metal bond partially restores structural integrity.

b. Solid-State Diffusion Healing in High-Temperature Alloys

In high-temperature steels and Ni-based superalloys, microcracks formed during creep or fatigue can heal through **grain boundary diffusion** if sufficient thermal energy is available. Elements like **boron or silicon**, which form mobile eutectic films at grain boundaries, have been used to enhance healing kinetics.

Example: Oxide-dispersion strengthened (ODS) steels have shown healing behavior at ~1000°C, where fine oxide particles anchor dislocations and aid in grain boundary diffusion.

c. Healing via Phase Transformation in SMAs

- Shape memory alloys such as **NiTi** and **Cu-Al-Ni** can exhibit reversible microstructural transformations that close cracks or voids during heating.
- The martensite-austenite phase change induces internal stresses that help realign and close damaged zones.

Example: Experiments on NiTi wires subjected to repeated bending demonstrated partial closure of surface cracks after thermal cycling across the transformation temperature.

d. Electrochemical or Redox-Based Healing

- Some metallic systems in corrosive environments show **autonomous healing of protective passive layers** (e.g., chromium oxide in stainless steel).
- Inspired by this, researchers are exploring **electrochemical healing agents** embedded in coatings or intergranular regions that can restore surface passivation after damage.

9.3 Limitations and Challenges

Despite promising early studies, numerous obstacles remain in realizing practical self-healing metal systems:

- **High activation energy:** Most healing processes require elevated temperatures, often incompatible with operational environments.
- **Healing vs. strength trade-off:** Incorporating healing agents or soft phases can compromise bulk mechanical properties.
- **Scaling difficulties:** Reproducibility and scalability of healing mechanisms across large components or complex geometries remain uncertain.

- **Healing reliability:** Repeatability and long-term fatigue behavior post-healing are not yet well understood.
- **Detection and actuation:** Autonomous triggering of the healing process without external cues is a major hurdle.

CONCLUSION

Advances in AM and materials informatics are opening new possibilities for engineered self-healing in metals:

- **Multi-material AM** enables spatial embedding of healing agents or soft phases in precise locations within a hard metal matrix.
- **Graded materials** can be designed with healing capability in critical regions (e.g., stress concentrators or crack-prone interfaces).
- **Nano-engineering** of grain boundaries and precipitates to promote healing under service temperatures is an active area of research.
- **In-situ monitoring** techniques using embedded sensors can help trigger active healing protocols in smart metal structures.

Abbreviations

Table 5.

Abbreviation	Definition
PBF	Powder Bed Fusion
DED	Directed Energy Deposition
ODS	Oxide Dispersion Strengthening
DIW	Direct Ink Writing
FFF	Fused Filament Fabrication
SLM	Selective Laser Melting
DSC	Differential Scanning Calorimetry
SMA	Shape Memory Alloys
MMA	Method of Moving Asymptotes
SIMP	Solid Isotropic Material Penalization
TO	Topology Optimization
FEA	Finite Element Analysis

continued on following page

Table 5. Continued

Abbreviation	Definition
LPBF	Laser-Powder Bed Fusion
FDM	Fused Deposition Modeling
CSAM	Cold Spray Additive Manufacturing
BJ	Binder Jetting
AM	Additive Manufacturing
FGM	Functionally Graded Materials

Declaration on the Use of AI

Artificial intelligence tools were used in the preparation of this chapter to assist with language editing, structuring, and drafting support. All technical content, interpretations, and conclusions are the author's own, and that AI tools were not used to generate original scientific results or fabricate references.. The author reviewed, edited, and verified the entire manuscript to ensure its accuracy, originality, and compliance with ethical and academic standards.

REFERENCES

Ashby, M. F. (2006). The properties of foams and lattices. *Philosophical Transactions of the Royal Society A: Mathematical, Physical and Engineering Sciences, 364*(1838), 15-30.

Baskaran, J. (2025). Metallic 3D-Printed Scaffolds with Functional Coatings: Advancing Tissue Regeneration through Bioactive and Antibacterial Surfaces. In *Innovations and Applications of Advanced Biomaterials in Healthcare and Engineering* (pp. 253-276): IGI Global Scientific Publishing.

Baskaran, J., Muthukannan, D., Shukla, R., & Gokuldoss, P. K. (2024). Manufacturability and deformation studies on a novel metallic lattice structure fabricated by Selective Laser Melting. *Vacuum, 222*, 113065. DOI: 10.1016/j.vacuum.2024.113065

Benafan, O., Noebe, R., Padula, I. I. S.II, Brown, D., Vogel, S., & Vaidyanathan, R. (2014). Thermomechanical cycling of a NiTi shape memory alloy-macroscopic response and microstructural evolution. *International Journal of Plasticity, 56*, 99–118. DOI: 10.1016/j.ijplas.2014.01.006

Blaiszik, B., Jones, A., Sottos, N. R., & White, S. R. (2014). Microencapsulation of gallium–indium (Ga–In) liquid metal for self-healing applications. *Journal of Microencapsulation, 31*(4), 350–354. DOI: 10.3109/02652048.2013.858790

Chen, C., Wang, X., Wang, Y., Yang, D., Yao, F., Zhang, W., Wang, B., Sewvandi, G. A., Yang, D., & Hu, D. (2020). Additive manufacturing of piezoelectric materials. *Advanced Functional Materials, 30*(52), 2005141. DOI: 10.1002/adfm.202005141

Cheng, X., Shen, Z., & Zhang, Y. (2024). Bioinspired 3D flexible devices and functional systems. *National Science Review, 11*(3), nwad314. DOI: 10.1093/nsr/nwad314

Chmielewska, A., Jahadakbar, A., Wysocki, B., Elahinia, M., Święszkowski, W., & Dean, D. (2022). Chemical polishing of additively manufactured, porous, nickel–titanium skeletal fixation plates. *3D Printing. Additive Manufacturing, 9*(4), 269–277.

Dutta, S., Sarma, D. K., Vora, J., Chaudhari, R., Bhowmik, A., Samal, P., & Khanna, S. (2025). A State-of-the-Art review on Micro-Machining of nitinol shape memory alloys and optimization of process variables considering the future trends of research. *Journal of Manufacturing Materials Processing, 9*(6), 183. DOI: 10.3390/jmmp9060183

Gibson, I., Rosen, D. W., & Stucker, B. (2015). *Additive manufacturing technologies* (2nd ed.). Springer. DOI: 10.1007/978-1-4939-2113-3

Jagadeesh, B., Duraiselvam, M., & Prashanth, K. (2023). Deformation behavior of metallic lattice structures with symmetrical gradients of porosity manufactured by metal additive manufacturing. *Vacuum, 211*, 111955. DOI: 10.1016/j.vacuum.2023.111955

Jagadeesh, B., Duraiselvam, M., & Prashanth, K. (2024). Deformation and energy absorption studies on FBCC and FBCCz lattice structures with symmetrical density gradients produced by L-PBF of Ti-6Al-4V alloy. *Materials Today: Proceedings*. Advance online publication. DOI: 10.1016/j.matpr.2024.02.008

Kannan, G. R., Sivakumar, M., Jagadeesh, B., & Balaji, N. (2024). Powder Bed Fusion Process–State of Art. In *Advances in Additive Manufacturing* (pp. 89-107). DOI: 10.1002/9781394238316.ch5

Khoo, Z. X., An, J., Chua, C. K., Shen, Y. F., Kuo, C. N., & Liu, Y. (2018). Effect of heat treatment on repetitively scanned SLM NiTi shape memory alloy. *Materials (Basel), 12*(1), 77. DOI: 10.3390/ma12010077

Li, M., Contreras-Almengor, O., Ordoño, J., Aguilar Vega, C., Zapata Martínez, R., Echeverry-Rendón, M., Díaz-Lantada, A., & Molina-Aldareguia, J. (2025). The microstructure and mechanical properties of nitinol manufactured by LPBF: Differences between Ni-rich and Ti-rich compositions. *Virtual and Physical Prototyping, 20*(1), e2476006. DOI: 10.1080/17452759.2025.2476006

Li, Y., Annan, C., Su, J., Li, Y., Zhang, Y., Li, Z., & Shi, Y. (2025). An overview on ceramic multi-material additive manufacturing: Progress and challenges. *International Journal of Extreme Manufacturing, 7*(4), 042005. DOI: 10.1088/2631-7990/adbc75

Machado, L., & Lagoudas, D. (2008). Thermomechanical constitutive modeling of SMAs. In *Shape Memory Alloys: Modeling and Engineering Applications* (pp. 121-187): Springer. DOI: 10.1007/978-0-387-47685-8_3

Paul, A. A., Nguyen, Q. H., & Shen, W. (2024). 3D printed magnetostrictive polymer composites (MPCs) for wireless stress sensing. *Materials & Design, 247*, 113402. DOI: 10.1016/j.matdes.2024.113402

Rozvany, G. (2000). *The SIMP method in topology optimization-theoretical background, advantages and new applications.* Paper presented at the 8th Symposium on Multidisciplinary Analysis and Optimization. DOI: 10.2514/6.2000-4738

Sames, W. J., List, F., Pannala, S., Dehoff, R. R., & Babu, S. S. (2016). The metallurgy and processing science of metal additive manufacturing. *J. International Materials Reviews, 61*(5), 315–360. DOI: 10.1080/09506608.2015.1116649

Tareq, S., Rahman, T., Poudel, B., Chung, H., & Kwon, P. (2024). Heat treatment protocol for additively manufactured nitinol shape memory alloys in biomedical applications. *Materials Science and Engineering A*, *897*, 146274. DOI: 10.1016/j.msea.2024.146274

Trenfield, S. J., Madla, C. M., Basit, A. W., & Gaisford, S. (2018). Binder jet printing in pharmaceutical manufacturing. In A. W. Basit & S. Gaisford (Eds.), *3D printing of pharmaceuticals* (pp. 41-54): Springer, Cham. DOI: 10.1007/978-3-319-90755-0_3

Varatharajulu, M., Duraiselvam, M., Pradeep, G., & Jagadeesh, B. (2025). Exploring Epitaxial Grain Growth and Marangoni-Induced Conversion in Columnar Dendrites: A Study of Inconel 718 Additive Manufacturing via Direct Laser Energy Deposition. *Surface Review and Letters*, *32*(1), 2450102–2450115. DOI: 10.1142/S0218625X24501026

Yan, C., Hao, L., Hussein, A., & Raymont, D. (2012). Evaluations of cellular lattice structures manufactured using selective laser melting. *International Journal of Machine Tools & Manufacture*, *62*, 32–38. DOI: 10.1016/j.ijmachtools.2012.06.002

Chapter 6
Advancements and Applications of Wire Arc Additive Manufacturing With Future Perspectives

Uzwalkiran Rokkala
https://orcid.org/0000-0001-9048-4384
Aditya University, Surampalem, India

Faseeulla Khan Mohammad
https://orcid.org/0000-0002-5193-3260
King Faisal University, Saudi Arabia

Saad Patel Md
https://orcid.org/0009-0003-3771-4509
Vellore Institute of Technology, Vellore, India

Akash Biradar
Dr. D.Y. Patil Institute of Technology, India

Nooruddin Ansari
Oryx Universal College, Liverpool John Moores University, Doha, Qatar

Prudvireddy Paresi
Federation University, Ballarat, Australia

Raffi Mohammed
National Institute of Technology Andhra Pradesh, India

ABSTRACT

Wire Arc Additive Manufacturing (WAAM) is a novel metal additive manufacturing process based on direct energy deposition destined to revolutionize the manufacturing sector in the next few decades. This process enables sequential deposition of metal on a substrate with the capability of creating intricate geometries with enhanced mechanical properties.This chapter discusses the basic concepts of WAAM, its categorizations, process planning strategies, control approaches, toolpath generation, material choice, defect avoidance and monitoring techniques. Recent

DOI: 10.4018/979-8-3373-9454-1.ch006

innovations, such as intelligent path optimization, real-time process monitoring, and simulation-driven control, highlight the rapid development of the technology. Industries like aerospace, automotive, and tooling are examined for critical applications, while future developments include hybrid production system integration, sustainability-driven practices, and the development of industry standards. WAAM presents an attractive solution for future manufacturing with its flexibility in design, cost-effectiveness and scalability.

1. INTRODUCTION

Industry 4.0 has brought revolutionary changes in the industries mainly in the advanced engineering sector of additive manufacturing (AM). With the primary objective of producing non-structural components for design reasons, the area of AM was founded over thirty years ago (Lu et al., 2015). When compared to traditional manufacturing technologies, this AM technique allows for the quick production of complicated shapes. The four main components of AM are as follows: (a) a digital model of the desired product that is created by converting a CAD model to an STL file, (b) the use of consolidated materials such as wires, powders, or liquid droplets to create the model, (c) provisions for material deposition support and (d) a digital control system that allows for precise material layering to create the desired shape (Beaman et al., 2020). AM is the right application for the perfect trifecta of simplified fabrication of complex geometries and of improved performance of the fabricated components with in the estimated cost.

AM as a growing technology has accelerated its innovation. Rapid prototyping, rapid tooling and 3D printing are subsets of AM. Due to its vast advantages like cost efficiency, minimal material wastage and power, faster and perfect replica of prototypes, hands on inventory are very less, recreation and changes in the manufacturing process at any point of time is very easy and has better product reliability (Bikas et al., 2016). Having all these advantages the AM was implemented into different industrial sectors like automobile, marine, shipbuilding, aerospace, biomedical and in to the construction fields. Currently, AM is gradually growing to commercial usages due to its vast advantages. The American Society for Testing and Materials (ASTM) group has given a standard classification of AM into seven different types based on the materials, methodologies and there working principals (Srivastava et al., 2022). The Fig.1 shows the seven types of AM methods.

Figure 1. Classification of additive manufacturing techniques

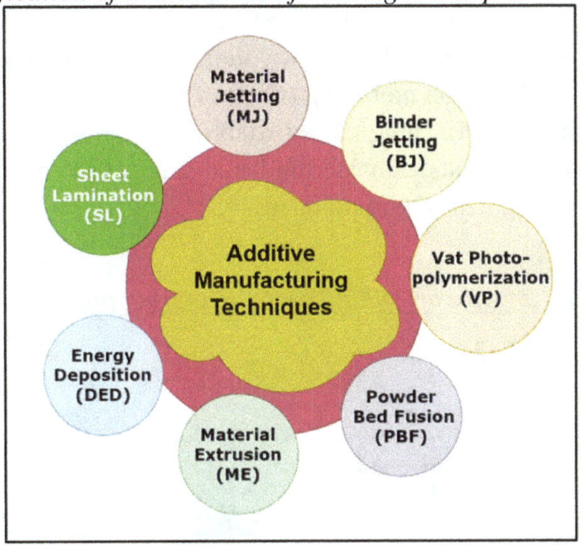

(Singh et al., 2024)

The first technique that was developed in the AM was VAT photo polymerization which uses photopolymers as the printing medium and a light source (UV) was used to bond the molecules there by it builds the required shapes and 3D models. Stereo lithography, Digital light processing, Continuous digital light processing is the subsets of the VAT photo polymerization (Sun et al., 2025). As this process have fine and good surface finish it was mostly used in jewellery making. Due to lack of photopolymer material choices and inadequate strength this process was confined to certain conditions. Second comes the material jetting process which uses polymers and plastics as their material to develop models of 2D shapes layer by layer in the form of droplets with the help of ultraviolent rays. As this process has benefits of higher accuracy of deposition the material wastage was very much low. This process was not suitable for materials other than plastics and polymers (Bezek & Williams, 2023).

The binder jetting process has the capability of printing the objects with colour and uses metals, polymers, and ceramics as the building materials. This process was usually fast when compared to the other AM process due to increase in the printing head holes that deposits material. This process allows two material method for varieties of different binder-powder combinations there by increases the chances of multiple fusions. The post processing of the parts that are build using this process takes time which impacts the overall production time. Then comes the material extrusion process in which fuse deposition modelling method was commonly used. This process uses plastics and polymers just as material jetting technique as their printing materials. The material extrusion process uses plastics and polymers in which acrylonitrile butadiene styrene (ABS) was widely used as their printing material. This method is best suitable for low volume production cases which is more economical than the injection moulding (Oleff et al., 2021).

WAAM, which is an AM technique used to fabricate large-scale metal components, medium to large components, considering its advantages in cost competitiveness, complexity, energy efficiency, and productivity.

Figure 2. Schematic of wire arc additive manufacturing process

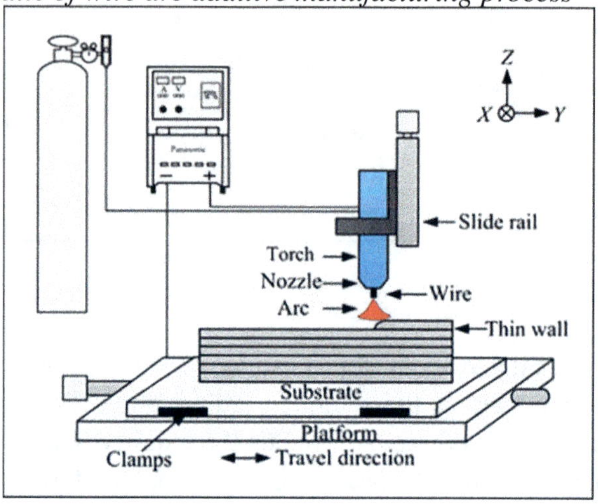

(Xiong et al., 2018)

The main component utilized in wire-feed additive manufacturing using Directed Energy Deposition (DED) is metal wire. Based on the energy source used for metal deposition, this technique can be divided into three different groups: arc welding, laser, and electron beam. In contrast to other AM techniques, WAAM melts and deposits the material layer by layer using an electric arc as a heat source. The origins of WAAM can be seen in a 1925 patent filing, and since about 1990, there has been considerable research into its potential for use in AM (Rodríguez-González et al., 2023). This technique is classified as an AM technology under the wire feed operation and DED category. Welding-based deposition, welding-based RP (Rapid Prototyping), 3D welding, and 3D micro welding are synonyms for WAAM. The greatest substitute for conventional subtractive manufacturing techniques is WAAM. This is due to WAAM's ability to create components with low buy-to-fly ratios and low to medium complex geometry. With the use of WAAM, a 58% cost reduction is achievable (Singh & Khanna, 2021).

WAAM process has many advantages, some of them are impressive building capabilities, high deposition rates, and compatibility with various welding equipment and filler materials. The process also requires minimal human intervention that has a quick setup time, and helps keep capital and material costs down. On the contrary, the process has some drawbacks as well. For instance, WAAM often needs post-processing, can have low surface accuracy, and might result in rough finishes. In addition to this, the heat build-up can lead to residual strains and distortions. Still, the process WAAM being used across a variety of industries, including construction, energy, aerospace, and automotive.

WAAM is a state-of-the-art additive manufacturing technology that employs an electric arc welding process for building metal parts layer by layer. It is a cost-effective and efficient method for producing large-scale structures with complex designs (Dinovitzer et al., 2019). The WAAM process begins with digital design, where a 3D model of the part to be produced is built using CAD software. Following is path planning, which concerns determining the optimal toolpath for the deposition head so that the material is deposited in the correct position. Material preparation is also a critical stage, where the proper feedstock wire is selected and surface conditions for the deposition are optimized. Throughout the layer-by-layer deposition stage, the wire is melted by an electric arc, and it is deposited on the substrate or the last layer, building up the component step by step (Su et al., 2019). Once each layer is deposited, it solidifies and cools to maintain the structural integrity of the material. Lastly, post-processing methods such as machining, heat treatment, or surface finishing may be employed to attain the required dimensional accuracy and mechanical properties. This rigorous process renders WAAM a versatile and scalable technology for multiple industrial uses.

To start off, a 3D CAD model of the object you want is either created or sourced. Then, specialized software takes a look at this CAD model and crafts a toolpath or trajectory for the welding torch to follow during the deposition process. This carefully planned path ensures that the material is deposited with precision and accuracy. Next, a metal wire, usually found on a spool, is chosen based on the specific properties needed and its compatibility with the welding process. This wire is then fed through a wire feeding mechanism into the welding torch. An electric arc is generated by a welding power source between the wire electrode and the work piece, melting the wire material in the process (Taşdemir & Nohut, 2021). The molten metal is then deposited exactly onto the work piece, bonding with the layers that came before it. A robotic arm or another motion control system guides the welding torch along the predetermined toolpath. As each layer is added, the molten metal cools quickly and solidifies, merging with the previous layers to create a solid structure. Depending on what's needed for the specific application, additional steps like machining, surface finishing, and heat treatment might be carried out to ensure the desired dimensional accuracy, surface quality, and material properties.

The recent research shows that the WAAM process is set to make even greater strides in areas like process control, surface quality, material properties, and overall reliability (Pan et al., 2024; Niu et al., 2024; Sasikumar & Oyyaravelu, 2024; Chen et al., 2025). Its ability to handle large-scale additive manufacturing makes it a truly exciting technology with bright future in the manufacturing sector.

2. PROCESS PLANNING, PROCESS ANALYSIS, PROCESS SELECTION AND MODELLING IN WAAM

2.1 Process Planning

Process planning in additive manufacturing, especially with wire arc additive manufacturing (WAAM), takes a methodical approach to outline the sequence of operations and parameters needed to create a part both efficiently and accurately. It is all about carefully selecting various factors to ensure top-notch performance, material integrity, and precise dimensions. This begins with design analysis, where the assessment of the parts is been done and also the geometric intricacies, and performance criteria's are figured out for the process feasibility (Omiyale et al., 2025; Nguyen et al., 2023). Next comes material selection, zero in on the ideal wire feedstock, considering its mechanical properties, weldability, and how well it meshes with the WAAM process to achieve the desired characteristics of the part. It is also needed to plan for support structures to boost stability during the deposition process, which helps reduce distortion and maintain geometric accuracy, particularly for those tricky overhangs or complex features.

The build strategy is a crucial part of the process, as it involves deciding the layer thickness, deposition paths, and orientation to make the most of the materials, reduce residual stresses, and enhance structural integrity. It is important to carefully select process parameters like arc current, voltage, travel speed, and wire feed rate to ensure the quality of deposition, the microstructure, and the mechanical properties. To maintain high standards, quality control measures such as in-process monitoring and post-build inspections are put in place to check dimensional accuracy, spot defects, and guarantee consistency. Moreover, planning for post-processing steps, like machining, surface finishing, and heat treatment is essential to meet the final specifications of the part. Lastly, keeping thorough documentation of all process parameters, build strategies, and quality checks is vital for ensuring repeatability, traceability, and ongoing process improvement. By focusing on these key elements, WAAM process planning paves the way for the efficient production of high-quality, functional metal components. Effective process planning in WAAM demands a solid understanding of part design, materials, equipment capabilities, and the desired outcomes. It takes a systematic approach to fine-tune build strategies, process parameters, and post-processing considerations to achieve efficient and top-notch production.

2.2 Process Analysis

Process analysis in WAAM is essential for enhancing productivity, minimizing defects, and ensuring consistent part quality (Barath Kumar & Manikandan, 2022). A thorough examination of the manufacturing workflow helps identify inefficiencies, allowing for targeted improvements in deposition strategies, material usage, and energy consumption. Researchers have extensively studied various aspects of WAAM to optimize its performance. For instance, Shen et al. investigated the effect of induction heat treatment on residual stress distribution of components fabricated by WAAM (Shen et al., 2022). Similarly, He et al. analysed thermal management strategies, showing that controlled interlayer cooling improves microstructural homogeneity and mechanical properties in titanium WAAM parts (He et al., 2024).

Material utilization and defect formation are critical areas of focus in process analysis. Martina et al. studied material flow efficiency in WAAM, proposing optimized wire feed mechanisms to minimize waste while maintaining deposition accuracy (Martina et al., 2019). Defect analysis, particularly porosity and lack-of-fusion imperfections, has been explored by Ding et al, who correlated process parameters such as travel speed and current settings with defect occurrence, providing guidelines for parameter optimization (Ding et al., 2021). Real-time monitoring and data-driven process control have emerged as key enablers for quality assurance. Rodrigues et al. developed an in-situ monitoring system using optical sensors to detect anomalies during deposition, allowing for immediate corrective actions (Rodrigues, et al. 2019). Additionally, Montevecchi et al. demonstrated the benefits of machine learning in predicting optimal process parameters, reducing trial-and-error iterations. Continuous improvement in WAAM relies on systematic feedback from post-process evaluations. Dhinakaran et al. highlighted the importance of post-deposition heat treatments in refining grain structure and relieving residual stresses in nickel-based super alloys (Dhinakaran et al., 2020). By integrating these findings into process planning, manufacturers can achieve higher repeatability and performance in WAAM-produced components. Through ongoing research and technological advancements, WAAM continues to evolve as a robust AM method, capable of producing high-quality, large-scale metal structures with optimized efficiency and reliability

2.3 Process Selection

The selection of an appropriate manufacturing process requires careful consideration of multiple technical and economic factors to ensure optimal production outcomes. When evaluating WAAM against conventional and other AM methods, recent research highlights several critical decision-making criteria. Part Design and Complexity play a fundamental role, as WAAM excels in producing large-scale, geometrically complex components with reduced material waste compared to subtractive methods. Recent work by Geng et al. demonstrated that WAAM's near-net-shape capabilities make it particularly advantageous for aerospace components with internal lattice structures, where machining would be prohibitively expensive (Geng et al., 2023).

Material Compatibility remains a key consideration, with WAAM supporting a growing range of weldable alloys. A comprehensive review of material-process interactions, showing that advanced wire chemistries (e.g., high-strength steels and nickel super alloys) can achieve mechanical properties comparable to wrought materials when proper thermal management is applied. For production volume and time constraints, studies revealed that WAAM becomes economically competitive for medium-batch production (50-500 units) of large parts, offering lead time reductions up to 60% compared to traditional forging including machining routes. However, Le et al. (2021) cautioned that post-processing requirements may offset these gains for high-precision applications. Cost Considerations must account for the total value chain. A cost model developed for showing WAAM's superiority for structural aerospace components exceeding 5 kg, where material savings outweigh higher energy costs. Meanwhile, Zhang et al. (2023) highlighted that hybrid manufacturing (WAAM + machining) optimizes costs for parts requiring tight tolerances. Regarding quality requirements, advances in real-time monitoring have significantly improved WAAM's reliability. Research by Colegrove et al. (2013) demonstrated that closed-loop control systems can achieve surface roughness below 25 μm and dimensional accuracy within ± 0.5 mm for large structures.

Finally, environmental impact assessments on WAAM technique shown favourably, with 30-40% lower CO_2 emissions compared to conventional manufacturing for weight-optimized designs, though energy intensity remains a concern for small parts. A holistic evaluation framework should integrate these recent findings with traditional decision factors. WAAM implementation requires matching the technology's strengths - large-part fabrication, design freedom, and material efficiency - with application-specific requirements. Emerging digital tools combining technical cost modelling with sustainability metrics now enable more informed process selection, ensuring WAAM is deployed where it delivers maximum value across technical, economic, and environmental dimensions.

This evolving body of research underscores that while WAAM may not universally replace conventional methods, it represents a transformative solution for specific applications where its unique capabilities align with production needs. Continued advancements in process control, material development, and hybrid approaches are expected to further expand WAAM's suitability across industrial sectors.

2.4 Modelling In WAAM

Recent advances in computational modelling have significantly enhanced the understanding and optimization of WAAM processes. Comprehensive modelling approaches are essential for predicting process behaviour, minimizing experimental trials, and improving final part quality (Hamrani et al., 2023). The current state of WAAM modelling encompasses several critical domains, each addressing distinct aspects of the deposition process.

2.4.1 Thermal Modelling

Thermal analysis remains fundamental to WAAM process optimization, with Ding et al. (2021) showing that accurate heat distribution predictions are crucial for controlling solidification behaviour. Finite element analysis (FEA) tools such as ANSYS, ABAQUS, and COMSOL are widely employed, as highlighted by Martina et al. (2016), who successfully predicted thermal cycles in large-scale titanium WAAM structures. Recent work by Wang et al. (2025) introduced machine learning-assisted thermal models that improve computational efficiency while maintaining accuracy.

2.4.2 Mechanical Modelling

Residual stresses and distortions pose significant challenges in WAAM, prompting extensive research into mechanical modelling. A thermo-mechanical FEA framework that accurately predicts distortion in multi-layer depositions, validated through experimental measurements. Further, these models enhanced by incorporating phase transformation effects in high-strength steels, demonstrating improved stress prediction capabilities.

2.4.3 Microstructure Modelling

Grain structure evolution directly impacts mechanical properties, making microstructure modelling a critical research area. Staroselsky et al. (2024) applied cellular automaton models to simulate grain growth in Inconel 718 WAAM deposits, correlating thermal history with microstructure development. More recently,

Wang et al. (2024) integrated phase-field modelling with thermal simulations to predict austenite-to-ferrite transformations in low-alloy steels, providing insights into mechanical property control.

2.4.4 Process Modelling and Parameter Optimization

The relationship between process parameters (e.g., wire feed rate, current, travel speed) and deposition quality has been extensively studied. Ou et al. (2021) used computational fluid dynamics to model melt pool dynamics, establishing optimal parameter windows for defect-free deposition. A hybrid analytical-numerical models were created to predict the bead geometry with high accuracy, reducing the need for trial-and-error experimentation. Recent efforts focus on multi-scale and multi-physics modelling approaches. Further, an integrated digital twin framework combining thermal, mechanical, and microstructure models developed for real-time process optimization. Mu et al (2024) demonstrated the effectiveness of Artificial Intelligence (AI)-driven surrogate models in accelerating simulation times while maintaining predictive accuracy

3. CHALLENGES IN WAAM

Recent studies highlight the complex dynamics of WAAM process control. It is demonstrated that the nonlinear coupling between heat input and deposition rate significantly affects bead geometry stability, particularly in multi-layer depositions. A real-time adaptive control system is developed using machine learning to maintain arc stability, showing a 40% improvement in deposition consistency. However, the process instabilities persist in large-scale builds due to cumulative thermal effects. Thermomechanical challenges remain a major research focus. The residual stresses in WAAM titanium components can reach 80% of yield strength without proper thermal management. An innovative interlayer cooling strategy is suggested to reduces distortion by 60% in steel components. Recent works introduced a hybrid in-situ rolling approach that simultaneously reduces residual stresses while improving mechanical properties.

Material limitations continue to constrain WAAM applications. Few systematic investigations on fifteen common engineering alloys were evaluated and identified only eight as currently suitable for WAAM without significant property degradation. Further, a novel wire chemistry was developed for high-strength aluminium alloys that overcome hot cracking tendencies. However, it is noted that material certification remains a bottleneck, with only five alloys currently approved for aerospace WAAM applications. Surface quality challenges persist despite advancements. An average WAAM surface roughness values were measured and resulted in between 50-200 μm, significantly higher than conventional manufacturing. It is demonstrated that hybrid WAAM-machining approaches can achieve R_a < 3.2 μm but increase production costs by 25 - 40%.

Defect mitigation remains critical for industrial adoption. WAAM defects are classified into six categories, with lack-of-fusion being most prevalent (occurring in 15-30% of builds). An in-situ monitoring system was developed combining infrared thermography and acoustic emission that detects 95% of critical defects (Alcaraz et al., 2024). However, the current NDT methods struggle with defect quantification s to improve large-scale WAAM productivity by 35% through optimized path planning. However, the energy consumption remains 2-3 times higher than conventional manufacturing for simple geometries.

4. RECENT ADVANCEMENTS

WAAM has witnessed significant technological progress in recent years, expanding its capabilities and industrial applicability. One of the most notable advancements lies in material diversity, with researchers exploring an increasingly broad range of alloys and functionally graded materials. Recent studies by Lu et al. (2024) have demonstrated successful deposition of high-entropy alloys and tailored material compositions, enabling component-specific property optimization. This material expansion has been complemented by innovative process techniques, particularly the integration of cold metal transfer technology and advanced robotic systems. These developments have substantially improved deposition precision and repeatability while reducing heat input and distortion.

Post-processing techniques have also seen remarkable innovation, addressing one of WAAM's traditional limitations. Research by Mansor et al. (2024) highlights integrated approach to WAAM optimization. The optimization of various wire deposition techniques and process parameters in the WAAM method is essential for producing high-quality AM metal components. Perhaps the most transformative advancement has been the implementation of AI and Machine Learning in WAAM processes. Fernández-Zabalza et al. (2025) developed AI-driven predictive modelling of homogeneous bead geometry for WAAM processes, The results confirm the feasibility of non-intrusive, signal-based anomaly detection, allowing for rapid identification of weld spalls and supporting scalable, automated WAAM process monitoring in industrial settings. These technological leaps are driving WAAM adoption across aerospace, energy, and defence sectors, particularly for large-scale, complex components promises to further accelerate WAAM's evolution. Emerging research directions include digital twin integration, advanced multi-material deposition strategies, and sustainable process optimization - all poised to unlock new applications and establish WAAM as a mainstream manufacturing technology in the coming decade. The convergence of these advancements suggests WAAM is transitioning from a prototyping tool to a viable production solution for high-value metal components.

5. FUTURE PROSPECTS

WAAM holds tremendous potential to revolutionize metal AM, with several promising developments on the horizon. The technology's inherent advantages including minimal energy consumption, reduced material waste, and shorter production lead times position it as a strong contender to dominate the AM market for large-scale metal components. One key area of future development involves the implementation of multiple heat sources, which could enable better thermal management, higher deposition rates, and improved microstructure control. Researchers are also focusing on applying Design for AM (DfAM) principles specifically tailored for WAAM processes to optimize part geometry, minimize residual stresses, and enhance mechanical performance.

Another critical research direction is the advancement of innovative defect detection and prevention strategies, leveraging real-time monitoring, machine learning, and advanced sensor technologies to ensure consistent quality in as-built components. Additionally, the development of advanced post-processing techniques—such as hybrid machining, laser polishing, and targeted heat treatments—will be essential to meet stringent surface finish and dimensional accuracy requirements. A particularly transformative opportunity lies in the integration of WAAM with other manufacturing technologies, such as subtractive machining, forging, or cold working, to create hybrid manufacturing systems that combine the strengths of multiple processes. This approach could unlock new possibilities for producing high-performance components with complex geometries and superior material properties. As WAAM technology continues to mature, it is expected to transition from a niche prototyping method to a mainstream manufacturing solution across aerospace, automotive, energy, and defence industries. With ongoing advancements in process control, material science, and digital integration, WAAM is poised to offer unprecedented design freedom, cost efficiency, and production flexibility for fabricating large, complex metal parts. The convergence of these innovations will likely establish WAAM as a cornerstone of next-generation manufacturing in the coming decade.

6. SUMMARY

WAAM stands at the forefront of manufacturing innovation, demonstrating remarkable potential to transform industrial production through its unique capabilities. This rapidly evolving technology combines the advantages of AM with the robustness of arc welding processes, offering manufacturers unprecedented design freedom while delivering substantial cost efficiencies and reduced production timelines. Recent technological breakthroughs have significantly enhanced WAAM's operational scope, particularly through developments in multi-material deposition techniques, advanced process control systems, and expanded material compatibility. These improvements have been complemented by sophisticated simulation and modelling approaches that enable more precise prediction and optimization of deposition outcomes.

Looking ahead, WAAM is poised for widespread industrial adoption across critical sectors including aerospace, automotive, energy infrastructure, defence applications, and specialized tooling production. The technology's future trajectory will be characterized by continued material science innovations, with researchers focusing on broadening the spectrum of compatible alloys and developing novel material combinations to produce components with precisely engineered mechanical and thermal properties. This expansion of material capabilities, coupled with ongoing refinements in process reliability and quality control, positions WAAM as a transformative manufacturing solution capable of meeting the increasingly complex demands of modern industrial production. As these advancements mature, WAAM is expected to transition from a specialized fabrication method to a main.

Funding Statement

This work was supported by the Deanship of Scientific Research, Vice Presidency for Graduate Studies and Scientific Research, King Faisal University, Saudi Arabia. Grant No. KFU251841.

REFERENCES

Alcaraz, J. Y.II, Foqué, W., Sharma, A., & Tjahjowidodo, T. (2024). Indirect porosity detection and root-cause identification in WAAM. *Journal of Intelligent Manufacturing, 35*(4), 1607–1628. DOI: 10.1007/s10845-023-02128-x

Barath Kumar, M. D., & Manikandan, M. (2022). Assessment of process, parameters, residual stress mitigation, post treatments and finite element analysis simulations of wire arc additive manufacturing technique. *Metals and Materials International, 28*(1), 54–111. DOI: 10.1007/s12540-021-01015-5

Beaman, J. J., Bourell, D. L., Seepersad, C. C., & Kovar, D. (2020). Additive manufacturing review: Early past to current practice. *Journal of Manufacturing Science and Engineering, 142*(11), 110812. Advance online publication. DOI: 10.1115/1.4048193

Bezek, L. B., & Williams, C. B. (2023). Process-structure-property effects of ultraviolet curing in multi-material jetting additive manufacturing. *Additive Manufacturing, 73*, 103640. Advance online publication. DOI: 10.1016/j.addma.2023.103640

Bikas, H., Stavropoulos, P., & Chryssolouris, G. (2016). Additive manufacturing methods and modelling approaches: A critical review. *International Journal of Advanced Manufacturing Technology, 83*(1-4), 389–405. DOI: 10.1007/s00170-015-7576-2

Chen, Z., Yuan, L., Pan, Z., Zhu, H., Ma, N., Ding, D., & Li, H. (2025). A comprehensive review and future perspectives of simulation approaches in wire arc additive manufacturing (WAAM). *International Journal of Extreme Manufacturing, 7*(2), 022016. DOI: 10.1088/2631-7990/ada099

Colegrove, P. A., Coules, H. E., Fairman, J., Martina, F., Kashoob, T., Mamash, H., & Cozzolino, L. D. (2013). Microstructure and residual stress improvement in wire and arc additively manufactured parts through high-pressure rolling. *Journal of Materials Processing Technology, 213*(10), 1782–1791. DOI: 10.1016/j.jmatprotec.2013.04.012

Dhinakaran, V., Ajith, J., Fahmidha, A. F. Y., Jagadeesha, T., Sathish, T., & Stalin, B. (2020). Wire Arc Additive Manufacturing (WAAM) process of nickel based superalloys–A review. *Materials Today: Proceedings, 21*, 920–925. DOI: 10.1016/j.matpr.2019.08.159

Ding, D., Zhao, R., Lu, Q., Pan, Z., Li, H., Wang, K., & He, K. (2021). A shape control strategy for wire arc additive manufacturing of thin-walled aluminium structures with sharp corners. *Journal of Manufacturing Processes*, *64*, 253–264. DOI: 10.1016/j.jmapro.2021.01.029

Dinovitzer, M., Chen, X., Laliberte, J., Huang, X., & Frei, H. (2019). Effect of wire and arc additive manufacturing (WAAM) process parameters on bead geometry and microstructure. *Additive Manufacturing*, *26*, 138–146. DOI: 10.1016/j.addma.2018.12.013

Fernández-Zabalza, A., Rodríguez-Díaz, Á., Veiga, F., Suárez, A., Uralde, V., Ballesteros, T., & Alfaro, J. R. (2025). AI-driven predictive modeling of homogeneous bead geometry for WAAM processes. *International Journal of Advanced Manufacturing Technology*. Advance online publication. DOI: 10.1007/s00170-025-16061-1

Geng, R., Cheng, Y., Chao, L., Wei, Z., & Ma, N. (2023). Microstructure and solute concentration analysis of epitaxial growth during wire and arc additive manufacturing of aluminum alloy. *Crystals*, *13*(5), 776. DOI: 10.3390/cryst13050776

Hamrani, A., Bouarab, F. Z., Agarwal, A., Ju, K., & Akbarzadeh, H. (2023). Advancements and applications of multiple wire processes in additive manufacturing: A comprehensive systematic review. *Virtual and Physical Prototyping*, *18*(1), e2273303. DOI: 10.1080/17452759.2023.2273303

He, B., Wang, Y., Wang, C., Ren, Y., & Yang, G. (2024). Effect of gradient decrease heat-input combined with inter-layer cooling on microstructure and mechanical properties of WAAM-Ti-6Al-2Zr-1Mo-1V alloy. *Materials Today. Communications*, *39*, 108628. DOI: 10.1016/j.mtcomm.2024.108628

Le, V. T., Mai, D. S., Doan, T. K., & Paris, H. (2021). Wire and arc additive manufacturing of 308L stainless steel components: Optimization of processing parameters and material properties. *Engineering Science and Technology, an International Journal*, *24*(4), 1015–1026. DOI: 10.1016/j.jestch.2021.01.009

Lu, B., Li, D., & Tian, X. (2015). Development trends in additive manufacturing and 3D printing. *Engineering (Beijing)*, *1*(1), 85–89. DOI: 10.15302/J-ENG-2015012

Lu, H., Shen, Q., Chen, X., Wen, M., & Jayalakshmi, S. (2024). Dual wire arc additive manufacturing of compositionally graded Alx-Co-Cr-Fe-Ni high-entropy alloy. *Journal of Materials Research and Technology*, *29*, 4052–4062. DOI: 10.1016/j.jmrt.2024.02.089

Mansor, M. S. M., Raja, S., Yusof, F., Muhamad, M. R., Manurung, Y. H., Adenan, M. S., & Ren, J. (2024). Integrated approach to Wire Arc Additive Manufacturing (WAAM) optimization: Harnessing the synergy of process parameters and deposition strategies. *Journal of Materials Research and Technology*, *30*, 2478–2499. DOI: 10.1016/j.jmrt.2024.03.170

Martina, F., Ding, J., Williams, S., Caballero, A., Pardal, G., & Quintino, L. (2019). Tandem metal inert gas process for high productivity wire arc additive manufacturing in stainless steel. *Additive Manufacturing*, *25*, 545–550. DOI: 10.1016/j.addma.2018.11.022

Martina, F., Roy, M. J., Szost, B. A., Terzi, S., Colegrove, P. A., Williams, S. W., Withers, P. J., Meyer, J., & Hofmann, M. (2016). Residual stress of as-deposited and rolled wire+arc additive manufacturing Ti–6Al–4V components. *Materials Science and Technology*, *32*(14), 1439–1448. DOI: 10.1080/02670836.2016.1142704

Mu, H., He, F., Yuan, L., Hatamian, H., Commins, P., & Pan, Z. (2024). Online distortion simulation using generative machine learning models: A step toward digital twin of metallic additive manufacturing. *Journal of Industrial Information Integration*, *38*, 100563. DOI: 10.1016/j.jii.2024.100563

Nguyen, V. T., Minh, P. S., Uyen, T. M. T., Do, T. T., Ngoc, H. V. T., Le, M. T., & Tien Nguyen, V. T. (2023). WAAM technique: Process parameters affecting the mechanical properties and microstructures of low-carbon steel. *Metals*, *13*(5), 873. DOI: 10.3390/met13050873

Niu, F., Bi, W., Liu, Z., Sun, X., Ma, G., & Wu, D. (2024). 304-Ni60B (NiCrBSi) composite components using WAAM-LC integrated hybrid manufacturing. *International Journal of Advanced Manufacturing Technology*, *132*(1-2), 463–474. DOI: 10.1007/s00170-024-13376-3

Oleff, A., Küster, B., Stonis, M., & Overmeyer, L. (2021). Process monitoring for material extrusion additive manufacturing: A state-of-the-art review. *Progress in Additive Manufacturing*, *6*(4), 705–730. DOI: 10.1007/s40964-021-00192-4

Omiyale, B. O., Ogedengbe, I. I., Olugbade, T. O., Rasheed, A. A., Ogbeyemi, A., & Farayibi, P. K. (2025). Cold metal transfer WAAM of aluminum alloys: Influence of processing parameters. *International Journal of Advanced Manufacturing Technology*, *136*(5), 1967–1987. DOI: 10.1007/s00170-024-14989-4

Ou, W., Knapp, G. L., Mukherjee, T., Wei, Y., & DebRoy, T. (2021). An improved heat transfer and fluid flow model of wire-arc additive manufacturing. *International Journal of Heat and Mass Transfer*, *167*, 120835. DOI: 10.1016/j.ijheatmasstransfer.2020.120835

Pan, Y., Yu, M., Xu, C., Zhang, J., & Geng, L. (2024). High-performance 2319 aluminum alloy via CMT-WAAM: Microstructure, porosity, and mechanical properties. *Metals*, *14*(7), 14. DOI: 10.3390/met14070797

Rodrigues, T. A., Duarte, V., Miranda, R. M., Santos, T. G., & Oliveira, J. P. (2019). Current status and perspectives on wire and arc additive manufacturing (WAAM). *Materials (Basel)*, *12*(7), 1121. DOI: 10.3390/ma12071121

Rodríguez-González, P., Ruiz-Navas, E. M., & Gordo, E. (2023). Wire arc additive manufacturing (WAAM) for aluminum-lithium alloys: A review. *Materials (Basel)*, *16*(4), 16. DOI: 10.3390/ma16041375

Sasikumar, C., & Oyyaravelu, R. (2024). Mechanical properties and microstructure of SS 316 L created by WAAM based on GMAW. *Materials Today. Communications*, *38*, 107807. DOI: 10.1016/j.mtcomm.2023.107807

Shen, H., Lin, J., Zhou, Z., & Liu, B. (2022). Effect of induction heat treatment on residual stress distribution of components fabricated by wire arc additive manufacturing. *Journal of Manufacturing Processes*, *75*, 331–345. DOI: 10.1016/j.jmapro.2022.01.018

Singh, G., Mehta, A., & Vasudev, H. (2024). Sustainability of additive manufacturing: A comprehensive review. *Progress in Additive Manufacturing*, *9*(6), 2249–2272. DOI: 10.1007/s40964-024-00579-z

Singh, S. R., & Khanna, P. (2021). Wire arc additive manufacturing (WAAM): A new process to shape engineering materials. *Materials Today: Proceedings*, *44*, 118–128. DOI: 10.1016/j.matpr.2020.08.030

Srivastava, M., Rathee, S., Patel, V., Kumar, A., & Koppad, P. G. (2022). A review of various materials for additive manufacturing: Recent trends and processing issues. *Journal of Materials Research and Technology*, *21*, 2612–2641. DOI: 10.1016/j.jmrt.2022.10.015

Staroselsky, A., Voytovych, D., & Acharya, R. (2024). Prediction of Ni-based alloy microstructure in wire arc additive manufacturing from cellular automata model. *Computational Materials Science*, *233*, 112721. Advance online publication. DOI: 10.1016/j.commatsci.2023.112721

Su, C., Chen, X., Gao, C., & Wang, Y. (2019). Effect of heat input on microstructure and mechanical properties of Al-Mg alloys fabricated by WAAM. *Applied Surface Science*, *486*, 431–440. DOI: 10.1016/j.apsusc.2019.04.255

Sun, H., Zou, B., Quan, T., Wang, X., Ma, X., Xing, H., Wei, S., & Huang, C. (2025). Multi-material ceramic hybrid additive manufacturing based on vat photopolymerization and material extrusion compound process. *Additive Manufacturing*, *97*, 104627. DOI: 10.1016/j.addma.2024.104627

Taşdemir, A., & Nohut, S. (2021). An overview of wire arc additive manufacturing (WAAM) in shipbuilding industry. *Ships and Offshore Structures*, *16*(7), 797–814. DOI: 10.1080/17445302.2020.1786232

Wang, F., Chen, W., Wang, D., Hou, H., & Zhao, Y. (2024). Phase-field modeling and Experimental investigation for rapid solidification in wire and arc additive manufacturing. *Journal of Materials Research and Technology*, *28*, 4585–4599. DOI: 10.1016/j.jmrt.2024.01.021

Wang, X., Sridar, S., Klecka, M., & Xiong, W. (2025). Rapid data acquisition and machine learning-assisted composition design of functionally graded alloys via wire arc additive manufacturing. npj. *Advances in Manufacturing*, *2*(1), 17. Advance online publication. DOI: 10.1038/s44334-025-00028-x

Xiong, J., Li, Y., Li, R., & Yin, Z. (2018). Influences of process parameters on surface roughness of multi-layer single-pass thin-walled parts in GMAW-based additive manufacturing. *Journal of Materials Processing Technology*, *252*, 128–136. DOI: 10.1016/j.jmatprotec.2017.09.020

Zhang, C., Zou, D., Mazur, M., Mo, J. P., Li, G., & Ding, S. (2023). The state of the art in machining additively manufactured titanium alloy Ti-6Al-4V. *Materials (Basel)*, *16*(7), 2583. DOI: 10.3390/ma16072583

Chapter 7
Analysis of Residual Stress on Commercially Pure Titanium Using Molybdenum Wire Machined on Reciprocated Travelling WEDM

Anoop Johny
KMCT College of Engineering, India

C. Thiagarajan
Saveetha School of Engineering, SIMATS, Saveetha University, India

N. Senthilkumar
https://orcid.org/0000-0002-2441-1061
Saveetha School of Engineering, SIMATS, Saveetha University, India

Deepanraj Balakrishnan
Prince Mohammad Bin Fahd University, Saudi Arabia

ABSTRACT

Wirecut Electrical Discharge Machining (WEDM) is a cost-effective and productive method of cutting electrically conductive hard materials. WEDM has the added benefit of being more accurate when it comes to manufacturing component forms.

DOI: 10.4018/979-8-3373-9454-1.ch007

In this research, the performance of reciprocated WEDM (RWEDM) on a titanium grade 2 alloy with a molybdenum wire electrode is evaluated by measuring the residual stresses present on the machined part for different conditions. Observations indicate that considerable residual stresses are seen in the specimen as watched from the X-ray diffraction test through measuring the residual stress, shear stress, D-spacing, full width at half maximum height, intensity, and breadth of stresses. Observations show that when feed rate of wire electrode in RWEDM and flow rate of dielectric is higher and higher variable frequency and higher current, residual stress tends to rise due to higher generation of spark density, which correspondingly removes higher quantity of material from the workpiece

1. INTRODUCTION

Residual stress (RS) measurement is crucial in assessing the mechanical integrity, performance, and longevity of machined parts, particularly in materials like Inconel 625 and titanium alloys, which exhibit high strength and work hardening tendencies (Outeiro, 2018). RSs, which are internal stresses retained in parts after fabrication or machining, can significantly influence the fatigue life, corrosion resistance, dimensional stability, and overall structural reliability of a component. Under cyclic loading, tensile RSs exacerbate the fracture start and propagation, ultimately resulting to failure sooner rather than later, whereas compressive RSs can enhance fatigue resistance by delaying crack growth. In machining, heat generation, plastic deformation, and tool-workpiece interaction contribute to RS formation (Chen et al., 2022; Senthilkumar et al., 2025). RS measurement techniques help identify internal stresses remaining in a material after manufacturing or processing. These stresses can significantly influence material performance, affecting fatigue life, dimensional stability, and susceptibility to cracking. Accurate measurement methods such as X-ray diffraction, hole-drilling, and neutron diffraction enable engineers to assess and manage these stresses, thereby improving structural integrity and service reliability. Controlling RSs is particularly critical in aerospace, biomedical, and high-performance applications, where components must withstand extreme mechanical and thermal conditions (Suresh et al., 2025). Proper RS management through post-machining treatments like shot peening, heat treatment, or cryogenic processing ensures improved wear resistance, dimensional accuracy, and overall durability of machined parts (la Monaca et al., 2021). Machining titanium alloys with wire electrical discharge machining (WEDM) develops RSs in response to short temperature cycles characterized by heating and cooling locally. Electrical sparks produce very high temperatures, which lead to the material's melting and vaporization, and then subsequent quenching rapidly, as dielectric fluid cools the workpiece (Johny & Thiagarajan, 2020; Kavimani et al., 2024). This rapid cooling induces tensile RSs on the machined surface due to thermal contraction, while subsurface layers may experience compressive stresses from the balancing effect of the tensile stresses above (Kumar et al., 2014). Additionally, recast layers and microstructural alterations, such as alpha-case formation and grain refinement, can further influence RS distribution (Bharathi et al., 2025; Perumal et al., 2025). The combination of these effects can impact fatigue performance, dimensional stability, and mechanical properties of titanium alloy components produced through WEDM (Arunnath et al., 2022; Chinna et al., 2025).

In a study, Soni et al. (Soni et al., 2018) showcases the impacts of WEDM using Taguchi's L_9 array on the shape memory alloy Ti50Ni45Co5. To achieve a greater material removal rate (MRR) and improved surface roughness (SR), the current research used hybrid optimization to determine the optimum circumstances as follows: 125μs pulse on time (Ton), 35μs pulse off time (Toff), and 40V servo voltage (SV). The machined surface shows signs of compressive RS after the WEDM process. Applying the Taguchi approach, Rao et al. (Srinivasa Rao et al., 2016) conducted a parametric investigation of the effects of WEDM variables on RSs during the milling of aluminum alloy. RSs ranged from 8.2 to 405.6 MPa, according to the data. Intermetallics like AlCu and AlCu3 were also affected by it throughout their development. Under no circumstances throughout the machining process cracks on the surfaces were not found, according to microscopic analysis. In a study conducted by Gaikwad et al. (Gaikwad et al., 2021), it was discovered that as the current and Ton levels are increased, the fatigue strength of EDM machined objects decreases and fatigue failure occurs early. At 10^6 cycles, the fatigue strength reached 397.88 MPa, which is lower than the work piece's ultimate strength of 465 MPa. It was discovered that the MRR grows as the current increases. When the current and Ton are increased, the fatigue strength of EDM machined products is reduced because fatigue failure occurs early. To determine how different WEDM process parameters affected the machining properties of Ti-6Al-4 V alloy, Jagdale et al. (Jagdale et al., 2025) conducted an evaluation. The degree of MRR is largely affected by peak current (72.75%) and pulse on time (11.68%), the two most important parameters for determining Ra. In real-world WEDM of Ti-6Al-4 V alloy, adjusting the peak current and pulse on time should be your top priority.

Machining and heating, according to Frame et al. (Frame et al., 2020), may cause titanium alloy parts to bend and introduce unwanted RSs. This research found that RSs on machined surfaces are heavily dependent on the tool path, and that solution treated and quenched samples of Ti 6Al-4V alloy behaved significantly differently from samples that were solution treated and aged. There is evidence of non-uniform stresses in the sample from the oscillatory connection shown in the matching d-spacing vs $\sin^2\varphi$ curve for the patterns. Titanium alloys are particularly difficult to machine because of the high tool wear and slow MRR. It is normal for all kinds of EDM procedures to result in the formation of recast layers, heat impacted zones, and tool wear. Depending on the kind of dielectrics, TiO_2 and TiC may be seen in the recast layer. Because of the heat phenomena linked to the mechanism of material removal, residual strains will always develop after an EDM operation. Due to the rapid cooling rate, limited thermal conductivity, and extensive temperature generation, the machined surfaces produced by EDM procedures include fractures, a recast layer, RS, and a heat impacted zone (Pramanik et al., 2020). When comparing the results of conventional electrical discharge machining with those of a powder-mixed dielectric with ultrasonic-assisted EDSM, Khosrozadeh (Khosrozadeh, 2024) discovered that the latter resulted in a thinner heat-altered layer with far fewer microcracks. Ultrasonic waves were shown to have lower RSs in contrast to other approaches, and their stress diagram stood out when compared to the others. Taking the L_9 array into account, Karataş & Biberci (Altin Karataş & Biberci, 2023) studied the effect of the WEDM parameters on Ti-6Al-4V: cutting width (kerf), MRR, and roughness. With the use of gray analysis, the optimal parameters are: 46 V voltage, 10 kg/cm^2 dielectric fluid pressure, and 8 m/min wire feed rate. The optimal machining settings resulted in a 4.22% improvement for kerf, a 54.65% increase for MRR, a 28.77% improvement for Ra, a 31.94% improvement for Rq, and a 35.24% improvement for Rz. The voltage's impact rate was 72.18% according to the ANOVA findings. It should be noted that there was no statistically significant influence on wire feed rate or dielectric fluid pressure. In their study of the WEDM process using Nickelvac HX, Manoj & Narendranath (Manoj & Narendranath, 2020) took into account the servo voltage, pulse on time, cutting speed override, and pulse off time. Raising the cutting speed override and pulse on-time caused the cutting rate and surface roughness to rise. Cutting rate and surface roughness were both affected by changes in the servo voltage and pulse off time. Because it regulates product quality, response surface optimization was used to optimize cutting rate and surface roughness.

At a low concentration of 0.1 M H_2SO_4 aqueous acidic electrolyte, Besekar & Bhattacharyya (Besekar & Bhattacharyya, 2022) studied the impacts of salient parameters on NiTinol micro-slit machining. They used low-frequency PZT vibration-assisted axial nozzle jet flushing and found that the effect had a limited amplitude. The slit width became wider as the electrolyte flow rate rose, a consequence of both the high flow pressure and the rising temperature, which led to a rise in the current flow density in the machining zone. The PZT tool vibration during milling had a low frequency and modest amplitude, which greatly decreased the slit width. It was also noted that when the input energy, including pulse voltage, frequency, and width, increased, the slit width also grew. In their study, Xu et al. (Xu, Xi, Zhang, et al., 2022) presents an innovative method for online workpiece height measurement in reciprocating moving wire-EDM. They do this by combining a pulse discharge state detector with discharge location detection (DLD). Stabilizing the feed rate in response to changes in workpiece height during machining is possible by modifying machining settings in accordance with the findings of workpiece height estimation. Machining efficiencies of a stepwise workpiece may be raised by 27.9% and a hollow workpiece by 74.5%, according to the verification trials of cutting two variable height workpieces. While working with Titanium Grade-5 alloy in WEDM, Sonawane & Wangikar (Ashok Sonawane & Sitaram Wangikar, 2022) suggested the PCA-TOPSIS method for optimizing multi-superiority reactions. The ideal values for the machine variables are 210 Amp, Time Off 55 μs, Gap voltage 20 V, Cable tension 2 g, and Wire feed rate 5 m/min. When looking at values for relative closeness coefficients, time is determined to be the most important component. At higher input energies, scanning electron microscopy study shows a rise in the surface inequality value (Ra = 2.47 μm) due to an excessive number of sphere-shaped droplets, large holes, and a lack of shine.

Previous studies on RS measurement in Wire Electrical Discharge Machining (WEDM) of titanium alloys have primarily focused on conventional setups with fixed parameters. These studies often reported tensile RSs due to high thermal gradients and inadequate cooling, which can adversely affect fatigue life and surface integrity (Ahmad et al., 2018; Das & Joshi, 2020). In contrast, the current study on reciprocating WEDM (RWEDM) of Titanium Grade 2 alloy explores a dynamic machining regime with optimized parameters: wire feed rate (8–12 mm/min), dielectric flow rate (5–15 g/sec), variable frequency (18–22 Hz), and discharge current (200–240 A). These tailored settings improved spark stability, heat dissipation, and surface cooling, leading to significantly lower tensile or even compressive RSs. This not only enhances dimensional accuracy but also improves fatigue resistance. Thus, the current study offers a more advanced and thermally controlled machining strategy compared to earlier WEDM-based investigations. Furthermore, the existing literature on WEDM of titanium grade 2 alloy has extensively explored process parameters, surface integrity, and material removal mechanisms. However, a significant gap remains in the comprehensive measurement and characterization of RSs induced during machining in RWEDM. While some studies discuss the presence of RSs, few provide quantitative assessments using advanced techniques like X-ray diffraction (XRD). Additionally, the influence of key WEDM parameters on RS distribution is not well understood. Addressing this gap is crucial for optimizing machining parameters to enhance fatigue life, mechanical performance, and dimensional stability of titanium grade 2 components in critical applications like aerospace and biomedical industries. The novelty of the present study is to measure the RSs developed in the machined specimen which can influence the performance and service life of the component.

2. MATERIALS AND METHODS

2.1 Material Selection

Titanium has superior corrosion resistance over stainless steel in most environments, with reduced weight. Of the four nearly pure titanium grades, Grade 2 is the alloy most often used in industrial settings due to its high ductility and corrosion resistance (Gao et al., 2024). Titanium Grade 2 is a popular choice due to its light weight and low density. Titanium Grade 2, due to its low density, is widely sought after for its light weight. Two main applications employing titanium alloys are medical and aerospace. The chemical, marine, and desalination industries may all benefit from titanium grade 2's strength and high corrosion tolerance (Marin & Lanzutti, 2023). The oil and gas sector often uses Grade 2 titanium alloy in a variety of components, including reaction and pressure vessels, heat exchangers, flue gas desulfurizing systems, tube or pipe systems, and many more. At room temperature, this grade 2 alloy acts as an alpha alloy, however it changes into a beta phase at 913 ±15°C and goes back to being an alpha phase when cooled down below (Totten, 2016).

Titanium is a reactive metal with a strong affinity towards oxygen, resulting in a particularly stable and adhesive oxide skin. Despite the metal's reactivity, the skin provides exceptional corrosion resistance. When exposed to the atmosphere, the oxide layer grows spontaneously and quickly. Among the many characteristics of grade 2 alloy are its 11.4 W/m°C thermal conductivity, 4.51 g/cc density, 103 GPa elastic modulus, 145 HV hardness, 344 MPa ultimate tensile strength, and 448 MPa yield strength (Leyens & Peters, 2006; Thirumalvalavan et al., 2025). Figure 1 shows a scanning electron micrograph, and Table 1 shows the chemical make-up of titanium grade 2 alloy along with the EDX spectrum showing the elemental composition, mainly consisting of titanium by 99.5.

Table 1. Titanium Alloy Grade 2: Composition

Element	Fe	Si	C	Ti
Composition (%)	0.088	0.050	0.002	99.86

Figure 1. SEM and EDX of Titanium Grade 2 alloy

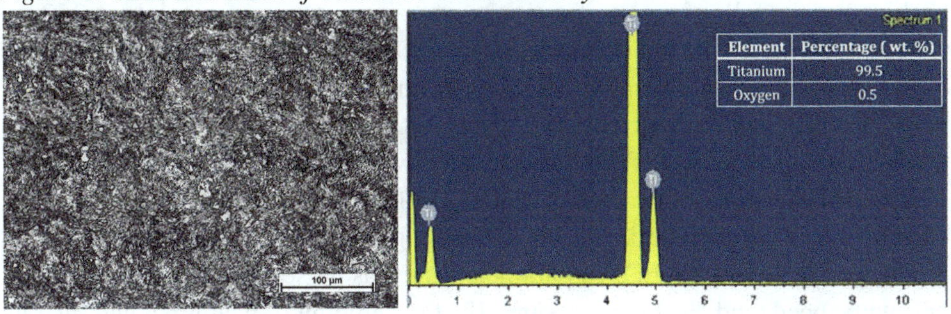

2.2 Travelling or Reciprocating WEDM

RWEDM differs from conventional WEDM primarily in the movement and configuration of the wire electrode. In conventional WEDM, the wire remains stationary in the vertical axis while the workpiece is moved in the X and Y directions to achieve the desired shape. The wire continuously feeds from a spool to maintain a fresh cutting edge, but its spatial position remains fixed during operation. In contrast, RWEDM introduces additional controlled motion in the wire itself usually in the lateral or horizontal direction along with or independent of the workpiece movement. This design enables more uniform wear distribution on the wire, improves surface finish, and enhances cutting precision, particularly when machining complex contours or thicker workpieces. Conventional WEDM continuously fed wire that travels in one direction from a supply spool to a take-up spool cutting the conductive workpiece through a series of rapid electrical discharges (sparks). This ensures that fresh wire is always present at the cutting zone, providing consistent performance, high precision, and good surface finish, though it results in higher wire consumption. In contrast, REDM reuses the same segment of wire by moving it back and forth in a reciprocating motion. This reduces wire wastage and operational costs but may lead to reduced cutting efficiency and surface quality over time due to wire wear. The mechanical setup of RWEDM includes additional guiding mechanisms and motorized drives to facilitate this wire motion, making the construction slightly more complex but beneficial in high-precision or large-scale industrial applications.

The machined surface is often subjected to a rough cut and a trim cut in a typical WEDM operation in order to attain a better surface finish. But in this research a one cut operation is done to study the actual surface finish and also the usage of single wire-molybdenum. This experiment is carried out on Sprint cut (ELPULS-40a DLX) manufactured by Electronica Machine Tool Limited, India using 0.18

mm molybdenum electrode wire as in Figure 2. The SH-300 model RWEDM CNC machine has the following specifications: a maximum workpiece weighing 350 kg, ±3° / 80 mm maximum taper/angle, Ø0.16 to Ø0.20 mm diameter of wire, accuracy of ±0.003 mm, and energy rating of 3KVA, with pure water as dielectric. The schematic illustration of conventional WEDM and RWEDM (Xu, Xi, Li, et al., 2022) is also presented in Figure 2. The relationship between machining parameters and RS is critical because these parameters directly influence the thermal and mechanical loads imparted to the material during processing. In processes like Wire Electrical Discharge Machining (WEDM), parameters such as pulse duration, discharge current, wire speed, and flushing pressure affect the amount of heat input and the rate at which it dissipates. High thermal gradients and rapid cooling can induce tensile RSs on the machined surface, potentially leading to microcracks and reduced fatigue life. Conversely, optimized parameters can promote compressive RSs, which are generally beneficial for enhancing fatigue resistance and dimensional stability. Therefore, careful control and optimization of machining parameters are essential to manage RSs and ensure superior surface integrity and material performance.

Figure 2. RWEDM setup with machined part and molybdenum wire and comparison between conventional WEDM and RWEDM

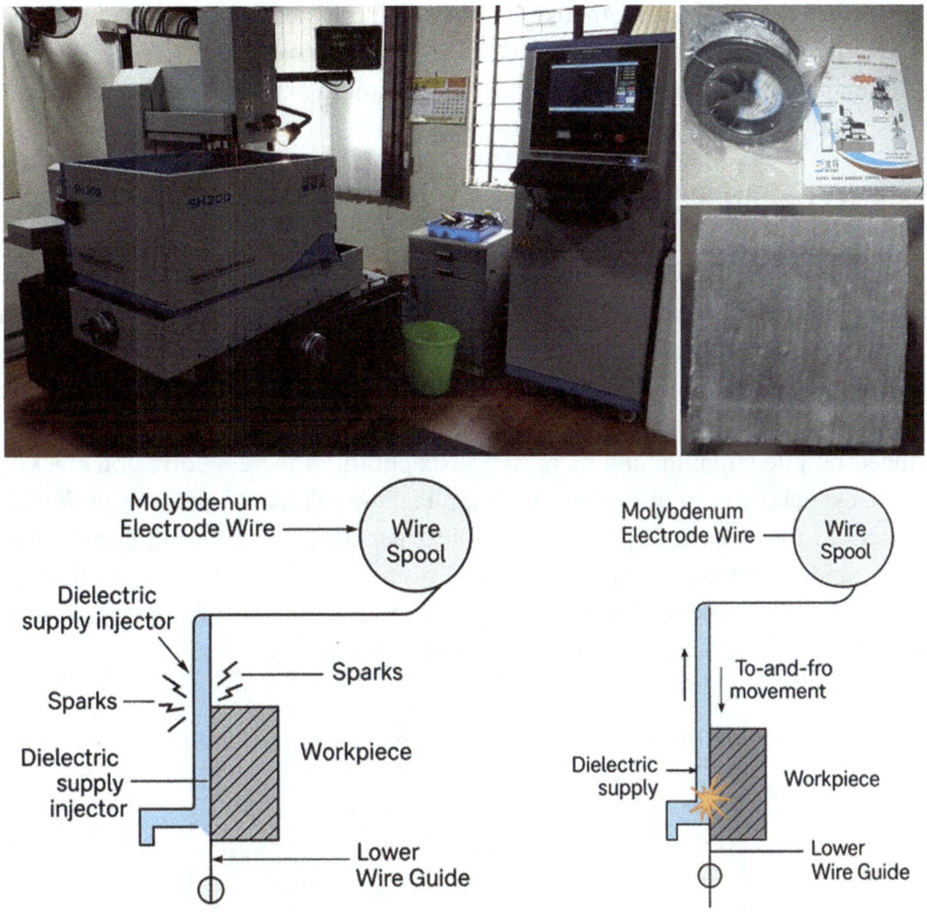

RWEDM is often preferred over conventional WEDM due to its cost efficiency and enhanced performance for thick materials. Unlike conventional WEDM, which uses a unidirectional wire that is discarded after a single pass, RWEDM employs bidirectional wire movement, allowing the same wire to be reused multiple times. This reduces wire consumption by up to 50–70%, significantly lowering operational costs. Additionally, RWEDM's controlled reciprocation improves flushing efficiency in deep cuts, enabling better machining of thick sections (e.g., aerospace components) with reduced wire breakage and improved surface finish. On the other hand, RS mapping in Grade 2 titanium is less reported primarily because of its low elastic modulus and high ductility, which make stress measurements challenging.

Unlike high-strength titanium alloys (e.g., Ti-6Al-4V), Grade 2 titanium's lower yield strength allows RSs to relax more easily during machining or testing, leading to inconsistent data. Additionally, its hexagonal close-packed (HCP) structure complicates diffraction-based methods like XRD, as it produces weaker peak intensities and overlapping reflections. Most studies focus on high-strength aerospace-grade titanium, leaving commercially pure Grade 2 titanium under-researched despite its widespread use in chemical and marine applications.

2.3 XRD based RS Measurements

Measuring RSs in WEDM machined components is crucial because WEDM is a thermal-based material removal process that induces significant tensile and thermal RSs owing to rapid heating and cooling cycles (Bhattacharya et al., 2019). These RSs can influence the machinable part's mechanical characteristics and functionality directly. In titanium alloys, excessive tensile RSs can lead to crack initiation, reduced fatigue strength, and increased susceptibility to stress corrosion cracking (SCC), especially in high-performance applications. Thus, RS analysis in WEDM is essential for ensuring optimal mechanical performance, preventing premature failure, and enhancing component durability in critical engineering applications such as aerospace, medical implants, and turbine components (Tabatabaeian et al., 2022). Figure 3 showcases the schematic of RS generation.

Figure 3. Schematic illustration of RS generation (Outeiro, 2018)

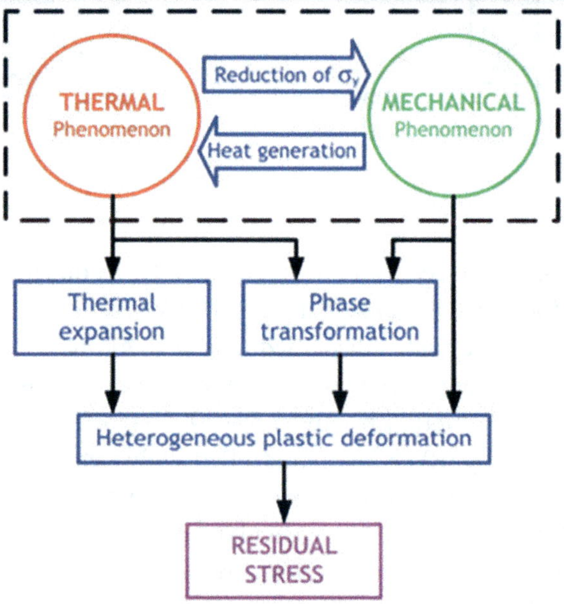

In X-ray diffraction (XRD) based RS measurement is a non-destructive technique that analyzes elastic strain in crystalline materials by measuring shifts in Bragg diffraction angles. In analysis, RS in a crystalline material is primarily determined using the $\sin^2\psi$ method, where ψ is the tilt angle of the sample. When a material is under stress, the atomic planes are distorted, altering the interplanar spacing (d-spacing), which is measured at different ψ angles. This change causes a shift in the 2θ angle according to Bragg's Law. By plotting the measured d-spacing against $\sin^2\psi$, the slope of the resulting line indicates the magnitude and nature (tensile or compressive) of the RS. Key advantages include non-destructiveness and phase-specific stress analysis, though it requires smooth surfaces and is limited to crystalline materials. Additionally, FWHM (Full Width at Half Maximum) and peak breadth are indicators of microstrain and crystal defects; broader peaks often signify higher internal stresses or strain gradients. Peak intensity may also vary under stress, though it is more influenced by factors like texture or grain orientation than by stress alone. Thus, RS correlates directly with changes in d-spacing and 2θ, and indirectly with peak shape and intensity in XRD data (Sarmast et al., 2023). Diffractometers designed to detect RS are essentially powder Diffractometers. Compared to a standard powder diffractometer, a RS diffractometer features multiple rotating axes. The sample may be tilted according to the requirements of the $\sin^2\psi$ approach. In particular, with RS diffractometers, 2θ and omega may be altered independently, or de-coupled. Omega is frequently physically set in a powder diffractometer at half the value of 2-theta (Dolabella et al., 2022). For RS assessment, the setup of a laboratory-type goniometer sensing a large 2θ angle is shown in Figure 4.

Figure 4. Directions of rotation and angles utilized to quantify RS (Fitzpatrick et al., 2002)

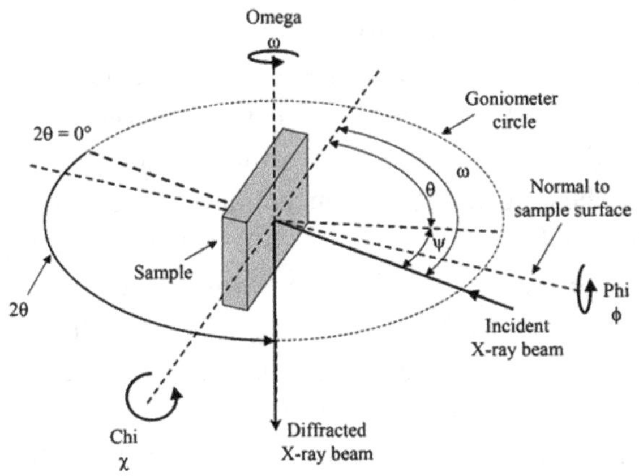

RSs act like hidden fingerprints left behind during manufacturing, silently shaping how everyday objects withstand wear and tear. In car suspension springs, beneficial compressive stresses introduced through shot peening help the metal endure millions of road bumps without cracking, effectively doubling its lifespan. Conversely, the welded joints in bicycle frames often harbor harmful tensile stresses; these combine with pedal forces to accelerate tiny cracks, explaining why fractures often appear near weld spots after years of use. Kitchen knives demonstrate this duality: honing creates compressive stresses at the edge for lasting sharpness, but improper grinding can leave tensile stresses that cause micro-chips over time. Even smartphone frames, forged from aluminum alloys, rely on carefully balanced RSs to survive daily drops without bending or cracking. Engineers combat these effects by using heat treatments (like annealing oven racks) or precision processes (laser-cutting surgical tools), proving that controlling these invisible forces determines whether everyday items become durable workhorses or premature failures.

3. RESULTS AND DISCUSSION

Machining on RWEDM is performed with varying parameters viz., 8-12 mm/min wire feed rate (WFR), 5-15 g/sec of dielectric flow rate (DFR), variable frequency (VF) of 18-22 Hz, and discharge current (DI) of 200-240 A. Experiments are planned and executed based on L_{27} orthogonal array of Taguchi's design and samples are tested to assess the amount of stress that has been maintained inside it (Anoop Johny & Thiagarajan, 2022; Pragadish et al., 2023). Simultaneously the outputs such as KW, SR, and MRR and then correlated with the measured RS. In RWEDM, specific machining settings significantly influence the RS levels due to the thermal and mechanical effects they impose on the workpiece. High discharge energy (from increased pulse duration or current) generates intense localized heating, followed by rapid cooling, which can cause steep thermal gradients and lead to high tensile RS on the surface. Wire feed rate affects the amount of fresh wire exposed to the cut; lower feed rates may result in worn wire generating inconsistent sparks, increasing localized heating and stress concentration. The dielectric flow rate plays a key role in cooling and debris removal insufficient flow can trap heat and debris, further increasing thermal stress. Variable frequency control affects the timing of discharges and cooling intervals, impacting how much heat builds up between pulses. Faster frequencies with shorter cooling times may result in higher surface stresses. Ultimately, fine-tuning these settings helps control heat input and cooling rates, which directly influences the type and magnitude of RSs formed in RWEDM.

3.1 Condition 1

The titanium grade 2 alloy is machined with a WFR of 8 mm/min, DFR of 5 g/sec, VF of 18 Hz and DI of 200 A that produces a SR of 4.77 μm, KW of 0.373 mm and MRR of 1.15 g/min. From XRD setup the RS measured on the machined specimen is +206.5 ± 18.9 MPa and shear stress of -33.5 ± 9.1 MPa. RSs and shear stress are interconnected through material deformation and thermal effects. During WEDM, high-temperature sparks cause localized melting and rapid cooling, resulting in residual strains on the machined surface that are either compressive or tensile. The mechanical characteristics of the material may be affected by these RSs, including its response to shear stress (Liu & Guo, 2016). Compressive RSs typically enhance shear strength by counteracting crack propagation, as RSs may weaken the material under shear loading. Additionally, variations in machining parameters, affect the RS magnitude and distribution, thereby influencing the material's shear stress resistance. The RSs are analyzed using different parameters (Childerhouse et al., 2022). The D-spacing (interplanar spacing, d) represents the distance between atomic planes in the crystal lattice, and variations in d-spacing indicate RSs; compressive stress decreases d-spacing, while tensile stress increases it. 2θ (diffraction angle) is the angle at which constructive interference occurs, shifting due to strain in the material (Almer & Winholtz, 2008). FWHM measures the peak broadening, which is influenced by factors like size of grain, dislocation density, and microstrain; broader peaks suggest higher microstrain or smaller grains (Mahrous et al., 2024; Vashista & Paul, 2012). Peak breadth is related to FWHM but considers the overall width of the peak profile, reflecting plastic deformation and defect structures. Intensity refers to the peak height and area, influenced by crystal orientation, phase composition, and material texture. To analyze these values, d-spacing variations are used to calculate RSs using Bragg's law and stress equations, while FWHM and peak breadth provide insights into material hardening and defects (Schubnell et al., 2023). Shifts in 2θ help quantify elastic strain, and intensity variations indicate preferred orientations or phase changes due to machining effects (Li et al., 2024).

The DSpacing, 2θ, FWHM, breadth and intensity are presented in Table 2. The peak profile is built from the observed data; the fitting of peak profile for specimen 1 is given in 2θ vs. Intensity plots as shown in Figure 5 separating K-α_1 and K-α_2. Figure 6 presents the relationship between the DSpacing and FWHM obtained during RS measurement in condition 1. It is observed that with higher values of Sin^{2}psi, DSpacing values tends to increase whereas FWHM tends to lower. The Dspacing value is around 0.815 Å which represent that RS exists in the machined sample.

Table 2. Outcomes of RS measurement in condition 1

Sin²psi	DSpacing	2Theta	FWHM	Breadth	Intensity
0.0109	0.814911	142.18	4.309	4.26	333.97
0.0000	0.815095	142.10	4.097	4.06	363.30
0.0090	0.815235	142.04	3.953	3.97	390.32
0.0428	0.815655	141.87	3.047	3.56	434.27
0.0965	0.815789	141.82	2.754	3.49	461.18
0.1060	0.815792	141.82	2.716	3.44	466.75
0.1159	0.815965	141.75	2.691	3.39	464.48
0.1930	0.816096	141.69	2.379	3.17	478.47
0.2895	0.816310	141.61	2.097	2.92	503.56
0.3790	0.816412	141.57	2.068	2.80	488.92
0.4826	0.816435	141.56	1.971	2.69	450.48

Figure 5. Plot of 2θ vs. Intensity for condition 1

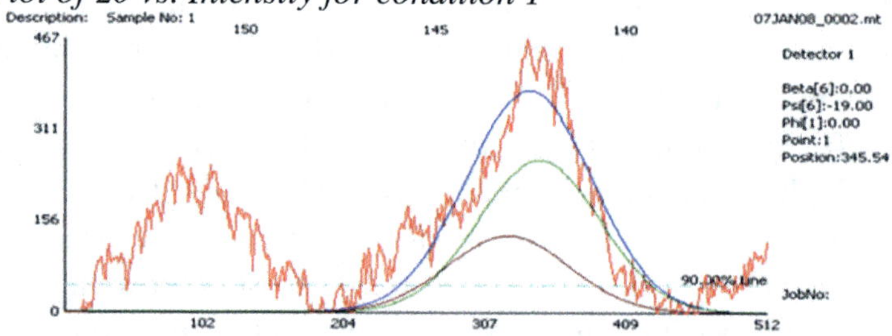

Figure 6. Sin²psi vs DSpacing and FWHM plots for condition 1

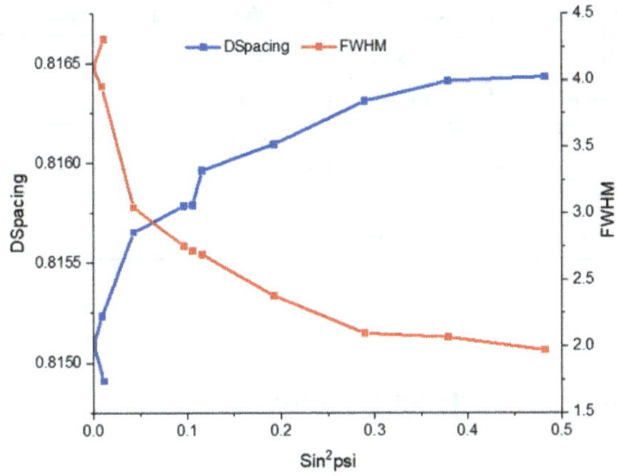

3.2 Condition 2

Machining with a WFR of 8 mm/min, DFR of 5 g/sec, VF of 22 Hz and DI of 240 A that produces a SR of 4.03 μm, KW of 0.397 mm and MRR of 0.97 g/min. The RS on the machined surface is measured as +444.5 ± 26.0 MPa and shear stress of -63.8 ± 12.6 MPa. The DSpacing, 2θ, FWHM, breadth and intensity are depicted in Table 3. Figure 7 shows the fitting of peak profile for condition 2 is given in 2θ vs. Intensity plot showing the profiles of breadth and FWHM. The relationship between the DSpacing and FWHM obtained during RS measurement in condition 2 as in Figure 8. It is observed that with higher values of Sin^2psi, DSpacing values tends to increase whereas FWHM tends to lower (Yazawa et al., 2020). The Dspacing value is around 0.817 Å which represent that RS exists in the machined sample.

Table 3. Outcomes of RS measurement in condition 2

Sin²psi	DSpacing	2Theta	FWHM	Breadth	Intensity
0.0109	0.815113	142.09	5.720	5.02	296.95
0.0000	0.815348	142.00	5.468	4.89	324.86
0.0090	0.815372	141.99	5.450	4.81	330.72
0.0428	0.815467	141.95	5.370	4.80	379.19
0.0965	0.815794	141.82	4.910	4.37	416.15
0.1060	0.816172	141.66	4.571	4.25	398.31

continued on following page

Table 3. Continued

Sin²psi	DSpacing	2Theta	FWHM	Breadth	Intensity
0.1159	0.816171	141.66	4.725	4.35	398.17
0.1930	0.816998	141.33	2.818	3.69	439.45
0.2895	0.817328	141.20	2.617	3.61	448.00
0.3790	0.817690	141.06	2.384	3.29	454.58
0.4826	0.818072	140.90	2.305	3.09	447.14

Figure 7. Plot of 2θ vs. Intensity for condition 2

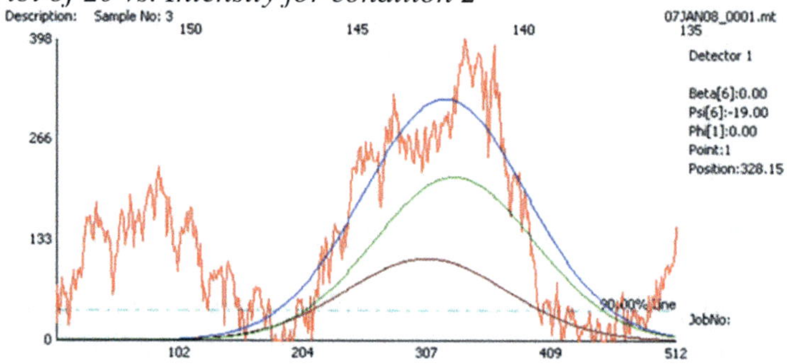

Figure 8. Sin²psi vs DSpacing and FWHM plots for condition 2

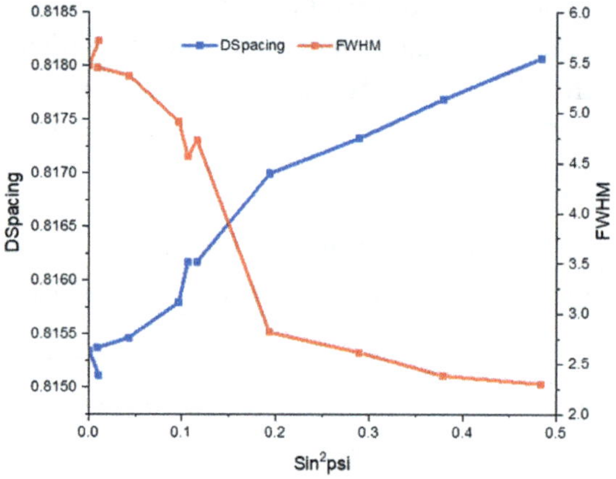

3.3 Condition 3

The condition machined with a WFR of 8 mm/min, DFR of 10 g/sec, VF of 18 Hz and DI of 200 A that produces a SR of 4.25 μm, KW of 0.378 mm and MRR of 1.04 g/min. The RS measured on the machined condition is +376.8 ± 18.7 MPa and shear stress of -46.1 ± 9.1 MPa. The DSpacing, 2θ, FWHM, breadth and intensity are presented in Table 4. Figure 9 shows the fitting of peak profile for condition 3 is given in 2θ vs. Intensity plot showing the profiles of breadth and FWHM. Figure 10 presents the relationship between the DSpacing and FWHM obtained during RS measurement in condition 3. It is observed that with higher values of Sin^2psi, DSpacing values tends to increase whereas FWHM tends to lower. The Dspacing value is around 0.817 Å which represent that RS exists in the machined sample.

Table 4. Outcomes of RS measurement in condition 3

Sin^2psi	DSpacing	2Theta	FWHM	Breadth	Intensity
0.0109	0.815072	142.11	4.546	4.62	294.06
0.0000	0.815193	142.06	4.503	4.53	332.14
0.0090	0.815339	142.00	4.598	4.41	351.58
0.0428	0.815371	141.99	4.222	4.20	395.52
0.0965	0.815932	141.76	3.321	3.81	417.37
0.1060	0.816258	141.63	3.767	3.78	425.06
0.1159	0.816101	141.69	3.077	3.77	423.64
0.1930	0.816553	141.51	2.683	3.55	448.02
0.2895	0.816724	141.44	2.313	3.26	464.80
0.3790	0.817264	141.22	2.024	2.97	471.55
0.4826	0.817517	141.12	1.985	2.80	440.40

Figure 9. Plot of 2θ vs. Intensity for condition 3

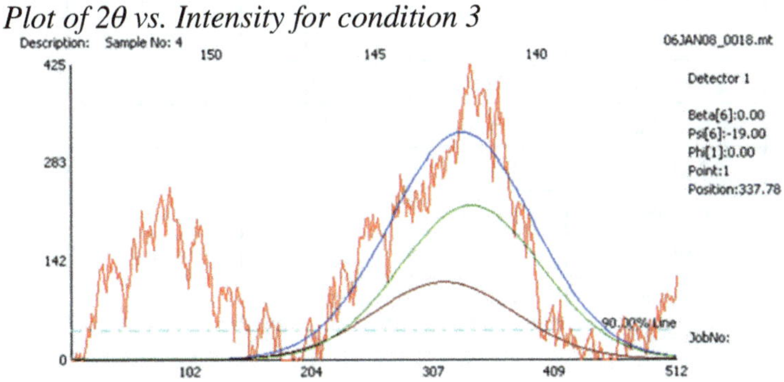

Figure 10. Sin²psi vs DSpacing and FWHM plots for condition 3

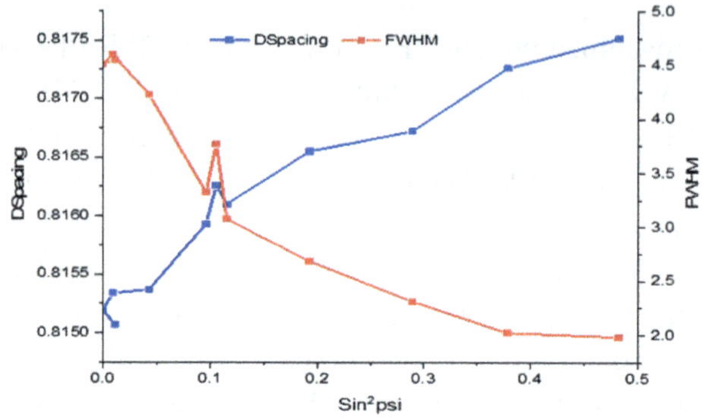

3.4 Condition 4

The specimen machined with a WFR of 8 mm/min, DFR of 15 g/sec, VF of 18 Hz and DI of 200 A that produces a SR of 3.68 μm, KW of 0.387 mm and MRR of 0.94 g/min. The RS is measured as +149.5 ± 22.2 MPa and shear stress of -20.1 ± 10.8 MPa. The DSpacing, 2θ, FWHM, breadth and intensity are depicted in Table 5. Figure 11 shows the fitting of peak profile for condition 4 is given in 2θ vs. Intensity plot showing the profiles of breadth and FWHM. Figure 12 presents the relationship between the DSpacing and FWHM obtained during RS measurement in condition 4. It is observed that with higher values of Sin²psi, DSpacing values tends to increase whereas FWHM tends to lower. The Dspacing value is around 0.816 Å which represent that RS exists in the machined sample.

Table 5. Outcomes of RS measurement in condition 4

Sin²psi	DSpacing	2Theta	FWHM	Breadth	Intensity
0.0109	0.815015	142.13	4.110	3.86	275.85
0.0000	0.815111	142.09	3.891	3.83	317.41
0.0090	0.815167	142.07	3.101	3.59	350.47
0.0428	0.815522	141.93	2.737	3.37	400.30
0.0965	0.815743	141.84	2.470	3.18	421.30
0.1060	0.815807	141.81	2.463	3.04	431.57
0.1159	0.815802	141.81	2.388	3.08	439.18
0.1930	0.815907	141.77	2.067	2.74	470.57

continued on following page

Table 5. Continued

Sin²psi	DSpacing	2Theta	FWHM	Breadth	Intensity
0.2895	0.816049	141.71	1.951	2.56	468.16
0.3790	0.816028	141.72	1.880	2.47	457.38
0.4826	0.816096	141.69	1.742	2.32	423.55

Figure 11. Plot of 2θ vs. Intensity for condition 4

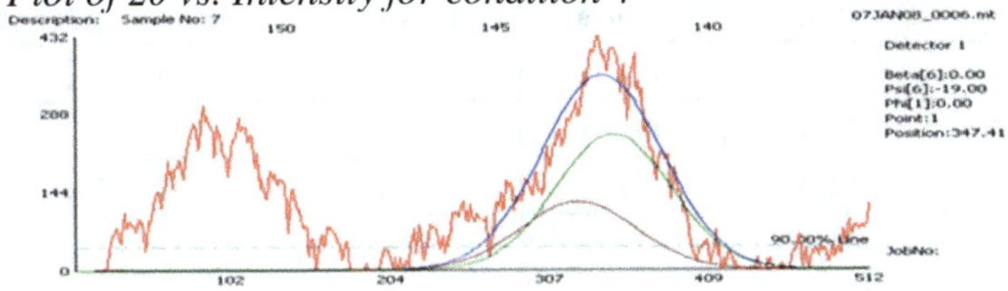

Figure 12. Sin²psi vs DSpacing and FWHM plots for condition 4

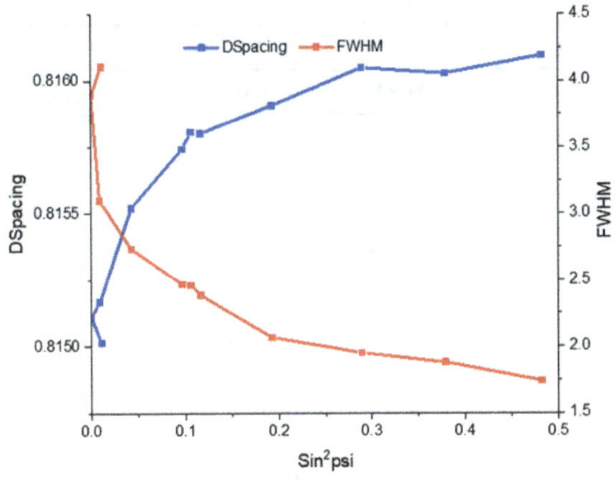

3.5 Condition 5

The machined condition subjected to RS measurement machined with a WFR of 8 mm/min, DFR of 15 g/sec, VF of 22 Hz and DI of 240 A produces a SR of 2.30 μm, KW of 0.381 mm and MRR of 0.98 g/min. From XRD setup the RS is measured as +235.7±20.4 MPa and shear stress of -47.0±9.9 MPa. The DSpacing, 2θ, FWHM) breadth and intensity are depicted in Table 6. Figure 13 presents the fitting of peak profile for condition 5 is given in 2θ vs. Intensity plot showing the profiles of breadth and FWHM. Figure 14 presents the relationship between the DSpacing and FWHM obtained during RS measurement in condition 5. It is found that with Sin^2psi higher values, DSpacing values tends to increase whereas FWHM tends to lower. The Dspacing value is around 0.817 Å which represent that RS exists in the machined sample.

Table 6. Outcomes of RS measurement in condition 5

Sin^2psi	DSpacing	2Theta	FWHM	Breadth	Intensity
0.0109	0.815137	142.08	4.248	4.30	313.47
0.0000	0.815154	142.08	4.243	4.29	352.61
0.0090	0.815273	142.03	3.689	4.04	376.51
0.0428	0.815544	141.92	3.088	3.73	403.70
0.0965	0.815914	141.77	2.588	3.35	437.63
0.1060	0.815735	141.84	2.474	3.27	443.48
0.1159	0.815887	141.78	2.453	3.35	455.42
0.1930	0.816328	141.60	2.221	2.97	464.87
0.2895	0.816548	141.51	2.092	2.76	494.53
0.3790	0.816692	141.45	2.127	2.85	518.82
0.4826	0.816894	141.37	2.081	2.62	482.04

Figure 13. Plot of 2θ vs. Intensity for condition 5

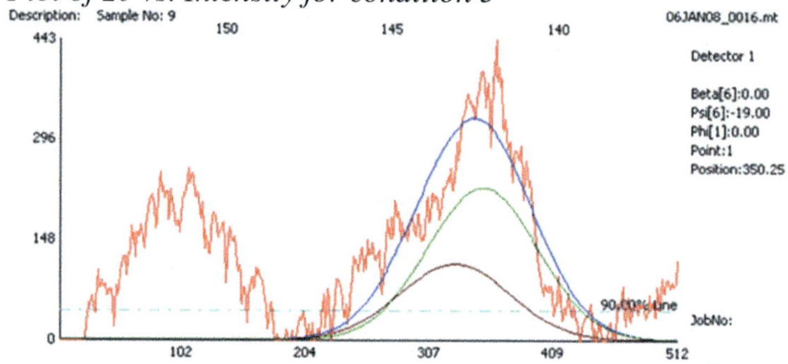

Figure 14. Sin²psi vs DSpacing and FWHM plots for condition 5

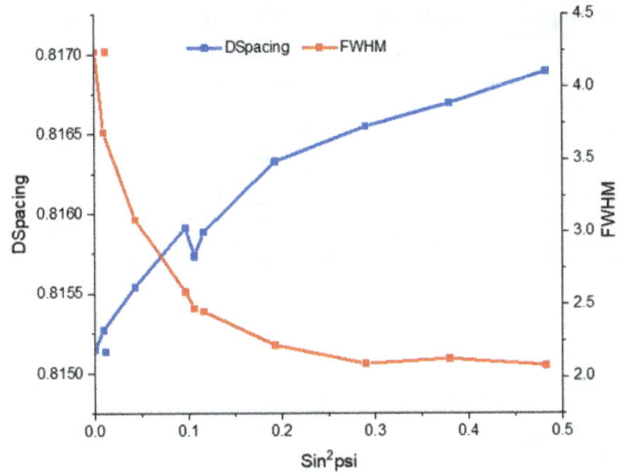

3.6 Condition 6

The specimen machined with a WFR of 10 mm/min, DFR of 5 g/sec, VF of 22 Hz and DI of 220 A that produces a SR of 4.68 μm, KW of 0.402 mm and MRR of 1.08 g/min. From XRD setup the RS measured is +479.5 ± 34.3 MPa and shear stress of -28.8 ± 16.6 MPa. The DSpacing, 2θ, FWHM, breadth and intensity are depicted in Table 7. Figure 15 illustrates the fitting of peak profile for condition 6 is given in 2θ vs. Intensity plot showing the profiles of breadth and FWHM. Figure 16 presents the relationship between the DSpacing and FWHM obtained during RS measurement in condition 6. It is observed that with higher values of Sin²psi, DSpacing values tends to increase whereas FWHM tends to lower. The Dspacing value is around 0.816 Å which represent that RS exists in the machined sample.

Table 7. Outcomes of RS measurement in condition 6

Sin²psi	DSpacing	2Theta	FWHM	Breadth	Intensity
0.0109	0.815201	142.06	5.852	5.12	263.48
0.0000	0.815667	141.87	5.486	4.95	281.13
0.0090	0.815202	142.06	5.194	4.65	351.57
0.0428	0.815324	142.01	5.164	4.79	392.96
0.0965	0.816093	141.69	4.899	4.33	397.71
0.1060	0.816124	141.68	4.848	4.22	396.86

continued on following page

Table 7. Continued

Sin²psi	DSpacing	2Theta	FWHM	Breadth	Intensity
0.1159	0.816082	141.70	4.924	4.25	391.64
0.1930	0.816307	141.61	4.681	4.08	428.30
0.2895	0.816692	141.45	2.510	3.74	459.17
0.3790	0.817441	141.15	2.215	3.39	461.49
0.4826	0.818049	140.91	2.049	3.07	458.56

Figure 15. Plot of 2θ vs. Intensity for condition 6

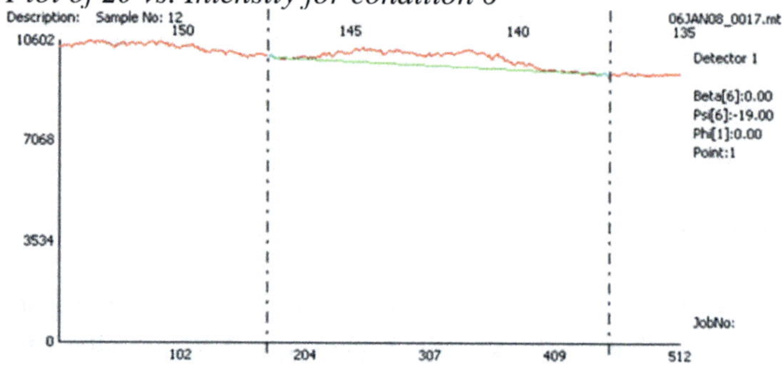

Figure 16. Sin²psi vs DSpacing and FWHM plots for condition 6

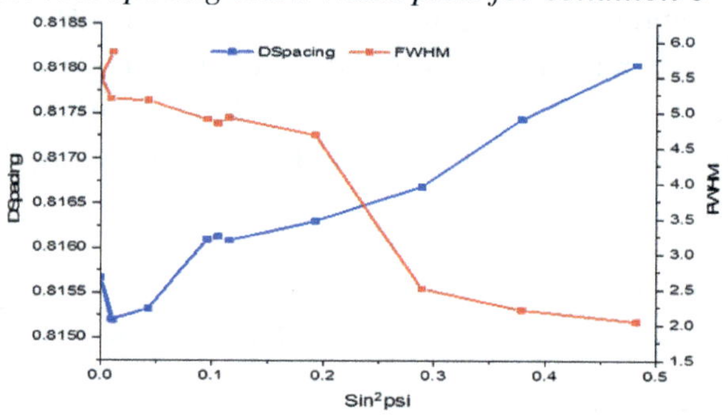

3.7 Condition 7

The machined specimen with a WFR of 10 mm/min, DFR of 10 g/sec, VF of 18 Hz and DI of 240 A that produces a SR of 4.52 μm, KW of 0.373 mm and MRR of

0.81 g/min. From XRD setup the RS is measured as +409.4±17.8 MPa and shear stress of -42.0±8.6 MPa. The DSpacing, 2θ, FWHM, breadth and intensity are depicted in Table 8. Figure 17 presents the fitting of peak profile for condition 7 is given in 2θ vs. Intensity plot showing the profiles of breadth and FWHM. Figure 18 presents the relationship between the DSpacing and FWHM obtained during RS measurement in condition 7. It is observed that with higher values of Sin^2psi, DSpacing values tends to increase whereas FWHM tends to lower. The Dspacing value is around 0.816 Å which represent that RS exists in the machined sample.

Table 8. Outcomes of RS measurement in condition 7

Sin^2psi	DSpacing	2Theta	FWHM	Breadth	Intensity
0.0109	0.815054	142.12	5.369	4.84	283.40
0.0000	0.815307	142.01	5.063	4.64	312.17
0.0090	0.815374	141.99	4.552	4.51	338.99
0.0428	0.815602	141.89	4.481	4.39	386.24
0.0965	0.816123	141.68	4.400	4.27	387.61
0.1060	0.816157	141.67	4.100	3.94	408.90
0.1159	0.815952	141.75	4.383	4.13	412.11
0.1930	0.816173	141.66	2.712	3.83	426.95
0.2895	0.816725	141.44	2.420	3.52	451.79
0.3790	0.817439	141.15	2.187	3.20	450.12
0.4826	0.817726	141.04	2.059	2.94	444.81

Figure 17. Plot of 2θ vs. Intensity for condition 7

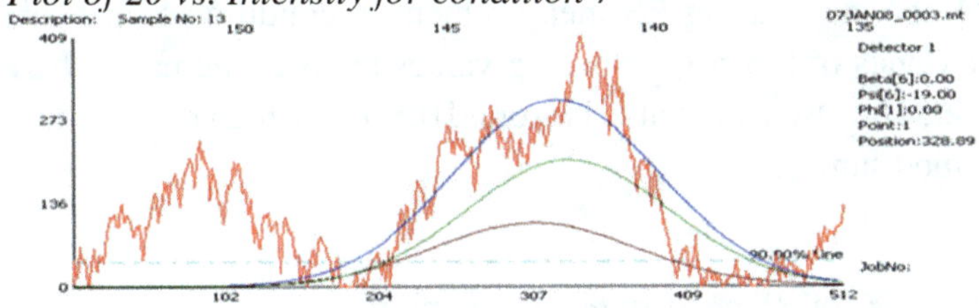

Figure 18. Sin²psi vs DSpacing and FWHM plots for condition 7

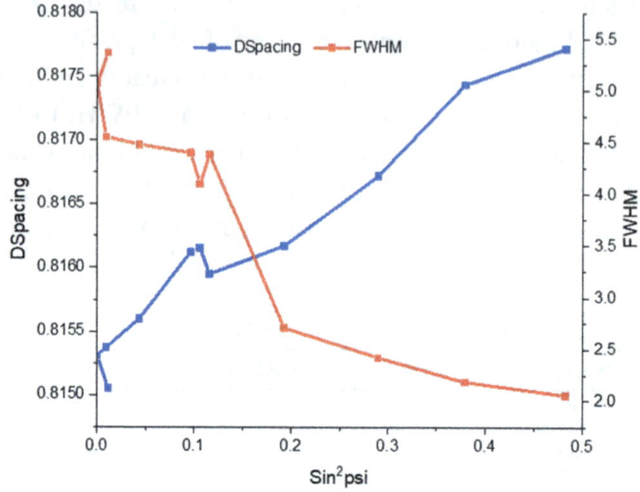

3.8 Condition 8

The specimen machined with a WFR of 10 mm/min, DFR of 15 g/sec, VF of 18 Hz and DI of 240 A that produces a SR of 4.32 μm, KW of 0.365 mm and MRR of 0.87 g/min. From XRD setup the RS measured on the machined condition is measured as +352.3 ± 27.4 MPa and shear Stress of -24.7 ± 13.3 MPa. The DSpacing, 2θ, FWHM, breadth and intensity are depicted in Table 9. Figure 19 presents the fitting of peak profile for condition 8 is given in 2θ vs. Intensity plot showing the profiles of breadth and FWHM. Figure 20 presents the relationship between the DSpacing and FWHM obtained during RS measurement in condition 8. It is observed that with higher values of Sin²psi, DSpacing values tends to increase whereas FWHM tends to lower. The Dspacing value is around 0.816 Å which represent that RS exists in the machined sample.

Table 9. Outcomes of RS measurement in condition 8

Sin²psi	DSpacing	2Theta	FWHM	Breadth	Intensity
0.0109	0.815461	141.95	4.119	4.05	248.85
0.0000	0.815091	142.10	4.248	4.28	275.02
0.0090	0.815134	142.08	4.320	4.23	304.68
0.0428	0.815566	141.91	4.171	3.97	346.74

continued on following page

Table 9. Continued

Sin²psi	DSpacing	2Theta	FWHM	Breadth	Intensity
0.0965	0.815835	141.80	4.103	3.86	366.76
0.1060	0.815933	141.76	2.972	3.84	374.15
0.1159	0.815883	141.78	2.945	3.83	362.32
0.1930	0.816361	141.59	2.444	3.36	393.48
0.2895	0.816695	141.45	2.201	3.23	417.93
0.3790	0.816901	141.37	2.224	3.14	429.95
0.4826	0.817220	141.24	1.924	2.83	421.99

Figure 19. Plot of 2θ vs. Intensity for condition 8

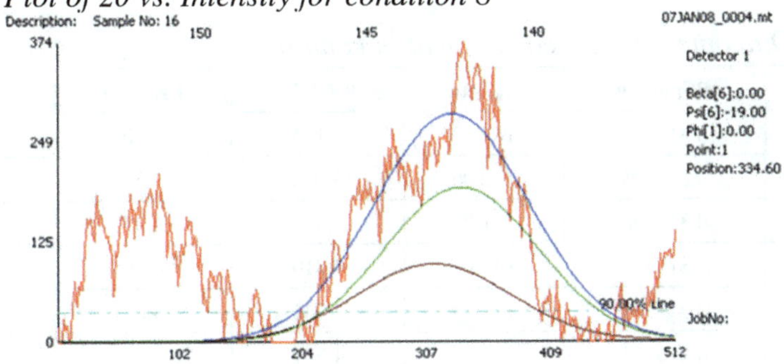

Figure 20. Sin²psi vs DSpacing and FWHM plots for condition 8

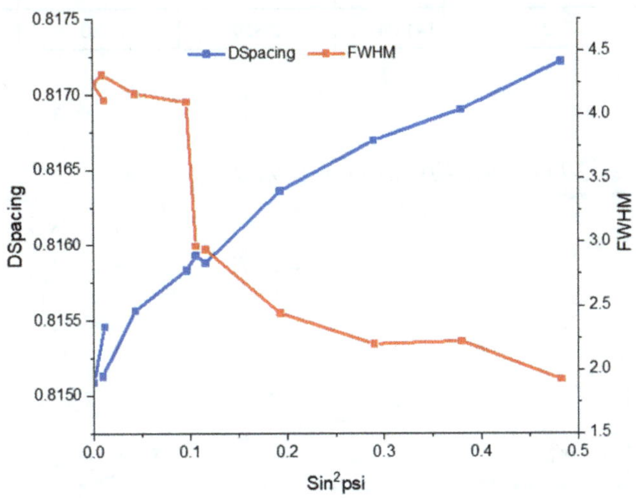

3.9 Condition 9

The machined surface with a WFR of 12 mm/min, DFR of 15 g/sec, VF of 20 Hz and DI of 240 A that produces a SR of 5.86 μm, KW of 0.375 mm and MRR of 1.16 g/min. From XRD setup the RS is measured as +483.2 ± 27.5 MPa and shear Stress of -17.4 ± 13.3 MPa. The DSpacing, 2θ, FWHM, breadth and intensity are depicted in Table 10. Figure 21 presents the fitting of peak profile for condition 9 is given in 2θ vs. Intensity plot showing the profiles of breadth and FWHM. Figure 22 presents the relationship between the DSpacing and FWHM obtained during RS measurement in condition 9. It is observed that with higher values of Sin^2psi, DSpacing values tends to increase whereas FWHM tends to lower. The Dspacing value is around 0.817 Å which represent that RS exists in the machined sample.

Table 10. Outcomes of RS measurement in condition 9

Sin^2psi	DSpacing	2Theta	FWHM	Breadth	Intensity
0.0109	0.814865	142.19	7.125	4.60	219.44
0.0000	0.815205	142.06	5.045	4.53	270.22
0.0090	0.815545	141.92	4.912	4.37	289.27
0.0428	0.815473	141.95	4.810	4.26	331.27
0.0965	0.815991	141.74	4.941	4.25	326.90
0.1060	0.816157	141.67	4.821	4.26	329.33
0.1159	0.816337	141.60	4.972	4.33	323.84
0.1930	0.816235	141.64	4.804	4.18	338.40
0.2895	0.816798	141.41	4.673	4.06	379.88
0.3790	0.817213	141.24	2.778	3.79	415.03
0.4826	0.817727	141.04	2.850	3.80	390.29

Figure 21. Plot of 2θ vs. Intensity for condition 9

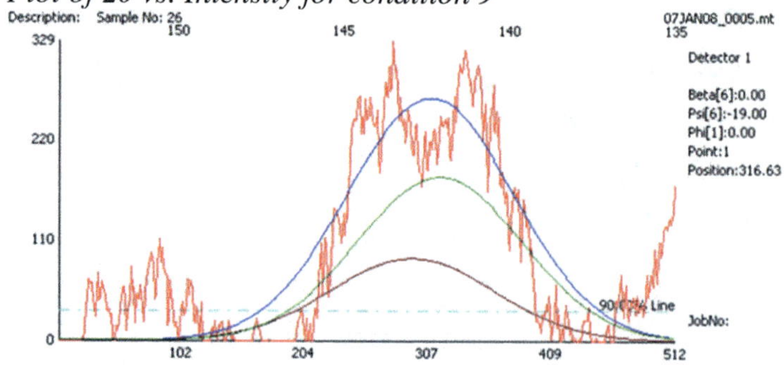

Figure 22. Sin²psi vs DSpacing and FWHM plots for condition 9

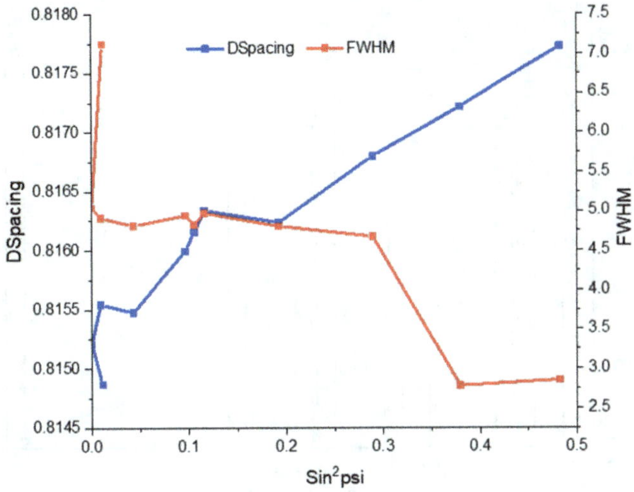

3.10 RS and Shear Stress on Machined Specimens

During XRD, the shear and RS present inside the machined specimen is measured for different specimens machined under different input machining conditions. Figure 23 presents the RS and Figure 24 presents the shear stress on the the machined specimens. It is observed that a considerable quantity of stresses is available in the specimens due to the material removal mechanism.

Figure 23. RS value for different specimens

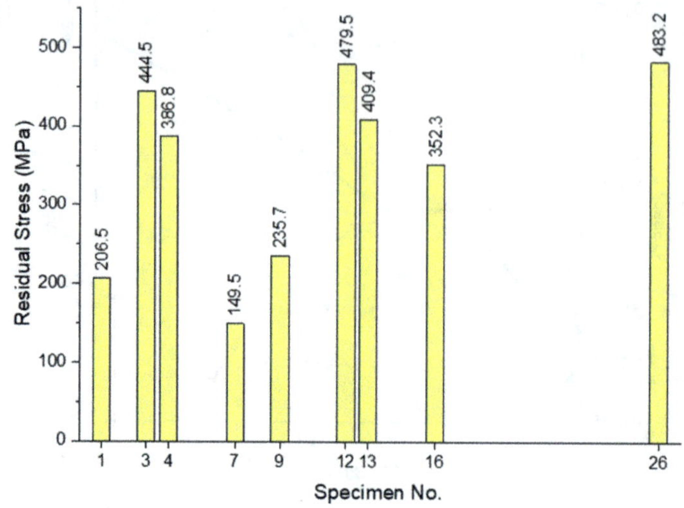

Figure 24. Shear stress value for different specimens

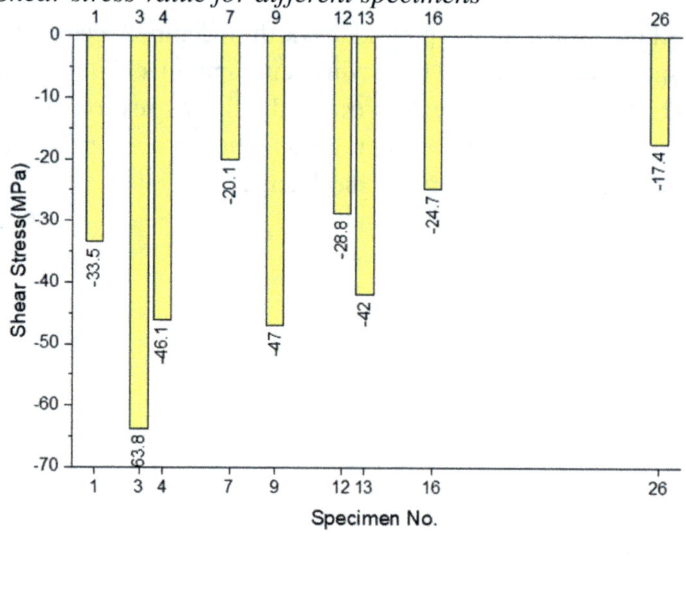

3.11 Input Parameter Effects on RSs

The effect of machining settings on the generated RS on the machined specimen is given as contour plot (2D plot) relating two input parameters and one output response in Figure 25. Observations show that when WFR in RWEDM and DFR is higher and higher VF and higher DI, RS tends to rise due to higher generation of spark density which correspondingly removes higher quantity of material from the workpiece (Ali et al., 2024). A higher WFR ensures a fresh cutting surface, reducing wire wear and stabilizing discharge conditions, which can minimize surface defects and lower tensile RSs (Salvati Korsunsky, 2020). The DFR is really important in flushing debris and controlling the cooling rate; inadequate flow may lead to excessive heat accumulation, increasing tensile stresses. The DI directly affects discharge energy, where higher currents generate intense localized heating, leading to deeper recast layers and increased tensile stresses (Hegde et al., 2024). Similarly, VF impacts spark distribution and energy input; lower frequencies may cause inconsistent discharges and localized overheating, promoting stress concentration, while higher frequencies ensure stable machining, reducing RS variations (Sharman et al., 2006).

Figure 25. Contour plot of RS vs. considered input parameters

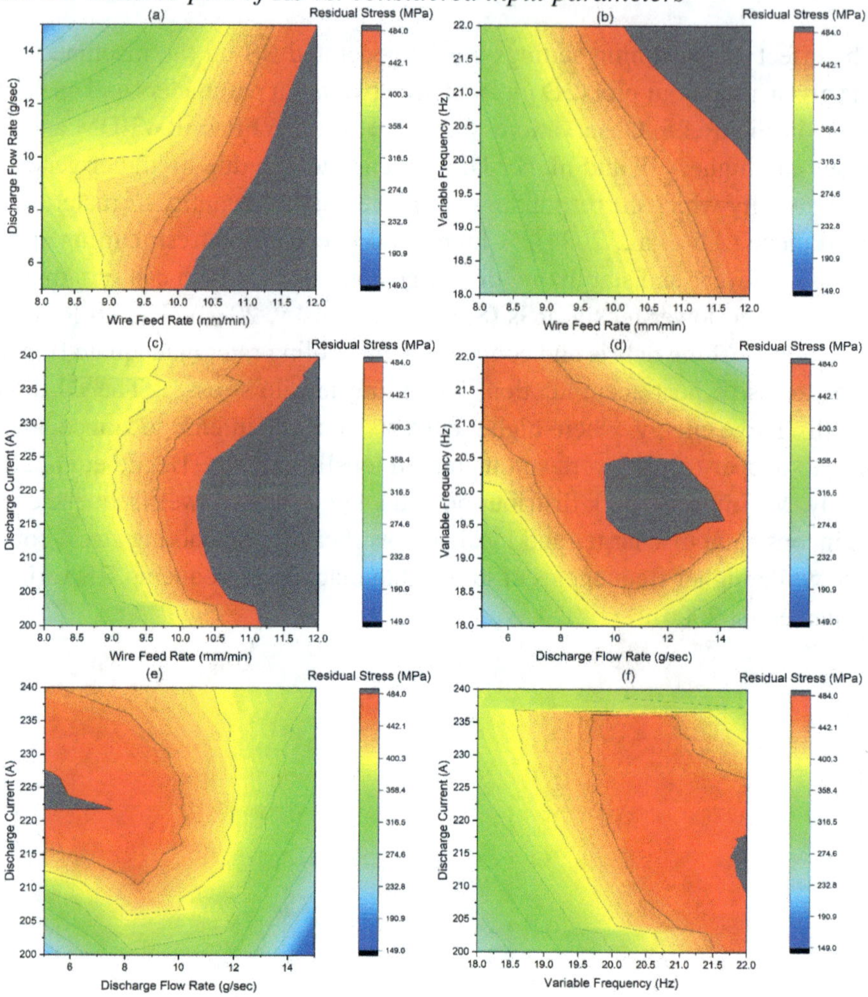

The contour plots in the image show that RS varies significantly by up to approximately 225% across the explored range of input parameters, from a minimum of 149 MPa to a maximum of 484 MPa. This variation is influenced most strongly by DI, WFR, and DFR, while VF also contributes but with less drastic effect. For instance, in Figure 25(c), RS increases from 160 MPa to over 450 MPa as DI and WFR increase, accounting for a variation of nearly 183%. Similarly, in Figure 25(a), changing the DFR and WFR leads to a variation of more than 200% in RS. On the other hand, in Figure 25(b) and (f), with variation in VF, the change in RS is more moderate ranging between 149 MPa to 400 MPa indicating a 168% variation. A key observation from Figure 25(d) is the existence of an optimal parameter window, where RS is minimized; deviation from this window causes a sharp rise in stress. Overall, controlling WFR and DI is crucial for managing RS, while precise tuning of DFR and VF can help in further stress optimization.

4. CONCLUSION

The RS induced inside the machined specimens are determined using XRD technique and analysis is done. The results of the experiment demonstrate that the MRR decreases as the WFR increases, and that the SR and KW increase when the DFR between the workpiece and electrode increases, while the MRR rises and the KW and SR fall. Raising the VF results in greater removal of material from the material as well as greater KW, but an improved surface quality. Impact of DI on the outputs presents that rising the DI lowers the KW, SR, and MRR. With increase in input parameter settings the RS tends to rise due to higher generation of spark density which correspondingly removes higher quantity of material from the workpiece. The tensile stresses rise due to the acting action and material removal mechanism.

REFERENCES

Ahmad, B., van der Veen, S. O., Fitzpatrick, M. E., & Guo, H. (2018). Measurement and modelling of residual stress in wire-feed additively manufactured titanium. *Materials Science and Technology, 34*(18), 2250–2259. DOI: 10.1080/02670836.2018.1528747

Ali, M. A., Mufti, N. A., Sana, M., Tlija, M., Farooq, M. U., & Haber, R. (2024). Enhancing high-speed EDM performance of hybrid aluminium matrix composite by genetic algorithm integrated neural network optimization. *Journal of Materials Research and Technology, 31*, 4113–4127. DOI: 10.1016/j.jmrt.2024.07.077

Almer, J. D., & Winholtz, R. A. (2008). *X-Ray Stress Analysis.*, DOI: 10.1007/978-0-387-30877-7_28

Altin Karataş, M., & Biberci, M. A. (2023). Statistical Analysis of WEDM Machining Parameters of Ti-6Al-4V Alloy Using Taguchi-Based Grey Relational Analysis and Artificial Neural Network. *Experimental Techniques, 47*(4), 851–870. DOI: 10.1007/s40799-022-00601-5

Anoop Johny, & Thiagarajan, C. (2022). Investigation, analysis, optimization of reciprocating wire-cut electrical discharge machining of titanium alloy with molybdenum electrode. *Journal of Ceramic Processing Research, 23*(4), 443–458.

Arunnath, A., Madhu, S., & Tufa, M. (2022). Experimental Investigation and Optimization of Material Removal Rate and Tool Wear in the Machining of Aluminum-Boron Carbide (Al-B4C) Nanocomposite Using EDM Process. *Advances in Materials Science and Engineering, 2022*, 1–11. DOI: 10.1155/2022/4254024

Ashok Sonawane, S., & Sitaram Wangikar, S. (2022). Multi-superiority response optimization of WEDM for titanium grade-5 using PCA based TOPSIS theory. *Materials Today: Proceedings, 62*, 841–847. DOI: 10.1016/j.matpr.2022.04.044

Besekar, N., & Bhattacharyya, B. (2022). Experimental investigation and characterization of NiTinol shape memory alloy during wire electrochemical machining. *Journal of Manufacturing Processes, 81*, 346–361. DOI: 10.1016/j.jmapro.2022.07.019

Bharathi, P., Sampath Kumar, T., & Anbuchezhiyan, G. (2025). Multioptimization analysis of machining characteristics on spark electrical discharge machining of Al/SiC and Al/SiC/B 4 C composites. *Engineering Research Express, 7*(1), 015518. DOI: 10.1088/2631-8695/ada723

Bhattacharya, S., Mishra, A., Singh, T., Bandyopadhyay, S., Kumar, S., Dey, G. K., & Kain, V. (2019). Simulation of residual stress using thermomechanical analysis on wire EDM cut surface. *Materials Today: Proceedings, 19*, 462–467. DOI: 10.1016/j.matpr.2019.07.636

Chen, S., Gao, H., Zhang, Y., Wu, Q., Gao, Z., & Zhou, X. (2022). Review on residual stresses in metal additive manufacturing: Formation mechanisms, parameter dependencies, prediction and control approaches. *Journal of Materials Research and Technology, 17*, 2950–2974. DOI: 10.1016/j.jmrt.2022.02.054

Childerhouse, T., M'Saoubi, R., Franca, L., Weston, N., Crawforth, P., & Jackson, M. (2022). The influence of machining induced surface integrity and residual stress on the fatigue performance of Ti-6Al-4V following polycrystalline diamond and coated cemented carbide milling. *International Journal of Fatigue, 163*, 107054. DOI: 10.1016/j.ijfatigue.2022.107054

Chinna, R., Sharma, P., & Fernandes, F. (2025). Machinability and surface integrity analysis of Ti-17 alloy using WEDC for advanced aero-engine application. *Heliyon, 11*(1), e41461. DOI: 10.1016/j.heliyon.2024.e41461

Das, S., & Joshi, S. N. (2020). Estimation of wire strength based on residual stresses induced during wire electric discharge machining. *Journal of Manufacturing Processes, 53*, 406–419. DOI: 10.1016/j.jmapro.2020.03.015

Dolabella, S., Borzì, A., Dommann, A., & Neels, A. (2022). Lattice Strain and Defects Analysis in Nanostructured Semiconductor Materials and Devices by High-Resolution X-Ray Diffraction: Theoretical and Practical Aspects. *Small Methods, 6*(2), 2100932. Advance online publication. DOI: 10.1002/smtd.202100932

Fitzpatrick, M., Fry, A., Holdway, P., Kandil, F. A., Shackleton, J., & Suominen, L. (2002). *NPL Good Practice Guide no. 52: determination of residual stresses by x-ray diffraction.*

Frame, L. D., Rambarran, I., Sala, K., & Sanders, C. (2020). Impacts of Machining and Heat Treating Practices on Residual Stresses in Alpha-Beta Titanium Alloys. *Journal of Materials Engineering and Performance, 29*(6), 3626–3637. DOI: 10.1007/s11665-020-04843-5

Gaikwad, M. U., Krishnamoorthy, A., & Jatti, V. S. (2021). Investigating the influence of electrical discharge machining process parameter on fatigue strength during machining of titanium grade-2. *Materials Today: Proceedings, 46*, 8951–8957. DOI: 10.1016/j.matpr.2021.05.367

Gao, K., Zhang, Y., Yi, J., Dong, F., & Chen, P. (2024). Overview of Surface Modification Techniques for Titanium Alloys in Modern Material Science: A Comprehensive Analysis. *Coatings*, *14*(1), 148. DOI: 10.3390/coatings14010148

Hegde, A., Shetty, R., Nayak, R., Shetty, S., & Shetty, S. V. (2024). Optimizing the Die-Sink EDM Machinability of AISI 316L Using Ti-6Al-4V-SiCp Electrodes: A Computational Approach. *Journal of Manufacturing and Materials Processing*, *8*(5), 202. DOI: 10.3390/jmmp8050202

Jagdale, M., Ambhore, N., Chaudhari, R., Kulkarni, A., & Abdullah, M. (2025). Experimental investigation of process parameters in Wire-EDM of Ti-6Al-4 V. *Scientific Reports*, *15*(1), 5652. DOI: 10.1038/s41598-025-90486-2

Johny, A., & Thiagarajan, C. (2020). Investigation of surface integrity and it's optimization on pure titanium using molybdenum wire by reciprocated travelling WEDM – A review. *Materials Today: Proceedings*, *33*, 2581–2584. DOI: 10.1016/j.matpr.2019.12.251

Kavimani, V., Gopal, P. M., Keerthiveettil Ramakrishnan, S., Giri, J., Alarifi, A., & Sathish, T. (2024). Predictive modelling and optimization of WEDM parameter for Mg–Li alloy using ANN integrated CRITIC-WASPAS approach. *Heliyon*, *10*(15), e35194. DOI: 10.1016/j.heliyon.2024.e35194

Khosrozadeh, B. (2024). Surface integrity of Ti-6Al-4V alloy in hybrid electrical discharge machining processes. *Proceedings of the Institution of Mechanical Engineers. Part E, Journal of Process Mechanical Engineering*, *238*(5), 2258–2267. DOI: 10.1177/09544089231160464

Kumar, A., Kumar, V., & Kumar, J. (2014). Surface integrity and material transfer investigation of pure titanium for rough cut surface after wire electro discharge machining. *Proceedings of the Institution of Mechanical Engineers. Part B, Journal of Engineering Manufacture*, *228*(8), 880–901. DOI: 10.1177/0954405413513013

la Monaca, A., Murray, J. W., Liao, Z., Speidel, A., Robles-Linares, J. A., Axinte, D. A., Hardy, M. C., & Clare, A. T. (2021). Surface integrity in metal machining - Part II: Functional performance. *International Journal of Machine Tools & Manufacture*, *164*, 103718. DOI: 10.1016/j.ijmachtools.2021.103718

Leyens, C., & Peters, M. (2006). *Titanium and Titanium Alloys: Fundamentals and Applications*. Wiley. Retrieved from https://books.google.co.in/books?id=okSA1N1xAxgC

Li, Y., Lin, J., & Zhao, Z. (2024). Phase field study on the grain size dependent electrical resistivity-strain response in superelastic nanocrystalline NiTi alloys. *Journal of Materials Research and Technology*, *32*, 4092–4105. DOI: 10.1016/j.jmrt.2024.08.215

Liu, J. F., & Guo, Y. B. (2016). Residual Stress Modeling in Electric Discharge Machining (EDM) by Incorporating Massive Random Discharges. *Procedia CIRP*, *45*, 299–302. DOI: 10.1016/j.procir.2016.02.060

Mahrous, M. A., Abdelghany, M. A., Bonney, C., Farag, H., & Jasiuk, I. M. (2024). Irradiation-induced microstrain and dislocation density in additively manufactured 316H stainless steel. *Journal of Materials Research and Technology*, *33*, 8306–8320. DOI: 10.1016/j.jmrt.2024.11.060

Manoj, I., & Narendranath, S. (2020). Variation and artificial neural network prediction of profile areas during slant type taper profiling of triangle at different machining parameters on Hastelloy X by wire electric discharge machining. *Proceedings of the Institution of Mechanical Engineers. Part E, Journal of Process Mechanical Engineering*, *234*(6), 673–683. DOI: 10.1177/0954408920938614

Marin, E., & Lanzutti, A. (2023). Biomedical Applications of Titanium Alloys: A Comprehensive Review. *Materials (Basel)*, *17*(1), 114. DOI: 10.3390/ma17010114

Outeiro, J. (2018). Residual Stresses in Machining Operations. In CIRP Encyclopedia of Production Engineering (pp. 1–13). Berlin, Heidelberg: Springer Berlin Heidelberg. DOI: 10.1007/978-3-642-35950-7_16811-1

Perumal, G., Senthilkumar, N., Satheshkumar, D., & Anandhakumar, S. (2025). Comparative analysis of solution treatment with short and long-term aging on mechanical, tribological, and corrosion potential of Ti-6Al-4V alloy. *Multiscale and Multidisciplinary Modeling. Experiments and Design*, *8*(3), 159. DOI: 10.1007/s41939-025-00759-6

Pragadish, N., Kaliappan, S., Subramanian, M., Natrayan, L., Satish Prakash, K., Subbiah, R., Kumar, T., & Ch, A. (2023). Optimization of cardanol oil dielectric-activated EDM process parameters in machining of silicon steel. *Biomass Conversion and Biorefinery*, *13*(15), 14087–14096. DOI: 10.1007/s13399-021-02268-1

Pramanik, A., Basak, A. K., Littlefair, G., Debnath, S., Prakash, C., Singh, M. A., Marla, D., & Singh, R. K. (2020). Methods and variables in Electrical discharge machining of titanium alloy – A review. *Heliyon*, *6*(12), e05554. DOI: 10.1016/j.heliyon.2020.e05554

Salvati, E., & Korsunsky, A. M. (2020). Micro-scale measurement & FEM modelling of residual stresses in AA6082-T6 Al alloy generated by wire EDM cutting. *Journal of Materials Processing Technology*, *275*, 116373. DOI: 10.1016/j.jmatprotec.2019.116373

Sarmast, A., Schubnell, J., Preußner, J., Hinterstein, M., & Carl, E. (2023). Residual stress analysis in industrial parts: A comprehensive comparison of XRD methods. *Journal of Materials Science*, *58*(44), 16905–16929. DOI: 10.1007/s10853-023-09069-z

Schubnell, J., Carl, E., Widerspan, V., & Collmann, M. (2023). Determination of Loading and Residual Stresses on Offshore Jacket Structures by X-ray Diffraction. *Journal of Marine Science and Engineering*, *11*(7), 1304. DOI: 10.3390/jmse11071304

Senthilkumar, N., Perumal, G., Palanikumar, K., Deepanraj, B., & Basha, M. S. (2025). Wear and mechanical characteristics of coated Ti-6Al-4 V alloy with an alumina-13 wt.% titania -40 wt.% yitrium stabilized zirconia ceramic composite via plasma spraying technique. *Results in Engineering*, *25*, 103777. DOI: 10.1016/j.rineng.2024.103777

Sharman, A. R. C., Hughes, J. I., & Ridgway, K. (2006). An analysis of the residual stresses generated in Inconel 718TM when turning. *Journal of Materials Processing Technology*, *173*(3), 359–367. DOI: 10.1016/j.jmatprotec.2005.12.007

Soni, H., S, N., & M R, R. (2018). Experimental Investigation on Effects of Wire Electro Discharge Machining of Ti50Ni45Co5 Shape Memory Alloys. *Silicon*, *10*(6), 2483–2490. DOI: 10.1007/s12633-018-9780-9

Srinivasa Rao, P., Ramji, K., & Satyanarayana, B. (2016). Effect of wire EDM conditions on generation of residual stresses in machining of aluminum 2014 T6 alloy. *Alexandria Engineering Journal*, *55*(2), 1077–1084. DOI: 10.1016/j.aej.2016.03.014

Suresh, S., Natarajan, E., Markandan, K., Sudhagar, S., Anbuchezhiyan, G., & Elayaraja, R. (2025). Strategic enhancement of joint strength in dissimilar Al/Mg alloys: A workable approach for aerospace application. *Discover Materials*, *5*(1), 23. DOI: 10.1007/s43939-025-00194-x

Tabatabaeian, A., Ghasemi, A. R., Shokrieh, M. M., Marzbanrad, B., Baraheni, M., & Fotouhi, M. (2022). Residual Stress in Engineering Materials: A Review. *Advanced Engineering Materials*, *24*(3), 2100786. Advance online publication. DOI: 10.1002/adem.202100786

Thirumalvalavan, S., Perumal, G., Senthilkumar, N., & Selvarasu, S. (2025). Enhancing Tribological Characteristics of Titanium Grade-5 Alloy through HVOF Thermal-Sprayed WC-Co Nano Coatings by TOPSIS and Golden Jack Optimization Algorithm. *Recent Patents on Nanotechnology*, *19*(4), 544–567. DOI: 10.2174/0118722105306841240808092616

Totten, G. E. (2016). *Heat treating of nonferrous alloys. Volume 4E*. ASM Handbook. ASM International., Retrieved from https://books.google.co.in/books?id=TtSZwgEACAAJ DOI: 10.31399/asm.hb.v04e.9781627081696

Vashista, M., & Paul, S. (2012). Correlation between full width at half maximum (FWHM) of XRD peak with residual stress on ground surfaces. *Philosophical Magazine*, *92*(33), 4194–4204. DOI: 10.1080/14786435.2012.704429

Xu, L.-Y., Xi, X.-C., Li, Z.-L., Ma, J.-Y., Gao, Q., & Zhao, W.-S. (2022). Wire breakage prevention for reciprocated traveling WEDM based on discharge location detection. *International Journal of Advanced Manufacturing Technology*, *123*(5–6), 1859–1875. DOI: 10.1007/s00170-022-10240-0

Xu, L.-Y., Xi, X.-C., Zhang, Y.-O., Li, Z.-L., Ma, J.-Y., Gao, Q., & Zhao, W.-S. (2022). Online workpiece height estimation method based on discharge location detection for reciprocated traveling WEDM. *Journal of Manufacturing Systems*, *65*, 317–329. DOI: 10.1016/j.jmsy.2022.09.018

Yazawa, K., Malay, A. D., Masunaga, H., Norma-Rashid, Y., & Numata, K. (2020). Simultaneous effect of strain rate and humidity on the structure and mechanical behavior of spider silk. *Communications Materials*, *1*(1), 10. DOI: 10.1038/s43246-020-0011-8

Chapter 8
Titanium-Based Implants in Additive Manufacturing:
Materials, Processing, and Future Prospects

Samrat Hazra
https://orcid.org/0009-0002-7796-1611
National Institute of Technology, Rourkela, India

Saurav Datta
National Institute of Technology, Rourkela, India

Kaushik Kumar
https://orcid.org/0000-0002-4237-2836
Birla Institute of Technology, India

Onur Mesut Şenaras
Oyak Renault Automotive Plants, Turkey

Arzu Eren Şenaras
https://orcid.org/0000-0002-3862-4551
Bursa Uludag University, Turkey

ABSTRACT

3D printing has transformed the manufacturing of medical implants by enabling precise and customized designs. This review explores key powder-based additive manufacturing (AM) methods, including Powder Bed Fusion (PBF) and Directed Energy Deposition (DED), used to create titanium-based implants. It also discuss-

DOI: 10.4018/979-8-3373-9454-1.ch008

es titanium powder production techniques such as gas atomization (GA), plasma atomization (PA), and the plasma rotating electrode process (PREP). The study highlights the benefits of additive manufacturing (AM), including its cost-effectiveness and design flexibility. However, it addresses significant challenges like powder quality, porosity, and contamination. The increasing adoption of titanium implants in biomedical applications underscores the urgent need for further research to improve material properties and enhance process efficiency. Future advancements in AM and powder production will be critical for improving the performance and accessibility of these implants.

1. INTRODUCTION

Rapid prototyping, or Additive Manufacturing (AM), is the process of creating objects from a digital model using successive layers of material, as shown in Figure 1. It is also called by many other names, such as on-demand manufacturing, digital generation, desktop manufacturing, solid layer manufacturing, layer construction, and direct manufacturing. The process involves depositing layers of thin material sections to form objects or components, as elaborated by (Mahamood et al., 2019). It is used to manufacture complex structures and shapes directly from 3D digital designs. The 3D process builds items layer by layer, adding a new layer on top of the last one each time. The technology was invented by Charles Hull in 1986 through the development of stereolithography (SLA). Since then, various innovative techniques such as powder bed fusion (PBF), fused deposition modelling, inkjet printing, and contour crafting (cc) have been developed (Dutta & Froes, 2017; Gill et al., 2017; Ngo et al., 2018). The international community pays much attention to this territory. It is relevant to a series of sectors where production is prioritized due to its advantages. It gives rise to the development of various complex structures that classical manufacturers cannot handle, offering innovative and quite effective solutions for time and cost saving at all levels. Each of the referred fields manifests its sign of academic or professional interest. They mostly find applications in the medical domain, like orthopaedic implants, dental implants, and heart devices (Xiong et al., 2016). Ti6Al4V is a unique alloy due to its strength, lightness, and high biocompatibility with the human body (Senthilkumar et al., 2025). It is used mainly in the manufacture of parts, which are strong and lightweight. Studies show that it is resistant to other forms of corrosion, durable, and integrates well into the human body, sometimes even sticking to bones. However, some studies show that it might leach some amounts of aluminium and vanadium, which may pose some problems and safety issues concerning medical uses (Haydar & Al-Deen, 2021). Ti64 is a distinctive generic titanium alloy that consists of two crystal structures, namely the α and β phases. These alloys have great strength, are light in weight, and are tough enough to withstand pressure without cracking or breaking. In addition, they are also resistant to corrosion, (very similar to austenite stainless steel filled welds (Sengottaiyan et al., 2025)) which makes them biocompatible. Due to these remarkable features, Ti64 is often applied in medicine, which is a suitable and effective application of this grade in practice (Liu & Shin, 2019a; Mahmud et al., 2020). The most usual way to produce components made of Ti6Al4V is first forging and then rolling large blocks, as well as casting. However, these technologies generally cause considerable cutting losses, higher expenses, and prolonged processing times (Huang et al., 2016; Lütjering & Williams, 2003). One of the key techniques in engineering is additive manufacturing (AM), which has many applications and provides a great amount of freedom. Many attempts have been made to fabricate metal components through directed energy deposition, selective laser melting, and electron beam melting(Gibson et al., 2021a; Y. Zhang et al., 2018)[REMOVED HYPERLINK FIELD]. Additive manufacturing, also known as 3D printing, is a technique that constructs objects by layering materials on top of each other (Ni et al., 2019; Popov et al., 2019) (Figure 1). The effectiveness of implants created through additive manufacturing (AM) largely depends on the quality of titanium (Ti) and its alloy powders. Prior to the emergence of the AM process, conventional techniques like powder metallurgy (PM) were frequently employed to produce nearly finished shape items from titanium and its alloys. These conventional approaches incorporate metal injection moulding and hot isostatic pressing. In additive manufacturing processes, powders must conform to rigorous physical and chemical standards.

Figure 1. Additive Manufacturing or 3D Printing Process Step

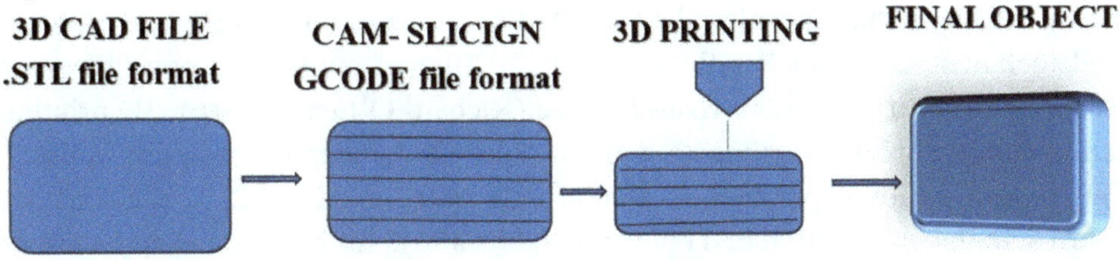

Figure 2. Impact Of the Complexity on The Cost of Additive Manufacturing and Traditional Manufacturing

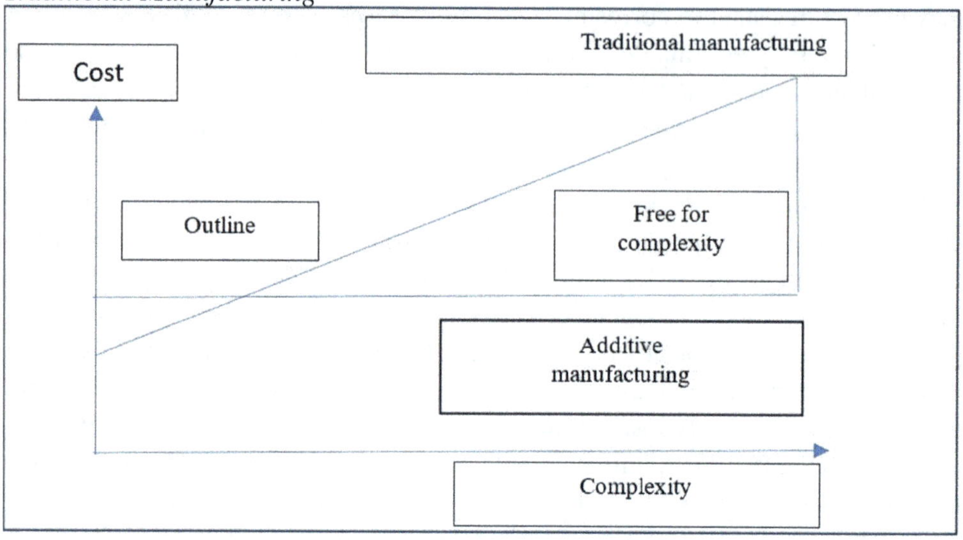

Typically, powders utilized in both powder bed and powder feed systems should possess excellent flow properties and elevated tap density to ensure the final product remains uniform (Sun et al., 2018; Tang et al., 2015). Compared to conventional production methods, the most notable advantage of AM is that it creates complex parts directly from raw material without exposing them to processes like extrusion, forging, casting, or additional machining to attain the end shape (Frazier, 2014; Shapiro et al., 2016). AM is also cost-effective since it creates parts in almost the end shape, reducing material waste. AM also accommodates a broad range of feedstock materials (Sanchez-Rexach et al., 2020), and any leftover powder can be reused, making it even more cost-effective (Richard O'Leary, 2016). 3D printing can handle complex components without seeing a comparable increase in costs. This representation explains the difference presented in Figure 2. This evaluation focuses on the purification of Ti powder and its alloys for biomedical applications by different metal 3D printing techniques. Table 1 below indicates the primary differences between conventional and additive manufacturing (AM)-based implant production. Table 1 below highlights the main distinctions between traditional and additive manufacturing (AM)-based implant manufacturing.

Table 1. Comparison between Traditional and Additive Manufacturing (AM)-based Implant Manufacturing

Aspect	Traditional Manufacturing	Additive Manufacturing (AM)
Design Flexibility	Cutting or shaping techniques have restrictions for traditional methods.	AM facilitates intricate and patient-specific designs.
Material Wastage	High.	Minimal; material is added layer by layer.
Customization	In traditional methods, customizing is hard and expensive.	AM makes it simple and personal.
Lead Time	Many steps are involved in traditional techniques.	AM advertised the use of direct digital printing for faster speeds.
Tooling Requirement	Tools are needed in Traditional techniques.	AM doesn't often require tools for every newly designed piece.
Production Volume Suitability	Mass production is Ideal for Traditional Methods.	AM excels in making a small number of intricate items due to the complexity of small items.
Surface Finish	A smoother surface finish is given by traditional methods.	AM requires additional steps in improving the surface quality.
Mechanical Properties	Most procedures of the Traditional methods have proven to be reliable.	AM outcomes depend on the complexity of the procedure and the materials used, causing results to differ.
Supply Chain Complexity	Multiple vendors and steps are needed for traditional.	AM, with its digital workflows, can be simpler.
Biocompatibility Optimization	Pores in traditional methods are restricted.	AM offers better control for integration with bone aiding.

2. POWDER-BASED 3D PRINTING PROCESSES

The use of powder-based additive manufacturing (AM) technology using titanium and titanium alloys has attracted a lot of interest in the biomedical field. This is because it can create customized products and save time, cost, and resources. Achievement in the implantation or scaffolding of products made using this method relies on different additive manufacturing processes and the purity of titanium and its alloy powder (Jang et al., 2020a). Recent studies show that powder 3D printing is useful for titanium alloys in medicine. Two important methods are selective laser melting and electron beam melting. These techniques help create precise and custom implants. They also offer great strength and compatibility with the body (Pesode & Barve, 2022). Beta-titanium alloys are of particular interest owing to the reduced elastic modulus, which could help reduce stress in implants (Sing, 2022). Beta-titanium alloys fabricated by PBF are stronger than the conventional ones but have less capacity to deform (Tong et al., 2017). Lattice structures can help alter elastic modulus to have better human bone matching (Chen et al., 2016). Despite the encouraging initial clinical outcomes, challenges such as optimization of process parameters, structural design, and long-term performance exist (A. B. Singh et al., 2024). PBF can be divided based on the source of heat used into infrared (IR), electron beam (EB), and laser beam (LB) categories. For example, in the case of processes such as selective laser sintering (SLS) and selective laser melting (SLM), lasers melt the powder either completely or partially. Electron beam-based techniques include electron beam melting (EBM), where a high-energy electron beam completely melts the material. EBM can be done with conductive materials such as metals alone (Shanthar et al., 2023).

2.1. Electron Beam Powder Bed Fusion of Metals

Electron beam powder bed fusion of Ti6Al4V for biomedical use highlights its ability to create personalized implants of intricate geometry and enhanced properties (Tamayo et al., 2021). There are also concerns about its long-term effect on the human body (Pesode & Barve, 2022). Beta-type titanium alloys fabricated using powder bed fusion have the potential to minimize the elastic modulus incompatibility between bone and materials, although improvements are necessary (Zhang & Chen, 2019). A new EB-PBF-Ti536 printed alloy exhibits an antibacterial attribute, eliminating the postoperative infection threat (Behjat et al., 2024). Despite promising early clinical outcomes, the regulatory approval processes for 3D-printed implants have not been able to keep up with the rapid advancements in technology (Pesode & Barve, 2022). Ongoing research focuses on optimising porous structures, surface coatings, and alloy compositions to enhance the long-term performance and biocompatibility of titanium implants produced by additive manufacturing. Electron beam powder bed fusion is a high-energy density powder bed method in which an electron beam is utilized to liquefy and agglomerate metal powder into fully dense, metallurgically sound components, as shown in Figure 3. PBF-EB/M is a variation of the PBF process; a high-power electron beam seizes to dither the metallic powder edge-wise using a layer-wise manner to end up complex three-dimensional components immediately from one PC-adapted style (CAD) file (Breuning et al., 2022)[REMOVED HYPERLINK FIELD]. Thus, PBF-EB/M is a unique subset of powder bed fusion (PBF) technologies, including also laser-based PBF (PBF-LB). The indicated difference pertains to the employed energy source being a laser (PBF-LB) or an electron beam (PBF-EB/M) (Roos & Rännar, 2024) [REMOVED HYPERLINK FIELD]. Electromagnetic coils steer the high-energy beam and control melting (Sharabian et al., 2022). The manufacturing process starts with the deposition of a powder layer on the build platform. Once this powder bed has been created, an electron beam is scanned according to a digital CAD model, enabling the beam to selectively melt the powder in the desired areas. The electron beam used for e-beam resistance derives from accelerated electrons focused by an electromagnetic lens into a high-energy beam (Heinl et al., 2008). A new layer of powder will then be distributed on top, and the process will continue. Each layer fuses with the layer below it, building the final product one layer at a time. After the auxiliary manufacturing process is done, the excess powder that has not been used is still left over within the build area (Gao et al., 2018a). Requirements for electron beam parameters in an additive manufacturing process are (Liu & Shin, 2019a) a single beam has a maximum power (≤3500 W), the beam spot size is between 200 and 1000 μm and typical scan speeds exceed 1000 m/s, while the layer thicknesses vary between 50 and 200 μm. It also enables an energy density of over 100 kw cm^2, relying on a beam power that exceeds≈60 kw, according to(Sing et al., 2016)[REMOVED HYPERLINK FIELD]. Especially in applications of orthopaedic materials, the Powder Bed Fusion Electron Beam/Melting (PBF-EB/M) and Powder Bed Fusion Laser Beam/Melting (PBF-LB/M) processes achieve different advantages and disadvantages. Parts made with the PBF-EB/M process have less residual stress than those from PBF-LB/M, which is advantageous for projects needing strong, stable materials (Cheng et al., 2012; Mumtaz & Hopkinson, 2010). Primarily, the problem with the PBF-EB/M technique is that the surface is rougher as a result of this issue. This occurs because PBF-LB/M uses finer powder particles, which leads to more details being brought out and more matte surfaces on the resulting parts. Furthermore, PBF-LB/M is energy-efficient and can use a great amount of materials. Because these are cost-cutting techniques, both of these cut down the expenses of PBF-LB/M to be used in situations where job cost is important. They are the most accurate methods for metal fabrication technology in terms of accuracy and detail. One of the advantages is better quality elements' manufacturing, which is a matter of importance; for instance, the accuracy of things like orthopedic implants depends on manufacturing (Roseti et al., 2017; X. Wang et al., 2016; C. Yan et al., 2015). Recent research has looked at different parts of PBF-EB/M processing. This includes creating complex structures (R. Yan et al., 2018; Yánez et al., 2016) and the enhancement of corrosion resistance (Acquesta & Monetta, 2020) and biocompatibility (Ataee et al., 2019; Chudinova et al., 2019)[REMOVED HYPERLINK FIELD]. PBF-EB/M has distinct noticeable characteristics that make the materials better. By strength and safety, they are woven together to serve strenuous medical requirements.

Process Step of Electron Beam Powder Bed Fusion of Metals:

Powder formation: A thin layer of metal powder is spread across the build platform

Melting: The electron beam accurately fuses the powder to a digital 3D blueprint.

Layering: Once the single layer has melted, the build stage is moved down, and a new layer of powder is poured on top of the previous one.

Repetition: It happens again since the whole part is constructed one layer at a time.

Post processing: The excess powder is stripped away, and the workpiece can be further processed, such as heat treatment or processing, to obtain the desired properties and finish on the surface.

Figure 3. Electron Beam Powder Bed Fusion of Metals

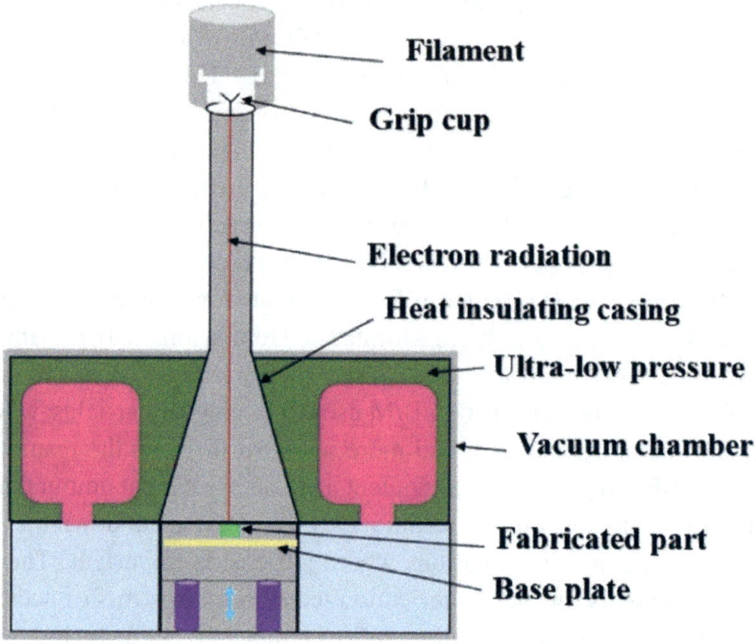

2.2. Binder Jetting (BJ):

Binder Jetting (BJ) is a new 3D printing technology that employs a liquid binder to bind Powder particles together layer by layer to create three-dimensional products (Ziaee & Crane, 2019). Researchers at MIT created BJ in the 1990s while conducting research on 3d printing. (Dawood et al.,2015). Researchers at MIT created BJ in the 1990s while conducting research on 3D printing. This has evolved over the years with firms simplifying the technology and creating new Applications. Initially, individuals utilized it to develop models with speed, but today it is also employed to produce end products (Lores et al., 2019).

Binder jetting (BJ) is an additive manufacturing (AM) process that has various advantages. It covers a wide range of powdered feedstocks, enabling the production of functionally graded materials. BJ systems also have relatively high build rates compared to methods such as material jetting, as they only need the printing of the binder, which represents a minor portion of the total part component volume. A print head with 100 nozzles has the capability to produce components at a rate of up to 200 cm^3/min (Bai & Williams, 2015). Metal parts can have build volumes of 780 mm × 400 mm × 400 mm while sand-casting machines can produce parts with widths of 850 mm, heights of 500 mm, and theoretically unlimited length(Nied et al., 1990)[REMOVED HYPERLINK FIELD]. However, BJ produces "green" parts, comprised of bonded powder particles that require post-processing to achieve final properties. The suitability of BJ for various applications depends on the quality of these final properties, which in turn relies on several process factors, including the raw materials, powder bed quality, construct parameters, and after-processing methods. Ongoing research continues to expand the potential applications of BJ technology (Halloran, 1999). Recent developments on Ti6Al4V alloy for medical use highlight its advantages, including high strength and low density, and biocompatibility (Mujahid et al., 2024). Nevertheless, challenges in the release of toxic aluminium and vanadium have generated research for the use of replacement alloying elements. BinderJet 3D printing is a very promising technique for the fabrication of Ti6Al4V implants with Process condition investigation and particle size optimization for the production of high-density Components with the desired mechanical and microstructure properties (Simchi et al., 2022). While binder jetting has a number of advantages over other additive manufacturing technologies, it continues to manufacture high-density components with minimal impurities (Simchi et al., 2023). Surface treatment technologies are also being researched to further improve biological activity, wear, and corrosion resistance of Ti alloys (L. Zhang & Chen, 2019). All these are included in ongoing research into the development of tailor-made biomaterials that are regulatory compliant for implantation.

The benefit of this mode of production is efficiency and economy. Other than a lack of residual stress. However, it has some disadvantages like sintering shrinkage, poor quality, and strength (Gao et al., 2018a; Pei et al., 2023).

Process Steps of Binder Jetting (BJ):

Design and preparation: Everything starts from a three-dimensional CAD model. The model is subsequently transformed into an STL file.

Slicing: The STL file is loaded into slicing software that slices the model into thin horizontal layers, depending on the selected layer thickness.

Printing: A printhead deposits a liquid binder onto a powder bed, layer by layer, to create the part. Each of these binder drops is about 50 microns across and is what binds the powder particles, combining to create every layer of the finished 3D model.

Layering: After the printing layer is completed, the powder bed is lowered, and a new layer of powder is applied over the previously printed layer, allowing the printhead to glide smoothly across it.

Post-Processing: Lastly, the printed piece is allowed to harden. Loose powder is blown away using pressurized air, and then the piece can be heated or filled with material to get the right qualities and smooth surface.

Figure 4. Binder Jetting (BJ)

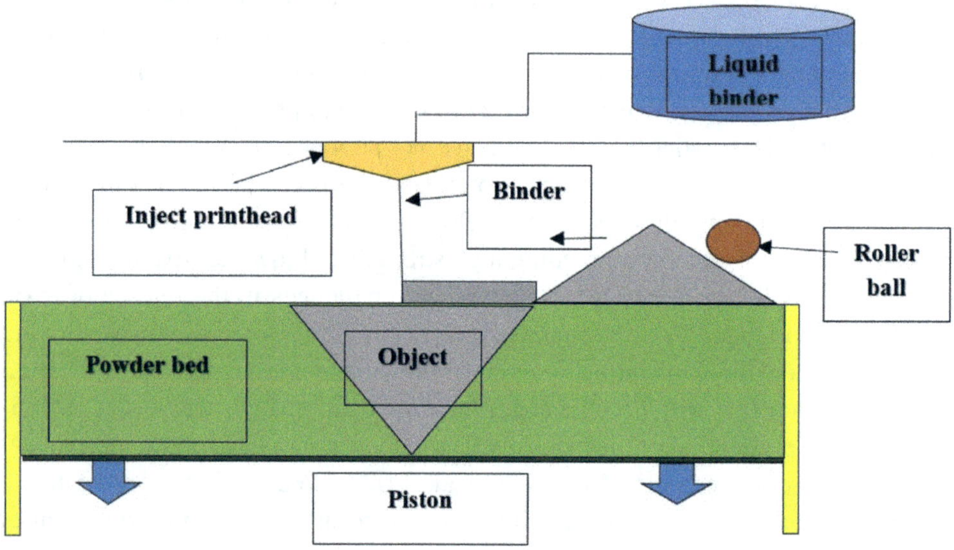

2.3. Directed Energy Deposition (DED)

This technique is used to repair parts and add material to existing components. It is also applied in developing a new part from a raw material base (Gibson et al., 2021a), [REMOVED HYPERLINK FIELD]shown in Figure 5. DED is very effective since it minimizes material consumption during production and can cut a variety of materials, for example, metals, ceramics, and polymers (Lalegani Dezaki et al., 2022). The process achieves high accuracy through concentrated thermal energy sources such as a laser, electron beam, or plasma arc. It does this in an effort to make geometrical shapes of different complexity and features, which would otherwise be difficult to accomplish using conventional manufacturing methods (Letov et al., 2021).

Another benefit of DED is its capacity to process vast numbers of builds, thereby making it highly suitable for large-scale parts manufacturing (Xu et al., 2018). The technology is employed for the manufacture of large Components rapidly, which is always time-consuming when attempting to make use of other types of 3D Printing (Wong & Hernandez, 2012). It is employed to repair old parts and build their Strength to make them long-lasting for a long life. It is a general technique that is used to Design elements for a variety of applications, including medical, aerospace, and automobile (Salmi,2021). The process is cost-effective and efficient, especially in handling other material Forms (Saboori et al., 2019). The process has a number of important constituents that work together to create High-quality products. One of the primary components is the source of energy: a laser, electron beam, or plasma arc. The energy source is a function of the material and product (A. Singh et al., 2020). The selected energy source provides the quantity of heat, which is adequate for heating the material that is used during the process. Material supplied to the DED system, which is referred to as feed material, can be powder or wire form, depending on the DED system's requirements and the product's nature (Svetlizky et al., 2021). The deposition head conveys the feed material to this site (Thompson et al., 2015). The control system is a vital component that oversees the movement of the tool used to deposit material and the platform upon which the object is built. It also regulates the settings of the source of energy. This system ensures that material is deposited correctly based on the digital design (Vallone et al., 2023). Oxygen and nitrogen are inert gases employed extensively in Direct Energy Deposition (DED) processes. The gases will undergo minimal reaction with most metals except for extremely reactive metals. Since they undergo minimal reactions, they prevent the surface from becoming dirty or corroded. This maintains the part in good health and strong. Therefore, each part remains whole, and the piece remains in the desired form (Everton et al., 2016). Additionally, some of the components of the building blocks of the DED process are sensors and monitoring systems that enable you to track conditions such as the rate of deposition. Feedback is given, and any adjustments during the building process can be made to achieve the desired result at the end (Chua et al., 2017). Much of the recent research into Directed Energy Deposition (DED) technology has involved application in biomedical implants that will withstand weight-bearing stresses (Balla et al., 2010; Marattukalam et al., 2015). For example, DED has been utilized to create structured materials, e.g., porous titanium-based implants, to allow for the simulation of the mechanical properties of human bone (Xue et al., 2007). This technology has recently been used to create titanium-silver alloy implants. These implants are safe to use inside the body and resistant to bacteria (Maharubin et al., 2019), analyzing their surface characteristics

(A Das et al., 2019; A, Mitun, et al., 2019), and examining the strength and load capacity of Ti6Al4V alloys (Bandyopadhyay et al., 2010; Meteyer et al., 2014).

Process Steps of Direct Energy Deposition (DED):

This blueprint determines the exact shape of the part.

Slicing: Finally, a slice of the STL is created through slicing software, and it's sliced into two vertical sections and then halved vertically so that all the elements within the horizontal section are displayed, as well as all the elements in the vertical section are displayed. These slices depict the parts of the machine and how the final parts are shaped to complete it.

Material Feeding: As the system works, the feed material, which can be in the form of powders or wires, is then sprayed out through a nozzle.

Energy Source Activation: The heat source may be a laser, an electron beam, or a plasma arc. Its main aim is to burn the material that is being used in the form it is supplied.

Layer Deposition: In this process, the model is divided into slices, and it is cut by the deposition head to get a thin cut. This is followed by the deposition of layers using a material similar to before, as mentioned above. When the melt is in contact with the air, it becomes a solid structure, which solidifies as it cools down.

Inert Gas Shielding: As for the protection of molten materials, one should be using an inert gas that would be similar in atomic number to oxygen and with a lower number, which is argon, to improve the standard and completion of the substance.

Monitoring and Control: Monitoring of parameters such as temperature, deposition rate, and any other factors that can be identified through the incorporation of sensors in the temperature and deposition rate measurement.

Post-Processing: The last step before the use of the fabricated part is the post-processing, where some of the manufacturing process takes place, for example, heat treating the part to reduce its temperature, machining the part to achieve the required standard, or giving the parts a surface finish to make them better.

Figure 5. Directed Energy Deposition (DED)

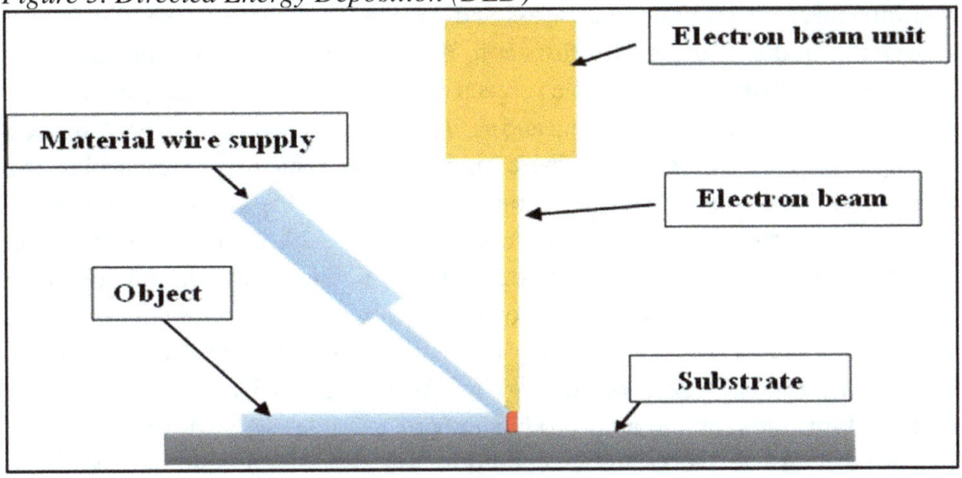

2.4. Laser-Based Powder Bed Fusion of Metals

Laser-based powder bed fusion of metals, the technique employed to generate the parts, combines a laser that melts and fuses the powders in layers, enabling parts fabrication with optimal mechanical and dimensional strength (Khorasani et al., 2020), shown in Figure 6. Such an approach received its attention in the biomedical area since its ability to produce Ti6Al4V components with a constant and random distribution of geometrical and dimensional variations makes possible the creation of parts designed to match the requirements of the patients and their operations as well as minimizing costs and time frames (Haglin et al. 2016). The process lets us create parts with special designs that control how much space is inside them and how they are built, like a lattice. This helps us make materials that are very similar to real bone. This can lower the problem of stress shielding (when implants are too stiff compared to bone) (Ahangar et al., 2019) and create surfaces on implants that help them bond better with bone, making the implant and bone work together more smoothly (Haglin et al., 2016b; Takizawa et al., 2018). While this method has a number of advantages, it also possesses some limitations. Studies have indicated that components of Ti6Al4V produced by this technology have relatively large Residual stress, are not very ductile, and may have holes on their surface or sub-surface. The layered Extensive temperature variations during the production process generate such issues (Mercelis & Kruth, 2006; Tan et al., 2017). Thermal residual stresses (TRS) can effectively minimize a material's life expectancy, leading to issues such as fatigue, cracking, and increased corrosion, particularly where cavities or holes exist.

It is particularly important in medical implants because they are placed in Severe environments such as prolonged stress and wet, corrosive conditions. To correct to perform failure analysis with respect to these problems, components produced by means of LPBF are sensitive be subjected to secondary treatments, such as Hot Isostatic Pressing (HIP), tempering and aging. These actions decrease heat stresses, making the material more pliable, increasing its shape, removing irregularities, and closing small openings. In-situ preheating methods are at times utilized to minimize TRS and anisotropy during production (Etesami et al., 2022). These methods can improve how materials are able to withstand wear and tear, and for how long they stabilise prior to failing. They don't, though, develop compressive residual stresses, which are crucial to ensuring the material lasts even longer. They also cannot completely remove small surface pits or slightly sub-surface pits. Due to this, additional steps are typically required after producing parts using laser powder bed fusion to improve the surface. Good surface conditions are highly important in making the bone cells adhere to the implant and for the implant to be well integrated into the bone (Hanawa, 2019; Vrancken et al., 2016).

Figure 6. Laser-Based Powder Bed Fusion of Metals

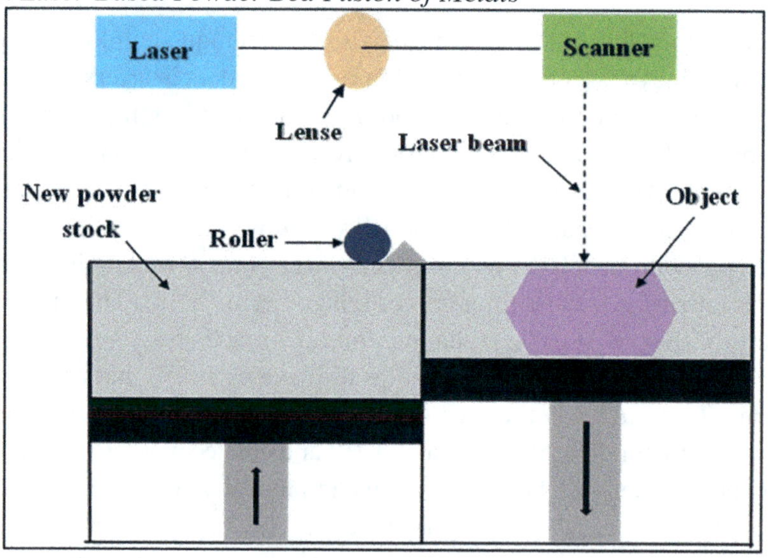

3. POWDER PRODUCTION TECHNIQUE

Titanium powders used industrially are typically produced by atomizing liquid metal into small, spherical particles (Jang et al., 2020).

The atomization process involves three main steps:

1. Melting metal can be achieved through vacuum induction, plasma arc, induction melting, or plasma heating.
2. Atomizing the molten metal into droplets using either high-pressure gas or centrifugal forces.
3. Solidifying these droplets in a cooling chamber filled with an inert gas, such as argon or helium, reduces oxygen absorption to levels between 100 and 500 ppm.

Even though the techniques are probably different, they all follow this basic system.

3.1 Gas Atomization (GA)

Gas atomization (GA) employs a high-speed gas, such as argon or nitrogen, to atomize Liquid metal into droplets. High-pressure gas is ejected at the rate of 350 kPa to 4 MPa (Mathias et al., 2024). The process works by undergoing three principal steps: melting, atomizing, and solidifying. A high-pressure reservoir releases an initially high-velocity blast of gas. Tank into an anesthetizer maintained at low pressure. When high-pressure compressed gas is expanded rapidly into the low-pressure chamber, the residual metal liquid deforms and is disintegrated into small, molten droplets due to the kinetic shock of the high-velocity gas stream (Mandal et al., 2022). Gas atomization has many advantages: the powder particles are spherical, increased Cleanliness and faster solidification, as well as high productivity. But therein lies the rub: the formation of satellite particles, tiny ones that stick to the bigger ones. The satellites are produced through particle impacts during the atomization and could lead the Particles to be less mobile. Gas atomization has three main techniques: free-fall gas atomization (FFGA), close-coupled gas atomization (CCGA), and electrode induction gas atomization (EIGA). These techniques differ in how close the nozzle is to the atomizer and how the metal is melted (Soong et al., 2023).

3.2 Plasma Rotating Electrode Process (PREP)

PREP is performed in a helium-filled stainless steel chamber (Jang et al., 2020). In this, an energy-rotating metal rod is heated up to a very high temperature at one of its ends. The molten metal droplets are produced, broken up, and ejected by centrifugal force. These droplets harden when they are in the air to form a spherical shape by the surface tension, yielding a powdered form (Sanaei & Fatemi, 2021). This method is known for producing particles with a more uniform spherical shape. However, it is associated with higher costs (Ju et al., 2020). Additionally, during the powder preparation process, solidification segregation may occur, potentially leading to the formation of impurity grains, such as oxides, at the particle boundaries due to diffusion during densification (Sanaei & Fatemi, 2021).

3.3 Plasma Ato

Plasma Atomization (PA) Utilizing pre-alloyed steel wire as a feedstock, it is simultaneously formed into droplets by melting it with high-speed plasma torches (Ju et al., 2020). Because the molten metal remains uncontaminated by solid impurities, there is no interaction with foreign particles, preventing the formation of satellite grains. However, the process can result in gas voids within the powder, caused by trapped gas during atomization. Additionally, PA is limited to wire-based feedstock, which increases production costs since wires are generally expensive. This limitation likewise hinders experimentation with non-conductive materials such as Ti_3Al(Jang et al., 2020c; Ju et al., 2020) [REMOVED HYPERLINK FIELD]. Parameters that can influence the size and distribution of particles include wire diameter, feed velocity, and gas pressure, in addition to the wire's distance and angle from the plasma outlet.

4. BIOMEDICAL APPLICATIONS OF METAL 3D PRINTING

Ti6Al4V (Ti64) is one of the most useful titanium alloys due to its durability, lightweight nature, biocompatibility, suitability for bone development, and corrosion resistance (Sanaei & Fatemi, 2021). Ti alloy implants consist of 80% of this material (Ju et al., 2020). The ideal AM processes with Ti are due to the phenomenon of allotropy and the diversity of the associated microstructures present in these alloys. The densities of Ti and Ti64 are 4.5 grams per cubic centimetre and 4.41 grams per cubic centimetre, respectively. Ti64, being the biomaterial mentioned, is used to manufacture clavicular, mandibular, dental, knee, and hip implants (Popov et al., 2018). The mechanical properties were enhanced by adding Zr to the Ti alloys. Zr and its biological properties, like Ti, are not toxic and are also good for biological reliability. Zr and Zr alloys belong to the family of bioactive metallic biomaterials, which is the ability for an apatite layer to form on their surfaces in vivo, comparable to that of bone. Besides, zirconium has large fracture toughness and mechanical strength, both of which are important properties that contribute to corrosion resistance. Due to them, they can be integrated as structural devices for the replacement of biological hard tissues(Nie et al., 2014)[REMOVED HYPERLINK FIELD]. The in vivo analysis and experimentation of Ti, Nb, and Ta, the elements, revealed their good biocompatibility & thus, they are the materials of choice for bone implantation (J.-X. Wang et al., 2009). (Matsuno, 2001; Yamanoglu et al., 2018) provides a list of 3D-printed titanium and its alloys that are currently utilized in the biomedical industry, as indicated in Table 2.

Table 2. Additive manufacturing of Ti-based alloys for biomedical uses

Materials	Applications
Ti6Al4V	Prosthetic valve, Vascular stent, orthopaedic fixation device.
Ti-29Nb-13Ta-4.6Zr	Crown, bridge, denture, implant.
Ti-6Al-7Nb	Crown, knee joint, hip joint, femoral, stem prosthesis.
CP-Ti	Screw and abutment
NiTi	Catheter, stent

5. KEY TECHNICAL CONSIDERATIONS IN MEDICAL 3D PRINTING

The regulatory landscape for 3D-printed medical devices presents unique challenges due to their customizability and manufacturing processes. Key considerations include design control, quality assurance, and post-printing processes like cleaning and sterilization (Morrison et al., 2015). The FDA's approval process for these devices requires addressing material properties, printing parameters, and build orientation to ensure quality (Di Prima et al., 2016). Standards development is crucial for demonstrating safety and efficacy across different geographic domains (Rafi et al., 2022). Critical factors in 3D printing for medical applications include process parameters, material properties, and regulatory requirements such as sterility and mechanical properties (Ng et al., 2024). The healthcare industry's early adoption of 3D printing technology has highlighted its potential for personalized medical devices and complex structures (Rafi et al., 2022). However, robust regulatory frameworks and standards are essential to provide confidence in the use of 3D-printed medical products (Ng et al., 2024; Rafi et al., 2022).

6. POST-PROCESSING AND QUALITY CONTROL

Post-processing techniques play a vital role in enhancing the quality of parts produced through additive manufacturing, addressing issues like poor surface finish and inadequate mechanical properties (Peng et al., 2021). Various methods, including grinding, drag finishing, and surface mechanical treatment, can significantly reduce surface roughness and enhance microhardness (Behjat et al., 2024). Laser shock peening and laser polishing have shown promising results in improving surface characteristics and mechanical properties, with laser polishing capable of reducing surface roughness by up to 95% (Mahmood et al., 2022). Thermal post-processing techniques can eliminate pores, enhance corrosion resistance, and improve mechanical properties (Mahmood et al., 2022). Pore removal, better corrosion resistance, and higher mechanical properties could be achieved using thermal post-processing methods (Mahmood et al., 2022). These techniques have proven to be effective in multiple materials as post-processed parts reveal improvements in surface roughness, microhardness, microstructure, mechanical strength, and wear resistance compared to as-built and Conventional conditions (Khan et al., 2021). Standardization of post-processing procedures for particular defects is needed for further Advancements.

7. AI-DRIVEN ADVANCEMENTS IN PATIENT-SPECIFIC IMPLANTS

Latest developments in the patient-specific implant (PSI) design, artificial intelligence (AI) and machine learning (ML) methods to improve efficiency and cut cost (Memon et al., 2021). The term 'biomaterialomics' embraces multidimensional data optimization and an AI approach to the development of biomaterials (Basu et al., 2022). Computational models and experimental databases are combined with advanced manufacturing (e.g., 3D printing) and experimental techniques to develop fourth-generation biomaterials and implants (Kilian & Moghe, 2016). Patient-specific total temporomandibular joint replacement (PS-TMJR) implants, for instance, fit better anatomically and have better fixation, but their designs, materials and manufacturing methods are not consistent (Mehrotra et al., 2021). Data science is central to the development of 'digital twins' that are used to model and predict clinical performance of materials and the processing-structure-property relationship of materials science and technology with biomaterials science (Basu et al., 2022). The use of artificial intelligence (AI) and machine learning (ML) tools in biomaterials science is becoming increasingly common, while the road toward providing in-depth training in data science will be instrumental in pushing this scientific community forward.

8. FAILURE ANALYSIS AND LONG-TERM PERFORMANCE

The titanium alloy Ti-6Al-4v, which is one of the materials that dentists and orthopedicians use, is affected by a variety of failure mechanisms. Corrosion is a major issue, as reported to occur as a selective attack on the microstructure of the alloy, especially in the case of a fluoride-rich atmosphere (X. Chen et al., 2020). Crevice corrosion within the modular junction can also result in oxide-induced stress corrosion cracking (OISCC), resulting in implant failure (Gilbert et al., 2012). Although in some conditions corrosion fatigue is not a Primary consideration, notches can reduce fatigue strength markedly. Fretting and fatigue have been demonstrated to be significant concerns with modular neck stems. Galvanic corrosion has been found to contribute to failure in mixed-alloy systems (Dolinar et al., 2023). These failure mechanisms can cause corrosion by-products to leach out into the surrounding biological environment, thereby leading to potential complications, such as peri-implant bone resorption (X. Chen et al., 2020). Knowledge of these mechanisms is important for the development of better implant designs and materials.

9. CHALLENGES IN CLINICAL ADOPTION

AM technologies such as selective laser melting (SLM) and electron beam melting (EBM) have shown great potential for the manufacture of Ti alloy medical implants, especially Ti6Al4V (Sing et al., 2016; Trevisan et al., 2018). These processes foster the generation of complex porous structures, which have been proven to be beneficial to osseointegration and mechanical properties, as well as at the same time feasible for patient-specific designs (Trevisan et al., 2018). Meanwhile, it is still a challenge to develop such that it can have the optimal mechanical properties, surface integration ability and antibacterial properties similar to anatomical bone (Aufa et al., 2022). Challenges of stress-shielding in load-bearing applications, the need for higher corrosion resistance remains (Aufa et al., 2022). Notwithstanding these difficulties, AM technologies have shown promise for transforming total joint arthroplasty by minimizing processing costs, reducing waste, and facilitating customized implants tailored to patient-specific information (Trevisan et al., 2018). Current research targets overcoming existing constraints to further propel the clinical acceptance of AM-made Ti6Al4V implants.

10. CURRENT TRENDS AND FUTURE OUTLOOK

Based on (Grigoriev et al., 2023), Figure 7 illustrates information on joint replacement procedures in adults from 2017 to 2020 in Moscow, Russia. In 2019, there were 136,442 surgeries, of which 76,849 were hip replacements (56.3%) and 54,720 were knee replacements (40.1%). The other joint replacements, including shoulder, elbow, ankle, and hand joints, represented 4,873 surgeries (3.6%). Between 2017 and 2019, surgeries steadily rose by 16.2%. There was a change in the trend in 2020.

Figure 7. Joint Replacement Surgeries in Russia (2017 To 2020)

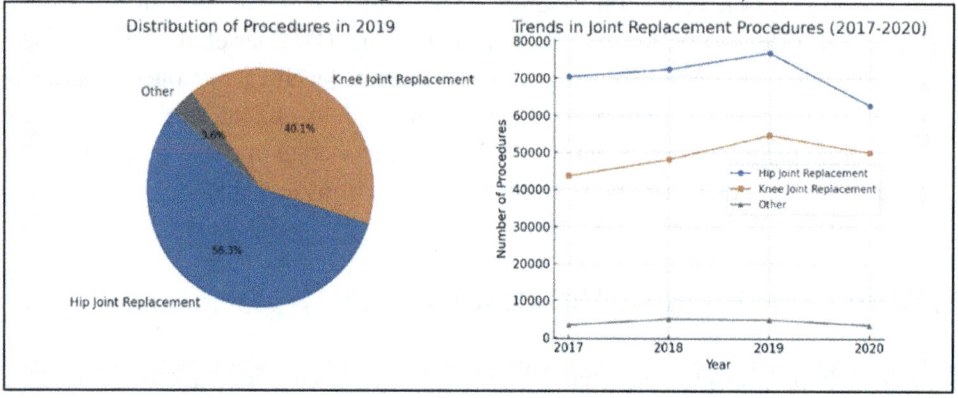

Figure 8 shows data on titanium implants [88]; a search in the Scopus database for titanium and implant from 1989 (when the first titanium implant patent was registered) found 11,968 publications. Over the past decade, the number of publications more than doubled, rising from 5,002 to 11,968.

Figure 8. Publications On Titanium Implants From 1989 To the Present.

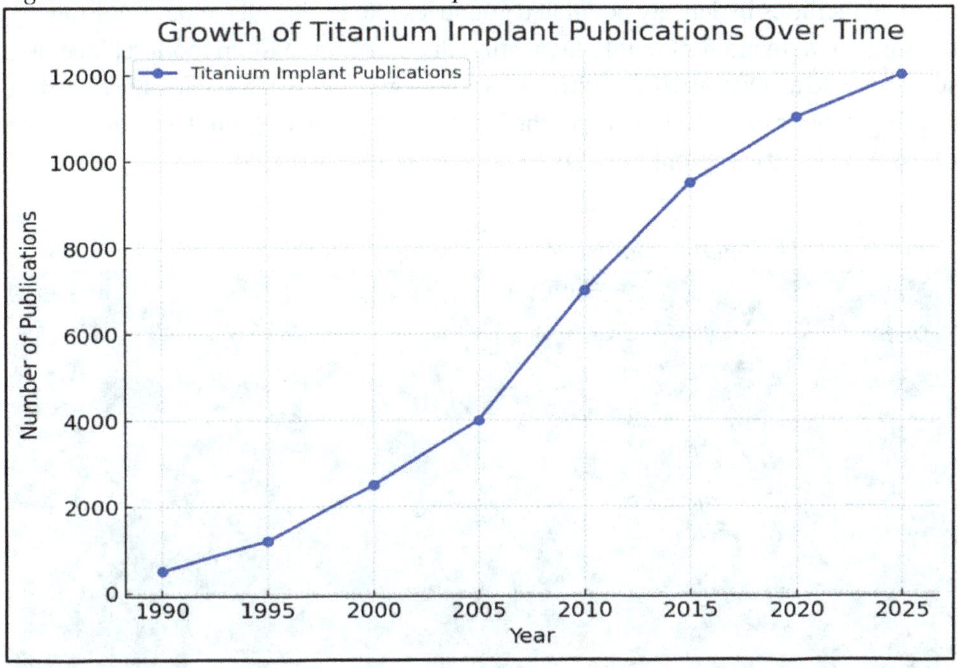

In 2020, medical implants were the leading sector, accounting for 32% of the total market share, and it is anticipated that this trend will persist in the future due to the increasing demand and need for implants. In 2020, the medical industry was valued at 1.34 billion dollars. The projected annual compound growth rate for the years 2021-2028 is 21.8%. The market sizes have expanded to 1.61 billion dollars and 6.44 billion dollars, respectively, between the years 2021 and 2028 (Sağbaş B & Gavcar B, 2022). Since 3D printing enables personalized implant production, the percentage of implants manufactured using this technology is expected to rise in the future. North America is anticipated to dominate the market with more than 36% revenue share in 2023 and is poised for considerable growth during the forecast period, shown in Figure 9 (Deepa Pandey, 2025). This expansion is fueled by key players in the area, such as GE Additive, 3D Systems, Inc., and Stratasys Ltd., who are making significant investments in research and development for innovative healthcare technologies. Additionally, U.S. government policies are fostering market growth. For example, on September 8, 2021, 3D Systems acquired Oqton, a software firm recognized for its smart, cloud-based Manufacturing Operating System (MOS). This acquisition will enable 3D Systems to establish more adaptable production environments that integrate sophisticated manufacturing and automated technologies, including AM capabilities. Conversely, the Asia Pacific region is projected to be the fastest-growing market, driven by emerging economies and the demand for advanced medical facilities in densely populated countries like India and China. Additionally, the surge in demand for dental 3D printing due to an increase in tooth replacement surgeries and an increase in arthritis cases, along with swift healthcare infrastructure development, is expected to propel the Healthcare Additive Manufacturing Market in the Asia Pacific region[REMOVED HYPERLINK FIELD].

Figure 9. U.S. Healthcare Additive Manufacturing Market Share, By Region, 2023 (%)

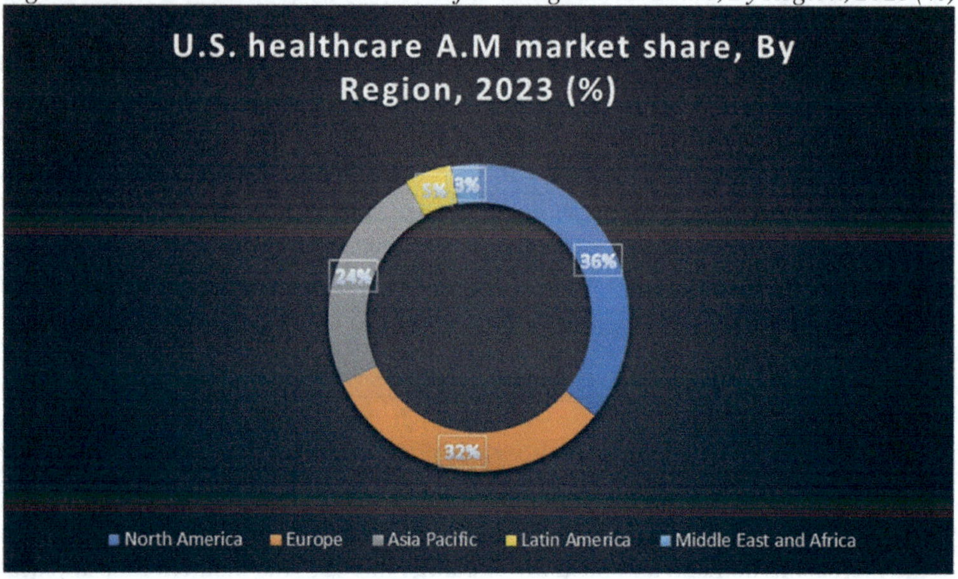

In India, the additive manufacturing market generated \$61.4 million in 2023 and is projected to reach \$314.0 million by 2030, with an impressive annual growth rate (CAGR) of 26.3% from 2024 to 2030. The hardware segment was the top revenue earner in 2023 and is set to grow the fastest in the coming years.

11. CONCLUSION

Additive Manufacturing (AM) has become a revolutionary method in the production of biomedical implants, especially those made of Ti6Al4V alloy, because of its higher strength-to-weight ratio, biocompatibility, and capacity to create complex, patient-specific shapes. Methods like Powder Bed Fusion (PBF), Binder Jetting (BJ), and Directed Energy Deposition (DED) bring unique benefits compared to conventional methods of production, such as the minimization of waste material and ease of design adaptability. However, in spite of these developments, there are still several issues that remain—e.g., residual stresses, surface roughness, and doubts regarding long-term biocompatibility. To address these challenges, post-processing treatments, such as heat treatment and surface modification, are required to provide the required mechanical and biological properties of additively manufactured implants. In addition, regulatory guidelines are required to ensure the safety, uniformity, and clinical effectiveness of these medical devices. Looking to the future, the future for additive manufacturing (AM) in biomedical applications looks highly promising. State-of-the-art research efforts are aimed at process parameter optimization, new alloy system exploration, and artificial intelligence-based design and simulation methodologies. As the need for individually designed, high-performance implants grows, AM will be a primary driver for the next generation of personalized healthcare devices. Intensive interdisciplinary cooperation between researchers, producers, and regulators will be necessary to maximize the potential of AM technologies in the biomedical field.

ACKNOWLEDGEMENT

The authors appreciate the encouragement provided and sincerely acknowledges the comments and suggestions of the editors and reviewers that have been instrumental for improving and upgrading the paper in its present form.

REFERENCES

Acquesta, A., & Monetta, T. (2020). As-Built EBM and DMLS Ti-6Al-4V Parts: Topography–Corrosion Resistance Relationship in a Simulated Body Fluid. *Metals, 10*(8), 1015. DOI: 10.3390/met10081015

Ahangar, P., Cooke, M. E., Weber, M. H., & Rosenzweig, D. H. (2019). Current Biomedical Applications of 3D Printing and Additive Manufacturing. *Applied Sciences (Basel, Switzerland)*, *9*(8), 1713. DOI: 10.3390/app9081713

Ataee, A., Li, Y., & Wen, C. (2019). A comparative study on the nanoindentation behavior, wear resistance and in vitro biocompatibility of SLM manufactured CP–Ti and EBM manufactured Ti64 gyroid scaffolds. *Acta Biomaterialia*, *97*, 587–596. DOI: 10.1016/j.actbio.2019.08.008

Aufa, A. N., Hassan, M. Z., & Ismail, Z. (2022). Recent advances in Ti-6Al-4V additively manufactured by selective laser melting for biomedical implants: Prospect development. *Journal of Alloys and Compounds*, *896*, 163072. DOI: 10.1016/j.jallcom.2021.163072

Bai, Y., & Williams, C. B. (2015). An exploration of binder jetting of copper. *Rapid Prototyping Journal*, *21*(2), 177–185. DOI: 10.1108/RPJ-12-2014-0180

Balla, V. K., Bodhak, S., Bose, S., & Bandyopadhyay, A. (2010). Porous tantalum structures for bone implants: Fabrication, mechanical and in vitro biological properties. *Acta Biomaterialia*, *6*(8), 3349–3359. DOI: 10.1016/j.actbio.2010.01.046

Bandyopadhyay, A., Espana, F., Balla, V. K., Bose, S., Ohgami, Y., & Davies, N. M. (2010). Influence of porosity on mechanical properties and in vivo response of Ti6Al4V implants. *Acta Biomaterialia*, *6*(4), 1640–1648. DOI: 10.1016/j.actbio.2009.11.011

Basu, B., Gowtham, N. H., Xiao, Y., Kalidindi, S. R., & Leong, K. W. (2022). Biomaterialomics: Data science-driven pathways to develop fourth-generation biomaterials. *Acta Biomaterialia*, *143*, 1–25. DOI: 10.1016/j.actbio.2022.02.027

Behjat, A., Sanaei, S., Mosallanejad, M. H., Atapour, M., Sheikholeslam, M., Saboori, A., & Iuliano, L. (2024). A novel titanium alloy for load-bearing biomedical implants: Evaluating the antibacterial and biocompatibility of Ti536 produced via electron beam powder bed fusion additive manufacturing process. *Biomaterials Advances*, *163*, 213928. DOI: 10.1016/j.bioadv.2024.213928

Breuning, C., Pistor, J., Markl, M., & Körner, C. (2022). Basic Mechanism of Surface Topography Evolution in Electron Beam Based Additive Manufacturing. *Materials (Basel)*, *15*(14), 4754. DOI: 10.3390/ma15144754

Chen, W.-M., Xie, Y. M., Imbalzano, G., Shen, J., Xu, S., Lee, S.-J., & Lee, P. V. S. (2016). Lattice Ti structures with low rigidity but compatible mechanical strength: Design of implant materials for trabecular bone. *International Journal of Precision Engineering and Manufacturing*, *17*(6), 793–799. DOI: 10.1007/s12541-016-0097-6

Chen, X., Shah, K., Dong, S., Peterson, L., Callagon La Plante, E., & Sant, G. (2020). Elucidating the corrosion-related degradation mechanisms of a Ti-6Al-4V dental implant. *Dental Materials*, *36*(3), 431–441. DOI: 10.1016/j.dental.2020.01.008

Cheng, X. Y., Li, S. J., Murr, L. E., Zhang, Z. B., Hao, Y. L., Yang, R., Medina, F., & Wicker, R. B. (2012). Compression deformation behavior of Ti–6Al–4V alloy with cellular structures fabricated by electron beam melting. *Journal of the Mechanical Behavior of Biomedical Materials*, *16*, 153–162. DOI: 10.1016/j.jmbbm.2012.10.005

Chua, Z. Y., Ahn, I. H., & Moon, S. K. (2017). Process monitoring and inspection systems in metal additive manufacturing: Status and applications. *International Journal of Precision Engineering and Manufacturing-Green Technology*, *4*(2), 235–245. DOI: 10.1007/s40684-017-0029-7

Chudinova, E. A., Surmeneva, M. A., Timin, A. S., Karpov, T. E., Wittmar, A., Ulbricht, M., Ivanova, A., Loza, K., Prymak, O., Koptyug, A., Epple, M., & Surmenev, R. A. (2019). Adhesion, proliferation, and osteogenic differentiation of human mesenchymal stem cells on additively manufactured Ti6Al4V alloy scaffolds modified with calcium phosphate nanoparticles. *Colloids and Surfaces. B, Biointerfaces*, *176*, 130–139. DOI: 10.1016/j.colsurfb.2018.12.047

Dawood, A., Marti, B. M., Sauret-Jackson, V., & Darwood, A. (2015). 3D printing in dentistry. *British Dental Journal*, *219*(11), 521–529. DOI: 10.1038/sj.bdj.2015.914

Deepa Pandey. (2025). Healthcare Additive Manufacturing Market.

Di Prima, M., Coburn, J., Hwang, D., Kelly, J., Khairuzzaman, A., & Ricles, L. (2016). Additively manufactured medical products – the FDA perspective. 3D Printing in Medicine, 2(1), 1. https://doi.org/DOI: 10.1186/s41205-016-0005-9

Dolinar, D., Gorenšek, M., Avsec, K., Šetina Batič, B., Hočevar, M., Godec, M., Žužek, B., Debeljak, M., Jenko, M., Grant, J. T., & Kocjančič, B. (2023). Mechanisms of Premature Fracture in Modular Neck Stems Made of CoCrMo/Ti6Al4V and Ti6Al4V/Ti6Al4V Alloy. *Coatings*, *13*(7), 1255. DOI: 10.3390/coatings13071255

Dutta, B., & Froes, F. H. (2017). The Additive Manufacturing (AM) of titanium alloys. *Metal Powder Report*, *72*(2), 96–106. DOI: 10.1016/j.mprp.2016.12.062

Etesami, S. A., Fotovvati, B., & Asadi, E. (2022). Heat treatment of Ti-6Al-4V alloy manufactured by laser-based powder-bed fusion: Process, microstructures, and mechanical properties correlations. *Journal of Alloys and Compounds*, *895*, 162618. DOI: 10.1016/j.jallcom.2021.162618

Everton, S. K., Hirsch, M., Stravroulakis, P., Leach, R. K., & Clare, A. T. (2016). Review of in-situ process monitoring and in-situ metrology for metal additive manufacturing. *Materials & Design*, *95*, 431–445. DOI: 10.1016/j.matdes.2016.01.099

Frazier, W. E. (2014). Metal Additive Manufacturing: A Review. *Journal of Materials Engineering and Performance*, *23*(6), 1917–1928. DOI: 10.1007/s11665-014-0958-z

Gao, C., Wang, C., Jin, H., Wang, Z., Li, Z., Shi, C., Leng, Y., Yang, F., Liu, H., & Wang, J. (2018a). Additive manufacturing technique-designed metallic porous implants for clinical application in orthopedics. *RSC Advances*, *8*(44), 25210–25227. DOI: 10.1039/C8RA04815K

Gibson, I., Rosen, D., Stucker, B., & Khorasani, M. (2021a). *Additive Manufacturing Technologies* (Vol. 17). Springer International Publishing., DOI: 10.1007/978-3-030-56127-7

Gilbert, J. L., Mali, S., Urban, R. M., Silverton, C. D., & Jacobs, J. J. (2012). In vivo oxide-induced stress corrosion cracking of Ti-6Al-4V in a neck–stem modular taper: Emergent behavior in a new mechanism of in vivo corrosion. *Journal of Biomedical Materials Research. Part B, Applied Biomaterials*, *100B*(2), 584–594. DOI: 10.1002/jbm.b.31943

Gill, S. S., Arora, H., Jidesh, , & Sheth, V. (2017). On the development of Antenna feed array for space applications by additive manufacturing technique. *Additive Manufacturing*, *17*, 39–46. DOI: 10.1016/j.addma.2017.06.010

Grigoriev, S., Sotova, C., Vereschaka, A., Uglov, V., & Cherenda, N. (2023). Modifying Coatings for Medical Implants Made of Titanium Alloys. *Metals*, *13*(4), 718. DOI: 10.3390/met13040718

Haglin, J. M., Eltorai, A. E. M., Gil, J. A., Marcaccio, S. E., Botero-Hincapie, J., & Daniels, A. H. (2016a). Patient-Specific Orthopaedic Implants. *Orthopaedic Surgery*, *8*(4), 417–424. DOI: 10.1111/os.12282

Haglin, J. M., Eltorai, A. E. M., Gil, J. A., Marcaccio, S. E., Botero-Hincapie, J., & Daniels, A. H. (2016b). Patient-Specific Orthopaedic Implants. *Orthopaedic Surgery*, *8*(4), 417–424. DOI: 10.1111/os.12282

Halloran, J. W. (1999). Freeform fabrication of ceramics. *British Ceramic Transactions*, *98*(6), 299–303. DOI: 10.1179/096797899680633

Hanawa, T. (2019). Titanium–Tissue Interface Reaction and Its Control With Surface Treatment. *Frontiers in Bioengineering and Biotechnology*, *7*, 170. Advance online publication. DOI: 10.3389/fbioe.2019.00170

Haydar, H. J., & Al-Deen, J. (1973). AbidAli, A. K., & Mahmoud, A. A. (2021). Improved performance of Ti6Al4V alloy in Biomedical applications - Review. *Journal of Physics: Conference Series*, *012146*(1). DOI: 10.1088/1742-6596/1973/1/012146

Heinl, P., Körner, C., & Singer, R. F. (2008). Selective Electron Beam Melting of Cellular Titanium: Mechanical Properties. *Advanced Engineering Materials*, *10*(9), 882–888. DOI: 10.1002/adem.200800137

Huang, R., Riddle, M., Graziano, D., Warren, J., Das, S., Nimbalkar, S., Cresko, J., & Masanet, E. (2016). Energy and emissions saving potential of additive manufacturing: The case of lightweight aircraft components. *Journal of Cleaner Production*, *135*, 1559–1570. DOI: 10.1016/j.jclepro.2015.04.109

Jang, T.-S., Kim, D., Han, G., Yoon, C.-B., & Jung, H.-D. (2020a). Powder based additive manufacturing for biomedical application of titanium and its alloys: A review. *Biomedical Engineering Letters*, *10*(4), 505–516. DOI: 10.1007/s13534-020-00177-2

Ju, J., Zhou, Y., Wang, K., Liu, Y., Li, J., Kang, M., & Wang, J. (2020). Tribological investigation of additive manufacturing medical Ti6Al4V alloys against Al2O3 ceramic balls in artificial saliva. *Journal of the Mechanical Behavior of Biomedical Materials*, *104*, 103602. DOI: 10.1016/j.jmbbm.2019.103602

Khan, H. M., Karabulut, Y., Kitay, O., Kaynak, Y., & Jawahir, I. S. (2021). Influence of the post-processing operations on surface integrity of metal components produced by laser powder bed fusion additive manufacturing: A review. *Machining Science and Technology*, *25*(1), 118–176. DOI: 10.1080/10910344.2020.1855649

Khorasani, A., Gibson, I., Veetil, J. K., & Ghasemi, A. H. (2020). A review of technological improvements in laser-based powder bed fusion of metal printers. *International Journal of Advanced Manufacturing Technology*, *108*(1–2), 191–209. DOI: 10.1007/s00170-020-05361-3

Kilian, K. A., & Moghe, P. V. (2016). High throughput strategies for the design, discovery, and analysis of biomaterials. *Acta Biomaterialia*, *34*, v–vi. DOI: 10.1016/j.actbio.2016.03.019

Lalegani Dezaki, M., Serjouei, A., Zolfagharian, A., Fotouhi, M., Moradi, M., Ariffin, M. K. A., & Bodaghi, M. (2022). A review on additive/subtractive hybrid manufacturing of directed energy deposition (DED) process. *Advanced Powder Materials*, *1*(4), 100054. DOI: 10.1016/j.apmate.2022.100054

Letov, N., Velivela, P. T., Sun, S., & Zhao, Y. F. (2021). Challenges and Opportunities in Geometric Modeling of Complex Bio-Inspired Three-Dimensional Objects Designed for Additive Manufacturing. *Journal of Mechanical Design*, *143*(12), 121705. Advance online publication. DOI: 10.1115/1.4051720

Liu, S., & Shin, Y. C. (2019a). Additive manufacturing of Ti6Al4V alloy: A review. *Materials & Design*, *164*, 107552. DOI: 10.1016/j.matdes.2018.107552

Lores, A., Azurmendi, N., Agote, I., & Zuza, E. (2019). A review on recent developments in binder jetting metal additive manufacturing: Materials and process characteristics. *Powder Metallurgy*, *62*(5), 267–296. DOI: 10.1080/00325899.2019.1669299

Lütjering, G., & Williams, J. C. (2003). *Titanium Matrix Composites.*, DOI: 10.1007/978-3-540-71398-2_9

Mahamood, R. M., Akinlabi, S. A., Shatalov, M., Murashkin, E. V., & Akinlabi, E. T. (2019). Additive Manufacturing / 3D Printing Technology: A Review. *Annals of Dunarea de Jos University of Galati Fascicle XII Welding Equipment and Technology*, *30*, 51–58. DOI: 10.35219/awet.2019.07

Maharubin, S., Hu, Y., Sooriyaarachchi, D., Cong, W., & Tan, G. Z. (2019). Laser engineered net shaping of antimicrobial and biocompatible titanium-silver alloys. *Materials Science and Engineering C*, *105*, 110059. DOI: 10.1016/j.msec.2019.110059

Mahmood, M. A., Chioibasu, D., Ur Rehman, A., Mihai, S., & Popescu, A. C. (2022). Post-Processing Techniques to Enhance the Quality of Metallic Parts Produced by Additive Manufacturing. *Metals*, *12*(1), 77. DOI: 10.3390/met12010077

Mahmud, A., Ding, D., Kiani, A., & Hasan, M. (2020). Corporate Social Responsibility Programs and Community Perceptions of Societal Progress in Bangladesh: A Multimethod Approach. *SAGE Open*, *10*(2), 2158244020924046. Advance online publication. DOI: 10.1177/2158244020924046

Mandal, S., Sadeghianjahromi, A., & Wang, C.-C. (2022). Experimental and numerical investigations on molten metal atomization techniques – A critical review. *Advanced Powder Technology*, *33*(11), 103809. DOI: 10.1016/j.apt.2022.103809

Marattukalam, J. J., Singh, A. K., Datta, S., Das, M., Balla, V. K., Bontha, S., & Kalpathy, S. K. (2015). Microstructure and corrosion behavior of laser processed NiTi alloy. *Materials Science and Engineering C*, *57*, 309–313. DOI: 10.1016/j.msec.2015.07.067

Mathias, L. E. T., Pinotti, V. E., Batistão, B. F., Rojas-Arias, N., Figueira, G., Andreoli, A. F., & Gargarella, P. (2024). Metal powder as feedstock for laser-based additive manufacturing: From production to powder modification. *Journal of Materials Research*, *39*(1), 19–47. DOI: 10.1557/s43578-023-01271-8

Matsuno, H. (2001). Biocompatibility and osteogenesis of refractory metal implants, titanium, hafnium, niobium, tantalum and rhenium. *Biomaterials*, *22*(11), 1253–1262. DOI: 10.1016/S0142-9612(00)00275-1

Mehrotra, D., Kumar, S., Mehrotra, P., Khanna, R., Khanna, V., Eggbeer, D., & Evans, P. (2021). Patient specific total temporomandibular joint reconstruction: A review of biomaterial, designs, fabrication and outcomes. *Journal of Oral Biology and Craniofacial Research*, *11*(2), 334–343. DOI: 10.1016/j.jobcr.2021.02.014

Memon, A. R., Li, J., Egger, J., & Chen, X. (2021). A review on patient-specific facial and cranial implant design using Artificial Intelligence (AI) techniques. *Expert Review of Medical Devices*, *18*(10), 985–994. DOI: 10.1080/17434440.2021.1969914

Mercelis, P., & Kruth, J. (2006). Residual stresses in selective laser sintering and selective laser melting. *Rapid Prototyping Journal*, *12*(5), 254–265. DOI: 10.1108/13552540610707013

Meteyer, S., Xu, X., Perry, N., & Zhao, Y. F. (2014). Energy and Material Flow Analysis of Binder-jetting Additive Manufacturing Processes. *Procedia CIRP*, *15*, 19–25. DOI: 10.1016/j.procir.2014.06.030

Morrison, R. J., Kashlan, K. N., Flanangan, C. L., Wright, J. K., Green, G. E., Hollister, S. J., & Weatherwax, K. J. (2015). Regulatory Considerations in the Design and Manufacturing of Implantable 3D-Printed Medical Devices. *Clinical and Translational Science*, *8*(5), 594–600. DOI: 10.1111/cts.12315

Mujahid, S., Boostani, A. F., Paudel, Y., Oppedal, A., Paliwal, B., Rhee, H., & El Kadiri, H. (2024). Examination of Novel Titanium-639 Alloy as a Means of Balancing Strength and Ductility through Molybdenum Addition Rather than Prolonged Aging Heat Treatment. *Metals*, *14*(7), 828. DOI: 10.3390/met14070828

Mumtaz, K. A., & Hopkinson, N. (2010). Selective Laser Melting of thin wall parts using pulse shaping. *Journal of Materials Processing Technology*, *210*(2), 279–287. DOI: 10.1016/j.jmatprotec.2009.09.011

Ng, W. L., An, J., & Chua, C. K. (2024). Process, Material, and Regulatory Considerations for 3D Printed Medical Devices and Tissue Constructs. *Engineering (Beijing)*, *36*, 146–166. DOI: 10.1016/j.eng.2024.01.028

Ngo, T. D., Kashani, A., Imbalzano, G., Nguyen, K. T. Q., & Hui, D. (2018). Additive manufacturing (3D printing): A review of materials, methods, applications and challenges. *Composites. Part B, Engineering*, *143*, 172–196. DOI: 10.1016/j.compositesb.2018.02.012

Ni, J., Ling, H., Zhang, S., Wang, Z., Peng, Z., Benyshek, C., Zan, R., Miri, A. K., Li, Z., Zhang, X., Lee, J., Lee, K.-J., Kim, H.-J., Tebon, P., Hoffman, T., Dokmeci, M. R., Ashammakhi, N., Li, X., & Khademhosseini, A. (2019). Three-dimensional printing of metals for biomedical applications. *Materials Today. Bio*, *3*, 100024. DOI: 10.1016/j.mtbio.2019.100024

Nie, L., Zhan, Y., Liu, H., & Tang, C. (2014). Novel β-type Zr–Mo–Ti alloys for biological hard tissue replacements. *Materials & Design*, *53*, 8–12. DOI: 10.1016/j.matdes.2013.07.008

Nied, H. F., Taylor, C. A., & Delorenzi, H. G. (1990). Three-dimensional finite element simulation of thermoforming. *Polymer Engineering and Science*, *30*(20), 1314–1322. DOI: 10.1002/pen.760302009

Pei, E., Bernard, A., Gu, D., Klahn, C., Monzón, M., Petersen, M., & Sun, T. (Eds.). (2023). *Springer Handbook of Additive Manufacturing*. Springer International Publishing., DOI: 10.1007/978-3-031-20752-5

Peng, X., Kong, L., Fuh, J. Y. H., & Wang, H. (2021). A Review of Post-Processing Technologies in Additive Manufacturing. *Journal of Manufacturing and Materials Processing*, *5*(2), 38. DOI: 10.3390/jmmp5020038

Pesode, P., & Barve, S. (2022a). Additive manufacturing of metallic biomaterials and its biocompatibility. *Materials Today: Proceedings*. Advance online publication. DOI: 10.1016/j.matpr.2022.11.248

Pesode, P., & Barve, S. (2022b). Additive manufacturing of metallic biomaterials and its biocompatibility. *Materials Today: Proceedings*. Advance online publication. DOI: 10.1016/j.matpr.2022.11.248

Popov, V. V.Jr, Muller-Kamskii, G., Katz-Demyanetz, A., Kovalevsky, A., Usov, S., Trofimcow, D., Dzhenzhera, G., & Koptyug, A. (2019). Additive manufacturing to veterinary practice: Recovery of bony defects after the osteosarcoma resection in canines. *Biomedical Engineering Letters*, *9*(1), 97–108. DOI: 10.1007/s13534-018-00092-7

Popov, V. V.Jr, Muller-Kamskii, G., Kovalevsky, A., Dzhenzhera, G., Strokin, E., Kolomiets, A., & Ramon, J. (2018). Design and 3D-printing of titanium bone implants: Brief review of approach and clinical cases. *Biomedical Engineering Letters*, *8*(4), 337–344. DOI: 10.1007/s13534-018-0080-5

Rafi, K., Zhonghong Liu, A., Di Prima, M., Bates, P., & Seifi, M. (2022). Regulatory and standards development in medical additive manufacturing. *MRS Bulletin*, *47*(1), 98–105. DOI: 10.1557/s43577-022-00266-w

Richard O'Leary, R. S. P. P. G. H. (2016). An Investigation into the Recycling of Ti-6Al-4V Powder Used Within SLM to Improve Sustainability. In Impact. *The Journal of Innovation Impact*, *8*(2), 377–388.

Roos, S., & Rännar, L.-E. (2024). Surface roughness reduction in electron beam powder bed fusion additive manufacturing of super duplex stainless steel 2507: Investigating optimisation techniques and face orientation-dependent irregularities. *Progress in Additive Manufacturing*. Advance online publication. DOI: 10.1007/s40964-024-00763-1

Roseti, L., Parisi, V., Petretta, M., Cavallo, C., Desando, G., Bartolotti, I., & Grigolo, B. (2017). Scaffolds for Bone Tissue Engineering: State of the art and new perspectives. *Materials Science and Engineering C*, *78*, 1246–1262. DOI: 10.1016/j.msec.2017.05.017

Saboori, A., Aversa, A., Marchese, G., Biamino, S., Lombardi, M., & Fino, P. (2019). Application of Directed Energy Deposition-Based Additive Manufacturing in Repair. *Applied Sciences (Basel, Switzerland)*, *9*(16), 3316. DOI: 10.3390/app9163316

Sağbaş, B., & Gavcar, B. (2022). Additive Manufacturing Of Titanium Alloys For Biomedical Applications. *Uluborlu Mesleki Bilimler Dergisi.*, *5*(1), 54–74.

Salmi, M. (2021). Additive Manufacturing Processes in Medical Applications. *Materials (Basel)*, *14*(1), 191. DOI: 10.3390/ma14010191

Sanaei, N., & Fatemi, A. (2021). Defects in additive manufactured metals and their effect on fatigue performance: A state-of-the-art review. *Progress in Materials Science, 117*, 100724. DOI: 10.1016/j.pmatsci.2020.100724

Sanchez-Rexach, E., Johnston, T. G., Jehanno, C., Sardon, H., & Nelson, A. (2020). Sustainable Materials and Chemical Processes for Additive Manufacturing. *Chemistry of Materials, 32*(17), 7105–7119. DOI: 10.1021/acs.chemmater.0c02008

Sengottaiyan, S., Shaisundaram, V. S., Shameer Basha, M., Deepanraj, B., & Senthilkumar, N. (2025). Effect of duplex and austenite stainless steel fillers on mechanical, microstructural, and corrosion characteristics of dissimilar welds using Cold Metal Transfer welding. *Results in Engineering, 26*, 104818. DOI: 10.1016/j.rineng.2025.104818

Senthilkumar, N., Perumal, G., Palanikumar, K., Deepanraj, B., & Shameer Basha, M. (2025). Wear and mechanical characteristics of coated Ti-6Al-4V alloy with an alumina–13 wt.% titania–40 wt.% yttrium stabilized zirconia ceramic composite via plasma spraying technique. *Results in Engineering, 25*, 103777. DOI: 10.1016/j.rineng.2024.103777

Shanthar, R., Chen, K., & Abeykoon, C. (2023). Powder-Based Additive Manufacturing: A Critical Review of Materials, Methods, Opportunities, and Challenges. *Advanced Engineering Materials, 25*(19), 2300375. Advance online publication. DOI: 10.1002/adem.202300375

Shapiro, A. A., Borgonia, J. P., Chen, Q. N., Dillon, R. P., McEnerney, B., Polit-Casillas, R., & Soloway, L. (2016). Additive Manufacturing for Aerospace Flight Applications. *Journal of Spacecraft and Rockets, 53*(5), 952–959. DOI: 10.2514/1.A33544

Sharabian, E., Leary, M., Fraser, D., & Gulizia, S. (2022). Electron beam powder bed fusion of copper components: A review of mechanical properties and research opportunities. *International Journal of Advanced Manufacturing Technology, 122*(2), 513–532. DOI: 10.1007/s00170-022-09922-6

Simchi, A., Petzoldt, F., Hartwig, T., Hein, S. B., Barthel, B., & Reineke, L. (2022). Binder Jet 3d Printing of Biomedical Grade Ti-6al-4v Alloy: The Role of Particle Size, Microstructural Features, and Mechanical Properties. SSRN Electronic Journal. https://doi.org/DOI: 10.2139/ssrn.4127679

Simchi, A., Petzoldt, F., Hartwig, T., Hein, S. B., Barthel, B., & Reineke, L. (2023). Microstructural development during additive manufacturing of biomedical grade Ti-6Al-4V alloy by three-dimensional binder jetting: Material aspects and mechanical properties. *International Journal of Advanced Manufacturing Technology*, *127*(3–4), 1541–1558. DOI: 10.1007/s00170-023-11661-1

Sing, S. L. (2022). Perspectives on Additive Manufacturing Enabled BetaTitanium Alloys for Biomedical Applications. *International Journal of Bioprinting*, *8*(1), 478. DOI: 10.18063/ijb.v8i1.478

Sing, S. L., An, J., Yeong, W. Y., & Wiria, F. E. (2016). Laser and electron-beam powder-bed additive manufacturing of metallic implants: A review on processes, materials and designs. *Journal of Orthopaedic Research*, *34*(3), 369–385. DOI: 10.1002/jor.23075

Singh, A., Kapil, S., & Das, M. (2020). A comprehensive review of the methods and mechanisms for powder feedstock handling in directed energy deposition. *Additive Manufacturing*, *35*, 101388. DOI: 10.1016/j.addma.2020.101388

Singh, A. B., Khandelwal, C., & Dangayach, G. S. (2024). Revolutionizing healthcare materials: Innovations in processing, advancements, and challenges for enhanced medical device integration and performance. Journal of Micromanufacturing. https://doi.org/DOI: 10.1177/25165984241256234

Soong, S. Z., Lai, W. L., & Kay Lup, A. N. (2023). Atomization of metal and alloy powders: Processes, parameters, and properties. *AIChE Journal. American Institute of Chemical Engineers*, *69*(11), e18217. Advance online publication. DOI: 10.1002/aic.18217

Sun, Y., Aindow, M., & Hebert, R. J. (2018). Comparison of virgin Ti-6Al-4V powders for additive manufacturing. *Additive Manufacturing*, *21*, 544–555. DOI: 10.1016/j.addma.2018.02.011

Svetlizky, D., Das, M., Zheng, B., Vyatskikh, A. L., Bose, S., Bandyopadhyay, A., Schoenung, J. M., Lavernia, E. J., & Eliaz, N. (2021). Directed energy deposition (DED) additive manufacturing: Physical characteristics, defects, challenges and applications. *Materials Today*, *49*, 271–295. DOI: 10.1016/j.mattod.2021.03.020

Takizawa, T., Nakayama, N., Haniu, H., Aoki, K., Okamoto, M., Nomura, H., Tanaka, M., Sobajima, A., Yoshida, K., Kamanaka, T., Ajima, K., Oishi, A., Kuroda, C., Ishida, H., Okano, S., Kobayashi, S., Kato, H., & Saito, N. (2018). Titanium Fiber Plates for Bone Tissue Repair. *Advanced Materials*, *30*(4), 1703608. Advance online publication. DOI: 10.1002/adma.201703608

Tamayo, J. A., Riascos, M., Vargas, C. A., & Baena, L. M. (2021). Additive manufacturing of Ti6Al4V alloy via electron beam melting for the development of implants for the biomedical industry. *Heliyon*, *7*(5), e06892. DOI: 10.1016/j.heliyon.2021.e06892

Tan, X. P., Tan, Y. J., Chow, C. S. L., Tor, S. B., & Yeong, W. Y. (2017). Metallic powder-bed based 3D printing of cellular scaffolds for orthopaedic implants: A state-of-the-art review on manufacturing, topological design, mechanical properties and biocompatibility. *Materials Science and Engineering C*, *76*, 1328–1343. DOI: 10.1016/j.msec.2017.02.094

Tang, H. P., Qian, M., Liu, N., Zhang, X. Z., Yang, G. Y., & Wang, J. (2015). Effect of Powder Reuse Times on Additive Manufacturing of Ti-6Al-4V by Selective Electron Beam Melting. *Journal of the Minerals Metals & Materials Society*, *67*(3), 555–563. DOI: 10.1007/s11837-015-1300-4

Thompson, S. M., Bian, L., Shamsaei, N., & Yadollahi, A. (2015). An overview of Direct Laser Deposition for additive manufacturing; Part I: Transport phenomena, modeling and diagnostics. *Additive Manufacturing*, *8*, 36–62. DOI: 10.1016/j.addma.2015.07.001

Tong, J., Bowen, C. R., Persson, J., & Plummer, A. (2017). Mechanical properties of titanium-based Ti–6Al–4V alloys manufactured by powder bed additive manufacture. *Materials Science and Technology*, *33*(2), 138–148. DOI: 10.1080/02670836.2016.1172787

Trevisan, F., Calignano, F., Aversa, A., Marchese, G., Lombardi, M., Biamino, S., Ugues, D., & Manfredi, D. (2018). Additive manufacturing of titanium alloys in the biomedical field: Processes, properties and applications. *Journal of Applied Biomaterials & Functional Materials*, *16*(2), 57–67. DOI: 10.5301/jabfm.5000371

Vallone, M., Damasio, M., Atzeni, E., Poggi, M., & Salmi, A. (2023). *Thermo-mechanical Modelling of the Directed Energy Deposition (DED).* Process for the Optimization of Deposition Strategies., DOI: 10.1007/978-3-031-28547-9_23

Vrancken, B., Buls, S., Kruth, J., & Van Humbeeck, J. (2016). Preheating of Selective Laser Melted Ti6Al4V: Microstructure and Mechanical Properties. In *Proceedings of the 13th World Conference on Titanium* (pp. 1269–1277). Wiley. https://doi.org/ DOI: 10.1002/9781119296126.ch215

Wang, J.-X., Fan, Y.-B., Gao, Y., Hu, Q.-H., & Wang, T.-C. (2009). TiO2 nanoparticles translocation and potential toxicological effect in rats after intraarticular injection. *Biomaterials*, *30*(27), 4590–4600. DOI: 10.1016/j.biomaterials.2009.05.008

Wang, X., Xu, S., Zhou, S., Xu, W., Leary, M., Choong, P., Qian, M., Brandt, M., & Xie, Y. M. (2016). Topological design and additive manufacturing of porous metals for bone scaffolds and orthopaedic implants: A review. *Biomaterials*, *83*, 127–141. DOI: 10.1016/j.biomaterials.2016.01.012

Wong, K. V., & Hernandez, A. (2012). A Review of Additive Manufacturing. *ISRN Mechanical Engineering*, *2012*, 1–10. DOI: 10.5402/2012/208760

Xiong, J., Yin, Z., & Zhang, W. (2016). Forming appearance control of arc striking and extinguishing area in multi-layer single-pass GMAW-based additive manufacturing. *International Journal of Advanced Manufacturing Technology*, *87*(1–4), 579–586. DOI: 10.1007/s00170-016-8543-2

Xu, J., Gu, X., Ding, D., Pan, Z., & Chen, K. (2018). A review of slicing methods for directed energy deposition based additive manufacturing. *Rapid Prototyping Journal*, *24*(6), 1012–1025. DOI: 10.1108/RPJ-10-2017-0196

Xue, W., Krishna, B. V., Bandyopadhyay, A., & Bose, S. (2007). Processing and biocompatibility evaluation of laser processed porous titanium. *Acta Biomaterialia*, *3*(6), 1007–1018. DOI: 10.1016/j.actbio.2007.05.009

Yamanoglu, R., Efendi, E., Kolayli, F., Uzuner, H., & Daoud, I. (2018). Production and mechanical properties of Ti–5Al–2.5Fe– x Cu alloys for biomedical applications. *Biomedical Materials (Bristol, England)*, *13*(2), 025013. DOI: 10.1088/1748-605X/aa957d

Yan, C., Hao, L., Hussein, A., & Young, P. (2015). Ti–6Al–4V triply periodic minimal surface structures for bone implants fabricated via selective laser melting. *Journal of the Mechanical Behavior of Biomedical Materials*, *51*, 61–73. DOI: 10.1016/j.jmbbm.2015.06.024

Yan, R., Luo, D., Huang, H., Li, R., Yu, N., Liu, C., Hu, M., & Rong, Q. (2018). Electron beam melting in the fabrication of three-dimensional mesh titanium mandibular prosthesis scaffold. *Scientific Reports*, *8*(1), 750. DOI: 10.1038/s41598-017-15564-6

Yánez, A., Herrera, A., Martel, O., Monopoli, D., & Afonso, H. (2016). Compressive behaviour of gyroid lattice structures for human cancellous bone implant applications. *Materials Science and Engineering C*, *68*, 445–448. DOI: 10.1016/j.msec.2016.06.016

Zhang, L., & Chen, L. (2019). A Review on Biomedical Titanium Alloys: Recent Progress and Prospect. *Advanced Engineering Materials*, *21*(4), 1801215. Advance online publication. DOI: 10.1002/adem.201801215

Zhang, Y., Wu, L., Guo, X., Kane, S., Deng, Y., Jung, Y.-G., Lee, J.-H., & Zhang, J. (2018). Additive Manufacturing of Metallic Materials: A Review. *Journal of Materials Engineering and Performance*, *27*(1), 1–13. DOI: 10.1007/s11665-017-2747-y

Ziaee, M., & Crane, N. B. (2019). Binder jetting: A review of process, materials, and methods. *Additive Manufacturing*, *28*, 781–801. DOI: 10.1016/j.addma.2019.05.031

Chapter 9

Exploring the Role of Natural, Synthetic, and Biodegradable Polymers in Biomedical Applications

Aditya Shrivastav
https://orcid.org/0009-0000-8461-8171
Ajeenkya D.Y. Patil University, India

Susanta Das
https://orcid.org/0000-0002-9314-3988
Ajeenkya D.Y. Patil University, India

ABSTRACT

Biomedical applications rely on polymers because these materials possess adjustable properties (chemical/physical). This chapter evaluates healing function of natural & synthetic polymers within wound healing applications & drug delivery systems alongside their implementation in tissue engineering & regenerative medicine. Biocompatibility along with degradability in nature found within polymers (collagen, chitosan & alginate) make them fit for use as scaffolds alongside drug carriers. Synthetic polymers (polylactic acid & polycaprolactone) deliver adjustable mechanical stability benefits which make them valuable for implants with drug delivery applications. Processing technologies have been improved through 3D printing & electrospinning to develop medical structures which match individual healthcare requirements precisely. Use of polymers for biomedical applications

DOI: 10.4018/979-8-3373-9454-1.ch009

still faces ongoing issues pertaining to biocompatibility as well as scalability and regulatory compliance requirements. Solving these challenges will enable polymer-based materials to succeed in clinical applications

1. INTRODUCTION

The biomedical field depends heavily on polymers since these large molecules made of monomers provide both specialized chemical behaviors and customizable functionality. The molecular structure of polymers enables the creation of functional materials for various medical applications such as wound healing and drug delivery systems and tissue engineering and implants and regenerative medicine (Dhaliwal, 2018; Halder et al., 2024; Oleksy et al., 2023). Figure 1 shows the schematic representation.

Figure 1. Overview of biomedical applications of polymers highlighting their tunable properties

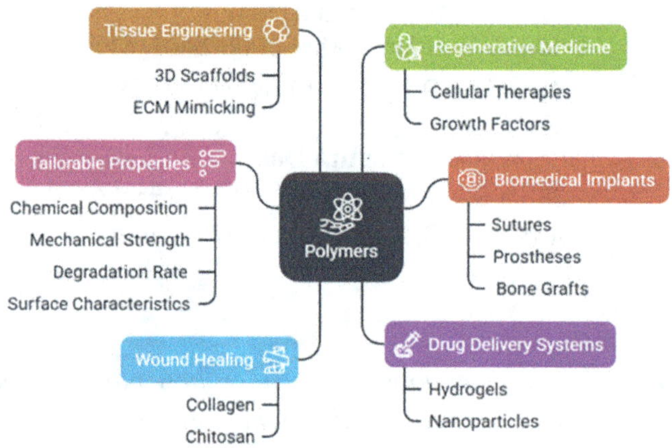

The field of wound healing makes use of both natural polymers (like hyaluronic acid, collagen, chitosan, gelatin, and alginate) alongside synthetic polymers (like polylactic acid, polyglycolic acid and polylactic-co-glycolic acid) to create scaffolds that promote cell proliferation while facilitating vascularization and controlled therapeutic agent delivery (Halder et al., 2024). Polymers in drug delivery take the form of hydrogels and nanofibers, and nanoparticles which allow developers to achieve specific drug release timing for therapeutic substances and bio factors leading to improved treatment outcomes with reduced adverse side effects (Dhaliwal, 2018; Halder et al., 2024). Through tissue engineering techniques polymers become essential components that develop three-dimensional scaffolds with extra-cellular matrix characteristics to help tissue cells anchor properly and expand and differentiate for tissue regeneration purposes (Oleksy et al., 2023). Polymers serve multiple roles in implanting applications and regenerative medicine by allowing their transformation into procedural components such as sutures and prostheses and bone grafts which provide structural integrity in addition to biologic utility (Dhaliwal, 2018; Oleksy et al., 2023).

Biocompatibility serves as an essential medical polymer requirement because it defines how well materials adapt to body contact without adverse responses. Medical devices and therapeutic devices need biocompatible polymer parts which harmonize with biological tissues through minimum immune reactions and inflammation and toxicity (Oleksy et al., 2023). The need for degradation properties becomes essential for both short-term medical implants together with scaffolds and drug delivery systems. The framework of biodegradable polymers makes them decompose within the biological environment through both enzymatic and hydrolytic processes before the body removes them through excretion. Biodegradable medical systems require no subsequent surgery for removal thereby minimizing long-term complications while conforming to sustainable healthcare practices (Dhaliwal, 2018; Nanda et al., 2022). The degradation products must be harmless and easily transformed by the body because undesirable tissue effects should be prevented (Dhaliwal, 2018). Biological tissues and organs require scaffolds that achieve their required functional mechanical properties which include strength along with elastic quality and degradation timing. Different biomaterial scaffolds need high compression strength together with fatigue resistance for bone regeneration applications and vascular grafts require flexible compliance for effective use. The possibility to precisely adjust these characteristics allows researchers to create polymeric materials which duplicate native tissue biomechanics for better functional integration and clinical outcomes (Dhaliwal, 2018; Oleksy et al., 2023). The biomedical application of natural, synthetic, and biodegradable polymers relies on meeting key standards biocompatibility, biodegradability, and mechanical performance. These properties enable treatment development along with safe and effective medical materials as well as new pathways for customized therapeutic approaches that represent a significant step forward in contemporary medicine.

2. NATURAL POLYMERS

Renewable biological substances transform into natural polymers which present intrinsic biocompatibility alongside complex structures for applications in biomedical fields (Hu et al., 2024). Their qualities to duplicate extracellular matrix elements together with adjustable physicochemical properties make them essential elements for tissue engineering scaffolds, drug delivery systems and regenerative medicine solutions (Puertas-Bartolomé et al., 2021). Collagen sourced from animal connective tissues represents one example of natural polymers along with chitosan obtained from crustacean exoskeletons and alginate extracted from brown algae which provide different biological functionalities (Jujjavarapu et al., 2020). To examine research trends researchers performed a bibliometric analysis through PubMed mapping key research areas involving natural polymers in biomedical applications. The key terminology "biocompatible materials," "tissue engineering," "drug delivery systems," "collagen," and "chitosan" function as core nodes according to Figure 2. Modern biomedical innovation heavily relies on natural polymers according to the identified research networks which focus on wound healing and regenerative medicine as well as nanoparticle-based drug delivery.

Figure 2. Keyword co-occurrence network related to natural polymers and their biomedical applications based on PubMed data (Source: PubMed database; analysis using VOSviewer; Range: 1930 - 2025).

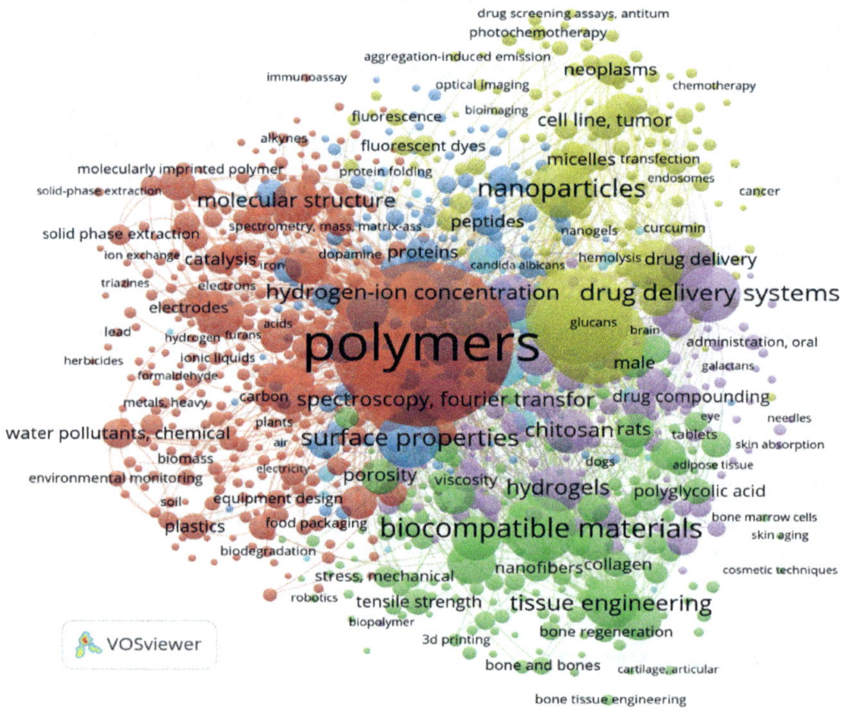

Building upon the insights from the bibliometric analysis, the following sections delve deeper into natural polymers, their sources, characteristics, and biomedical applications, with a focus on key materials such as collagen, chitosan, and alginate.

2.1 Sources and General Characteristics

Biological polymers derive from plants and animals and microbes serving as the three fundamental biological groups. The natural polymers cellulose and starch and pectin's derive from plants where they maintain strong mechanical stability while being readily available (Jujjavarapu et al., 2020; Wypij et al., 2023). The polymers that originate from animals like collagen, gelatin, and hyaluronic acid integrate bioactive signaling elements for effective cell bonding and tissue healing properties (Hu et al., 2024; Puertas-Bartolomé et al., 2021). Two types of microbial polymers including bacterial cellulose and polyhydroxyalkanoates (PHAs) present important characteristics of high purity and precise biosynthetic pathways (Puertas-Bartolome et al., 2021; Wypij et al., 2023). Natural polymers exhibit key properties which contribute to their value for biomedical uses despite their various sources of origin. Natural polymers seamlessly integrate with host tissues because their inherent biocompatibility and native biochemical signals trigger minimal immune responses (Hu et al., 2024; Puertas-Bartolome et al., 2021). Synchronized tissue regeneration rates are possible through enzymatic degradation mechanisms that contribute to controlled biodegradability (Hu et al., 2024). The range of structural complexity in natural polymers extends from linear molecular chains to intricate fibrillar networks which aid researchers in developing complex scaffold designs that replicate the extracellular matrix structure (Jujjavarapu et al., 2020; Puertas-Bartolome et al., 2021). Natural polymers showcase extensive functional adaptability because they enable easy chemical crosslinking and conjugation processes as well as nanoparticle and synthetic polymer integration to improve their properties (Wypij et al., 2023). Natural polymers demonstrate superior qualities as unmatched candidates against synthetic alternatives because they possess ecological sustainability and biomimetic properties together with adaptability for applications requiring biological interaction. To achieve their complete clinical potential research needs to solve complex issues regarding batch-to-batch reproduction as well as mechanical performance optimization (Puertas-Bartolome et al., 2021).

2.2 Key Natural Polymers in Biomedical Use

Biomedical applications require natural polymers as critical materials because they possess three vital features: biocompatibility, biodegradability and structural versatility. Collagen together with chitosan and alginate have become prominent natural polymers because of their special characteristics and their ability to fulfil a wide range of therapeutic requirements. The research analyses the biological origins and chemical frameworks and physicochemical attributes alongside biomedical applications of these significant polymers (Liu et al., 2019; Alemu et al., 2023; Periayah et al., 2016).

2.2.1 Collagen

Humans contain collagen which serve as the major structural protein within their body using connective tissue elements consisting of skin and tendons and bones. The structure consists of three left-handed polyproline II helices arranged into a right-hand triple helix where hydrogen bonds and hydroxyproline residues create stabilization. Such a three-dimensional shape grants the molecule extreme tensile strength alongside elastic properties. Recent cryo-electron microscopy analysis uncovered unanticipated structural variations in collagen which manifest through non-standard molecular packing patterns that enable novel molecular contacts (Liu et al., 2019).

Through purification of collagen enables medical applications through its compatibility with biological systems and minimal immune response that would make it perfect for tissue engineering solutions and wound healing and drug delivery needs. The enzyme-activated breakdown mechanism of collagen can be regulated through crosslinking approaches which include glutaraldehyde treatment or UV irradiation. Native collagen possesses subpar thermal stability and enzymatic degradation rates which leads to the need for chemical modifications for improving its mechanical stability. Medical applications of collagen involve using it for skin grafts and bone scaffolds and corneal implants because its ability to attract cells during regeneration provides therapeutic benefits. Clinical benefits remain compromised by issues arising from inconsistent batch qualities and remaining impurities despite current therapeutic efforts. The production of recombinant collagen using modern techniques provides versatile structural capabilities and enhanced reliability according to Liu et al. (2019).

2.2.2 Chitosan

Chitosan exists as a linear polysaccharide obtained through deacetylating chitin and remains the most prevalent material derived from crustacean exoskeletons together with fungal cell walls. The cationic behavior of chitosan creates efficient electrostatic binding between anionic biomolecules thus enhancing its potential as a delivery agent for drugs and genes (Alemu et al., 2023; Periayah et al., 2016).

Chitosan possesses three key physicochemical features that include pH-sensitive dissolution and natural antimicrobial effect related to its membrane-damaging properties together with its high biocompatibility properties. Chitosan shows extensive use for wound dressings since it speeds up wound clotting while speeding up skin cell growth and functions as a tissue engineering scaffold to boost cell growth specifically for osteoblasts (Alemu et al., 2023). The use of chitosan as a standalone material is limited by its low mechanical strength and its inability to dissolve at the pH levels of biological systems. The material needs additional blending with synthetic polymers or chemical modifications to increase its operational capabilities because of its fundamental weaknesses. Current research aims to optimize the degree of deacetylation (DD) and molecular weight to enhance cohesion and achieve controlled drug release kinetics particularly for nasal and oral drug delivery systems (Periayah et al., 2016).

2.2.3 Alginate

Alginate functions as an anionic copolymer which occurs naturally in seaweed brown variety or develops from bacterial fermentative processes. The two different units β-D-mannuronate (M) and α-L-guluronate (G) form structural blocks within alginate which determine its physicochemical properties based on M/G composition. Gels formed by high-G alginates become rigid and brittle with divalent cation exposure such as Ca^{2+} but high-M alginates develop flexible hydrogels. The gelation behavior of alginate can be controlled through its M/G ratio which renders it appropriate for usage in 3D bioprinting and wound dressings as well as controlled drug delivery systems (Lee & Mooney, 2011).

The therapeutic properties of alginate including biocompatibility and low toxicity have strong evidence, but raw extract impurities may induce immune responses in some cases. Wound management becomes more effective because alginate quickly hydrates and generates protective gel formations when it interacts with wound exudate because of its water-attracting properties. The topographical characteristics of extracellular matrices are replicated by scaffolds made from alginate which find applications in regenerating cartilage and bone tissue. The absence of bioactive motifs along with inconsistent molecular weight and M/G ratio between different batches create significant problems regarding cell adhesion. The field evolved with RGD peptide attachment through covalent bonds to boost cell connections and researchers combined alginate with collagen or chitosan to create hybrid structures that increased mechanical strength (Lee and Mooney, 2011).

Merging knowledge from material science and biomedical innovation has produced these natural polymers as an example. The table below provides a comparison of the features along with biomedical uses and major constraints among these natural polymers (Table 1).

Table 1. Comparative summary of collagen, chitosan, and alginate in biomedical applications.

Characteristic	Collagen	Chitosan	Alginate
Key Properties	Biocompatibility, mechanical strength	Biocompatibility, antimicrobial activity	Biocompatibility, gel-forming ability
Applications	Tissue engineering, wound healing	Wound healing, drug delivery	Tissue engineering, drug delivery
Challenges	Batch variability, mechanical weakness	Batch variability, poor cell adhesion	Mechanical weakness, rapid degradation
References	(Lee & Mooney, 2011)	(Alemu et al., 2023)	(Liu et al., 2019)

2.3 Biomedical Applications

2.3.1 Wound Healing

Wound healing applications depend heavily on natural polymers including collagen, chitosan and alginate because these substances show excellent compatibility with human tissue and degrade naturally while duplicating the extracellular matrix structure. Collagen which functions as one of the major extracellular matrix proteins, acts as a tissue framework by allowing cell movement and growth thus enabling wound closure and renewal of tissue structures. The haemostatic as well as antimicrobial and film-forming properties of chitin-derived chitosan make this substance very effective at reducing wound infections and accelerating wound closure. When alginate derived from brown seaweed encounters wound exudate it transforms into hydrogel while creating a moist healing environment which promotes autolytic debridement (Zhang et al., 2023). Research has confirmed the effectiveness of these polymers when applied to modern wound care bandages. Modern chronic wound treatment relies on chitosan hydrogels with integrated antimicrobial agents and growth factors which boost healing performance. The use of collagen dressings in diabetic ulcer management leads to faster granulation tissue formation accompanied by improved treatment outcomes. The high exudate-wicking capability of alginate dressings makes them optimal for treating wounds which generate large amounts of fluid (Zhang et al., 2023). The main benefits of these materials consist of low immunological reactions and adjustable breakdown times which researchers can convert them into different format structures (Zhang et al., 2023). Multiple challenges exist for natural polymers such as batch variability together with weak mechanical properties especially for collagen and alginate structures while additional materials may be necessary for performance improvements (Zhang et al., 2023).

2.3.2 Drug Delivery Systems

The drug delivery field takes advantage of natural polymers because these materials create drug delivery systems which release therapeutic compounds while achieving specific drug release times (Lopes et al., 2024). Drug protection against degradation and the delivery of sustained release occurs through encapsulation methods that use collagen along with chitosan and alginate. The cationic character of chitosan permits electrostatic binding with negatively charged drugs and genetic material which improves tissue absorption and sticking ability to biological tissue. The gelation characteristics of alginate enable drug encapsulation in microspheres and beads while allowing for pH-oriented and ion-depending release profiles (Lopes et al., 2024). Scientific studies show chitosan nanoparticles improve oral insulin bioavailability by protecting insulin from chemical breakdown and at the same time increasing accessibility. Research on drug delivery shows that microcapsules made from alginate protect both probiotics and chemotherapeutics while they travel through the gastrointestinal tract until reaching disease sites where these drugs selectively release. Microspheres made from collagen allow clinicians to use growth factors in regenerative medicine therapies (Lopes et al., 2024). The main benefits of drug delivery systems built with natural polymers include their ability to work well with biological systems and the ease of modifying them and the capacity to customize release patterns (Lopes et al., 2024). Technology faces two main drawbacks which include physical instability in specific biological environments and limited structural strength and difficult manufacturability on a mass scale (Lopes et al., 2024).

2.3.3 Tissue Engineering

Natural polymers form an essential component of tissue engineering applications because they mirror native extracellular matrix structures to support cellular binding as well as growth and specialization processes. The structural support and biological signals delivered by collagen scaffolds regulate effectively the regeneration of bone tissue as well as cartilage tissue and skin cells. The utilization of chitosan occurs in nerve and cartilage repair because this material resembles glycosaminoglycans structurally and effectively encourages cell binding. The medical field uses injectable alginate hydrogel matrices for encapsulating cells through which medical professionals support the regeneration of soft tissue structures (Hu et al., 2024). The field of tissue engineering has progressed through the creation of composite scaffolds which merge collagen with bioactive ceramics for treating bones as well as chitosan-based hydrogels with embedded stem cells for cartilage healing applications. 3D bioprinting using alginate materials allows the development of intricate tissue structures that achieve cell-specific arrangement leading to improved integration and operational capacities (Hu et al., 2024). The advantages of natural polymer scaffolds include excellent biocompatibility and enzymatic degradability together with the capability to modify chemicals for bioactive factor addition and mechanical property regulation (Hu et al., 2024). The functional uses of natural polymers in load-bearing applications face challenges because their mechanical strength is inadequate while their in vivo degradation rate is rapid even when enhanced crosslinking methods are utilized to improve their functionality and stability (Hu et al., 2024).

The biomedical sector relies on natural polymers including collagen chitosan and alginate because these elements provide essential biological capabilities while offering adaptable functionality for wound healing and drug delivery systems together with tissue engineering applications. The ongoing research focuses on improving natural polymers through chemical modification while also forming composites and using advanced fabrication techniques to enhance their potential in biomedical applications (Hu et al., 2024).

3. SYNTHETIC POLYMER

The manufacturing process of synthetic polymers consists of transforming monomers into extended chain structures and networks utilizing controlled polymerization systems. This procedure produces engineered macromolecules. The laboratory production of synthetic polymers differs from biopolymer origins that come from natural sources because synthetic polymers mostly derive from petrochemical feedstocks (LibreTexts, 2024). The four primary groups of synthetic polymers include thermoplastics, thermosets, elastomers along with synthetic fibres depending on their structure and application needs. Four commonly used synthetic polymers are polyethylene, polypropylene, polyvinyl chloride (PVC) and polystyrene, nylon and polyurethanes. All synthetic polymeric backbones derive from carbon elements, but polyolefins use exclusively carbon while polyesters and silicon's along with polyamides have backbones composed of oxygen, nitrogen, or silicon heteroatoms (LibreTexts, 2024).

Figure 3. Bibliometric visualization of synthetic polymer research related to biomedical applications (Source: PubMed database; analysis using VOSviewer; Range: 1945 - 2025).

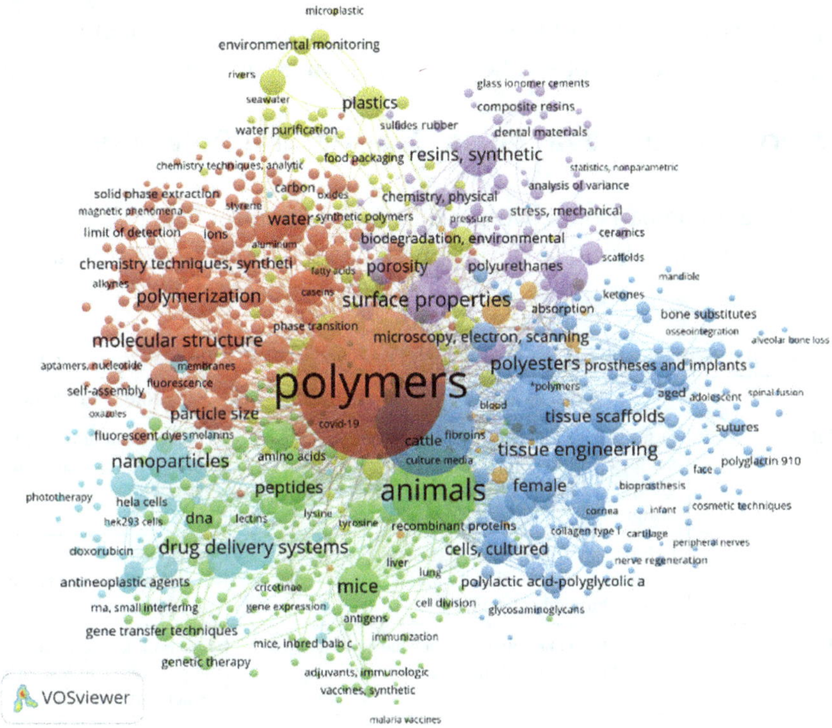

Synthetic polymers undergo manufacturing through three main polymerization procedures that consist of chain-growth addition polymerization and step-growth condensation polymerization following advanced processes utilizing ring-opening and metathesis polymerizations. Through different manufacturing approaches scientists gain extensive control to design polymers with targeted molecular designs and properties. Researchers employ modifications to monomers and catalysts alongside reaction conditions to generate polymers featuring high levels of customized mechanical properties and chemical/biological attributes (LibreTexts, 2024). Synthetic polymers hold a great importance when used in biomedical applications. Synthetic polymers serve biomedical applications as essential components because they provide control over their composition while blending flexibility with adjustable degradation properties for drug delivery vessels and tissue engineering scaffolds and prostheses and sutures and implantable devices. Advanced manipulation of polymer attributes leads to increased compatibility with biological systems while facilitating their integration into living environments (LibreTexts, 2024). The VOSviewer visualization of PubMed bibliometric data (2020-2024) shows increasing research interest in synthetic polymers for biomedical applications (shown in Figure 3). The bibliometric map showcases four main clusters of terms which include "polymerization", "nanoparticles", "tissue scaffolds", and "drug delivery systems" that represent the developing scientific role of synthetic polymers in modern medical technologies.

3.1 Overview and Advantages Over Natural Polymers

A systematic comparison between synthetic and natural polymers shown in Table 2 reveals the following advantages favoring synthetic polymers in biomedical contexts:

Table 2. Comparative Analysis of Synthetic and Natural Polymers (LibreTexts, 2024)

Property	Synthetic Polymers	Natural Polymers
Tunability	Offer precise control over molecular weight, composition, degradation rate, and functional group modification, enabling customized material design.	Limited tunability due to biological constraints and variability.
Mechanical Properties	Exhibit superior and adjustable mechanical strength, elasticity, and toughness, suited for load-bearing biomedical devices.	Often display inconsistent or insufficient mechanical performance.
Stability	Possess high chemical and thermal stability, facilitating sterilization and long-term storage.	Susceptible to enzymatic degradation and instability under variable conditions.
Scalability	Can be mass-produced with consistent quality using well-established industrial processes.	Limited scalability due to sourcing and extraction complexities.
Batch-to-Batch Reproducibility	High reproducibility owing to controlled synthesis protocols.	Variability inherent to natural extraction methods.

Natural polymers face processing difficulties when going to large scale because biological elements introduce variability and inconsistent mechanics (LibreTexts, 2024). The engineering capabilities for precise creation of synthetic polymers produce superior results than natural polymers. Advanced biomedical applications require these advantages since material performance along with reliability and regulatory compliance and scalability are essential factors (LibreTexts, 2024).

3.2 Major Synthetic Polymers and Their Biomedical Applications

The biomedical field uses three important biodegradable synthetic polymers which include polylactic acid (PLA) and polyglycolic acid (PGA) together with polycaprolactone (PCL) because of their dissimilar structural characteristics and functional abilities.

3.2.1 Poly(lactic acid) (PLA):

The synthetic process of PLA production involves ring-opening polymerization (ROP) for achieving high molecular weights from aliphatic polyester lactic acid monomers. The material shows excellent tensile strength together with modulus values but becomes brittle unless chemical modifications occur (Hamad et al., 2015). PLA undergoes thermal transformations between 50-80 °C glass transition (Tg) and melts from 170-180 °C (Hamad et al., 2015). The degradation of PLA materials occurs through water-induced cleavage of its ester bonds at rates determined by both material crystallinity level and molecular weight. PLA serves as a popular biomedically approved material for creating orthopaedic screws and pins because it breaks down gradually inside the body. The substance functions as a network structure to support drug delivery systems which release medications over extended periods. PLA serves in scaffold engineering for developing 3D-printed porous systems which aid in bone and cartilage tissue regeneration (Hamad et al., 2015).

3.2.2 Poly(glycolic acid) (PGA):

PGA occurs as a linear polyester made of glycolic acid units through glycoside ring-opening polymerization or thermally driven polycondensation reactions. PGA has high tensile strength but degrades in a short period of weeks to months from its use. The linear polyester PGA exhibits melting characteristics at 220-230°C and glass transition characteristics between 35-40°C according to Hosseini et al. (2023). The fast hydrolysis rate of PGA occurs because its highly crystalline structure surpasses PLA in degradation speed. PGA serves as the main component for medical sutures known as Dexon and wound closure meshes. The material serves as a delivery vessel for short-term medications for antibiotic drugs as well as anti-inflammatory agents. The application of PGA in tissue engineering includes its usage for creating fibrous mats which serve both vascular grafts and skin regeneration purposes (Hosseini et al., 2023).

3.2.3 Polycaprolactone (PCL):

The hexanoate repeating unit defines PCL as a semi-crystalline aliphatic polyester which undergoes copolymerization modifications for practical applications. The material shows high elasticity and toughness which qualifies its use in load-bearing biomedical applications. The thermal processing characteristics of PCL material become accessible because it shows a melting point of about 60°C and a glass transition temperature at around -60°C (Hosseini et al., 2023). The hydrophobic characteristics lead to PCL degradation at a pace of around 2-4 years. The manufacturing of tracheal stents alongside contraceptive devices makes use of PCL as an implant material. Researchers utilize PCL for building both micelles and nanospheres to encapsulate hydrophobic drugs. The bioactive ceramic composition of hydroxyapatite is frequently combined with PCL to develop 3D-printed composites for bone tissue regeneration according to Hosseini et al. (2023).

4. BIODEGRADABLE POLYMERS

A bibliometric analysis was done using VOSviewer with PubMed keywords to understand current research trends. The study distinguished three primary research groups which included environmental biodegradation science (red cluster) and biomedical applications of polymers (green cluster) and drug delivery systems (blue cluster). The environmental cluster demonstrates how biodegradable plastics effectively fight chemical pollutants while fighting microplastics through their crucial role. The biomedical cluster demonstrates robust networking between pharmaceutical components including polylactic acid and collagen because these substances find application in tissue engineering and absorbable implant development. The drug delivery cluster focuses on researching nanoparticle systems along with drug carriers and methods for controlled drug release. Biodegradation appears along with polymers and environmental aspects and animals while drug delivery systems emerge as top keywords according to a keyword analysis network (Figure 4). The assembled keyword distribution indicates extensive interdisciplinary focus on translating green materials into biomedical and environmental uses.

4.1 Definition and Importance

Biodegradable polymers break down through biological processes as microorganisms transform them into non-harmful substances which include water, carbon dioxide, methane, and biomass based on the degradation environment (Hu & Zhou, 2024). Their importance emerges because they solve the worldwide plastic waste disaster through offering an eco-friendly substitute to permanent synthetic materials. The biomedical industry benefits from their safe degradation inside biological environments because it enables temporary medical equipment that avoids invasive surgical removal procedures while saving patients both healthcare expenses and potential risks (Kurowiak et al., 2023).

4.2 Degradation Mechanisms and By-Products

Two main pathways exist when biodegradable polymers degrade through hydrolytic and enzymatic mechanisms that lead to their breakdown. The breakage of polymer chains triggered by water molecules defines hydrolytic degradation primarily because aliphatic polyesters including PLAs and PGAs are particularly sensitive to water (Hu & Zhou, 2024). Microbial production of specific enzymes creates the basis for enzymatic degradation through which microorganisms break down PHAs alongside natural polymers including chitosan and collagen. The complete degradation pathways using these processes yield nontoxic end products that consist of water along with carbon dioxide under aerobic conditions and methane under anaerobic conditions as well as biomass which allows safe biological reabsorption (Hu & Zhou, 2024).

4.3 Key Biodegradable Polymers and Their Biomedical Applications

Biomedical applications use prominent biodegradable polymers because of their adjustable degradation rates combined with biocompatibility and appropriate mechanical properties (Kurowiak et al., 2023). PLA and PGA primarily appear in absorbable sutures and orthopaedic fixation devices and drug delivery systems because their specific hydrolytic degradation patterns prove beneficial. Through its slower decomposition properties Polycaprolactone (PCL) performs in both long-term implant applications and tissue engineering scaffold technologies (Kurowiak et al., 2023). Microbial fermentation produces Polyhydroxyalkanoates (PHAs) which have demonstrated uses in wound dressings and controlled drug release systems because they exhibit favorable biocompatibility along with enzymatic degradability. Regenerative medicine and tissue engineering applications frequently use natural polymers like collagen, gelatin and chitosan because these materials exhibit both bioactivity and cell-response properties (Kurowiak et al., 2023). Through careful engineering processes scientists transform these materials into various forms that include fibres, films, hydrogels and 3D-printed scaffolds for applications in tissue regeneration, drug delivery systems and medical implant systems (Kurowiak et al., 2023).

4.4 Advantages and Challenges in Clinical Translation

The primary benefit that biodegradable polymers offer to clinical applications is their dual role as temporary support devices before their safe breakdown into body-compatible products that eliminate the requirement for extra medical procedures. Biodegradable polymers permit engineers to shape devices with intricate designs along with programmed drug delivery methods through rate controls matching human tissue renewal times (Kurowiak et al., 2023). However, several challenges impede their widespread clinical translation. The precise manipulation of degradation time frames remains crucial because improper timing leads to two issues: it causes early mechanical failures from degradation, and it results in acidic byproducts which cause local inflammation. The expansion of manufacturing scale together with quality control and reproducibility remains a barrier along with the need to prevent material-induced inflammatory reactions (Hu & Zhou, 2024). The approval process for new biodegradable polymers undergoes lengthy assessment by regulatory agencies who need to prove their safety as well as their functionality as well as long-term effects inside living systems. The clinical applications of biodegradable polymers show promising developments because of continuous progress in material science and engineering approaches (Hu & Zhou, 2024). Figure 4 shows keyword co-occurrence map of biodegradable polymer research using VOSviewer.

Figure 4. Keyword co-occurrence map of biodegradable polymer research

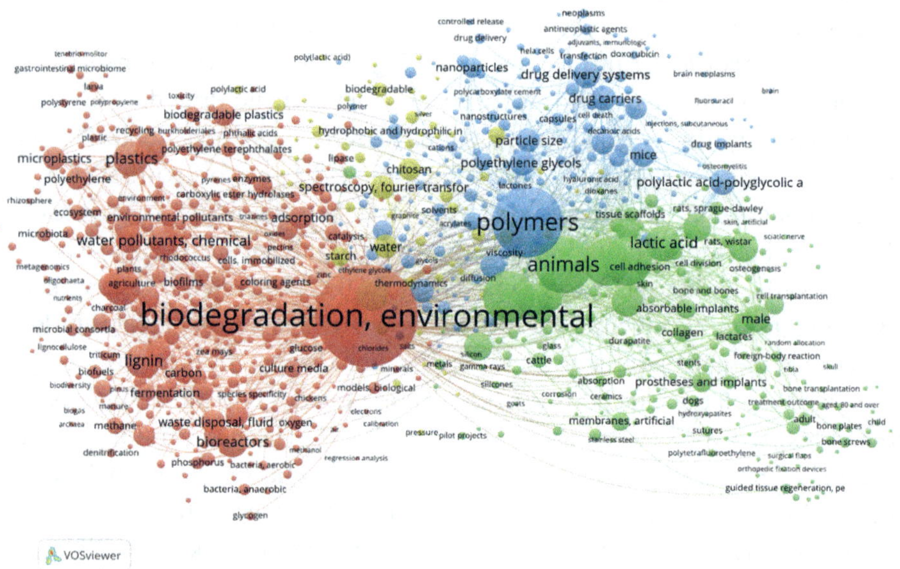

Source: PubMed database; analysis using VOSviewer; Range: 1945 - 2025.

5. ADVANCES IN POLYMER PROCESSING TECHNOLOGIES

5.1 3D Printing of Biomedical Polymers

Additive manufacturing or 3D printing has developed as a revolutionary manufacturing technology which alters the process of biomedical device production. Figure 5 visually shows how by growing time the research had grown in 3D Printing for Biomedical Polymers. The technology allows users to directly produce complex and customized medical devices from digital models through which rapid prototyping becomes feasible, and the creation of intricate geometries becomes achievable that traditional manufacturing methods cannot produce (Arefin et al., 2021). 3D printing technology developed its initial surgical planning applications into modern uses for implants and tissue engineering scaffolds as well as diagnostic tools and drug delivery systems (Chia & Wu, 2015; Arefin et al., 2021). The selection of materials represents an essential element in biomedical 3D printing as polymers need to demonstrate biocompatibility alongside mechanical properties that are suitable for fabrication while maintaining processability. Different types of polymers including artificial and natural materials are employed according to their particular usage as scaffolds or implants or biosensor applications (Agueda et al., 2021). Biocompatibility presents a vital challenge in biomedical printing as the printing process specifically heat and ultraviolet light exposure changes properties of materials and biological functionality (Agueda et al., 2021). Developments in material science have produced polymers and composite materials with flexible mechanical and biological properties as well as combinations of different materials elements (Arefin et al., 2021).

3D printing technology creates patient-specific devices as its main strength when applied to biomedical applications. Patient-specific implants and prosthetics derived from CT or MRI scan data result in better fit and functionality and improved clinical results according to research by Chia & Wu (2015) and Arefin et al. (2021). The tissue engineering process achieves better cell growth and tissue regeneration by using 3D printing technology to create scaffolds which duplicate the structural hierarchy of natural tissues (Arefin et al., 2021). Through 3D printing technology healthcare professionals can merge different materials and bioactive elements into solitary constructs which speed up development of advanced biomedical devices and medications delivery platforms (Chia & Wu, 2015; Assad et al., 2023). The field of biomedical 3D printing witnessed several innovative advancements which extended its operational potential. The actual fabrication of tissue-like structures through precise cell placement with combined biomaterials and growth factors happens thanks to modern bioprinting techniques according to Assad et al. (2023). Researchers have developed sophisticated tissue engineering scaffolds that achieve desired porosity while maintaining sufficient strength and displaying proper bioactivity (Arefin et al., 2021). Multiple materials printing methods allow for integrated placement of different polymer and composite components which manage biological and mechanical device demands (Arefin et al., 2021). The emerging field of personalized pharmaceuticals combines 3D-printing technology to create drug formulations which are designed with specialized dosages and custom shapes while also controlling release profiles. Four-dimensional printing techniques enable smart biomedical devices through polymers which adapt their form and functionalities according to environmental stimuli.

Figure 5. Trend of PubMed-indexed publications in 3D Printing for Biomedical Polymers.

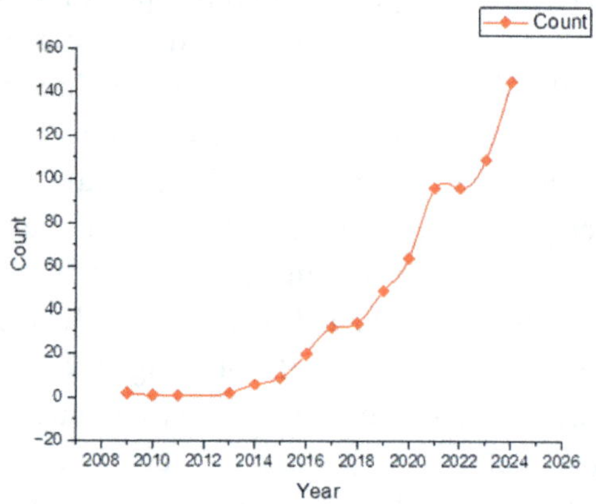

The revolutionary biomedical device fabrication process through 3D printing now delivers rapid and affordable production techniques with superior customizable capabilities. Polymers continue to advance for biomedical applications through continuous studies of biocompatible materials which remains the primary developmental focus. The technology demonstrates great potential through its capability to develop specialized patient implants alongside complex tissue scaffolds and multifunctional therapeutic systems. The technologies of bioprinting and multi-material deposition and 4D printing serve to expand biomedical applications further while making additive manufacturing the key foundation for future healthcare advancement (Chia & Wu, 2015; Arefin et al., 2021; Assad et al., 2023). Table 3 shows several polymers used in 3D printing for biomedical applications.

Table 3. Polymers Used in 3D Printing for Biomedical Applications

Material/Polymer Used	3D Printing Technique	Polymer Formed After Printing	Biodegradation	Cytocompatibility	Applications	References
Collagen	Robotic Dispensing	Hydrogel scaffolds, aligned wall composite fibers	Biodegradable	High	Skin, cartilage, heart valves, vocal cords, spinal cord	Saroia et al., 2018
Alginate (with methylcellulose)	Robotic Dispensing (Coaxial nozzle)	Hollow filament scaffolds with built-in micro-channels	Biodegradable	High (>90% cell viability)	Bone, cartilage, tissue scaffolds	Saroia et al., 2018
Gelatin-Chitosan (with PCL)	Robotic Dispensing	3D biodegradable patches with thin PCL-layer core	Biodegradable	High	Congenital heart defect patches	Saroia et al., 2018
Alginate	Inkjet Bioprinting	3D scaffolds sprayed with $CaCl_2$ for crosslinking, with uniform pores	Biodegradable	High (good endothelial cell adhesion)	Endothelial tissue scaffolds, tubular or branched structures	Saroia et al., 2018
Chitosan-Hydroxyapatite	Robotic Dispensing (Pneumatic system)	Layered scaffolds with good adhesive interlayer bonding	Biodegradable	Good adhesion	Bone tissue engineering scaffolds	Saroia et al., 2018
Polyethylene Glycol (PEG) Hydrogels	Laser-Assisted Bioprinting (Stereolithography SLA)	Cross-linked photo-active PEG-dma polymer scaffolds	Biodegradable	High	Soft tissue scaffolds	Saroia et al., 2018

continued on following page

Table 3. Continued

Material/Polymer Used	3D Printing Technique	Polymer Formed After Printing	Biodegradation	Cytocompatibility	Applications	References
Poly(ethylene glycol-co-depsipeptide) (PEG-co-PDP) & PEG acrylate (RGD-functionalized)	SLA with visible light source	Photo-crosslinked scaffolds with variable stiffness 3±1 to 38±13 kPa	Biodegradable	High (enhanced proliferation and adhesion)	Semilunar cartilage structures, tissue scaffolds	Saroia et al., 2018
Polycaprolactone (PCL) and chitosan	Robotic Dispensing and Electrospinning (combined)	Composite scaffolds with improved protein observing capacity	Biodegradable	High	Skin, ligament and tendon tissues	Saroia et al., 2018
Injectable Hydrogels	Injectable methods (chemical, physical)	Smart hydrogels with tunable properties	Biodegradable	High	Bone and cartilage tissue engineering	Dutta et al., 2020
Alginate	3D Printing (Micro-extrusion)	3D printed scaffolds with porous structures	Biodegradable	Moderate	Cartilage tissue engineering	Dutta et al., 2020
Collagen	3D Printing (Micro-extrusion)	3D printed collagen scaffolds	Biodegradable	High	Bone and cartilage tissue engineering	Dutta et al., 2020
Polycaprolactone (PCL)	3D Printing (Extrusion)	PCL scaffolds with enhanced mechanical properties	Biodegradable	Moderate	Bone tissue engineering	Dutta et al., 2020
Hyaluronic Acid	3D Printing (SLA, Inkjet)	Hydrogels with improved mechanical properties	Rapidly biodegradable	High	Cartilage repair, drug delivery	Dutta et al., 2020

continued on following page

Table 3. Continued

Material/Polymer Used	3D Printing Technique	Polymer Formed After Printing	Biodegradation	Cytocompatibility	Applications	References
Fibrin	3D Printing (Inkjet)	Injectable fibrin-based hydrogels	Biodegradable	High	Cartilage regeneration	Dutta et al., 2020
Chitosan	3D Printing (Micro-extrusion)	Composite hydrogels with improved properties	Biodegradable	High	Bone tissue engineering	Dutta et al., 2020
Programmable Hydrogels	3D Printing (Various techniques)	Hydrogels with reversible and sequential property changes	Biodegradable	High	Tissue engineering, diagnostics	Dutta et al., 2020
Alginate	3D Printing (DIW)	Hydrogel scaffolds with ionic crosslinking	Biodegradable	Moderate	Tissue engineering	Dutta et al., 2020
Gelatin	3D Printing (DIW)	Hydrogel scaffolds with thermally reversible properties	Biodegradable	High	Tissue engineering	Dutta et al., 2020
CNCs	3D Printing (DIW)	Reinforcing agent in hydrogel scaffolds	Biodegradable	High	Tissue engineering	Dutta et al., 2020
Polycaprolactone (PCL)	Selective Laser Sintering (SLS), Fused Deposition Modeling (FDM), Extrusion Bioprinting	3D printed scaffolds with enhanced mechanical strength	Biodegradable	High	Bone and cartilage tissue engineering	Fan et al., 2019

continued on following page

Table 3. Continued

Material/Polymer Used	3D Printing Technique	Polymer Formed After Printing	Biodegradation	Cytocompatibility	Applications	References
Polypropylene Fumarate (PPF)	Stereolithography (SLA), Digital Light Processing (DLP)	3D printed scaffolds with tunable mechanical properties	Biodegradable	High	Bone and cartilage tissue engineering	Fan et al., 2019
Polylactic Acid (PLA)	Fused Deposition Modeling (FDM), Stereolithography (SLA)	3D printed scaffolds with enhanced mechanical strength	Biodegradable	High	Bone tissue engineering	Fan et al., 2019
Polyether-Ether-Ketone (PEEK)	Selective Laser Sintering (SLS), Fused Deposition Modeling (FDM)	3D printed scaffolds with enhanced mechanical strength	Non-biodegradable	High	Bone and dental implants	Fan et al., 2019
Polyethylene Glycol (PEG)	Extrusion Bioprinting, Stereolithography (SLA)	3D printed hydrogels with tunable mechanical properties	Biodegradable	High	Tissue engineering and drug delivery	Fan et al., 2019
Polydimethylsiloxane (PDMS)	Stereolithography (SLA), Two-Photon Polymerization (TPP)	3D printed scaffolds with enhanced mechanical strength	Non-biodegradable	High	Tissue engineering and microfluidics	Fan et al., 2019

continued on following page

Table 3. Continued

Material/Polymer Used	3D Printing Technique	Polymer Formed After Printing	Biodegradation	Cytocompatibility	Applications	References
Hyaluronic Acid (HA)	Extrusion Bioprinting, Stereolithography (SLA)	3D printed hydrogels with tunable mechanical properties	Biodegradable	High	Tissue engineering and drug delivery	Fan et al., 2019
Gelatin (GEL)	3D Extrusion Printing	3D printed composite scaffolds with decellularized bone particles (dbPTs)	Biodegradable	High	Bone tissue engineering	Kara et al., 2022
Decellularized Bone Particles (dbPTs)	3D Extrusion Printing	Reinforcement agent in gelatin scaffolds	Biodegradable	High	Bone tissue engineering	Kara et al., 2022
Microbial Transglutaminase (mTG)	Crosslinking	Crosslinked gelatin scaffolds	Biodegradable	High	Bone tissue engineering	Kara et al., 2022
Polycaprolactone (PCL)	Fused Deposition Modeling (FDM)	3D printed scaffolds with antimicrobial properties	Biodegradable	High	Bone tissue engineering	Radhakrishnan et al., 2020
Enzymatic degradation (Lipase)	In vitro degradation study	Degradation of PCL/AgNps scaffolds	Biodegradable	High	Bone tissue engineering	Radhakrishnan et al., 2020
Water-Based Polyurethane (PU)	Digital Light Processing (DLP)	3D printed scaffolds for cartilage tissue engineering	Biodegradable	High	Cartilage tissue engineering	Shie et al., 2017

continued on following page

Table 3. Continued

Material/Polymer Used	3D Printing Technique	Polymer Formed After Printing	Biodegradation	Cytocompatibility	Applications	References
Hyaluronic Acid (HA)	Incorporated in PU for 3D printing	Hybrid scaffolds with enhanced properties	Biodegradable	High	Cartilage tissue engineering	Shie et al., 2017
Oxidized Alginate-Gelatin (ADA-GEL)	3D Printing (Direct Extrusion)	3D printed hydrogel scaffolds with electrical conductivity	Biodegradable	High	Cartilage tissue engineering	Distler et al., 2021
Polypyrrole (PPy)	In situ polymerization in hydrogel	Conductive hydrogel scaffolds	Biodegradable	High	Electroactive tissue engineering	Distler et al., 2021
Polystyrenesulfonate (PSS)	Incorporated in hydrogel for conductivity	Conductive hydrogel scaffolds	Biodegradable	High	Electroactive tissue engineering	Distler et al., 2021

5.2 Electrospinning for Nanofiber Production

The electrospinning method enables the production of ultra-fine fibres through an efficient procedure that generates fibres with dimensions from nanometer scale to micrometer scale. A high-voltage electric field connects a droplet of polymer solution (or melt) through a grounded collector to initiate the manufacturing process. Electric forces exceeding the surface tension of the polymer droplet cause the droplet to stretch into a thin jet that becomes thinner while travelling through the air. The solvent evaporates while the fibres travel through the air during which time solid nanofibers form that get collected based on the design of the collector (Cho et al., 2025). An electrospinning setup holds four essential components such as a high-voltage power supply, a syringe pump, a metallic needle and a collector. Fibre morphology together with properties get their direction from critical electrospinning parameters of applied voltage and polymer solution flow rate and concentration and viscosity and environment conditions (Cho et al., 2025). The electrospinning process operates with numerous polymers from both synthetic and natural origins to perfectly serve biomedical applications. Figure 6 shows the distribution of PubMed-indexed publications, comparing two research areas: general polymer processing through electrospinning and the more specific polymer electrospinning for nanofiber production. The narrowing gap between the two trends in recent years also suggests increasing specialization toward nanofiber applications. The biomedical field utilizes synthetic polymers like PCL and PLA along with PGA and PLGA because of their mechanical properties and ability to degrade in nature. Researchers also prefer PEO and PVA because of their positive mechanical characteristics and degradability properties. Natural polymers including gelatin as well as collagen and chitosan and silk fibroin and hyaluronic acid find widespread use in the field due to their superior biocompatibility and bioactivity features. Manufacturers combine natural and synthetic polymers to create materials which merge synthetic mechanical strength with natural biological properties while retaining adequate synthetic degradation times (Wang et al., 2024). The various biomedical applications which utilize electro spun nanofibers demonstrate significant impacts in medical practices. The structure of electro spun nanofiber mats duplicates the features of the extracellular matrix by creating a supportive framework that enables cells to attach and multiply while moving throughout the structure. Their Altaic structure enables proper gas exchange along with exudate absorption and their antimicrobial additives and growth factors function as healing boosters (Wang et al., 2024).

Figure 6. Trend of PubMed-indexed publications on Advances in Polymer Processing through electrospinning and Electrospinning for Nanofiber Production

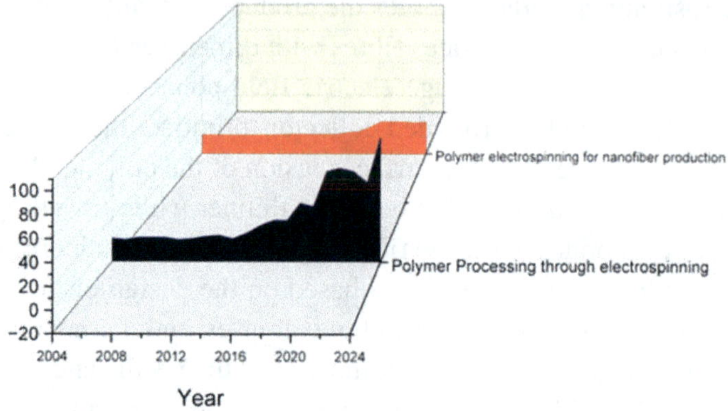

Nanofibers operate as vehicles for therapeutic drug delivery which enables regulated and extended agent distribution through customized design. Therapeutic compounds remain shielded through core-shell electrospinning methods until they arrive at their target destination (Wang et al., 2024). Tissue engineering benefits from electro spun scaffolds which duplicate native tissue structure for supporting cell attachment while promoting cell proliferation and differentiation. The applications of electrospinning have resulted in successful development of engineered skin tissue and bone structures and cartilage tissue and blood vessels together with nerve tissue while proper fibre alignment alongside nuanced bioactive cues proves effective for tissue regeneration (Wang et al., 2024). The electrospinning field made recent progress through the development of intricate fibre systems featuring hollow constructions along with core-shell configurations and porous structures and aligned fibrillar forms. New opportunities emerge through the combination of electrospinning processes with 3D printing and microfluidic systems for producing adaptive nanofiber materials. The technique has improved its industrial viability through developments including AI-based parameter optimization process and utilization of sustainable solvent methods (Cho et al., 2025; Wang et al., 2024). Electrospinning represents a revolutionary technology within polymer processing because it provides exceptional control of nanofiber composition and structure which creates numerous biomedical applications across wound healing and drug delivery and advanced tissue engineering (Wang et al., 2024).

5.3 Emerging Fabrication Techniques

Modern improvement in polymer processing relies on the integration of microfluidics with bioprinting and solvent casting solutions. The biomedical field today experiences significant progress due to innovative processing techniques which develop tailor-made advanced devices and materials targeted for drug delivery systems and tissue engineering applications and wound healing applications (Alexandre-Franco et al., 2024). Figure 7 shows the research development of new and advance fabrication techniques of polymers.

Figure 7. Trend of PubMed-indexed publications on Advances in Polymer Processing through Emerging Fabrication Techniques

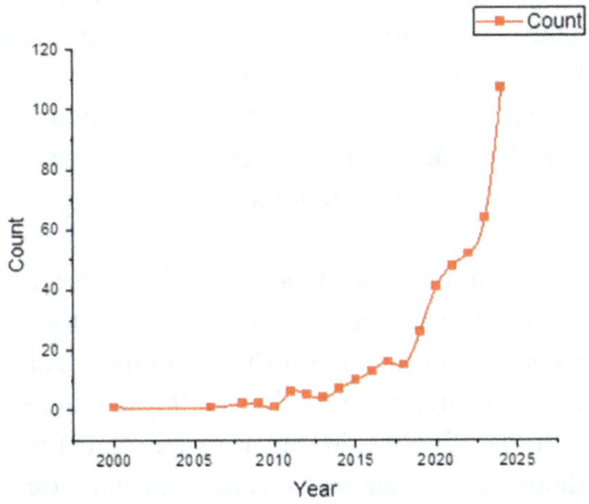

The main technology in biomedical polymer processing uses microfluidics since this technology allows precise fluid volume control in channels with micrometer dimensions. Advanced polymer fabrication methods such as replica moulding and microcontact printing and solvent-assisted moulding enable the production of microfluidic devices. Microfluidics provides precise monitoring capabilities to enable scientists who develop point-of-care testing methods and organ-on-chip platforms and screening diagnostic systems. The analysis between 2020 and 2024 demonstrates that polymer microfluidic systems achieve scalable functionality and multitasking capability that enables both biomedical diagnostic tests and precise drug delivery. The technological progress of three-dimensional (3D) printing introduced fresh methods to create microfluidic devices and accelerated novel designs which match specific patient requirements (Alexandre-Franco et al., 2024). Through automated layered depositions of biological components with biomaterials and bioactive molecules the creation of bioinks is possible. Biological components become easier to arrange with precision because researchers use this method which leads to enhanced development of complex tissue structures and organ models. Through modern bioprinting innovations scientists can develop vascularized tissues together with cartilage tissue and skin equivalents that help advance customized therapeutic strategies and modern regenerative medicine. The construction of scaffolds with bioinks containing hydrogel polymers like GelMA and alginate promotes cell survival while enabling robust cell proliferation as well as specified differentiation patterns (Alexandre-Franco et al., 2024). The solvent casting method functions as an established flexible approach which produces polymeric films and scaffolds for biomedical applications. A polymer solution poured into a mould leads to the formation of a solid structure after solvent evaporation. The main benefits of solvent casting include its convenient nature and budget-friendly approach coupled with the straightforward addition of bioactive substances within the polymer material. The fabrication method enables the production of drug delivery systems and wound dressings and biodegradable implants. Current research attempts to enhance solvent processes alongside mechanical film properties while adding nanoparticles which improve drug delivery patterns and biological tissue compatibility (Alexandre-Franco et al, 2024).

Recent fabrication techniques work together to drive biomedical polymer processing toward a substantial innovation milestone. The performance of diagnostic therapeutic devices improves through microfluidic miniaturization along with spatial control of cells in tissue engineering achieved through bioprinting and solvent casting that provides drug delivery solutions for wound treatment. The distinctive benefits of these methods allow scientists to create high-functioning patient-tailored biomedical solutions which were impossible to achieve through traditional processing techniques. Advanced fabrication tools synergize to enhance biomedical polymer capabilities and speed up laboratory-to-clinical transfers of advanced therapeutic devices (Alexandre-Franco et al., 2024).

6. ENGINEERING PERSPECTIVES ON POLYMER USE IN BIOMEDICAL APPLICATIONS

Biomedical engineering relies on polymer-based materials, however, using them in practice is not without issues. Biocompatibility is very important, as polymers should not activate the immune system and should work smoothly with the tissue around them (Pappalardo et al., 2019; Hu et al., 2024). If the mechanical properties of an implant and the surrounding tissue do not match well, the result may be minor injuries, tissue scarring or device failure mostly in areas with large movements like the heart or the bones (Mazza & Ehret, 2015; Satchanska et al., 2024). Another important problem is making sure the scaffold breaks apart at the same rate as tissue growth, to avoid it disappearing too soon or hanging around for too long (Mndlovu et al., 2024; Samir et al., 2022).

In response to these difficulties, choosing which engineering techniques to use now depends on the individual polymer. People use natural polymers like collagen and chitosan a lot because of their excellent compatibility in the body and inherent activity. Silk fibroin composites are handy in promoting healing of wounds in burn patients, while applied alginate hydrogels in dressings ensure a moist environment for faster tissue growth (Oliveira et al., 2023). At the same time, Farjaminejad et al. (2024) have demonstrated its beneficial properties. However, because of low strength (usually between 1 and 10 MPa), uncertain lifespan inside the human body and involved purification, natural polymers continue to face some difficulties. Because of their irregular results from one batch to another such materials are often strengthened with different chemical agents or plastic components (Reddy et al., 2021; Kalirajan et al., 2021).

The mechanical and surface qualities of synthetic polymers can be adjusted to meet the standards set for use in structure and orthopedics. They can give more strength (usually within the range of 50 to 3000 MPa) and are made with high precision. When polyether ether ketone (PEEK) is made more compatible with bone by adding hydroxyapatite and texturing, it is found to have a 40% greater success in hip surgery. Combining graphene oxide in polyurethane meshes can raise the strength of hernia repair by up to 300% but the flexibility is not decreased (Oliveira et al., 2023; Satchanska et al., 2024). Nevertheless, lots of synthetic polymers are unsafe for humans and won't biodegrade, so they must be handled carefully for safe disposal.

Biodegradable polymers have properties from both natural and manufactured materials. Adjusting the lactide-to-glycolide ratio in PLGA makes the scaffold degrade as the tissue grows. Generally, the strength is moderate (ranging from 20 to 200 MPa) and they provide more stable decay compared to natural rubber. PLGA-based stents help maintain the openness of the vessel for six to nine months and contribute to artery fixation and PCL scaffolds are designed to be in the bones from twelve to twenty-four months, helping to restore bone tissue (Mndlovu et al., 2024; Samir et al., 2022; Ulery et al., 2011). Implanting a large number of discs at once remains a problem, accompanied by increased expense and the acid surplus that can lower pH by more than 1.5 which slows osteogenesis in the spine (Pappalardo et al., 2019; Mndlovu et al., 2024).

Even now, implementing these materials in real-life clinical practice remains problematic for many reasons. Because natural polymers have bioactivity, the immune system usually flags them and the digestive enzymes in the body rapidly destroy them. It was demonstrated that chitosan scaffolds lose 60% of their original mass within two weeks in a person, whereas this occurs only in eight weeks in an artificial environment (Reddy et al., 2021; Kalirajan et al., 2021). Wearing synthetic fibres such as polyethylene may produce microplastic particles which could cause chronic inflammation in up to 15% of joint implant patients (Liang et al., 2023; Ulery et al., 2011). Even though PLGA implantation is safer than most polymers, its molecule breakdown can still cause trouble for the proper integration of tissues. This points out that polymers that can change in response to the surroundings would ensure more safety and reliability. Tumours in hyperthermia heated to high temperatures show accurate droplets from hydrogels with poly(N-isopropylacrylamide) (PNIPAAm), according to several studies (Purohit et al., 2022; Tanaka et al., 2014). However, medical systems using AI and machine learning can only be popularized if there are major improvements in technology. Using new algorithms, researchers can now predict how polymers will interact with organisms up to 85 percent correctly, based only on the behavior of macrophages in a test tube (Wolf et al., 2014; Seah et al., 2022).

At this point, every polymer that makes a great biomedical material has its own strengths and weaknesses. This observation is pushing the area to explore new material systems and designs that merge the advantages of biology, industry and the ability to degrade. Traditional approaches in materials science are now aligning with bio-fabrication technologies and AI optimization which is changing how we obtain clinically viable devices (Wolf et al., 2014; Seah et al., 2022; Purohit et al., 2022).

7. EMERGING TRENDS, CHALLENGES AND IMPLICATIONS OF POLYMER-BASED MATERIALS IN BIOMEDICAL APPLICATIONS

Biomedical polymers come with notable dangers of toxicity and immunological reactions in medical applications. Human bodies respond to polymer insertion through immune reactions which might result in tissue inflammation and rejection while producing various harmful effects. Biocompatible materials demand a priority status because a potent immune reaction caused by polymers will compromise the functionalities of medical devices and drug delivery systems and minimize their future safety and effectiveness (Yin & Luan, 2016; Jiang et al., 2025). The main problem resides in making sure polymers avoid releasing toxic by-products during their degradation process. It is essential to prevent consistent inflammation and immune rejection because these factors determine the long-term performance of these materials. Medical personnel face ongoing difficulties in establishing stable blood and tissue and cell connections while avoiding negative biological responses according to research by Jiang et al (2025) and Yin and Luan (2016). The successful clinical implementation of polymeric nanocomposites for drug delivery needs to prove both toxic free operations and non-activation of immune responses (Jiang et al., 2025). Tissue regeneration requirements determine the proper degradation rates of biodegradable polymers that must also produce safe decomposition products (Li et al., 2024). Achieving large-scale manufacturing of biomedical polymers faces multiple implementation hurdles. The production of advanced polymers through methods like solvent casting electrospinning and 3D printing enables customization but these techniques require extended cycle durations along with significant energy use. The approaches also struggle to generate advanced designs with personalized characteristics (Li et al., 2024). The important obstacle to overcome relates to quality management. Achieving constant high-quality product purity together with functionality at commercial-scale production becomes hard under conditions requiring strict mechanical performance and biocompatibility standards (Jiang et al., 2025). High production costs alongside limited access to particular natural polymers restricts the availability of polymer-based medical products thus decreasing their adoption potential within healthcare settings (Li et al., 2024). The extended cost of medical devices due to manufacturing complexities decreases their practical use particularly in areas with limited resources.

Regulatory guidelines represent multiple hurdles which biomedical polymers need to overcome for proper market approval. Material products meant for medical devices complete extensive testing for biocompatibility and toxicity alongside performance assessment to satisfy required standards (Yin & Luan, 2016). Regulatory approval schedules become prolonged because biological system interaction complexity meets with requirements to study extended-term safety properties. The evaluation process for new polymers and innovative nanocomposite formulations extends longer than usual because of concerns regarding their long-term performance within tissues. The process of gaining regulatory approval demands complete safety and efficacy documentation from agencies which lengthens and increases the financial burden of introducing new polymer technologies to the market (Yin & Luan, 2016).

Biomedical products made from polymers offer cost-saving advantages compared to traditional materials because such products facilitate minimally invasive surgical procedures while sometimes eliminating the requirement for additional surgeries. The high expenses involved in developing and producing and obtaining regulatory clearances for polymer materials specifically advanced and personalized types present challenges to their overall cost-effectiveness (Li et al., 2024). Biodegradable polymers used in medical devices succeed in decreasing healthcare plastic waste volumes which represent a major environmental problem in medical fields (Li et al., 2024). The usage of sustainable polymers helps decrease product-related environmental damage of medical single-use items, yet continued challenges emerge from raw material acquisition and degradation consistency and performance sustainability trade-offs (Li et al., 2024). The potential of polymers to transform biomedical applications remains restricted because of restrictions from manufacturing scalability along with biocompatibility issues, toxicity concerns, regulatory barriers, and economic and environmental factors. To achieve safe and sustainable medical solutions the industry needs interdisciplinary collaboration between researchers along with innovative material engineering and regulatory partnership (Yin & Luan, 2016; Jiang et al., 2025).

Modern medicine is benefiting greatly from biomedical polymers, as developments in materials, sustainability and technology are helping them make a difference in the health industry. New developments in smart polymers, environmentally friendly manufacturing and AI assisted design are rapidly making the field transformative. The use of smart polymers which sense changes in temperature, pH, light or mechanical signals, is advancing drugs for delivery, tissue engineering and regenerative medicine. With these materials, changes in their structure or function can be easily reversed, making it possible to accurately regulate drug release, repair implants on demand and build scaffolds similar to natural body tissues (Hunter & Moghimi, 2016). For example, injectable hydrogels are being designed to gel at body temperature in response to stimuli, enabling accurate drug delivery and improving patient compliance (Hunter & Moghimi, 2016). They are being used in wound repair and used for implantable devices where their healing ability by itself decreases surgical steps and can make them last longer. Along with how well a material works, sustainability is now an important factor in developing biomedical polymers. Moving to biomaterials and green chemistry is changing medical-grade polymer processes and cutting down on their effect on the environment. Now, renewable plant oils, lignocellulosic biomass and left-over farming residues are turned into biodegradable plastics with less environmental impact (Papageorgiou, 2018). Methods based on enzymes and catalysts that work under eco-friendly settings help develop materials that can be tuned for breaking down and have minimized toxic danger. By using these methods, companies both help the world reach its sustainable goals and comply with regulations for safer and eco-friendly biomaterials (Papageorgiou, 2018).

At the same time, using AI in materials design is helping with the search for and improved performance of biomedical polymers. Advances in machine learning allow these models to predict important features of materials like biocompatibility, strength and rate of breakdown, so researchers now depend on fewer experiments (Hang et al., 2025; Chen et al., 2023; Butler et al., 2018). AI has greatly improved 3D bioprinting by predicting which formulas of polymers will create functional bioinks with a high level of accuracy and this has quickened the development of specialized scaffolds and implants. AI is now boosting efforts to personalize medical equipment and treatments for each patient, so they cause fewer adverse effects (Hang et al., 2025). These new discoveries in polymers have not led to easy transition into actual medical uses. The FDA, EMA and ISO 10993 standards have made it necessary for medical polymers to satisfy demanding rules about biocompatibility, risk assessment and acceptable sterilization (Kurowiak et al., 2023). When applying for premarket approval, FDA requires thorough information on how the product is created, its toxicological safety and how stable its materials are. In the same way, the EMA Medical Device Regulation (MDR) requires strict measures for labeling, tracking and banning hazardous ingredients. Although ISO 10993 supports biological testing, problems arise because there is not one clear definition of a medical-grade material, and the list of regulated materials keeps changing (Curtis, 2018).

Most polymer-based devices are brought to market only after passing through both preclinical screening and human testing in clinical trials. Several products made from polyglycolide have obtained approval for use, these include biodegradable sutures (Dexon), orthopedic implants and drug-loaded stents in cardiology all have demonstrated success and safety in patients (Kurowiak et al., 2023). Even so, integrating the medical device market is often complicated by polymer composition differences in different batches, the cost of production, sterilization-related challenges and unknowing effects of degradation products when used with the body (as reported by Hua et al., 2018; Kurowiak et al., 2023). To illustrate, making larger batches of polymers designed to be controlled in both bioactivity and degradation is not easy and maintaining product sterility is still a big problem even after constant improvement (Hua et al., 2018; Kurowiak et al., 2023).

A comparison of different polymer types stresses the point of choosing the appropriate material for each task. Building materials made from collagen and chitosan are very good at repairing and healing soft tissues. Yet, there are difficulties because of weak mechanics and the variability between each batch which means standardization is hard (Puertas-Bartolomé et al., 2021; Hu et al., 2024). Synthetic polymers, for example polyethylene glycol (PEG), polylactic acid (PLA) and polycaprolactone (PCL), offer predictable strength and the ability to break down, can be manufactured on a large scale and are consistent in structure, making them suitable for use in implants and as drug carriers (Hu et al., 2024; Kurowiak et al., 2023). Still, these problematic features are reasons to rethink using these substances. This group of polymers which can be made from either nature or in the lab, gives temporary scaffolding and helps control drug delivery, eliminating the need for surgeons to operate. Even though their ability to break down benefits patients, the difficulty of regulating such implants and possible harmful effects from the breakdown products are holding back newer adoptions (Kurowiak et al., 2023; De Souza et al., 2022).

The shift in biomedical polymers is being driven by improvements in smart technology, green chemistry and using AI in materials science (Butler et al., 2018). Every polymer group, for example natural, synthetic and biodegradable, comes with its strengths and weaknesses during the translation to commercial use. Improvements here will rely on teamwork between experts in materials, biomedical engineering and healthcare. Integrating modern design methods with strict ways of bringing polymers to market and complying with rules promises more reliable, handy and eco-friendly healthcare solutions in the future (Hunter & Moghimi, 2016; Hang et al., 2025; Kurowiak et al., 2023).

8. CONCLUSION

Material science has undergone significant advancements through biomedical applications which utilize polymers to produce innovative medical solutions. The biocompatible natural polymers collagen together with chitosan and alginate demonstrate unique properties by reconstructing extracellular matrices thus serving as optimal materials for wound healing along with drug delivery and tissue engineering applications. Natural polymers provide structural adaptability which enables researchers to create advanced scaffolds that enhance cell growth while fostering tissue regeneration. The advancements of synthetic polymers have been offset by essential barriers to implementation including batch inconsistency together with material structural constraints and the necessity of chemical modifications for enhanced functional capability. The dedicated control of mechanical performance and degradation dynamics and simple production methods of synthetic polymers including polylactic acid and polyglycolic acid and polycaprolactone allows medical engineers to use these materials for different biomedical applications. Synthetic polymers demonstrate adjustable capabilities allowing engineers to fabricate materials that match precise functional needs rendering them suitable for biological operations. The implementation of synthetic polymers poses challenges because it entails intensive regulatory procedures while presenting potential toxic substance risks.

Modern polymer processing technologies such as electrospinning with 3D printing allows biomedical device manufacturers to produce complex shapes alongside customized medical solutions. Medical researchers currently develop flexible drug delivery systems and tissue-scaffolds by uniting these technical approaches to produce personalized healthcare products.

The transformation from research discoveries to clinical application entails significant hurdles. Biocompatibility together with immunological response reduction must be fundamental considerations for establishing durable success using polymer-based materials. The process of increasing, manufacturing volume, faces significant barriers because it demands continuous maintenance of product quality and uniformity levels. Manufacturers face extensive testing requirements when introducing new polymer technologies to market since they must obtain regulatory approval of safety and effectiveness. The wider adoption of polymers for biomedical purposes remains restricted due to multiple obstacles which include toxicity issues and challenges of maintaining biocompatibility along with scaling up production capacity and facing regulatory hurdles and economic limitations. The successful adoption of polymers for biomedical applications demands interdisciplinary teamwork along with innovative material engineering and sustainability assurance. The complete realization of polymer materials in medical applications requires ongoing research efforts in polymer science which will eventually generate safer medical solutions that improve clinical outcomes effectively and sustainably.

REFERENCES

Agueda, J. R. H. S., Chen, Q., Maalihan, R. D., Ren, J., Da Silva, Í. G. M., Dugos, N. P., Caldona, E. B., & Advincula, R. C. (2021). 3D printing of biomedically relevant polymer materials and biocompatibility. *MRS Communications*, *11*(2), 197–212. DOI: 10.1557/s43579-021-00038-8

Alemu, D., Getachew, E., & Mondal, A. K. (2023). Study on the Physicochemical Properties of Chitosan and their Applications in the Biomedical Sector. *International Journal of Polymer Science*, *2023*, 1–13. DOI: 10.1155/2023/5025341

Alexandre-Franco, M. F., Kouider, R., Al-Karany, R. K., Cuerda-Correa, E. M., & Al-Kassir, A. (2024). Recent Advances in polymer science and fabrication processes for enhanced microfluidic applications: An Overview. *Micromachines*, *15*(9), 1137. DOI: 10.3390/mi15091137

Arefin, A. M. E., Khatri, N. R., Kulkarni, N., & Egan, P. F. (2021). Polymer 3D Printing Review: Materials, process, and design Strategies for medical applications. *Polymers*, *13*(9), 1499. DOI: 10.3390/polym13091499

Assad, H., Assad, A., & Kumar, A. (2023). Recent developments in 3D Bio-Printing and its biomedical applications. *Pharmaceutics*, *15*(1), 255. DOI: 10.3390/pharmaceutics15010255

Butler, K. T., Davies, D. W., Cartwright, H., Isayev, O., & Walsh, A. (2018). Machine learning for molecular and materials science. *Nature*, *559*(7715), 547–555. DOI: 10.1038/s41586-018-0337-2

Chen, H., Liu, Y., Balabani, S., Hirayama, R., & Huang, J. (2023). Machine learning in predicting printable biomaterial formulations for direct ink writing. *Research*, *6*, 0197. Advance online publication. DOI: 10.34133/research.0197

Chia, H. N., & Wu, B. M. (2015). Recent advances in 3D printing of biomaterials. *Journal of Biological Engineering*, *9*(1), 4. Advance online publication. DOI: 10.1186/s13036-015-0001-4

Cho, Y., Beak, J. W., Sagong, M., Ahn, S., Nam, J. S., & Kim, I. (2025). Electrospinning and nanofiber technology: Fundamentals, innovations, and applications. *Advanced Materials*, *37*(28), 2570190. Advance online publication. DOI: 10.1002/adma.202570190

Curtis, B. (2018). *Regulatory considerations for medical polymers*. Foster Corporation. https://www.fostercomp.com/regulatory-considerations-for-medical-polymers/

De Souza, A. A., De Cássia Braga Granja, R., De Melo, E. C. C., De Oliveira, W. F., De Albuquerque, P. B. S., Correia, M. T. D. S., & Carneiro-Da-Cunha, M. D. G. (2022). Biodegradable synthetic polymers in biomedical application: A review. *Brazilian Journal of Health Review, 5*(6), 24628–24644. DOI: 10.34119/bjhrv5n6-217

Dhaliwal, K. (2018). Biodegradable Polymers and their Role in Drug Delivery Systems. *Biomedical Journal of Scientific & Technical Research, 11*(1). Advance online publication. DOI: 10.26717/BJSTR.2018.11.002056

Distler, T., Polley, C., Shi, F., Schneidereit, D., Ashton, M. D., Friedrich, O., Kolb, J. F., Hardy, J. G., Detsch, R., Seitz, H., & Boccaccini, A. R. (2021). Electrically conductive and 3D-Printable oxidized Alginate-Gelatin Polypyrrole:PSS hydrogels for tissue engineering. *Advanced Healthcare Materials, 10*(9), 2001876. Advance online publication. DOI: 10.1002/adhm.202001876

Dutta, S. D., Hexiu, J., Patel, D. K., Ganguly, K., & Lim, K. (2020). 3D-printed bioactive and biodegradable hydrogel scaffolds of alginate/gelatin/cellulose nanocrystals for tissue engineering. *International Journal of Biological Macromolecules, 167*, 644–658. DOI: 10.1016/j.ijbiomac.2020.12.011

Fan, D., Staufer, U., & Accardo, A. (2019). Engineered 3D polymer and hydrogel microenvironments for cell culture applications. *Bioengineering (Basel, Switzerland), 6*(4), 113. DOI: 10.3390/bioengineering6040113

Farjaminejad, S., Farjaminejad, R., Hasani, M., Garcia-Godoy, F., Abdouss, M., Marya, A., Harsoputranto, A., & Jamilian, A. (2024). Advances and Challenges in Polymer-Based Scaffolds for Bone Tissue Engineering: A path towards Personalized Regenerative Medicine. *Polymers, 16*(23), 3303. DOI: 10.3390/polym16233303

Halder, T., Barot, H., Kumar, B., Kaushik, V., Patel, H., Bhut, H., Saha, B., Poddar, S., & Acharya, N. (2024). An Insight into Biodegradable Polymers and their Biomedical Applications for Wound Healing. *Current Pharmaceutical Design, 30*(31), 2425–2444. DOI: 10.2174/0113816128295935240425101509

Hamad, K., Kaseem, M., Yang, H. W., Deri, F., & Ko, Y. G. (2015). Properties and medical applications of polylactic acid: A review. *Express Polymer Letters, 9*(5), 435–455. DOI: 10.3144/expresspolymlett.2015.42

Hang, R., Yao, X., Bai, L., & Hang, R. (2025). Evolving Biomaterials Design from Trial and Error to Intelligent Innovation. *Acta Biomaterialia, 197*, 29–47. Advance online publication. DOI: 10.1016/j.actbio.2025.03.013

Hosseini, A., Shahverdi, M., Sourani, Z., Habibi, R., Ahangari, Z., Shirian, S., & Zafari, F. (2023). A Review on Synthesis, Characterization and Applications of Polycaprolactone as a Novel Drug Delivery System and Tissue Engineering. *Journal of Inflammatory Diseases*, *27*(3). Advance online publication. DOI: 10.69107/jid-150907

Hu, T., Fang, J., Shen, Y., Li, M., Wang, B., Xu, Z., & Hu, W. (2024). Advances of naturally derived biomedical polymers in tissue engineering. *Frontiers in Chemistry*, *12*, 1469183. Advance online publication. DOI: 10.3389/fchem.2024.1469183

Hu, Z., & Zhou, R. (2024). Review on some important research progresses in Biodegradable Plastics/Polymers. *Recent Progress in Materials*, *06*(02), 1–19. DOI: 10.21926/rpm.2402015

Hua, S., De Matos, M. B. C., Metselaar, J. M., & Storm, G. (2018). Current Trends and Challenges in the Clinical translation of nanoparticulate Nanomedicines: Pathways for translational development and Commercialization. *Frontiers in Pharmacology*, *9*, 790. Advance online publication. DOI: 10.3389/fphar.2018.00790

Hunter, A. C., & Moghimi, S. M. (2016). Smart polymers in drug delivery: A biological perspective. *Polymer Chemistry*, *8*(1), 41–51. DOI: 10.1039/C6PY00676K

Jiang, M., Fang, H., & Tian, H. (2025). Latest advancements and trends in biomedical polymers for disease prevention, diagnosis, treatment, and clinical application. *Journal of Controlled Release*, *380*, 138–174. DOI: 10.1016/j.jconrel.2025.01.063

Jujjavarapu, S. E., Chandrasekhar, K., Naik, S., Toppo, A. L., & Thakur, V. (2020). Sources of Natural Polymers from Plants with Green Nanoparticles. In *CRC Press eBooks* (pp. 81–102). https://doi.org/DOI: 10.1201/9781351045155-4

Kalirajan, C., Dukle, A., Nathanael, A. J., Oh, T., & Manivasagam, G. (2021). A Critical review on Polymeric Biomaterials for Biomedical applications. *Polymers*, *13*(17), 3015. DOI: 10.3390/polym13173015

Kara, A., Distler, T., Polley, C., Schneidereit, D., Seitz, H., Friedrich, O., Tihminlioglu, F., & Boccaccini, A. R. (2022). 3D printed gelatin/decellularized bone composite scaffolds for bone tissue engineering: Fabrication, characterization and cytocompatibility study. *Materials Today. Bio*, *15*, 100309. DOI: 10.1016/j.mtbio.2022.100309

Kurowiak, J., Klekiel, T., & Będziński, R. (2023). Biodegradable Polymers in Biomedical Applications: A Review-Developments, Perspectives and Future Challenges. *International Journal of Molecular Sciences*, *24*(23), 16952. DOI: 10.3390/ijms242316952

Lee, K. Y., & Mooney, D. J. (2011). Alginate: Properties and biomedical applications. *Progress in Polymer Science, 37*(1), 106–126. DOI: 10.1016/j.progpolymsci.2011.06.003

Li, F., Chen, C., & Chen, X. (2024). Tremendous advances, multifaceted challenges and feasible future prospects of biodegradable medical polymer materials. *RSC Advances, 14*(44), 32267–32283. DOI: 10.1039/D4RA00075G

Liang, S., Xu, F., Wu, J., & Deng, X. (2023). Editorial: Biomedical applications of water-soluble synthetic polymers. *Frontiers in Bioengineering and Biotechnology, 11*, 1235931. Advance online publication. DOI: 10.3389/fbioe.2023.1235931

Libretexts. (2024). *30.1: Chain-Growth polymers*. Chemistry LibreTexts. https://chem.libretexts.org/@go/page/474086?pdf

Liu, X., Zheng, C., Luo, X., Wang, X., & Jiang, H. (2019). Recent advances of collagen-based biomaterials: Multi-hierarchical structure, modification and biomedical applications. *Materials Science and Engineering C, 99*, 1509–1522. DOI: 10.1016/j.msec.2019.02.070

Lopes, R. C., Nossa, T., Lustri, W. R., Lombardo, G., Errea, M. I., & Trovatti, E. (2024). Renewable polymers in biomedical applications: From the bench to the market. *Journal of Renewable Materials, 12*(4), 643–666. DOI: 10.32604/jrm.2024.048957

Mazza, E., & Ehret, A. E. (2015). Mechanical biocompatibility of highly deformable biomedical materials. *Journal of the Mechanical Behavior of Biomedical Materials. Journal of the Mechanical Behavior of Biomedical Materials, 48*, 100–124. DOI: 10.1016/j.jmbbm.2015.03.023

Mndlovu, H., Kumar, P., Du Toit, L. C., & Choonara, Y. E. (2024). A review of biomaterial degradation assessment approaches employed in the biomedical field. *NPJ Materials Degradation, 8*(1), 66. Advance online publication. DOI: 10.1038/s41529-024-00487-1

Nanda, H. S., Yang, L., Hu, J., Mao, H., & Jiang, S. (2022). Editorial: Biodegradable Polymers for Biomedical Applications. *Frontiers in Materials, 9*, 944755. Advance online publication. DOI: 10.3389/fmats.2022.944755

Oleksy, M., Dynarowicz, K., & Aebisher, D. (2023). Advances in Biodegradable Polymers and Biomaterials for Medical Applications - A Review. *Molecules (Basel, Switzerland), 28*(17), 6213. DOI: 10.3390/molecules28176213

Oliveira, C., Sousa, D., Teixeira, J. A., Ferreira-Santos, P., & Botelho, C. M. (2023). Polymeric biomaterials for wound healing. *Frontiers in Bioengineering and Biotechnology, 11*, 1136077. Advance online publication. DOI: 10.3389/fbioe.2023.1136077

Papageorgiou, G. Z. (2018). Thinking Green: Sustainable Polymers from Renewable Resources. *Polymers, 10*(9), 952. DOI: 10.3390/polym10090952

Pappalardo, D., Mathisen, T., & Finne-Wistrand, A. (2019). Biocompatibility of Resorbable Polymers: A Historical perspective and Framework for the future. *Biomacromolecules, 20*(4), 1465–1477. DOI: 10.1021/acs.biomac.9b00159

Periayah, M. H., Halim, A. S., & Saad, A. Z. M. (2016). Chitosan: A promising marine polysaccharide for biomedical research. *Pharmacognosy Reviews/Bioinformatics Trends. Pharmacognosy Reviews, 10*(19), 39. DOI: 10.4103/0973-7847.176545

Puertas-Bartolomé, M., Mora-Boza, A., & García-Fernández, L. (2021). Emerging Biofabrication Techniques: A review on natural polymers for biomedical applications. *Polymers, 13*(8), 1209. DOI: 10.3390/polym13081209

Purohit, A., Jain, S., Nema, P., Vishwakarma, H., & Jain, P. K. (2022). Intelligent or smart polymers: Advance in novel drug delivery. *Journal of Drug Delivery and Therapeutics, 12*(5), 208–216. DOI: 10.22270/jddt.v12i5.5578

Radhakrishnan, S., Nagarajan, S., Belaid, H., Farha, C., Iatsunskyi, I., Coy, E., Soussan, L., Huon, V., Bares, J., Belkacemi, K., Teyssier, C., Balme, S., Miele, P., Cornu, D., Kalkura, N., Cavaillès, V., & Bechelany, M. (2020). Fabrication of 3D printed antimicrobial polycaprolactone scaffolds for tissue engineering applications. *Materials Science and Engineering C, 118*, 111525. DOI: 10.1016/j.msec.2020.111525

Reddy, M. S. B., Ponnamma, D., Choudhary, R., & Sadasivuni, K. K. (2021). A comparative review of natural and Synthetic biopolymer composite scaffolds. *Polymers, 13*(7), 1105. DOI: 10.3390/polym13071105

Samir, A., Ashour, F. H., Hakim, A. A., & Bassyouni, M. (2022). Recent advances in biodegradable polymers for sustainable applications. *NPJ Materials Degradation, 6*(1), 68. Advance online publication. DOI: 10.1038/s41529-022-00277-7

Saroia, J., Yanen, W., Wei, Q., Zhang, K., Lu, T., & Zhang, B. (2018). A review on biocompatibility nature of hydrogels with 3D printing techniques, tissue engineering application and its future prospective. *Bio-Design and Manufacturing, 1*(4), 265–279. DOI: 10.1007/s42242-018-0029-7

Satchanska, G., Davidova, S., & Petrov, P. D. (2024). Natural and synthetic polymers for biomedical and environmental applications. *Polymers, 16*(8), 1159. DOI: 10.3390/polym16081159

Seah, I., Ong, C., Liu, Z., & Su, X. (2022). Polymeric biomaterials in the treatment of posterior segment diseases. *Frontiers in Medicine, 9*, 949543. Advance online publication. DOI: 10.3389/fmed.2022.949543

Shie, M., Chang, W., Wei, L., Huang, Y., Chen, C., Shih, C., Chen, Y., & Shen, Y. (2017). 3D Printing of Cytocompatible Water-Based Light-Cured Polyurethane with Hyaluronic Acid for Cartilage Tissue Engineering Applications. *Materials (Basel), 10*(2), 136. DOI: 10.3390/ma10020136

Tanaka, M., Sato, K., Kitakami, E., Kobayashi, S., Hoshiba, T., & Fukushima, K. (2014). Design of biocompatible and biodegradable polymers based on intermediate water concept. *Polymer Journal, 47*(2), 114–121. DOI: 10.1038/pj.2014.129

Ulery, B. D., Nair, L. S., & Laurencin, C. T. (2011). Biomedical applications of biodegradable polymers. *Journal of Polymer Science. Part B, Polymer Physics, 49*(12), 832–864. DOI: 10.1002/polb.22259

Wang, C., Su, Y., & Xie, J. (2024). Advances in electrospun nanofibers: Versatile materials and diverse biomedical applications. *Accounts of Materials Research, 5*(8), 987–999. DOI: 10.1021/accountsmr.4c00145

Wolf, M. T., Vodovotz, Y., Tottey, S., Brown, B. N., & Badylak, S. F. (2014). PredictingIn VivoResponses to biomaterials via CombinedIn VitroandIn SilicoAnalysis. *Tissue Engineering. Part C, Methods, 21*(2), 148–159. DOI: 10.1089/ten.tec.2014.0167

Wypij, M., Trzcińska-Wencel, J., Golińska, P., Avila-Quezada, G. D., Ingle, A. P., & Rai, M. (2023). The strategic applications of natural polymer nanocomposites in food packaging and agriculture: Chances, challenges, and consumers' perception. *Frontiers in Chemistry, 10*, 1106230. Advance online publication. DOI: 10.3389/fchem.2022.1106230

Yin, J., & Luan, S. (2016). Opportunities and challenges for the development of polymer-based biomaterials and medical devices. *Regenerative Biomaterials, 3*(2), 129–135. DOI: 10.1093/rb/rbw008

Zhang, H., Lin, X., Cao, X., Wang, Y., Wang, J., & Zhao, Y. (2023). Developing natural polymers for skin wound healing. *Bioactive Materials, 33*, 355–376. DOI: 10.1016/j.bioactmat.2023.11.012

ADDITIONAL READINGS

Feldman, D. (2008). Polymer history. *Designed Monomers and Polymers, 11*(1), 1–15. DOI: 10.1163/156855508X292383

Gil, E., & Hudson, S. (2004). Stimuli-reponsive polymers and their bioconjugates. *Progress in Polymer Science, 29*(12), 1173–1222. DOI: 10.1016/j.progpolymsci.2004.08.003

Klemm, D., Schumann, D., Kramer, F., Hebler, N., Hornung, M., Schmauder, H., & Marsch, S. (2006). Nanocelluloses as innovative polymers in research and application. In *Advances in polymer science* (pp. 49-96). https://doi.org/DOI: 10.1007/12_097

Mohanty, A. K., Wu, F., Mincheva, R., Hakkarainen, M., Raquez, J., Mielewski, D. F., Narayan, R., Netravali, A. N., & Misra, M. (2022). Sustainable polymers. *Nature Reviews. Methods Primers, 2*(1), 46. Advance online publication. DOI: 10.1038/s43586-022-00124-8

Olatunji, O. (2015). Natural polymers. In *Springer eBooks*. https://doi.org/DOI: 10.1007/978-3-319-26414-1

KEY TERMS AND DEFINITIONS

Alginate: The polysaccharide obtained from brown seaweed occurs naturally in its original state. Salts of chitosan find extensive use in biomedical areas through their ability to form gels and their biocompatible properties.

Biocompatibility: A material demonstrates biocompatibility if it functions appropriately with suitable bodily responses after its intended application while producing no toxic nor immunological body reactions.

Biomedical: Relating to the application of natural sciences, especially biology and physiology, to clinical medicine. Biomedical sciences encompass medical research, diagnostics, and the development of therapies and medical devices.

Chitosan: Natural polymer chitosan originates from the chitin material which exists inside shrimp and crab shells. The application of this material in drug delivery among other biomedical applications benefits from its biodegradable nature and also its compatibility with biological systems.

Collagen: Natural proteins exist in animal connective tissues where skin, bones and tendons also contain them. The medical field utilises this substance due to its biological compatibility and its support for promoting cell growth.

Natural Polymers: It consists of substances produced by living organisms that automatically occur in nature. The three main types of natural polymers are proteins along with nucleic acids and the polysaccharides cellulose and starch. Such polymers show both their ability to decompose naturally and their compatibility with human tissue.

Polymers: It consists of many monomers which function as repeated subunits. These compounds find extensive use throughout different markets since they exist as synthetic or natural materials for packaging and construction as well as biomedical applications.

Synthetic Polymers: The synthesis of artificial polymers from petroleum-derived monomers constitutes their definition. The common synthetic polymer group consists of polyethylene together with polystyrene followed by nylon. The medical and industrial sectors widely utilise these polymers in their applications.

Tissue Engineering: The interdisciplinary field known as tissue engineering focuses on integrating cells with engineering principles alongside materials to support tissue maintenance or perform tissue repair while replacing broken biological tissues.

Chapter 10
IoT-Driven Wearable Devices Based on Flexible and Stretchable Materials

Dankan Gowda V.
https://orcid.org/0000-0003-0724-0333
Department of Electronics and Communication Engineering, BMS Institute of Technology and Management, Bangalore, India

Prashant A. Athavale
https://orcid.org/0000-0003-3084-6405
Department of Electrical and Electronics Engineering, BMS Institute of Technology and Management, Bangalore, India

P. Ramesh Naidu
https://orcid.org/0000-0003-4941-9569
Department of Computer Science and Engineering, Nitte Meenakshi Institute of Technology, Nitte University, Bangalore, Bangalore, India

Maria Lapina
North Caucasus Federal University, Stavropol, Russia

K. D. V. Prasad
https://orcid.org/0000-0001-9921-476X
Symbiosis Institute of Business Management, Symbiosis International University, Pune, India

ABSTRACT

The emergence of Internet of Things (IoT) panned a lot of wearables that has a broad scope of applications, especially in health, fitness and personal monitoring.

DOI: 10.4018/979-8-3373-9454-1.ch010

However, traditional electronics have been made of hard materials which have strong limitations on sensations of feeling, flexibility, durability whereas all of those are required for wearable devices. With the development of IoTdriven wearables, flexible and stretchable material integration into the device could be a new opportunity to break through these obstacles. These advanced materials, which include conductive polymers, graphene, and elastomers, can be used to fabricate wearable electronics for curving, stretching and matching the human body. In this chapter, we review the use of such materials for designing and developing the ubiquitous wearable IoT devices described above by discussing their characteristics, benefits, and processing methods. Issues such as flexible electronics integration, sensor accuracy, power management, and long-term durability are also addressed.

INTRODUCTION

The acceleration of the Internet of Things (IoT) technology has enabled a new type of wearable devices which are changing the way of human interaction with technology, especially in the domains of healthcare, fitness and smart environments. These tools can range from smartwatches and fitness trackers to health monitoring touch points to a plethora of connected devices, providing real-time health information and real-time feedback. IoT-enabled wearables can collect vast amounts of data, process it in real-time, and send it to other devices or cloud-based platforms to help deliver actionable insights (Jing, Lai, and Sun, 2024). In the healthcare field, these devices help personalized health monitoring to detect diseases at the early stage, monitor chronic diseases, and manage health in real-time. In the fitness sector, they track and analyze physical and biometric metrics of activity and rest to enhance fitness results (Khadse and Hajare, 2025). Furthermore, wearables are becoming more and more connected to smart environments, in order to cooperate to IoT eco-systems, providing open deployment of smart homes and cities, that is, where the devices cooperate togheter to provide energy, security and so on.

Wearable device operation relies on IoT infrastructure. (IoT-based) wearables rely on cloud platform, where data are created, stored and processed from devices, offer insight capabilities to the user and conduct long-term analysis (Madunuri, Chitta, Bonam, Vangoor, Yellepeddi, and Ravi, 2024). Response Time Several models have shown that the cloud can be used to store and retrieve huge amounts of data and is capable of rapid elasticity which make it ideal for data- centric application with continuous data flow and processing. In IoT wearables, edge computing (a form of computing where processing is done closer to the data) serves to bring down latency, facilitate real-time decision-making, and minimize the need for cloud servers for time-sensitive applications (Tôn, Lê, and Nguyen, 2021). This structure guarantees that these wearable devices are not just responsive, but can operate consistently with little disruptions, making them more dependable for sustained on-the-go use.

What is seamless continuous monitoring: its importance in medical and fitness wearables. Wearable medical devices for heart rate, glucose, ECG etc require to work continuously without interruption to provide meaningful insights to healthcare professionals and patients (Chophla, Sharda, Madan, and Chopra, 2024). Likewise, fitness wearables need to monitor the users' physical activity in real-time and provide feedback on performance, all of which demands for real-time data collection and transmission. In both cases, reliable and accurate sensors, low power consumption as well as efficient communication protocols are necessary to guarantee that data is continuously gathered and transmitted without missing any information.

In spite of the recent advancements in the wearable devices based on IoT, number of challenges are existed in designing and deployment of such devices (Thangam, Joshith, Santhosh, Harish, and Kumari, 2024). One of the main restrictions of the conventional wearable devices (e.g., fitness tracker and smartwatch) is the rigid constituent materials. Most of these devices work, but none provide a comfortable or flexible enough fabric to wear for warranty coverages. These types of devices can be difficult to wear for long periods of time, and that can be particularly true when worn on sensitve areas of the body, such as the wrist or chest. Moreover, they are rigid, threatening the comfort of the user, because wearables are expected to move to the body shape and size.

Emerging applications such as wearable devices require the development of wearable materials that are flexible, stretchable, lightweight, and comfortable to wear, and at the same time, strong, and durable for function (Ronald, Raman, Vekariya, Baskar, Rahul, and Srinivasan, 2024). Flexible and stretchable materials may follow contours of the body and not interfere with body motion, and may be worn for extended periods without discomfort. These materials also make it possible to make the device smaller and lighter, without sacrificing its performance or capability. For instance, stretchable electronics will keep conducting electricity while being deformed, so it is suitable for wearable sensors which should maintain their performance while being bended or stretched or compressed." These kind of materials could change how wearable technology is used, leading to improved user experiences, longer-lasting wear and perhaps new kinds of wearables.

While flexible and stretchable materials can be included in wearable devices, there is a potential for further developments of more comfortable and more versatile wearable devices that can be more tailored for use by users across various fields. They enable the production of comfortable, lightweight wearables with superior performance (J. and D., 2025). In addition, they are enabling new applications in smart textiles and healthcare monitoring, in which wearables effortlessly mold to the body, for example as accessories integrated into garments or as epidermal electronics, without compomising comfort and motion.

The main focus of this chapter is to discuss the incorporation of flexible, stretchable materials in wearable IoT technologies. They are critical for tackling the comfort, durability, and versatility challenges in wearable technologies. A study of the use of these advanced materials will show how they contribute to the design, performance, and use of wearables, for a better long term use. This article also discusses how stretchable electronic, sensors, and energy harvesting devices can collaborate in wearable technology to bring out better functionality and user-friendly. Some of the organic materials used here has been summarized and the work of other flexible materials can be found in (Jing, Lai, and Sun, 2024). It is anticipated that the use of the flexible and stretchable materials into wearable IoT system can drive the IoT-enabled wearables toward more efficient, durable, and user-friendly devices that can be easily integrated into our daily lives.

LITERATURE SURVEY

Wearable Electronics and IoT Applications

Wearable electronics are an important segment in the Internet of Things (IoT) landscape that is ripe with potential in areas such as health monitoring, fitness tracking, and a diverse range of smart applications. Personal devices, including smartwatches, fitness trackers, medical diagnostic wearables and health-monitoring sensors, have become important for users and healthcare professionals to collect and receive data in real-time. For example, smartwatches integrate the features of regular watches with functionality such as activity recording, personal heart rate measurement, and instant notifications, while in contrast, fitness trackers are dedicated to monitor physical activities and related physiological measures (e.g., the numbers of taken steps, distance, and caloric burned by activities). Now wearable medical diagnostic devices such as electrocardiogram (ECG) monitors and glucose sensors are able to monitor patients' health around-the-clock, presenting the possibility to practice proactive healthcare (Liu, et al., 2019).

Usually, the devices to be tested make use of different sensors and actuators, as well as different communication modules to acquires, process and transmit data. It is necessary to collect and recognize the physical and physiological changes that happen in real-tim ethrough sensors embedded into wearable devices. Popular sensors in wearables include accelerometers, gyroscopes, temperature sensing, optical heart rate sensing, and electrochemical sensing for glucose and other biomarkers. Actuators like Vibration motors or Haptic feedback systems, physically respond to user interaction and alarms enriching the user experience (Redwan, Ahmed, Turjo, Munna, Hossain, Akter, and Al Sohan, 2024). Communication modules (e.g., Bluetooth, Wi-Fi, Zigbee, etc.) are important for the transfer of data between wearables and other devices such as smartphones, cloud platforms, and patient monitoring systems. One of the most popular is Bluetooth Low Energy (BLE), which is especially suitable for wearable devices that need to operate in low-power mode for longer periods of time on small batteries.

The incorporation of sensors, actuators, and communication capabilities allows these wearables to collect, process, and send/receive health and activity information while becoming part of the IoT environment (Rajaa, Manee, Mohammed, Mohammed, and Sharif, 2024). This connection gives users up-to-the-minute tracking of their health data for improved health monitoring, and more individualized treatment, especially for those with chronic conditions. Nevertheless, these devices have brought a number of challenges, in terms of user comfort, power consumption, and information security, to the development and deployment of wearable IoT services.

Materials for Flexible and Stretchable Electronics

Flexible and stretchable electronics have evolved as the underlying technology for applications in wearable technology with the demand for high comfort levels, easy fit and long-term wear. These materials are flexible bendy and stretchy enough to the shape of the human body so perfect for wearables. Conductive polymers, graphene, carbon nanotubes, and silver nanowires, metal meshes, silicone, elastomers, and organic electronics are some of the main substances used for the fabrication flexible and stretchable electronic devices (Mani, Singh, and Litoriya, 2024). Such materials possess distinctive properties that make them promising in various wearable device applications.

Conducting Polymers are one of the most popular materials for flexible electronics. Conductive polymers including polyaniline (PANI) and PEDOT:PSS are well-known polymers with superior electrical conductivity, good capability of flexibility and processability. The polymers have especially high value in devices such as sensors, organic light-emitting diodes (OLED), and thin-film transistors, that need flexible substrates. Conducting polymers also have the potential for inexpensive fabrication and are highly scalable, which makes them well-suited for large scale manufacturing of wearable devices.

Other important materials are Graphene and Carbon nanotubes (CNTs) because of their extraordinary properties such as high electrical conductivity, excellent mechanical strength and stretchability (Saravanan, Parvathy, Meenakshi, GaneshBabu, and Meenakshi, 2025). Graphene, consisting of a single layer of carbon atoms arranged in a honeycomb structure, has good flexibility and still inherits high conductance, and is thus a promising material for stretchable applications as electrodes and sensors. Other promising carbon-based materials, like carbon nanotubes — rolled-up cylinders of carbon atoms — also boast impressive strength, flexibility and conductivity. These materials have been widely investigated to study its potential applications such as transparent conductor, stretchable circuit, and flexible display (Anjum, 2023).

Stretchable circuits and transparent electrodes are two product applications of Silver Nanowires and Metal Meshes. Nanowires of silver provide high electrical conductivity and good mechanical flexibility, which makes them suitable for stretchable electronics. Metal meshes, especially made out of silver or copper, are employed to design flexible and transparent electrodes, which can be embedded into wearables, such as touchscreens and sensors among other electronic elements (Aslam, Aslam, Wang, Feng, and Zhang, 2025). They are indispensable for the development of stretchable and transparent electronics, which can maintain their electrical performance in the deformed state.

Silicone and Elastomers are used as flexible substrates in wearables devices. These materials stretch easily and are extremely rugged, which is useful for many wearable applications in which the device needs to survive bending (Kaur, Malik, Krishna, Kumar, Aluvala, and Singh, 2025). Silicone materials, such as polydimethylsiloxane (PDMS), are frequently employed in the development of wearable sensor, actuator, and flexible circuitry due to their biocompatibility and easy processability of high resolution.

Organic Electronics are the class of low-cost, stretchable devices that have been developed using semi-conductive organic materials that are based on carbon. Organic semiconducting polymers like poly(3-hexylthiophene) (P3HT) and polyfluorene, that are commonly employed in flexible transistors, organic solar cells, and OLEDs. These materials are attractive for wearables because of their flexibility, low cost, and simple processability by solution processing, such as inkjet printing.

Integration of IoT with Wearable Devices

Wearable devices implement IoT communication technologies to help in bringing about the effortless connectivity and the object-data interaction in the real-time way. Low-power wireless The most common forms of communication used in wearable IoT devices are Near-Field Communication (NFC), Bluetooth Low Energy (BLE), and a wide range of low-power wireless protocols such as Zigbee and LoRaWAN. NFC, providing near-field communication, can also be used for communication between two devices when in close proximity to each other, such as for payment systems or secure identification. On the other hand, BLE has also been preferred to apply to wearables since it possesses low power consumption and effective data transmission which would enable an extended duration of operation with a single battery charging (Jamalpur, Peddinti, Pawar, Zagade, Podduturi, and Thakur, 2025).

The use of energy is the primary consideration in the context of wearables for IoT, because to maintain the long hours of use between charges that wearables require. Energy harvesting approaches (like solar energy, piezoelectric energy, and thermoelectric energy harvesting) are considered as a potential solution for constant power of wearable devices (Cruz, et al., 2022). These techniques harvest ambient energy from body motion, temperature differences or light (for example) to provide energy to recharge batteries or to power sensors, thereby reducing the reliance on traditional sources of battery power.

Challenges and Opportunities

Although there have been big advances in wearable electronics, there are still a few challenges to overcome when scaling down flexible and stretchable IoT gadgets. The durability of the materials under loading is one of the major issues. Wearable devices are under long-term bending, stretching and flexing, which may cause the damage of material and the failure of electronics. To keep flexible materials functioning and intact during many deformations is needed for long-term use.

Interconnecting electronics and the human body is a difficult problem. Wearable sensors should be comfortable to wear and nonintrusive in data acquisition. Creating electronics that can interface non-invasively with the skin is of critical importance. Additionally, it is paid much attention to the biocompatibility and safety, due to the fact that wearable devices are usually in direct contact with the human body for a long time. Materials that are used in the devices should not induce any negative skin reactions (e.g., irritation or allergic reactions) over prolonged skin contact.

Although these are substantial barriers, they are also opportunities for creative approaches. The creation of self-healing materials, better energy efficient electronics and more adaptable wearable designs offers new avenues to develop the wearable internet of things.

FLEXIBLE AND STRETCHABLE MATERIALS FOR IOT WEARABLES

Material Properties for Wearable Electronics

The performance of IoT wearables highly relies on the incorporation of materials with particular properties to provide sustained and consistent performance. Conductivity, stretchability, flexibility, biocompatibility, and durability are the most desired material characteristics to be fulfilled for wearable electronics applications. These properties help to ensure that the devices function in any environment and always feel good to the user.

Conductivity is one of the most important properties for the materials applied in wearable electronics, it guarantees well signal transmission, energy storage and device communication. Conductive materials, conductive polymers, metallic nanowires, carbon-based materials (for example graphene), etc., are used to keep a desirable performance for wearable sensors and circuits even at a stretched status. It is useful for the wearables that undergo such continuous movement depending on their electrical conducting property even under the application of tensile or bending force.

Stretchability is also an important property that can enable wearable devices to conform to human body dynamics. The material also needs to maintain its electronic properties even when stretched, bent or deformed. Such compliance is indispensable for wearables including fitness trackers, medical monitors, and smart textiles, which need to naturally integrate with the body. Those materials, like elastomers, and silicone, and carbon nanotubes, are usually selected for how well they stretch without breaking and while remaining conductive.

Flexibility Follows Stretchability, And Is Just As Important For Wearables. Flexible material allows the device to be used on body contours, increasing comfort and wearability. Bendability makes garments, accessories, and other flexible substrates infinitely more wearable. Flexible materials such as thin films of graphene, metals, and conductive polymers can be used to fabricate flexible circuits and structural components.

Good biocompatibility is required for WEs being worn directly on skin. The materials must be biocompatible over time: meaning that no irritation or nocompatibility reactions may occur. So far, silicone, elastomers and some conductive polymers are the popular functional materials because their good biocompatibility and they are applicable for skin-touch applications. Wearables must also be non-toxic and comfortable to be worn continuously.

Finally, the robustness of these wearable devices when they are exposed to a multitude of physical and environmental factors should be well addressed. Flexible and stretchable materials should be resistant to mechanical deformation, wear, and use without degradation. Long-term reliability also depends on the capacity to preserve performance after multiple cycles of stretching, bending or exposure to environmental conditions like humidity, temperature or sweat.

Figure 1. Integration of Flexible and Stretchable Materials in IoT Wearables

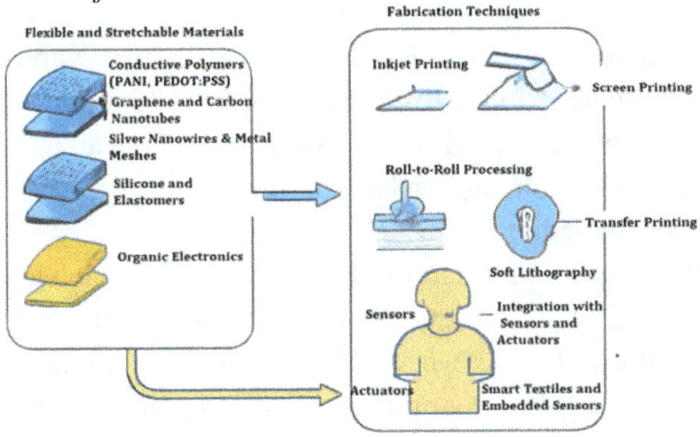

Figure.1. The integration of flexible and stretchable materials into IoT-enabled wearables is shown, breaking down the process into three major areas: (1) flexible and stretchable materials, (2) fabrication strategies, and (3) integration with sensing and actuation applications.

The first part discusses such materials as flexible transparent conductive films, stretchable conductive materials, and ultraflexible organic electronics. These materials\textemdash including conductive polymers (e.g., PANI, PEDOT:PSS), graphene, carbon nanotubes, silver nanowires, metal meshes, silicone, elastomers and others \textemdash are essential for device performance. These components must be selected to preserve electrical conductive behaviour, stretchability and flexibility - but above all skin compatibility and consequent durability if in skin contact over prolonged periods.

The second part of the figure demonstrates the technology description for flexible electronics. Inkjet printing, screen printing, roll-to-roll processing, transfer printing, soft lithography, and other techniques are used for the manufacturing of stretchable circuits, sensors, and actuators. The methods can result in well-defined and wafer-scale processes for the manufacturability of flexible electronics, including the integration of electronic devices on a substrate having bending, stretching, or conforming properties for use in the human body.

The last part of the figure provides an insight in the embedding of sensors and actuators in the wearable devices. Flexible sensors monitoring a variety of physiological signals, including temperature, pulse, pressure and movement, are incorporated into the flexible substrate, allowing real-time acquisition of thermal signals. The devices may include actuators that provide feedback to the user, such as vibration or haptic feedback. Also indicated are smart textiles, with sensors and actuators incorporated throughout fabrics and a constant, non-intrusive health monitoring with comfort and wearability.

Fabrication Techniques for Stretchable Electronics

Making stretchable electronics demands processes that allow materials to be bent into form without breaking or losing their ability to function. Various advanced fabrication technologies are used for such purposes for flexible and stretchable electronic devices, with comparisons at appropriate scales, from small prototype quantities to mass production.

Printing Technologies are often used to fabricate flexible electronics for their advantage in cost and scale. Techniques inkjet printing, screen printing, and roll to roll processing allows for the every say production of soft circuits, sensors, and other electronic apparatus. For example, inkjet printing makes it possible to deposit conductive inks at small scale onto a variety of flexible underlayers, providing a basis for tailored circuits and sensors. Screen printing is a more stable technique and may be used to print thicker conductive layers suitable for industrial applications. Roll-to-roll processing is a high-speed continuous printing process that is well suited for the high-throughput production of high volumes of flexible and stretchable devices, such as wearable sensors, directly on to flexible substrates.

Another advanced method for transfer of thin films of electronic materials onto flexible substrates is called Transfer Printing. This approach is based upon transferring prefabricated films – like semiconductor layers or electrodes – from a donor onto a flexible receiver. Transfer printing offers the advantages of accurate positioning of electronic devices on stretchable surfaces and integrating high-performance materials that are difficult to directly process on soft surfaces. This strategy provides a facile means of fabricating sophisticated electrics devices with high-resolution patterns and structures on flexible substrates, which are highly desire for wearables and smart textiles.

Soft Lithography is a versatile fabrication technique for microstructures and nanoscale patterns on stretchable platforms. It uses a flexible elastomeric stamp to print patterns on a substrate and enables the fabrication of sophisticated microelectronic devices, such as sensors and actuators, on flexible substrates. Soft lithography can be utilized to create wearable devices with high functionality such as flexible microcircuits and stretchable MEMS with high throughput and low cost to fabricate complex stretchable electronics.

Integration with Sensors and Actuators

One of the key role of flexible/stretchable materials in IoT enabled wearables refers to ability to smoothly cooperate with both sensors and actuators. A variety of sensors monitoring a range of physiological signals (i.e., temperature, heartbeats, pressure, motion, electrochemicals) are needed for wearable devices. These sensors need to perform at high sensitivity and be sensitive even during the deformation and movement of the device, while it is worn.

Deformable materials, eg conductive polymers and carbon nanotubes, are attractive for embedding sensors in wearables because of their conductive properties during stretch or bend. Such sensors can sense physical changes such as temperature differences on the skin, movement, pressure, etc., and convert it to an electrical signal that is then processed and transmitted by the wearable. For instance, the temperature sensors incorporated within flexible materials can be employed to sense the body temperature in real time, whereas the pressure sensors can be used to detect blood pressure or sense the physical activity level. Heart Rate There are already some methods for monitoring the heart rate using both optical and electrochemical sensing principle by integrating such sensors into a flexible substrate, therefore continuously monitoring without limiting the user's comfort or accuracy.

Actuators, giving physical feedbacks according to the sensors as input, are also important on wearable devices. Such elements are capable of providing feedback to a user including, for example, vibration notifications, haptic feedback, or stimulation based on changing monitored data. Flexible materials allow actuators, such as micro-vibration motors or electroactive polymers, to be attached to wearable devices such that they are both flexible and responsive. Such actuators can readily be incorporated into the structure of the device because of its small size, suitable for wearable applications.

A relevant advancement of wearable technology are smart textiles that embed sensors and actuators directly into textile structures. Such fabrics can track health data such as heart rate, body temperature and body movements, while also managing the device's operations in response to the user's input or the environmental conditions. For example, smart shirts that contain embedded sensors can assess heart rate and respiration, and smart socks can measure foot pressure and gait. Pliable substances like conductive fibers and stretchable polymers are employed in the fiber design, enabling the implantation of sensors into textiles and ultimately generating clothing that act as wear-and-forget health indicators.

When sensors and actuators are combined with flexible materials, however, the design space for wearables become limitless. Such devices can be worn for long periods of time in a comfortable fashion while still affording continuous, real-time monitoring. In addition, the advancement of stretchable and wearable electronic devices and systems allows us to build unprecedented smart textiles and wearables that can sample vast numbers of physiological and environmental signals, and makes the next generation of IoT-wearables much closer to daily life.

RESULTS AND DISCUSSIONS

Performance of IoT Wearables with Flexible Materials

The use of flexible and stretchable materials in IoT wearables has greatly enhanced their practicability in real world scenarios. Conformable materials enable wearable electronics to remain functional, and comfortable, on and around the human body. Figure 2 represents the response of a stretchable sensor; conductivity variation with strain. It can be seen that, with increasing strain, the conductivity follows a trend of decreasing, which is essential to wearables monitoring health and fitness. Conductive polymers, carbon nanotubes and graphene have better flexibility but do not deteriorate their electrical properties and make the device work stably under mechanical deformation as well.

Conventional rigid electronic devices and their counterparts on flexible/stretchable devices are also compared and wearable devices have distinct advantages. There are known wearable devices, with the Best ease of introduction toppunkt of rigid parts, which may be uncomfortable and limiting when the apparatus is worn, such as during prolonged wearing and/or exercise. Flexible electronics, on the other hand, make it possible for the wearables to bend and fit the body, which means that they're more comfortable to wear for longer stretches of time. Materials with high mechanical flexibility, such as elastomer, silicone, and conductive polymer, allow wearables to maintain functionality when bent, stretched and/or compressed. For example, Figure. 3 presents monitoring of heart rate over time and demonstrates real-time data acquisition by using flexible sensors embedded in the wearables with high accuracy even during the period of motion.

Figure 2. Sensor Response to Stretching (Conductivity vs. Strain)

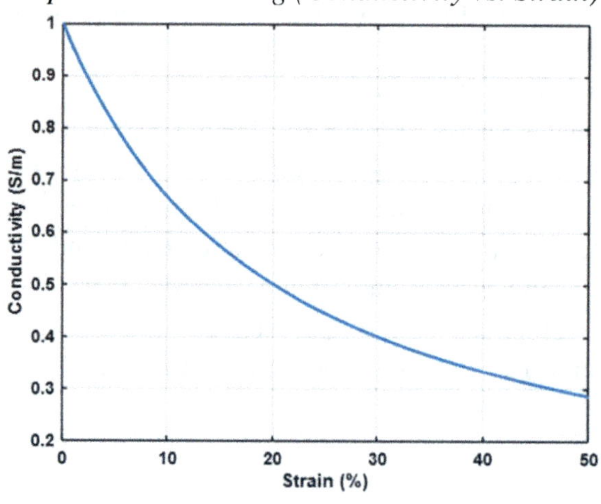

Figure 3. Heart Rate Monitoring (Heart Rate vs. Time)

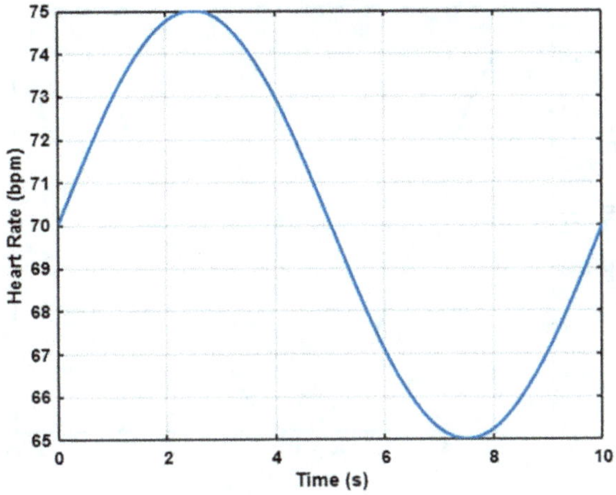

Figure 4. Power Consumption of Wearable Device (Power vs. Time)

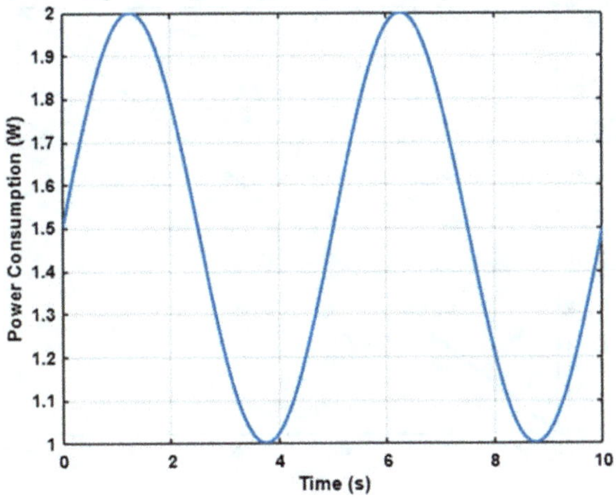

Figure 5. Temperature Change Monitoring (Temperature vs. Time)

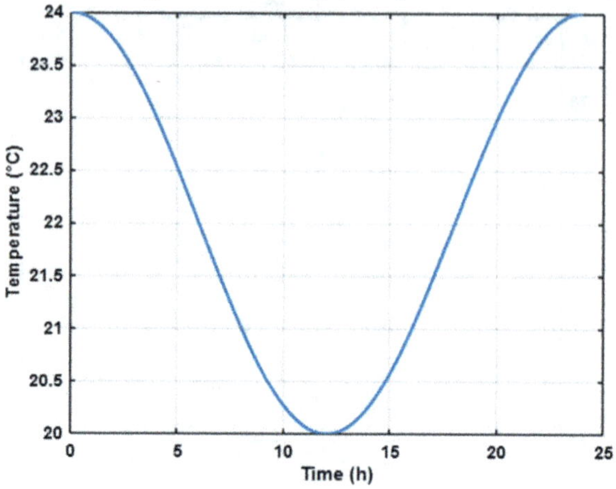

Figure 6. Strain vs. Resistance in Flexible Electronics

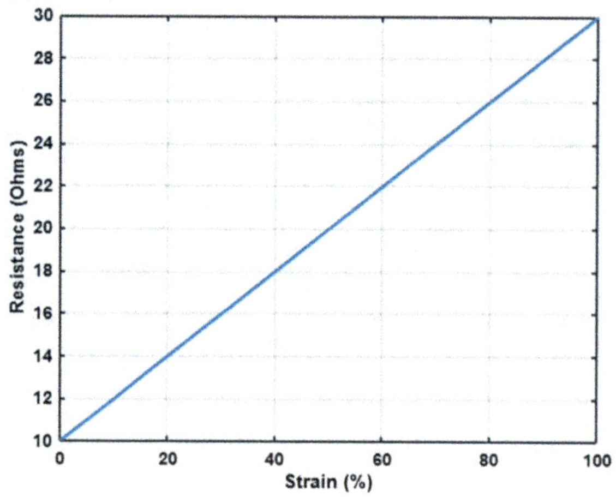

Data Accuracy and Reliability

One of the primary issues for the cases of IoT wearables is the accuracy of sensor with respect to the real-time health monitoring. Wearables use sensors to monitor a range of people's physical information, including heart rate, temperature, pressure, and movement. Precision sensing also is critical to applications such as health monitoring because the data collected can be used to catch fluctuations in health or early signs of illness, and to monitor patients with chronic conditions. For instantaneous measurement, the heart rate sensor can continuously measure over time as depicted in Figure 3, where the fluctuation is low and matches closely (instantaneous measurement) even with movement of the wearer:

But, for these sensors, the extra loads, such as tension and bending, may also affect the sensitivity. The stretching of electroconductive materials as in the case of Figure.2. may lower electric conductivity of such materials, which is not desirable in terms of the accuracy of a sensor. Nevertheless, compliant materials have been engineered to counteract these effects with rugged designs and compositions that maintain their electrical properties under mechanical loading. In addition, state-of-art wearable sensors use signal-processing algorithms that correct for minor deformation and noise errors, resulting in acquisition of reliable data during the dynamic activity. As medical wearables become increasingly more integrated with health care patient monitoring, the ability to deliver accurate and consistent data is still paramount to its success.

Power Management and Energy Harvesting

Power consumption is a primary concern for wearable IoT devices. These devices are expected to work for long periods without repeated recharging and thus should have efficient power management. Wearable devices have dynamic power consumption for activity, usage of sensors, and communication protocols (as shown in Figure 4). Fuel-efficiency is essential in preserving the stamina of the battery and at the same time continue to perform at a good standard.

To overcome power constraints, power harvesting systems, including solar, thermoelectric, and piezoelectric systems, are being incorporated in the wearables. Solar energy scavenging It is possible to prolong the battery's life by scavenging the ambient light source for solar power, while the thermoelectric and piezoelectric materials convert the body heat and mechanical movement into electricity. These approaches are particularly interesting for wearable systems in the exercise and healthcare fields where ongoing physical activity can produce sufficient energy to operate sensors and communication devices. For example, in Figure.4., a drop in the power is observed, that might be overcome to a certain extent employing energy harvesting technologies leading to a sustainable and robust energy supply for wearables. The development of flexible energy harvesting materials will increasingly lead the way and cut traditional battery consumption in the future.

Durability and Longevity of Flexible Wearable Devices

The durability and lifetime of flexible WEA is the key for its sustainable and long-lasting day-to-day use. Such devices are subjected to different environmental influences such as dampness, heating, and mechanical vulnerability, which may influence their performance during usage. Temperature variations are monitored as shown in Figure 5 for wearable sensors to withstand an exposure under varying environmental conditions. It is crucial to maintain a material's environment in WTDs for the sustained and accurate operation of sensors in fast changing conditions.

Furthermore, mechanical fatigue of soft materials needs to be taken into account, particularly in wearables that experience frequent bending, stretching, or compressing. This repeated mechanical deformation will result in the device used enough and some wear and tear, so the deterioration of its performance. It has recently been demonstrated that the flexible polymers such as silicone and elastomers are able to survive such long-term mechanical stress and hence, excellent material choices for wearable devices. Figure.6. indicates that the enhancement of resistance by means of strain is essential for the flexible electronics, thus highlighting the necessity of the material durability for the sensor long-last lifetime under physical deformation. In addition, it should be water (sweat and dirt) proof and washable to protect the wearables from environmental hazards if used repeatedly.

Biocompatibility and Safety

Since wearable electronics are in close contact with the human body in application, their biocompatibility and safety are also very important. The materials for wearable devices need to be bio-compatible, that is, not be harmful when coming in contact with the skin for extended periods, so that there is no itching or allergic reaction. Silicone and elastomer are popular materials used in wearables as they are biocompatible; unlike metals or rigid plastics, the likelihood of their skin irritative or allergic reaction is low. Also, wearables that include electronic devices need to comply with safety standards to prevent hazards such as electrical shock, electrical fire, or overheating where the wearables are close to a human body.

There are also potential issues, over the long term, with wearables impact on skin health and the body's natural functions. Based on current literature, most wearable materials are considered to be safe for extended wear; however, continued research should be conducted to investigate the long-term impact of incorporating electronic devices in close proximity to the body. Wearables need to be built for safety including top priorities such as electrical safety and avoiding that materials creep out harmful substances as time goes by. And finally, wearables that can be thrown in the wash without ruining the embedded electronics or sensors are necessary to keep everyone clean and the wearable alive.

In summary, the adoption of flexible and stretchable materials in IoT wearables has dramatically enhanced the performance, wearability, and capability of these devices. In the case of sensor performance under stress, power consumption and material longevity, much work has been done to improve wearability, efficiency and reliability. With the advances in materials, energy harvesting and sensors, the future of IoT wearables is bright and fitness, health applications are high yielding fields where real-time, continuous medical monitoring is of importance for customized health care.

FUTURE TRENDS AND EMERGING TECHNOLOGIES

Advanced Materials

Advanced materials science will be a game-changer in the next wave of wearable IoT devices. Materials such as graphene, MXenes, and other 2D materials provide advantages in electrical and mechanical performance of devices and functionality of the same. Graphene, a single layer of carbon atom in a hexagonal lattice, becomes so well-known for its high electrical conductivity, strong mechanical property and large mechanical flexibility. It renders the wearable device to be extremely thin, lightweight, stretchable and with superior performance. All of these properties render graphene a candidate material for stretchable electronics and soft displays, while also in transparent electrode for wearable sensors.

MXenes, a class of 2D materials, also have great prospects in the field of wearable electronics. MXenes are highly conductive strong and flexible materials, which makes them applicable for a wide range of technologies from energy storage and conversion devices to sensors. There materials possess superior thermoelectric and electrical conductivity, which are highly desirable for the power-driven wearable devices that involve energy harvesting and real-time data computation. In addition, 2D materials such as transition metal dichalcogenides (TMDs) provide flexibility and scalability which can be used to construct flexible devices/circuits/sensors/transistors etc., and can be integrated into wearable devices without losing function and comfort. These advanced materials, once integrated into a wearable system will significantly increase the performance (eg., function, robustness, reliability) of wearable products to enable more efficient continuous using in a dynamic world.

Artificial Intelligence and Machine Learning Integration

Wearable IoT devices are increasingly taking advantage of the capabilities and performance improvements afforded by Artificial Intelligence (AI) and Machine Learning (ML). With the help of AI and ML algorithms, wearables can learn from enormous amounts of real-time data captured by sensors, to deliver smarter and more personalized user insights. These are technologies that allow wearables to decide on their own, make inferences, and predict things based on data they collect. In health-monitoring scenarios, AI can be used to identify early symptoms of diseases, such as arrhythmia or diabetes, by analyzing the continuous data from several wearable sensors.

AI-driven algorithms can, for instance, leverage heart rate variability, body temperature, and other biometric data to predict the risk of health issue, thus, facilitating timely actions. Machine learning could also allow the device to improve its behavior over time, learning from its users routines and adjusting its own behavior, including operations like adjusting activity recommendations, or being able to detect abnormal behavior. AI and ML are moving forward, as are wearable devices, wearable devices will become smarter as well, resulting in on-the-fly health diagnostics and predictive analytics evolving healthcare and wellness sector.

Next-Generation Energy Solutions

Effective energy utilization continues to be a vital challenge for the wearable IoT. As wearables are increasingly being adopted in everyday life, there will be an increasing need for low-power, wearable technology. One of the strategies to address this issue is fabrication of wearable energy storage devices, namely, flexible batteries and supercapacitors, whose physical architecture could be directly incorporated into the device. Flexible batteries (e.g., lithium-ion batteries or solid-state batteries) also manifest lightweight and high-power properties, and may be disposed in the wearable device itself to provide prolonged operation life without adding any discomfort and load.

Another viable alternative are supercapacitors, which store energy electrostatically and have the capability to charge and discharge relatively fast, which makes them also a greener power source. By incorporating these energy solutions in wearables, they remain charged often enabling longer operational times and thus may become more usable for continuous monitoring. In addition, energy harvesting such as solar, piezoelectric, and thermoelectric systems will continue to develop to allow wearables to capture ambient environmental energy, like body heat, motion etcetera, to prolong battery life even further.

Self-Healing Materials

Self-healing materials are shaping up to be one of the more exciting applications of wearables. These materials can heal themselves when torn and thus vastly extend the longevity of wearables. This property makes it possible for self-healing polymers and composites to repair themselves when and if they are scratched, punctured, or stretched, allowing the wearable to remain in ideal working condition for its lifespan, without having to go for repairs or be replaced.

Self-healing materials are vital for wearables that are subject to harsh conditions including sweat, physical activity and environmental stress. They also minimize the effects of mechanical fatigue, which in turn extends the life and the aesthetic appeal of wearables. This advancement will make wearable devices more sustainable and lower the requirement for replacements.

Wireless Communication

Progress of Wireless Technologies also Need to Support IoT Wearables Improvements of wireless technologies are indispensable to make IoT wearables connected more closely and work more effectively. (Related: With 5G, Real-Time Data Could Change the Game for Wearables) 5G networks, when implemented, will provide ultra-low latency and high rate data transfer, which will help aid wearables in transmitting instantaneously and seamlessly. This is critical in application domains such as remote health monitoring where sensor data needs to be transmitted in a timely and reliable manner for analysis. When the 5G kicks in, wearables can support the flow of significantly more data, which will allow more sophisticated sensors to be included, and the user interaction with the device will be more seamless.

Moreover, low power wide-area networks (LPWANs), such as LoRaWAN and Narrow Band Internet of Things (NB-IoT), provide power-efficient communication mechanisms for wearables with long distance connectivity with low power usage. When it comes to wearables that have to be used for long periods without constant recharging, these technologies are especially valuable. "With the evolution of wireless technology, wearables will become able to connect to larger IoT environments, to form part of more complete and smart devices that can communicate with a range of smart systems, from healthcare networks to smart homes and cities.

CONCLUSION

Flexible materials have played a critical role in revolutionizing the wearable IoT scene. These materials, such as conductive polymers, graphene, and MXenes, improve the electrical response, flexibility, and mechanical strength of wearables so that they can better conform to the human body. With the breakthroughs in the material, the sensor integration, and the communication IoT technology, wearables now play a central role in providing real-time health status monitoring, physical fitness input, and instant data transfer. Today's performance of the wearable IoT devices gains better comfort, reliability, and accuracy such that there can appear lots of advanced applications in different fields. However, despite the progress in development, current limitations persist in terms of durability, power consumption, and user-friendly characteristics. The challenge remains to safeguard long term use of flexible materials, especially under mechanical strain, and minimize power consumption for the operation of the device over long periods. In the future, more efforts will be dedicated to increasing the flexibility, sustainability, multilinearity of wearables, especially by decreasing the reliance on traditional energy sources and the lifetime of the devices. Wearable IoT for personal health monitoring and preventive medicine. By monitoring useful biometrics it is possible to use the information these devices provide to assist users in becoming aware of long term concerns and focus on the areas where they can track progress.

REFERENCES

Anjum, M. N. (2023). eNext: An IoT and AI driven solution to the plugged-ear pandemic. *IEEE Internet of Things Journal, 10*(13), 11940–11941. DOI: 10.1109/JIOT.2023.3244800

Aslam, S., Aslam, S., Wang, T., Feng, D., & Zhang, S. (2025). A novel and secure machine learning-based Hyperledger blockchain for IoT healthcare. *IEEE Internet of Things Journal, 12*(15), 31126–31140. Advance online publication. DOI: 10.1109/JIOT.2025.3574471

Chophla, E., Sharda, S., Madan, S., & Chopra, A. (2024). IoT-driven healthcare: A new era of connected care. In *2024 2nd International Conference on Advances in Computation, Communication and Information Technology (ICAICCIT)* (pp. 585–592). DOI: 10.1109/ICAICCIT64383.2024.10912296

Cruz, A., Villena, M. L., Castro, I. M., Cervantes, J. P., Ondra, S. A., Quijano, R., & Alipio, M. (2022). Vitaband: IoT-driven health monitoring system for home quarantine COVID-19 patient using LoRaWAN. In *2022 IEEE International Conference on Internet of Things and Intelligence Systems (IoTaIS)* (pp. 351–357). IEEE. DOI: 10.1109/IoTaIS56727.2022.9975992

. J., R., & D., N. (2025). MediBot: IoT-driven healthcare for remote communities. In *2025 3rd International Conference on Advancements in Electrical, Electronics, Communication, Computing and Automation (ICAECA)* (pp. 1–7).

Jamalpur, B., Peddinti, D. R., Pawar, S. A., Zagade, A. D., Podduturi, M. R., & Thakur, J. (2025). IoT-driven data analytics: Transforming digital marketing campaigns with real-time consumer insights. In *2025 International Conference on Pervasive Computational Technologies (ICPCT)* (pp. 875–879). DOI: 10.1109/ICPCT64145.2025.10940271

Jing, J., Lai, H., & Sun, L. (2024). Research on AI and IoT-driven monitoring in military physical fitness training. In *2024 6th International Conference on Frontier Technologies of Information and Computer (ICFTIC)* (pp. 1668–1672). https://doi.org/DOI: 10.1109/ICFTIC64248.2024.10912941

Kaur, N., Malik, P. K., Krishna, G., Kumar, M., Aluvala, S., & Singh, P. (2025). An IoT-driven workers health monitoring system utilizing LoRa technology for real-time vital signs collection, tracking and cloud-based data logging at industrial entry points. In *2025 International Conference on Intelligent Control, Computing and Communications (IC3)* (pp. 423–429). DOI: 10.1109/IC363308.2025.10956658

Khadse, C., & Hajare, S. (2025). Sensor based smart wearable devices for women: A review. In *2025 International Conference on Machine Learning and Autonomous Systems (ICMLAS)* (pp. 1933–1939). DOI: 10.1109/ICMLAS64557.2025.10968950

Liu, C., Zhang, X., Zhao, L., Liu, F., Chen, X., Yao, Y., & Li, J. (2019). Signal quality assessment and lightweight QRS detection for wearable ECG SmartVest system. *IEEE Internet of Things Journal*, *6*(2), 1363–1374. DOI: 10.1109/JIOT.2018.2844090

Madunuri, R., Chitta, S., Bonam, V. S. M., Vangoor, V. K. R., Yellepeddi, S. M., & Ravi, C. S. (2024). IoT-driven smart healthcare systems for remote patient monitoring and management. In *2024 Asian Conference on Intelligent Technologies (ACOIT)* (pp. 1–7). DOI: 10.1109/ACOIT62457.2024.10939717

Mani, K., Singh, K. K., & Litoriya, R. (2024). IoT-driven revolution: Transforming elderly healthcare through dynamic heart health monitoring and analysis. In *2024 International Conference on Advances in Computing Research on Science Engineering and Technology (ACROSET)* (pp. 1–8). DOI: 10.1109/ACROSET62108.2024.10743704

Rajaa, J. M., Manee, M. J., Mohammed, M. A., Mohammed, M. A., & Sharif, S. F. A. (2024). IoT-driven physical training and fitness management in colleges. In *2024 International Conference on Emerging Research in Computational Science (ICERCS)* (pp. 1–6). DOI: 10.1109/ICERCS63125.2024.10895309

Redwan, K., Ahmed, M., Turjo, K. I. H., Munna, M. M., Hossain, S. S. A., & Al Sohan, M. F. A. (2024). Smart IoT-driven shopping assistance system for cognitive impairment patients: Ensuring safe and independent shopping experiences. In *2024 10th International Conference on Computing, Engineering and Design (ICCED)* (pp. 1–6).

Ronald, B. S., Raman, R., Vekariya, D., Baskar, V. V., Rahul, J., & Srinivasan, C. (2024). SVM-enhanced dynamic audiology screening with IoT-driven monitoring for personalized hearing health management. In *2024 2nd International Conference on Self Sustainable Artificial Intelligence Systems (ICSSAS)* (pp. 1028–1033). DOI: 10.1109/ICSSAS64001.2024.10760912

Saravanan, S. K., Parvathy, P. R., & Meenakshi, R., K, P., GaneshBabu, T. R., & Meenakshi, B. (2025). Predictive analytics for ear health management using IoT and ensemble learning. In *2025 International Conference on Intelligent Control, Computing and Communications (IC3)* (pp. 319–324). DOI: 10.1109/IC363308.2025.10957573

Thangam, S., Joshith, G., Santhosh, S., Harish, V., & Kumari, J. J. J. (2024). Smart steps to sporting success: IoT-driven footwear innovations. In *2024 15th International Conference on Computing Communication and Networking Technologies (ICCCNT)* (pp. 1–6).

Tôn, L.-P., Lê, L.-S., & Nguyen, M.-S. (2021). Micraspis: A computer-aided proposal toward programming and architecting smart IoT wearables. *IEEE Access : Practical Innovations, Open Solutions*, *9*, 105393–105408. DOI: 10.1109/ACCESS.2021.3096749

Chapter 11
Smart Materials for IoT-Enabled Structural Health Monitoring Using Image Processing

Mandeep Kaur
https://orcid.org/0009-0001-6997-9531
Department of Electronics and Communication Engineering, Punjabi University, Patiala, India

ABSTRACT

Safety and service life of infrastructures including bridges, buildings, and industrial plants can be validated by the structural health monitoring (SHM) techniques. Combining Internet of Things (IoT) technology with advanced image processing can present a solution for real-time, non-destructive, automated structure health monitor. In this chapter, we review smart solutions that integrate IoT-enabled sensor networks with image-based data acquisition for structural defect detection, classification, and analysis. A recent overview for the developments in IoT and image processing for SHM is summarized, such as machine learning assisted crack detection, and cloud-edge based computation model. Case studies show the functionality of these integrated systems allows for improved accuracy of monitoring, maintenance expenditures, and the use of prognostic maintenance tactics. Challenges such as environmental variability, data handling, and computation limitations are addressed, and the future direction of more reliable and scalable SHM is presented.

DOI: 10.4018/979-8-3373-9454-1.ch011

INTRODUCTION

Structural health monitoring (SHM) is crucial to the safety, reliability, and durability of critical infrastructures like bridges, buildings, tunnels, dams, and industrial facilities. Continuous monitoring of structural health can enable early detection of damage, deterioration or failure, which can avert catastrophic events, lower maintenance expenses, and prolong its service life (Shereen, Shahid, and Tanveer 2025). Conventional SHM strategies are typically based on regular manual monitoring using the hard-mounted sensors, which are difficult and time-consuming, costly and not capable of covering all parts of the infrastructure. In addition, human vision is subjective and sensitive to human factors, and conventional detection technologies cannot generate complete surface details like crack propagation and corrosion.

The advent of the Internet of Things (IoT) has revolutionized monitoring due to the possibility of implementing networks of sensors and devices that gather and remotely transmit real-time data constantly. The SHM systems based on IoT enable automated monitoring and data collection that improves the status of tracking during the different environmental and loading conditions (Rumi, Pavel, Islam, Shakir, and Hossain 2019). These systems offer better scalability, that is, the possibility for large-scale deployments on existing infrastructures (Sharma and Pathak 2025), and support for anywhere, anytime data access. The fusion between IoT, cloud computing and edge computing architectures improves even more the processing, storage and analysis of large amounts of sensor data, to the benefit of an increasingly smarter decision making and predictive maintenance (P. S., S. K. T., S. P. M., D. S. V., S. V., and R. A. V. K. 2024).

Image processing has been used widely in1,7,9,24–26 structural health monitoring (SHM) applications. Constant 3D coverage can be achieved in the case without contact by measuring the defectslie in the main focus while the image processing isperformed in the set of conjugate planes. In contrast with the sensors that measure indirect parameters, image processing methods analyze the visual data recorded by cameras, drones or other visualization tools for the detection of surface anomalies such as cracks, spalling, corrosion, and deformation in a large spatial detail (Balaji, Kalvinathan, Dheepanchakkravarthy, and Muthuvel 2021). Advanced image analysis techniques make it possible to extract features, classify defects, and provide quantitative estimations about the degree of damage. Image processing, combined with artificial intelligence and machine learning algorithms, allows for the accurate defect recognition and real-time anomaly detection that drives away reliance on manual interpretations for reliable and repeatable inspection results (Lopes, Pinho, Marques, Abreu, Carvalho, and Ferreira 2021).

This chapter presents intelligent methods to harmonize IoT systems and image processing solutions to improve the efficiency of structural health monitoring applications. Combining these two domains together you get the best of both worlds: IoT's real-time, scalable data collection and the specific, detailed learnings from image based analysis (Choudhary, Jaiswal, and Das 2021). These sensing systems allow monitoring of structural health in real-time as well as automated defect detection and classification, which are beneficial for proactive maintenance and timely intervention. The content of this chapter is limited to a complete overview of the latest developments in IoT-based image processing applications for SHM that include sensor network (Thangaraj, Ponmalar, and Anuradha 2015), image acquisition technique, processing algorithm, and data routing protocol. Moreover, some issues, including environmental fluctuation, data handling, computational burden, and implementation of such method in practice, are discussed (Priya, Al-Fatlawy, Khare, Mahalakshmi, and Ganesh 2023).

Objectives of the Smart SHM Solutions Incorporating IoT and Image Processing-This Chapter The purposes of this chapter are to browse the recent popular state-of-the-art related with smart SHM solutions, which incorporate both IoT and image processing, and also to point the significant methodologies and technologies to push such trends, to introduce the real-world cases to see how better these novel solutions are used for real applications. In addition, the chapter intends to identify the gap that currently exists and to suggest further research and development in this multidisciplinary area. In providing a comprehensive perspective on IoT-enabled image processing for SHM, this work helps to enhance the knowledge base and facilitate the acceptance of smart monitoring systems which enhance infrastructure resilience and security.

LITERATURE SURVEY

For decades, structural health monitoring (SHM) of infrastructure has mainly considered the combination of visual inspections with sensor based techniques to assess infrastructure condition. The traditional SHM methods are manual visual inspections carried out by trained professionals that rely on the identification of surface cracks, corrosion, deformation, or any other observable damage (Tyagi, Gupta, and Sisodia 2025). Visual inspections offer in situ assessment, although being subjective, both physically demanding, and restricted to the access and environmental conditions. In parallel, sensor-based SHM approaches have incorporated different sensors types, such as strain gauges, accelerometers, acoustic emission sensors (Suthar, Chithra, Abhinaya, and Harikumar 2023), and temperature probes, in order to record quantitative data regarding the response of the structural members to different loads and conditions. Such sensing systems allow for continual monitoring, however are often concerned with localised parameter monitor- ing and may miss surface- level defects, or may be spatially distributing damage. Although traditional SHM methods are popular, scalability is a bottleneck in practical applications, as many sensors are required for a large infrastructure and much human labor is needed. Furthermore, absence of the real-time processing and well draw spatial can contribute to delaying the detection of the crucial problems which are often critical to the prompt maintenance (Goyal, Yadav, and Mukherjee 2021).

Recent developments in the Internet of Things (IoT) have greatly influenced SHM by offering the possibility of the interconnected smart sensor networks and real-time data acquisition and monitoring at a remotely-located computer system. IoT-based SHM systems enable a real-time structural evaluation through the aggregating sensors measuring strain, vibration, acoustic emission, thermal shrinking, and environment etc. Data transmission from distributed sensor nodes to central processing units or cloud platforms can be realized by employing wireless communication technologies including Wi-Fi, ZigBee, and cellular networks (Tej, Potharaju, Jakkidi, Sharma, and Reddy 2024). Such a connection provides more scalability as it allows to scale all over a complex infra structure. Additionally, IoT platforms help integrating different data source types, which allow monitoring structural health on a world view (Khan, 2020). This data can be processed locally using edge computing or transmitted to the cloud for further analysis, promoting predictive maintenance and preventing system downtime." These developments provide markedly enhanced safety and value for infrastructure operations (Prathyusha, Krishnan, and Deepak 2024).

The development of image processing techniques is given great attention as a complementary method to sensor-based SHM, since it can provide detailed visual information of structural surfaces. Acquisitions can be carried out with different acquisition devices, from static cameras installed over structures, to low cost drones unmanned aerial vehicles (UAVs) with high resolution cameras up to handheld wearable acquisition devices as smartphones. In particular, drones have extended the capability of image-based inspections, having access to hard-to-reach or dangerous places, and forming complete visual records (R., Prasad, and S. K. C. 2024). After images are obtained, the images are analyzed using a variety of processing technies in order to determine the types and amounts of structural defects in such asphalt samples. Edge detection techniques are used for crack boundary enhancement and pattern recognition emerges in order to identify crack defects in term of their shape and texture. Sophisticated image segmentation procedures can separate the damaged parts from the background and allow an accurate analysis of crack widths, lengths, and the surface deformation. Image-based SHM techniques have been grafted with ML and DL algorithms and this has improved the merits of image-based SHM by automating the defect detection and classification procedures. Such as convolutional neural networks (CNNs), CNNs have been applied successfully for crack inspection and surface defect recognition, and they have also demonstrated a competitive accuracy and robustness even under various lights and environmental conditions (Ilangaweera, Naramada, Sajith, Aashiq, and Hinas 2024).

With the incorporation of IoT and image processing in SHM systems, a real-time and intelligent infrastructure monitoring has been achieved, which is a potential breakpoint in SHM systems. Image data can be transferred into IoT networks for real-time imaging data to achieve uninterrupted monitoring and instantaneous structural anomaly discovery (Veerabomma, S. K., Rajeswari, Murthy, and Selvan 2024). Cloud computing offers scalable resources for storing and analyzing massive quantities of image and sensor data whereas edge computing devices facilitate real-time, on-premise analysis, diminishing reliance on high-bandwidth networks. Some recent research works have shown the successful applications of IoT based image processing systems for SHM. For instance, drones and wireless IoT sensor networks have been used for bridge inspection, where the drone can take high-resolution images and the sensors could also get data simultaneously. In the cloud, we use machine learning models to process this data to predict early stage damage, and send alerts to the maintenance team (Liu, Briones, Zhou, and Magotra 2017). Unrivalled traditional glass fracture data and compatibility with graphic software can seamlessly analyse breaking via cameras various stationary graphic networks, such as vibration and acoustic sensors, and multi modal signal fusion, as well as comprehensive building inspection and monitoring.

In spite of these promising progressions, several limitations and challenges still persist for the acceptance of IoT and image processing in SHM. Data privacy and security are major concerns when developing sensitive infrastructure data transferred and archived in the cloud. Secure communication networks and building a defence against cyber-attacks are some of the critical issues in which the encryption and the authentication protocols have significant roles. Another hurdle is computation, as high resolution image processing and machine learning inference require fast processing but is not always possible on resource-constrained, low power consumption, lightweight devices located at the edge (Awais, Khan, Akram, and Nam 2025). Environmental effects like change in light conditions, levels of occlusions, weather conditions can result in poor image quality and hence on the defect detection which is a function of the input image. Integrating multi-source data from diverse sensors and image streams, also brings challenges for the synchronization, calibration and fusion of data for the generation of robust health indicators. Furthermore, the definition of standards and benchmarks in image-based SHM is still a challenge to be addressed, which is a barrier for comparison and evaluation among various algorithms.

In brief, the literature shows a progression of SHM strategies starting from manual inspection and discrete sensors up to IoT-based, image-centric systems. The combination of real-time IoT data acquisition and sophisticated image processing with the help of AI paves the way for accurate, scalable, and automated structural health monitoring. To the broader community, addressing the open issues such as data privacy, computational efficiency, environmental robustness, and multi-modal data fusion is imperative to further promote these smart SHM solutions towards practical deployment.

SMART SOLUTIONS FRAMEWORK

In this respect, the smart solutions of SHM for IoT-based can be achievable through efficient integration of a system architecture with all the indispensable components to make the data acquisition, transmission, analysis and decision more efficient using image for processing. Key components of this structure are IoT objects, ranging from high-resolution cameras to a variety of sensors such as strain gauges, accelerometers, temperature sensors, and acoustic sensors. They sense visual and tactile properties of a structure that is in view. Data transmission infrastructures, columnated around wireless standards (e.g., Wi-Fi, ZigBee and LoRaWAN) or installed as communication compatible with the cellular networks support the transmission of raw data gathered with the help of distributed devices to processing units (centralized or decentralized) without any hiccups. Image processing units can reside either in edge devices (next to the data acquisition source) or in the cloud (according to the desired latency and computation capabilities). While cloud platforms provide an elastic capacity to store large amount of data and execute complex analytics, edge computing provides such agility for fast processing and fast response by analyzing data close to the source, and thus reducing the need for continuous connection to the network. The complete process first involves taking pictures or video of a scene using fixed or flying cameras and thereafter transmitting the picture and sensor data. Image processing algorithms then process the visual information to identify structural issues while integrating with sensor information to provide full assessments for maintenance decisions.

Figure 1. Smart Solutions Framework for IoT-Enabled Structural Health Monitoring Using Image Processing

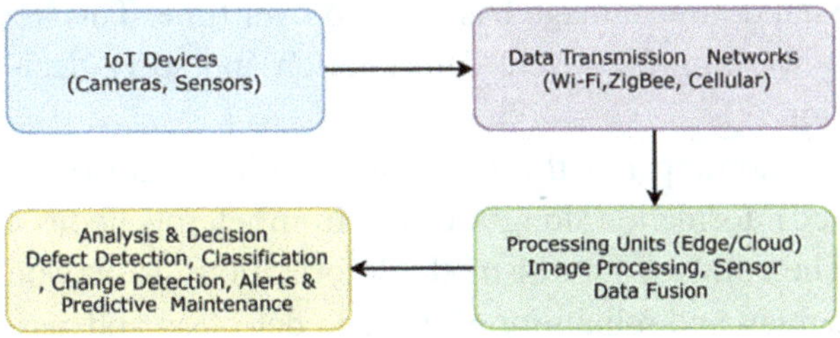

The Figure. 1. depicts the overall architecture of the intelligent SHM system by IoT devices and photograph technology. The pipeline starts from IoT devices, for instance, cameras and other sensors, installed on the structure to collect visual and physical data. Such data is sent over wireless networks, including Wi-Fi, ZigBee, and cellular, and is received by processing units at the edge or in the cloud, near the data source. The processing Π units carry out image pre-processing, feature extraction, defect classification with machine-learning based models and sensor data fusion to improve the damage detection sensitivity. The system allows real-time monitoring, change over time detection, alert generation of anomalies and predictive maintenance planning. Such integration simplifies the monitoring of health structure in automatic, continuous and large scale to timely maintain and enhance the safety of infrastructure.

Image processing pipeline is the infrastructure in the smart SHM framework which includes multiple sequential stages. First, pre-processing methods are used to enhance the quality of the captured images. These range from noise reduction algorithms to eliminate artifacts that are due to the imperfections in sensors or environmental interference, to image enhancement techniques such as intensity stretching or filtering to emphasize critical features. After the preprocessing, feature extraction methods can be used to find regions of interest such as cracks, corrosion spots or surface distortions due to structural damage. Edge detectors aid in defining the limits of the cracks, texture analysis and morphologic operations guide identification of corrosion patterns and material degradation. After feature extraction, defect classification is implemented through machine learning models. Both supervised and unsupervised learning methods are applied based on labeled data being available. The ML models, such support vector machines (SVM), random forests and convolutional neural networks (CNN) are trained with annotated data and can successfully classify various defect types. Unsupervised approaches, e.g., clustering techniques, do not rely on pre-existing damage conditions and allow for the identification of new or unknown damage modes. One of the core challenges in the pipeline is image time series analysis and change detection, where the user needs to compare the current images and sensor information with historical information to understand how damage has evolved over time. This temporal analysis can enable early warning for severe deterioration and serve decision making for timely intervention.

The data integration as part of the IoT is a key enabler of holistic approach to SHM, where image based information along with data from heterogeneous sensor networks get combined. The combined multi-modal information fusion would significantly improve the accuracy and reliability of damage detection and assessment. As an example, a crack identified in image processing can be confirmed visually, based on structural stress or vibration data present in the same area (and vice versa), giving a complete knowledge of the structural health. The combined solution offers immediate alerts based on pre-set limits as well as an anomaly-detection engine, which means workers receive alerts when maintenance action is required. Furthermore, predictive maintenance techniques use historical information and machine learning techniques to predict the future state of a structure, providing advanced warning of when repairs may be required before failures actually do occur. Such a method minimizes downtime and maintenance expenditures, enhances safety, and increases lifespan of structures. Thereby, the smart solutions framework is a combination of IoT connectivity, sophisticated image processing and intelligent analytics, which provides an efficient, competitive and scalable solution for contemporary SHM.

RESULTS AND DISCUSSION

This section presents the simulation results and pilot testing of image processing based Internet of Things (IoT) based Structural Health Monitoring (SHM) systems. In the experiments, we use simulated data sets of structural surfaces with artificial features, including those created by cracks, corrosion or deformations. These synthetic datasets are meant to simulate common real-world scenes such as bridges, buildings, and industrial infrastructure as would be captured using imaging devices including drones, stationary cameras, and mobile phones. The simulations integrate visual data and sensor information to maintain a realistic SHM surrounding. Figure 2 illustrates the utility of edge detection, especially the Canny algorithm, for crack pattern recognition in images of structures. This picture demonstrates how the edge-detection approach senses important information from noisy backgrounds and gives a clear demarcation of crack boundaries crucial for early stage" damage detection. This method serves as the basis of automated defect recognition.

Crack width measurements along the length of a simulated structural specimen are illustrated in Figure 3. This graph demonstrates that crack width is likely to increase gradually over distance with small wiggles representing real-life stochasticity, and only measurement error. The ability to identify the defect severity is important in prioritizing necessary maintenance. Comparison of the classification accuracy of different machine learning models, including SVM, random forest (RF), CNN, KNN, and Naive Bayes, is shown in Figure. 4. It can be seen that the recently, deep learning results, especially CNN, achieve the best classification accuracy on defect detection, above 95%, suggesting that the modern AI technologies have a good potential to achieve automation and enhancement of SHM trustworthiness. "Machine learning models help to decrease errors from human analysis, in addition to that they save time compared to the manual evaluation.

The temporal component of SHM is represented in Figure 5, where the evolution of damage is estimated over time based on a damage index measure. The exponential trend of growth, which we followed in the course of the development demonstrates the significance of ongoing monitoring to identify of an accelerated phase of the ischemia. This temporal analysis paves the way for predicting structural failures and planning for promptly repairing the structures, this reveals predictive maintenance due to integration of IoT and image processing systems. Real-time image processing through thresholding and defect highlighting on a structural surface is demonstrated in Figure.6. The red overlays show the regions as potential defects, thus supporting the applicability of visual segmentation methods for real-time on-site inspections.

Experimental results indicate that the image processing scheme with IoT has high accuracy in the defective detection and classification, based on the quantitative results in Figures 3 and 4. The inclusion of edge detection and machine learning improves the ability of system to identify minor and complicated defects, failures which even normal sensor data alone could overlook; data transmission latency and reliability of IoT connectivity is also crucial for real-time monitoring in the context of getting device data stored in cloud, simulated wireless communications proves the feasibility of transmitting high res- olution of images and sensor data with negligible delay particularly when pre-processing data before uploading into the cloud using edge computing. This strategy minimizes the usage of bandwidth and the round trip times, important for time critical infrastructure safety applications. IoT-based image processing methods scale better than the conventional manual inspection because they can be performed on a large scale for an extended period of time - independently of the available workforce. Automated measurements also enforce uniform criteria between the evaluations, thus minimizing the subjective factor in damage assessment.

Figure 2. Crack Detection using Edge Detection (Canny)

Figure 3. Feature Extraction - Crack Width Estimation (Simulated data)

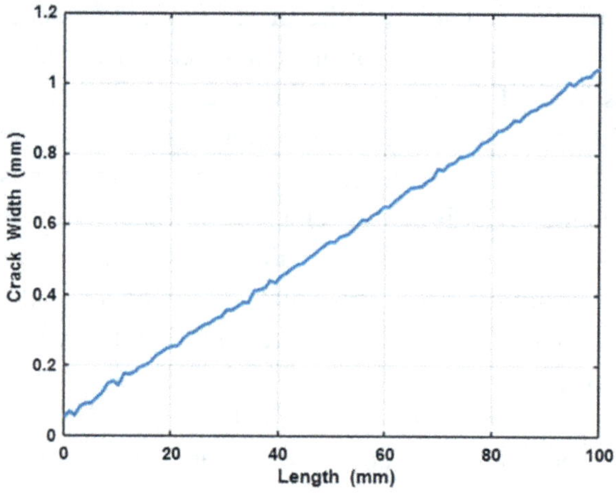

Figure 4. Classification Accuracy of Defect Detection Models

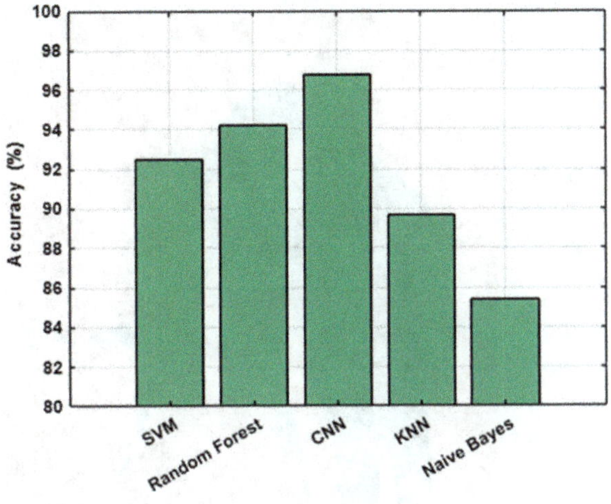

Figure 5. Change Detection Over Time (Simulated Damage Growth)

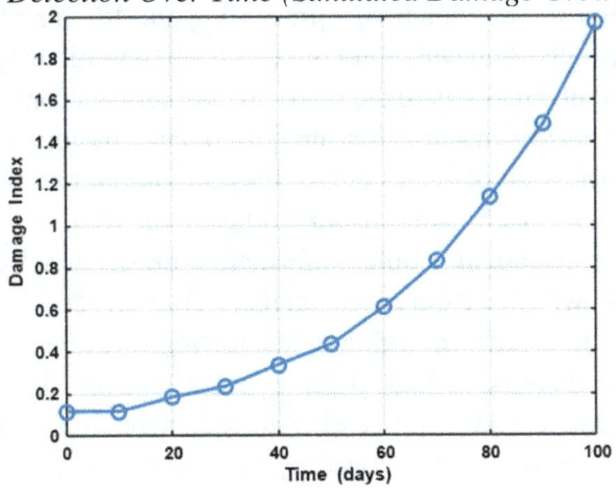

Figure 6. Real-time Image Processing - Thresholding and Defect Highlighting

A few strengths to combine image processing and IoT for SHM are discussed according to the results. And with non-contact image capture, images can be taken in dangerous or difficult areas where the safety of workers is a concern. Classification automation, with use of machine learning, automates the inspection process and allows for early detection of issues before they are driven out of hand. The multi-modal data fusion of images and sensor readings improves the reliability of health evaluations and brings about detailed information on structural conditions. Sadly, there are some drawbacks. System and quality variations can result from environmental factors such as ambient lighting, shadowing, weather and surface debris. These factors introduce noise that to some extent blur the defect characteristics, thus requiring sophisticated preprocessing and adaptive algorithms. The minimum detectable defect size is also influenced by the resolution of the imaging system, which affects the overall sensitivity. To tackle these issues, further efforts are needed in developing robust image enhancement and adaptive learning techniques.

Scalability and cost efficiency are essential for practical deployment. IOT image processing based systems could be scalable to monitor vast networked infrastructures using a collection of sensor and camera nodes provided with wireless connectivity. Cloud and edge computing environments offer elastic resources capable to scale according to monitoring requirements. Even though hardware and software development requires initial investment in equipments like cameras and handling system, being able to perform inspections automatically without manual intervention, and to keep the system up and running with nominally maintenance work for a long period of time gives us an interval in return on investment that is attractive. The modularity of the framework can result in a step-by-step improvement and also an easy transition to the existing self-health monitoring systems, making it practicable for real environments.

The presented results and simulations demonstrate that intelligent solutions combining IoT with image processing show great potential in the domain of structural health monitoring. When integrated, these enable the acquisition of data in real-time, with advanced image analysis and machine learning-based defect classification allowing for more precise, fast and preemptive infrastructure maintenance. In future, challenges need to be directed towards mitigation of environmental effects, to enhance computational efficiency, and to develop benchmark standards to ensure ease of access and use of such technologies in a variety of SHM applications.

REAL-TIME APPLICATIONS

In addition, the adoption of IoT technologies for image processing in SHM has enabled a wide range of real-time applications in the domain of critical infrastructure and urban areas. In the field of infrastructure monitoring, bridges, roads, tunnels and dams are important bondage which structural soundness has a direct impact on the public welfare and economic development. By keeping these structures under constant surveillance, this technique helps to prevent crack propagation, corrosion and material degradation from reaching a point where they become a problem. You could, for example, visually monitor fatigue crack propagation in bridge decks and support beams while at the same time autonomously collecting ultra-high resolution imaging and sensor data capturing intricate stress and deformed patterns. Also, by means of image processing a change in colour and hence in texture, of a metal surface, as for instance tunnel lining or dam gates, can be early detected, so that without prior physical inspection the maintenance can be realized before a potential breakdown. In such cases, access to data using IoT networks, combined with real-time data processing, can give structural health managers real-time intelligence and enable them to remain situationally aware in order to immediately act on developing threats.

Another important domain in which smart and SHM solutions could be helpful is building safety. Tall structures, particularly those in seismic zones, need close monitoring to ensure safety to occupants. Moreover historical monuments, in many cases made of material of limited resistance and subjected to environmental erosion through the centuries, require tailor-made surveillance to ensure their structural conservation. In an IoT-based image processing system, a detailed inspection is enabled and subtle appearance changes are checked such as shifts, cracks, and spalling of materials after an earthquake, typhoon, or other natural catastrophe. These systems can quickly pinpoint weak areas that need shoring up making it possible to target repairs. Preventing further degradation, reducing risks of delayed maintenance, and making informed decisions by monitoring damage have been possible with the use of early detection and monitoring techniques. Moreover, the benefit of the non-invasive methodology of image-based monitoring is of utmost importance for cultural heritage buildings: in this case results without physical intervention and at the same time allow the largest preservation level.

The industrial establishments, i.e., the pipeline, storage tank, the manufacturing complexes could also enjoy the noteworthy benefits, because of the installation of IoT Incorporated image processing for SHM. Oil, gas, or chemical pipes that would need to be continuously monitored to ensure no leakage, corrosion, or mechanical damages that could lead to hazards in the environment, or operation down time. On the other hand, storage tanks, and machinery equipment, have to be monitored for any structural shortcomings likely to cause breakdowns, or any triggering accidents. Image-based defect detection offers non-contact, scalable means to inspect large industrial structures in a continuous manner. When used in conjunction with IoT sensor networks that collect pressure, temperature, and vibration information, the image processing system improves overall assessment accuracy. Preventive maintenance regimes based on these integrated data streams reduce the chances of unexpected breakdowns, maximise repairs scheduling and fulfil safety legislation. Real-time alarms at the instant of defect detection provide quick response, protecting the employees, the environment and assets.

The idea of smart cities however, takes the use of IoT-based image processing for SHM into citywide management perspectives by including structural monitoring in larger urban management systems. Wireless networks operating across the city enable IoT to connect complimentary sets of infrastructure, from road and bridges and public buildings through to utilities. In this environment, SHM on an ongoing basis supports early warning systems to recognise structural anomalies or danger, before they actually develop. Real-time responses may also be automated to inform enabled parties, implement emergency responses (such as dispatching police, fire and/or ambulance forces), or modify traffic patterns to reduce hazard. For instance, image processing can be used to determine if cracks are growing rapidly on a bridge, and if they are, the system can provide instant warnings and suggest shutting down the road in order to avert accidents. Such integration contributes to an urban's overall resilience by enabling real-time situation awareness and enabling co-located emergency response. These solutions are scalable and flexible readily adapting to the complex and dynamic nature of urban infrastructure, promoting safer, more sustainable cities.

To summarize, applications of the smart- SHM systems are varied and showing a significant potential as real-time solution combining IoT and image processing. The ability to continually monitor infrastructure, identify flaws early, and proactively maintain assets means improved safety, lower operating expenses, and greater longevity of assets. As new technologies are developed, deeper and deeper integration with concepts like artificial idents computing, edge computing, augmented reality, and others will extend the capabilities and the impact of these systems, leading to a safer, smarter, and more resilient infrastructure across the globe.

CONCLUSION

The combination of Internet of Things (IoT) and advanced image processing technologies provides a novel platform for structural health monitoring (SHM) and is expected to make the infrastructure smart and safe. This chapter demonstrated the vital necessity of monitoring these human-made structures, e.g., bridges, buildings, tunnels, and industrial plants automatically and continuously, so as to remain operative, to be safe to the public and long-lasting. However, conventional SHM techniques may not be able to achieve scalability, real-time response, and surface-level focused examination. The merger of IoT and image analysis is able to address these issues by acquiring data in real time, monitoring remotely and detecting defects accurately and automatically. The implementation of ML models help to increase the accuracy of defect classification, which supports predictive and preventative maintenance and reduces downtime.

The potential of image processing in SHM using IoT can be applicable in numerous scenarios, such as in infrastructure, civil structure safety, asset management in industry, city system and so on. This holistic concept provides early warning of damage, permanent health monitoring and predictive maintenance, and thereby enhances safety and reduces costs. Further, the large-scale nature of the IoT networks in conjunction with the wealth of visual information that can be found in the image processing, leads to an increase in the understanding of structural health beyond 'sensor -only orthodoxy'.

However, despite the recent progress made, several challenges stand ahead such as privacy preserving, computational complexity, environmental influence on acquired images quality, and fusion of different source of cues. The development towards more secure and more standardized algorithmic, communication and processing techniques is still under continuous research and development. The future also holds potential in the use of new technology, including edge computing, artificial intelligence, and augmented reality to improve SHM technology.

In conclusion, the compounding potential of the IoT and image processing technologies presents great opportunity for smart SHM, and subsequently the future resilient, efficient, and intelligent infrastructure systems globally.

REFERENCES

Awais, M., Khan, W., Akram, T., & Nam, Y. (2025). Energy-efficient discrete cosine transform architecture using reversible logic for IoT-enabled consumer electronics. [or access via IEEE Xplore]. *IEEE Access : Practical Innovations, Open Solutions, 13*, 12292–12307. DOI: 10.1109/ACCESS.2025.3528261

Balaji, V. R., Kalvinathan, V., Dheepanchakkravarthy, A., & Muthuvel, P. (2021). IoT enabled smart irrigation system. In *2021 International Conference on Advancements in Electrical, Electronics, Communication, Computing and Automation (ICAECA)* (pp. 1–6).

Choudhary, O. P., Jaiswal, S., & Das, S. (2021). IoT enabled condition monitoring of low voltage motors using fuzzy inference system. In *2021 IEEE 5th International Conference on Condition Assessment Techniques in Electrical Systems (CATCON)* (pp. 132–133). DOI: 10.1109/CATCON52335.2021.9670486

Goyal, V., Yadav, A., & Mukherjee, R. (2021). Thermographic camera-based E-IoT enabled architecture for smart poultry farm. In *2021 5th International Conference on Electrical, Electronics, Communication, Computer Technologies and Optimization Techniques (ICEECCOT)* (pp. 772–775).

Ilangaweera, L. S., Naramada, S. E. M. H., Sajith, N. T. M., Aashiq, M. N. M., & Hinas, M. N. A. (2024). IOT based smart plant growing and caring platform for urban environments. In *2024 9th International Conference on Information Technology Research (ICITR)* (pp. 1–6). DOI: 10.1109/ICITR64794.2024.10857784

. Khan, J., Li, J.-P., Ahamad, B., Parveen, S., Haq, A. U., Khan, G. A., & Khan, M. A. (2020). SMSH: Secure surveillance mechanism on smart healthcare IoT system with probabilistic image encryption. *IEEE Access, 8*, 15747–15767. https://doi.org/ (Note: Publication year appears as 2020 based on sources, despite input listing 2022.) DOI: 10.1109/ACCESS.2020.2969534

Liu, Y., Briones, J., Zhou, R., & Magotra, N. (2017). Study of secure boot with a FPGA-based IoT device. In *2017 IEEE 60th International Midwest Symposium on Circuits and Systems (MWSCAS)* (pp. 1053–1056). DOI: 10.1109/MWSCAS.2017.8053108

Lopes, S. I., Pinho, P., Marques, P., Abreu, C., Carvalho, N. B., & Ferreira, J. (2021). Contactless smart screening in nursing homes: An IoT-enabled solution for the COVID-19 era. In *2021 17th International Conference on Wireless and Mobile Computing, Networking and Communications (WiMob)* (pp. 145–150).

. P. S., S. K. T., S. P. M., D. S. V., S. V., & R. A. V. K. (2024). Intelligent waste disposal: An IoT enabled automated system for dumpster monitoring and medical segregation. In *2024 International Conference on Power, Energy, Control and Transmission Systems (ICPECTS)* (pp. 1–6).

Prathyusha, G., Krishnan, M. B. M., & Deepak, S. P. K. (2024). Integration of IoT-enabled UAVs in smart agriculture: Types, applications, trends, challenges. In *2024 7th International Conference on Circuit Power and Computing Technologies (ICCPCT)* (pp. 1124–1130).

Priya, S. S., Al-Fatlawy, M. H., Khare, N., Mahalakshmi, V., & Ganesh, S. S. (2023). Machine and deep learning classifications for IoT-enabled healthcare devices. In *2023 4th International Conference on Computation, Automation and Knowledge Management (ICCAKM)* (pp. 1–7). DOI: 10.1109/ICCAKM58659.2023.10449617

. R., S., Prasad, K., & C., S. K. (2024). Optimal bandwidth utilization techniques for IoT device networks using edge computing. In *2024 4th Asian Conference on Innovation in Technology (ASIANCON)* (pp. 1–7).

Rumi, R. I., Pavel, M. I., Islam, E., Shakir, M. B., & Hossain, M. A. (2019). IoT enabled prescription reading smart medicine dispenser implementing maximally stable extremal regions and OCR. In *2019 Third International Conference on I-SMAC (IoT in Social, Mobile, Analytics and Cloud) (I-SMAC)* (pp. 134–138). DOI: 10.1109/I-SMAC47947.2019.9032709

Sharma, K., & Pathak, A. K. (2025). Transforming radiology with IoT-enabled smart wireless solutions in healthcare infrastructure. In *2025 3rd International Conference on Disruptive Technologies (ICDT)* (pp. 806–809).

Shereen, Y., Shahid, M. F., & Tanveer, M. H. (2025). Self-aware AI-based anomaly detection for AI IoT enabled smart healthcare system. In *2025 4th International Conference on Computing and Information Technology (ICCIT)* (pp. 406–413). https://doi.org/ (or access via IEEE Xplore: https://ieeexplore.ieee.org/document/10989421)DOI: 10.1109/ICCIT.2025.10989421

Suthar, S. B., Chithra, R., Abhinaya, G. K., & Harikumar, H. (2023). Smart waste segregation and collection system with IoT-enabled monitoring and analytics. In *2023 International Conference on Recent Advances in Electrical, Electronics, Ubiquitous Communication, and Computational Intelligence (RAEEUCCI)* (pp. 1–6). DOI: 10.1109/RAEEUCCI57140.2023.10134423

Tej, Y. V., Potharaju, V. S., Jakkidi, P. R., Sharma, A., & Reddy, N. S. (2024). Deep learning techniques for image recognition in IoT-enabled surveillance systems. In *2024 Asian Conference on Intelligent Technologies (ACOIT)* (pp. 1–8).

Thangaraj, M., Ponmalar, P. P., & Anuradha, S. (2015). Internet of Things (IoT) enabled smart autonomous hospital management system - A real world health care use case with the technology drivers. In *2015 IEEE International Conference on Computational Intelligence and Computing Research (ICCIC)* (pp. 1–8). DOI: 10.1109/ICCIC.2015.7435678

Tyagi, L. K., Gupta, A., & Sisodia, V. S. (2025). Machine and deep learning classifications for IoT-enabled healthcare devices. In *2025 International Conference on Pervasive Computational Technologies (ICPCT)* (pp. 582–587). DOI: 10.1109/ICPCT64145.2025.10940415

Veerabomma, S. K. S., Rajeswari, G., Murthy, B. S., & Selvan, R. S. (2024). Optimising IoT-healthcare systems with deep learning: Sensors data fusion, prognostic modelling, and smart management. In *2024 Second International Conference Computational and Characterization Techniques in Engineering & Sciences (IC3TES)* (pp. 1–6). DOI: 10.1109/IC3TES62412.2024.10877558

Chapter 12

Influence of Fiber Hybridization on the Mechanical Behavior of Kenaf-Sisal Epoxy Composites

Kodidasu Pranathi
https://orcid.org/0009-0007-5811-1714
MLR Institute of Technology, India

Abdul Gafoor
MLR Institute of Technology, India

Shaik Mohammed Adib Ayub
MLR Institute of Technology, India

M. Arthi
Saveetha Institute of Medical and Technical Sciences, India

Emad M. Elsehly
https://orcid.org/0000-0001-8811-275X
Faculty of Science, Damanhour University, Egypt

Shijo Thomas
Christ University, India

ABSTRACT

The natural composites are widely preferred nowadays in versatile application due to easily degradable and moderate mechanical properties when compare to polymer matrix composites. The kenaf fiber are used as reinforcement in composites wherever high flexural strength and modulus is required whereas sisal fiber is preferred for tensile strength, impact and wear resistance. In our work hybrid composite comprising kenaf and sisal as reinforcement and epoxy as matrix is fabricated and its mechanical properties are explored and compared with the conventional composite fabricated using only kenaf or sisal as reinforcement. The hybrid composite shows

DOI: 10.4018/979-8-3373-9454-1.ch012

excellent flexural strength but inferior tensile strength when compare to conventional composite. The failure of the specimens are investigated using scanning electron microscopy (SEM). In this chapter detailed insight is presented for emphasizing the need for tailored fiber selection and processing to achieve optimal mechanical properties.

1. INTRODUCTION

In recent years, there has been a significant rise in research on natural fiber-reinforced hybrid composites, particularly focusing on combining fibers with diverse properties to achieve enhanced mechanical performance (Nguyen et al., 2020). In the aerospace and automobile industries, synthetic fiber-reinforced polymers (FRPs) are widely used due to their high strength, stiffness, and durability (Muthu, 2020). Their eco-friendly nature and potential for sustainability have further driven interest in the research and development of advanced composite materials (Saha et al., 2025). Natural fibers like sisal and kenaf shown in figure 1 have high mechanical properties like tensile strength and good flexural behavior, in order to achieve a composite material which is good and both properties, hybrid composites are made (Karimah et al., 2023; Dhinakarraj et al., 2025). The kenaf fibre has good mechanical properties and specific strength, whereas the sisal fibre has high tensile strength and stiffness, making it useful in increasing the mechanical properties of the hybrid composite(Elfaleh et al., 2023, Deepanraj et al., 2025).This process of hybridization overcomes the limitation of single-fiber composites and also optimizes the performance characteristics(Saba et al., 2015). Recent publications highlight that achieving desired properties using natural fiber composites is a challenging thing (Solahuddin, 2022; Senthilkumar et al. 2024) and also the adhesion between the fiber matrix interface plays a crucial role in enhancing the propertries, this is improved by using some chemical treatment and coupling agents (Rana et al., 2023).

In the process of fiber hybridization mixture of two or more fibers in to single or more matrix is done to improve the characteristics of composites material (Hao et al., 2018). Mixing the sisal-kenaf fibre as reinforcement with the epoxy resin as matrix can increase the mechanical property of the composite material (Li et al., 2000). Bonding of fiber and matrix is called as fiber-matrix interface, which plays an important role enhancing the properties (Lin et al., 2025).The main reason behind the mechanical properties of natural fiber reinforced composites is quantity of fiber used, how the fiberis aligned in terms of direction, other thing is properties of matrix material (Abbood et al., 2021). The tensile strength and flexural strength of hybrid composite, will be reported and compared to the single-fibre composite(Velmurugan et al., 2020). These specimens are again studied under scanning electron microscopy for knowing the failure modes.

Figure 1. Natural fibres (a) Sisal fiber, (b) Kenaf fibre

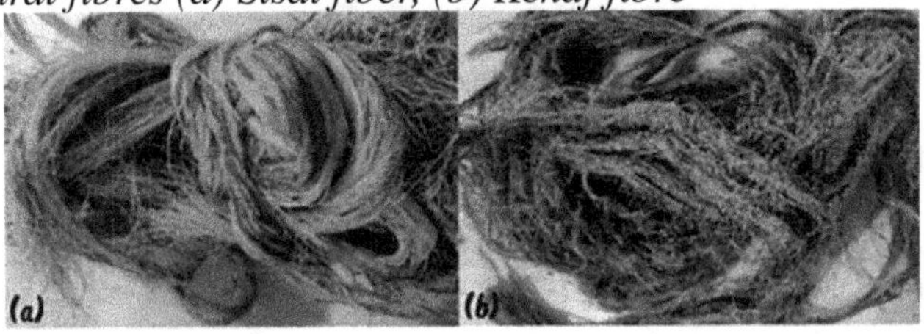

Out of all resins epoxy resin is a unique resin for its superior properties. The high strength and stiffness of epoxy resin providegreateroutputs in terms of mechanical behavior of composite (Abdellaoui et al., 2019). This resin is also good at adhesion properties, by which the strong bounding is made between the fibers (Fiore and Valenza, 2013). Besides, it has good chemical resistance, can be used at any conditions, with low shrinkage while curing making it ideal for usage in all industries (Skalková et al., 2021).In this study a commercial clear epoxy resin supplied by clearresin.in, India was used as matrix material and clear hardener as curing agent mixed in a 2:1 weight ratio as per manufacturer instruction.

Table 1. Properties of the epoxy resin

Property	Value
Density	1.1–1.2 g/cm^3
Tensile Strength	60–70 MPa
Flexural Strength	90–100 MPa
Curing Time (at room temperature)	24 hours
Glass Transition Temperature (Tg)	~80–90°C

In the industry of aerospace or automobilewhere composites are used for lightweight and load bearing components, like aircraft structural components and interior panels there this material can be used (Phiri et al., 2024). Generally natural fiber composites have distinct properties like strength, stiffness making them suitable for all industries (Al-Azad et al., 2021). Other industries like automobile can also be profited by using the natural fiber composites using sisal and kenaf fiber (Khan et al., 2025). These composites can be used to make ecofriendly, lightweight and less cost composite components in various parts, such as interior panel, dashboard components, door trims etc.By using the composite material in these areas, the vehicle weight will be reduced, and there by efficiency increases as fuel consumption decreases and provides less cost. The other main usage of these composite is in construction industry for roofing sheets, insulation material, wall panels, due to the unique property of thermal and sound proof nature. Promoting the composites leads to sustainable environmental impact and increase energy efficiency(Fornés et al., 2022).

2. EXPERIMENTAL METHODS

2.1 Fibre Preparation

Fibres used here are sisal and kenaf as reinforcement, combined with resin used is epoxy as the matrix. First, the fibres are cut into short pieces to ensure uniform treatment and are clean of any impurities (Mahmud et al., 2025). The epoxy resin is prepared in a ratio of 2:1 (resin to hardener), while the fibre to resin ratio is set at 70:30 (fibre to resin). The fibres are thoroughly coated with the resin to achieve even saturation and wetting. This process is carried out for 5 to 15 minutes, ensuring the mixture is well combined (Rashid et al., 2024) shown in figure 2.

Figure 2. Photographs of fibre preparation process:(a) fibre cutting, (b) resin preparation, (c) chopped fibre and resin mixing.

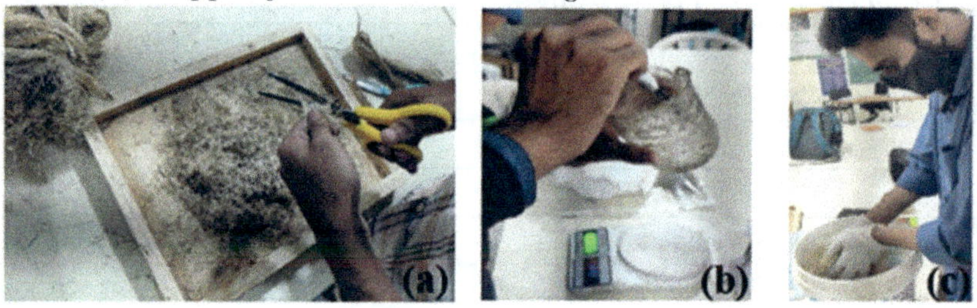

In this study, untreated kenaf and sisal fiber were used. The decision to use untreated fibers was made to investigate the baseline mechanical performance without the influence of chemical modification.

2.2 Composite Molding

The composite laminate is made using a hand mixing technique for each individual laminate (Middleton, 2016). A plastic sheet is placed on an iron plate, followed by the uniform random spread of chopped wet fiber. An additional iron plate is placed on top, and this mold is subjected to a compression load of 10 kN at room temperature (20–25°C) for approximately 24 hours. After the curing period, the iron plates are removed, and the fabricated composite laminate is cut with the cutting machine according to ASTM standards for each test (Thirumurugan et al., 2016). The photographs of compression mold and wetted fibre pacing is shown in figure 3.

Figure 3. (a) Fibre layering, (b) mold under compression.

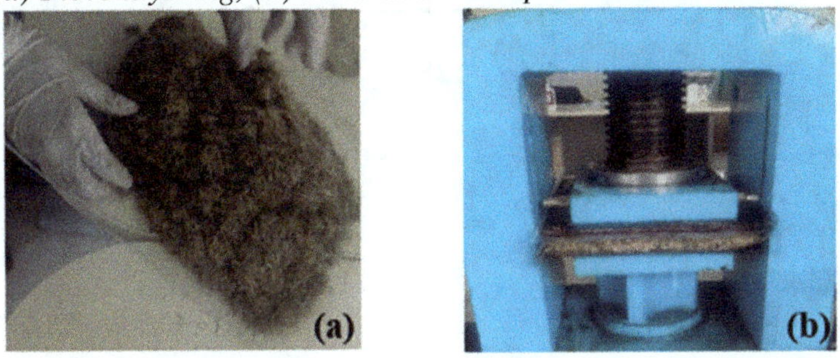

2.3 ASTM Standards

ASTM D3039 is a grade need to be followed while testing the tensile properties of polymer matrix composite (Hariharan & Rajeshkumar, 2024). This test includes the standard dimensions of the test specimen to perform tensile test (Link et al., 2005). It reveals the strength and stiffness of the tested composite. The specimen is hold using tensile grips, uniaxial load is applied, the following parameters are obtained while testing (Pranathi et al., 2024).

Figure 4. ASTM D3039 Standard specimen dimensions

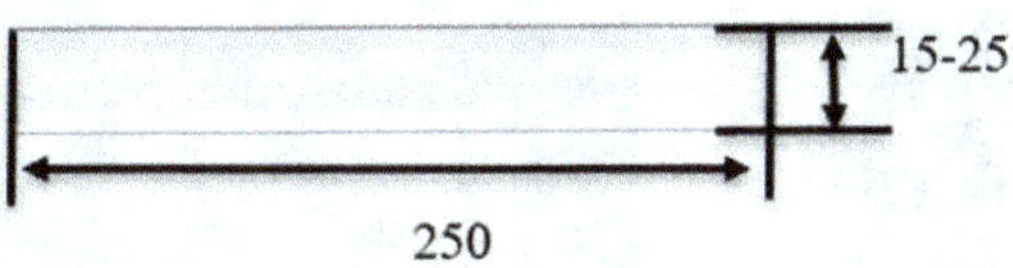

ASTM D7264 is a grade need to be followed while testing the flexural properties of polymer matrix composite (Ogunleye et al., 2022). This test includes the standard dimensions of the test specimen to perform tensile test. It reveals the bending performance and resistance to damage of the tested composite (Tang et al., 2023).

The specimen is supported at its both ends, point load is applied at the centre as shown in figure 4 and figure 5, defection is measured at mid span, and the following parameters are obtained while testing.

Figure 5. ASTM D7264 Standard specimen dimensions

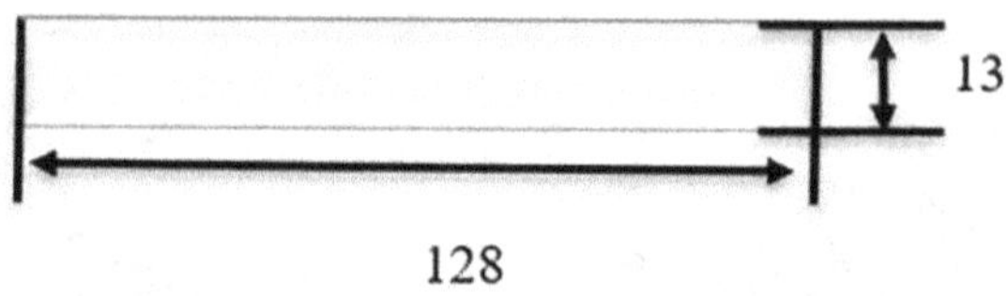

2.4 Tensile test

2.4.1 Sisal fiber

The tensile test is performed on the composite laminate sample according to ASTM standard D3039 using Universal Testing Machine (ASTM D3039). Before the test, the composite laminate is cut into the dimensions 250 mm x 5mm x 25 mm under the load of 2kN as shown in figure 6.

Figure 6. Images of tensile test (a) kenaf, sisal test specimens, (b) tensile testingin UTM

2.4.2 Kenaf fibre

The tensile test is performed on the composite laminate sample according to ASTM standard D3039 using the Universal Testing Machine (Pavani and Shashidhar, 2024). Before the test, the composite laminate is cut to the dimensions 250mm x 5mm x 25mm and tested under the load of 2 kN.

2.5 Flexural test

2.5.1 Sisal fibre

The flexural test is conducted on a composite laminate sample following ASTM standards D7264. Before the test, the fabricated laminate is shaped to the specified dimensions of 128 mm x 5 mm x 13 mm and was tested under a force of 2 kN (ASTM D3039

2.5.2 Kenaf fiber

The flexural test is conducted on a composite laminate sample following ASTM standards D7264. Before the test, the fabricated laminate is shaped to the specified dimensions of 128 mm x 5 mm x 13 mm and is tested under a force of 2 kN (Pavani and Shashidhar, 2024) as shown in figure 7.

Figure 7. Photographs of flexural test (a) kenaf, sisal test specimens(b)flexural testing in UTM

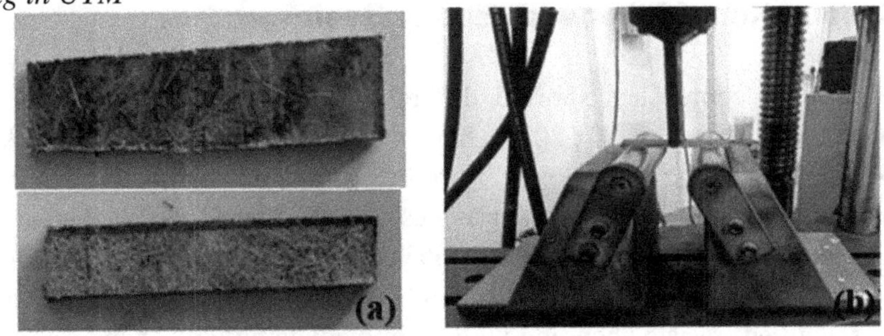

3. RESULTS AND DISCUSSION

3.1 Case I: Sisal Fiber with Epoxy Resin

The composite laminate reinforced solely with sisal fibers demonstrated high tensile strength and excellent elongation capacity (Gebremichael et al., 2025). The test specimens are shown in figure 8. When subjected to tensile testing, these specimens exhibited consistent resistance to axial loading, confirming sisal's natural strength and ability to withstand stretching forces (Zhu et al., 2011). The load-bearing capacity peaked at 2.320 kN, highlighting sisal's excellent fiber-to-matrix bonding and surface compatibility with epoxy resin (Ichim et al., 2024).

Figure 8. Images of tensile and flexural test sisal specimen

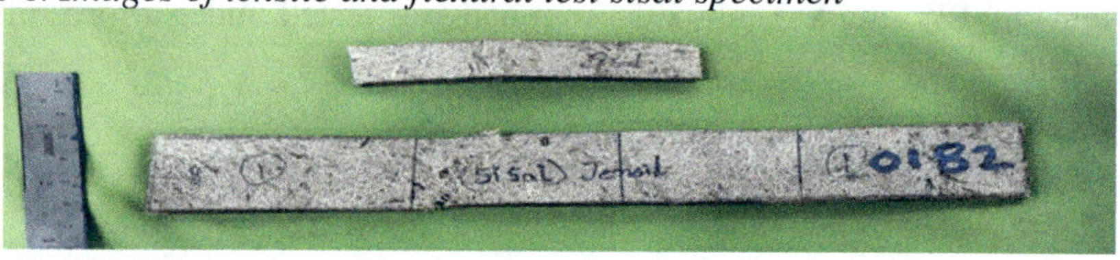

During flexural tests, however, the sisal composite showed moderate performance (Porwal et al., 2023). While it resisted initial bending, premature cracking was observed at higher deflections, suggesting that while sisal is strong in tension, it may not offer the same stiffness in flexure (Huang & Rodrigue, 2022).

3.2 Case II: Kenaf Fiber with Epoxy Resin

Kenaf-based composites showed different mechanical behavior than sisal-reinforced ones (Yorseng et al., 2020). In tensile testing, they achieved a peak load of 5.340 kN, indicating better load absorption but slightly lower tensile strength due to brittleness and fiber dispersion issues (Mahjoub et al., 2014). However, in flexural testing, the kenaf composites displayed superior performance (Zampaloni et al., 2007). The flexural strength was higher compared to sisal-only laminates, and the specimens resisted bending without immediate failure (Thandavamoorthy et al., 2024). The specimen used for testing is shown in figure 9.

Figure 9. Images of tensile and flexural test kenaf specimen

The increased flexural strength can be attributed to kenaf's inherent stiffness and the structural arrangement of the fibers during molding (Abbas et al., 2022). This allowed better load distribution along the length of the sample under bending (Salem et al., 2017).

3.3 Case III: Hybrid Kenaf–Sisal Fiber with Epoxy Resin

The hybrid composite, consisting of both kenaf and sisal fibers, presented a balanced mechanical response in both tensile and flexural testing (Neto et al., 2022). While the tensile strength was slightly lower than that of pure sisal laminates, it was still within a competitive range (Olanrewaju et al., 2025). More importantly, the hybrid composite displayed the highest flexural strength among all three cases, proving that combining fibers with complementary properties can significantly improve performance in multi-axial loading conditions (Khalil et al., 2012). The fabricated specimen is shown in figure 10.

Figure 10. Images of tensile test hybrid specimen

The presence of both flexible (sisal) and rigid (kenaf) fibers created a synergistic effect (El Hawary et al., 2023). Under tension, the sisal fibers absorbed strain, while the kenaf added stiffness and minimized excessive elongation as shown in figure 11. Under bending, the kenaf provided core structural rigidity while sisal helped in distributing the stress uniformly (Mohammed et al., 2024).

Table 2. Tensile and flexural properties of the three cases (test results)

Composite	Tensile Strength (MPa)	Flexural Strength (MPa)	Elongation (%)
Sisal-Epoxy	13.11	38.14	3.21
Kenaf-Epoxy	31.13	36.25	3.02
Hybrid	6.82	92.72	5.90

Figure 11. Tensile and flexural traits of the three cases

3.4 Morphological Observations via SEM

Scanning Electron Microscopy (SEM) is used to analyze the fracture surfaces of tested composite specimens (Mohammed and Abdullah, 2018). Magnification in SEM refers to the ratio of the size of an observed object in an image to its actual size. Unlike optical microscopes, SEM achieves significantly higher magnification, typically ranging from tens to hundreds of thousands of times (Pednekar et al., 2017). It allows observation of the fiber-matrix interface, fiber pull-outs, and crack propagation at high magnification also provide critical failure mechanisms which are affecting the mechanical properties of composite, the key failure modes are identified using SEM and discussed below.

Figure 12 presents SEM images of flexural test specimens at 200× magnification for (a) kenaf and (b) sisal composites. The kenaf specimen (Figure 12a) shows clear evidence of fiber pull-out and matrix cracking. Under bending, the applied stress causes fibers to detach from the resin, resulting in pull-out and the formation of gaps, which reduce load transfer efficiency between the fiber and matrix. In contrast, the sisal specimen (Figure 12b) exhibits fiber breakage along with the presence of matrix debris. The sisal fibers fracture under high stress, and weak fiber–matrix bonding produces fragmented surfaces, indicating regions of brittle failure.

Figure 12. Photographs of SEM images of flexural test specimens (a) kenaf test specimens (b) sisal test specimen (at magnification 200 X)

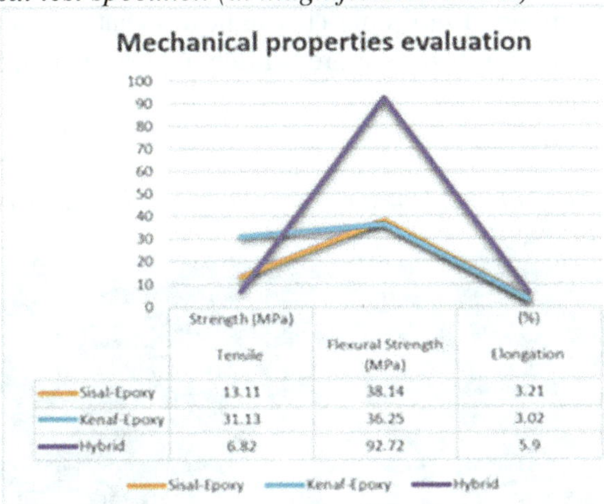

Figure 13 presents SEM images of flexural test specimens at 25,000× magnification for (c) kenaf and (d) sisal composites. In the kenaf specimen (Figure 13c), more pronounced matrix cracking and interfacial debonding are evident. Repeated stress concentrations during bending cause cracks to propagate along the fiber–matrix interface, thereby weakening the structural integrity. In the sisal specimen (Figure 13d), fiber fracture and delamination are clearly visible. The sisal fibers exhibit sharp fracture surfaces, and interlayer separation (delamination) occurs due to weak fiber–matrix adhesion combined with bending-induced stresses.

Figure 13. Photographs of SEM image of flexural test (c) kenaf test specimens(d) sisal test specimen (at magnification 25.00 K X)

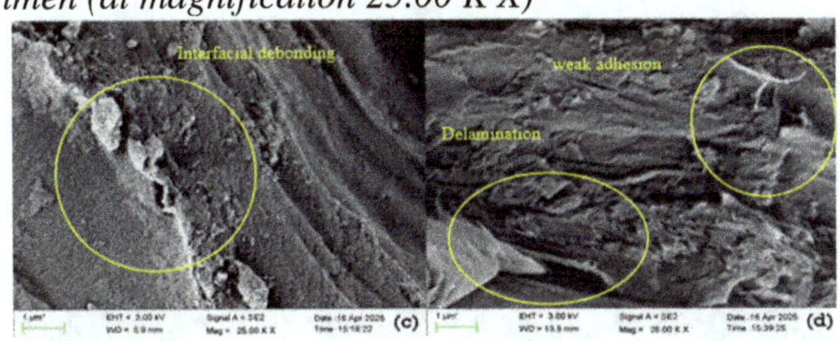

Figure 14 shows SEM images of tensile test specimens at 1,000× magnification for (e) kenaf and (f) sisal composites. The kenaf specimen (Figure 14e) exhibits fiber pull-out and matrix tearing, indicating that under tensile loading, fibers are pulled from the matrix without complete breakage, which suggests relatively weaker fiber–matrix adhesion. In contrast, the sisal specimen (Figure 14f) displays clean fiber fractures with minimal pull-out. This indicates that sisal fibers have better bonding with the matrix but behave more brittlely under tension, breaking sharply rather than pulling out.

Figure 14. Photographs of tensile test (e) kenaf test specimens(f)sisal test specimen (at magnification 1.00 K X)

Figure 15 shows SEM images of tensile test specimens at 1,000× magnification for (g) kenaf and (h) sisal composites. The kenaf specimen (Figure 15g) exhibits rough fiber surfaces and matrix cracking, indicating fiber bridging where some fibers resist fracture longer but eventually cause matrix cracking and rough break surfaces. In contrast, the sisal specimen (Figure 15h) shows brittle fiber fracture accompanied by microvoids. These microvoids represent weak spots in the matrix that act as initiation sites for early fracture under tensile loading, indicating that brittle failure dominates in the sisal composite.

Figure 15. Photographs of tensile test (g) kenaf test specimens(h)sisal test specimen (at magnification 5.00 K X)

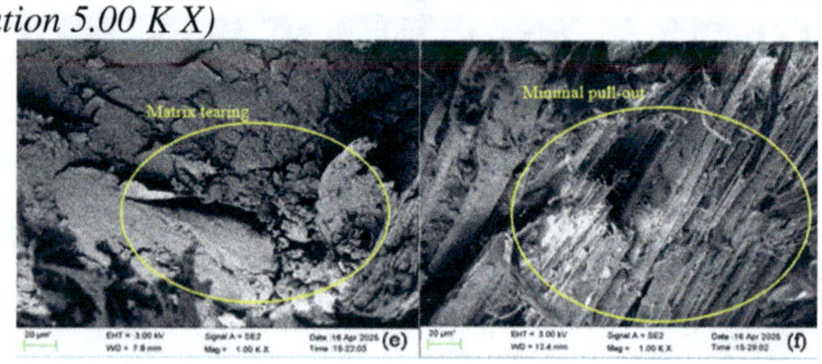

3.5 Comparative Summary

The findings confirm that fiber hybridization can optimize certain mechanical characteristics, depending on the loading conditions (Zuccarello et al., 2022). The hybrid kenaf-sisal composite did not outperform sisal in tensile strength but surpassed both individual composites in flexural performance (Gangwar et al., 2024). This validates the strategic advantage of hybridization when a balance between multiple loading modes is required (Selvan et al., 2021).

Table 3. Table showing comparative analysis key findings

Composite	Interface Bonding	Fiber Failure	Fracture Features
Sisal/Epoxy	Strong, minimal gaps	Clean break, slight pull-out	Smooth surface, minor cracks
Kenaf/Epoxy	Moderate, visible gaps	Brittle break, debonding	Rough surface, pull-out zones
Hybrid (Kenaf–Sisal)	Mixed, mostly good bonding	Mixed break and pull-out	Bridging, delayed cracks

4. CONCLUSIONS

This study evaluated the mechanical behavior of kenaf–sisal hybrid epoxy composites, with a specific focus on tensile and flexural characteristics. According to experimental findings, fiber hybridization is essential for customizing the composite's strength properties depending on the intended use. Reported that composites fabricated with sisal fiber provided improved tensile strength due to resistance to axial loading and strong interfacial bonding between layers of sisal composite, kenaf-based composites performed better under flexural loading, mainly because of their structural stiffness and resistance to bending.

The hybrid composite, made with both kenaf and sisal fibers, achieved a balance in mechanical properties, yielding increased flexural strength compared to other pure fiber composites, but still it reported a lower value for tensile strength compared to the pure sisal composite. The mixing of the fibers proved that a well-designed hybrid composite can offer desired, various performance characteristics to may load bearing applications subjected to different loading conditions based on the requirement.

REFERENCES

Abbas, A.-G. N., Aziz, F. N. A. A., Abdan, K., Mohd Nasir, N. A., & Norizan, M. N. (2022). Kenaf Fibre Reinforced Cementitious Composites. *Fibers (Basel, Switzerland), 10*(1), 3. Advance online publication. DOI: 10.3390/fib10010003

Abbood, I. S., Odaa, S., Hasan, K. F., & Jasim, M. A. (2021). Properties evaluation of fiber reinforced polymers and their constituent materials used in structures – A review. *Materials Today: Proceedings, 43*(Part 2), 1003–1008. DOI: 10.1016/j.matpr.2020.07.636

Abdellaoui, H., Raji, M., Bouhfid, R., & Qaiss, A. el K. (2019). Investigation of the deformation behavior of epoxy-based composite materials. In Failure Analysis in Biocomposites, Fibre-Reinforced Composites and Hybrid Composites (Woodhead Publishing Series in Composites Science and Engineering), pp. (chapter 2). Woodhead Publishing, 29-49. DOI: 10.1016/B978-0-08-102293-1.00002-4

Al-Azad, N., Asril, M. F. M., & Shah, M. K. M. (2021). A review on development of natural fibre composites for construction applications. *Journal of Materials Science and Chemical Engineering, 9*(7), 1–9. DOI: 10.4236/msce.2021.97001

ASTM D3039, Standard Test Method for Tensile Properties of Polymer Matrix Composite Materials.

Deepanraj, B., Saravanan, A. M., Senthilkumar, N., Afzal, A. A., & Afzal, A. R. (2025). Drilling characteristics optimization of polymer composite fortified with eggshells using Box–Behnken design and zebra optimization algorithm. *Results in Engineering, 25*, 104102. DOI: 10.1016/j.rineng.2025.104102

Dhinakarraj, C. K., Senthilkumar, N., Palanikumar, K., & Deepanraj, B. (2025). Machinability evaluation of magnesium composite using response surface methodology and nature-inspired metaheuristic algorithms. *International Journal on Interactive Design and Manufacturing, 19*, 2607–2627. DOI: 10.1007/s12008-024-01866-3

El Hawary, O., Boccarusso, L., Ansell, M. P., Durante, M., & Pinto, F. (2023). An overview of natural fiber composites for marine applications. *Journal of Marine Science and Engineering, 11*(5), 1076. DOI: 10.3390/jmse11051076

Elfaleh, I., Abbassi, F., Habibi, M., Ahmad, F., Guedri, M., Nasri, M., & Garnier, C. (2023). A comprehensive review of natural fibers and their composites: An eco-friendly alternative to conventional materials. *Results in Engineering, 19*, 101271. DOI: 10.1016/j.rineng.2023.101271

Fiore, V., & Valenza, A. (2013). Epoxy resins as a matrix material in advanced fibre-reinforced polymer (FRP) composites. In Advanced Fibre-Reinforced Polymer (FRP) Composites for Structural Applications (pp. 88–121). Woodhead Publishing. DOI: 10.1533/9780857098641.1.88

Gangwar, A., Kumar, V., Yaylacı, M., & Panda, S. K. (2024). Computational modelling and mechanical characteristics of polymeric hybrid composite materials: An extensive review. *Archives of Computational Methods in Engineering*, *31*, 3901–3921. DOI: 10.1007/s11831-024-10097-4

Gebremichael, G., Mekonone, S. T., & Baye, T. M.. (2025). Mechanical and microstructural characterization of sisal fiber-reinforced polyester laminate composites for improved durability in automotive applications. *Journal of Materials Science: Composites*, *6*(1), 1. Advance online publication. DOI: 10.1186/s42252-024-00064-4

Hao, L. C., Sapuan, S. M., Hassan, M. R., & Sheltami, R. M. (2018). Natural fiber reinforced vinyl polymer composites. In Sapuan, S. M. (Ed.), *Natural fibre reinforced vinyl ester and vinyl polymer composites* (pp. 27–70). Woodhead Publishing., DOI: 10.1016/B978-0-08-102160-6.00002-0

Hariharan, V., & Rajeshkumar, G. (2024). Investigation of mechanical properties of sisal fiber and sugar palm fiber reinforced hybrid composites. Qeios, January, 1–9. https://www.qeios.com/read/JT0ETV

Huang, J., & Rodrigue, D. (2022). Stiffness behavior of sisal fiber reinforced foam concrete under flexural loading. *Journal of Natural Fibers*, *19*(15), 12251–12267. DOI: 10.1080/15440478.2022.2054896

Ichim, M., Mure an, E. I., & Codău, E. (2024). Natural-fiber-reinforced polymer composites for furniture applications. *Polymers*, *16*(22), 3113. DOI: 10.3390/polym16223113 PMID: 39599204

Karimah, A., Ridho, M. R., Munawar, S. S., Adi, D. S., Ismadi, R. D., Damayanti, R., Subiyanto, B., Fatriasari, W., & Fudholi, A. (2023). A review on natural fibers for development of eco-friendly bio-composite: Characteristics and utilizations. *Journal of Materials Research and Technology*, *13*, 2442–2458. DOI: 10.1016/j.jmrt.2021.06.014

Khalil, H. A., Tehrani, M., Davoudpour, Y., Bhat, A., Jawaid, M., & Hassan, A. (2012). Natural fiber reinforced poly(vinyl chloride) composites: A review. *Journal of Reinforced Plastics and Composites*, *32*(5), 330–356. DOI: 10.1177/0731684412458553

Khan, F., Hossain, N., Mim, J. J., Rahman, S. M. M., Iqbal, M. J., Billah, M., & Chowdhury, M. A. (2025). Advances of composite materials in automobile applications – A review. *Journal of Engineering Research, 13*(2), 1001–1023. DOI: 10.1016/j.jer.2024.02.017

Li, Y., Mai, Y.-W., & Ye, L. (2000). Sisal fibre and its composites: A review of recent developments. *Composites Science and Technology, 60*(11), 2037–2055. DOI: 10.1016/S0266-3538(00)00101-9

Lin, S., Ye, B., Peng, C., Ran, M., Jie, D., Du, Y., & Yin, D. (2025). Enhancing interfacial bonding and mechanical properties of CF/EP composites through low-temperature jet plasma treatment. *International Journal of Adhesion and Adhesives, 141*, 104067. DOI: 10.1016/j.ijadhadh.2025.104067

Mahjoub, R., Mohamad Yatim, J., Mohd Sam, A. R., & Hashemi, S. H. (2014). Tensile properties of kenaf fiber due to various conditions of chemical fiber surface modifications. *Construction & Building Materials, 55*, 103–113. DOI: 10.1016/j.conbuildmat.2014.01.036

Mahmud, M. Z. A., Rabbi, S. F., Islam, M. D., & Hossain, N. (2025). Synthesis and applications of natural fiber-reinforced epoxy composites: A comprehensive review. *SPE Polymers, 6*(1), e10161. DOI: 10.1002/pls2.10161

Middleton, B. (2016). Composites: Manufacture and application. In Design and Manufacture of Plastic Components for Multifunctionality: Structural Composites, Injection Molding, and 3D Printing (Chapter 3, pp. 53–101).

Mohammed, A., & Abdullah, A. (2018) Scanning Electron Microscopy (SEM): A Review. *Proceedings of 2018 International Conference on Hydraulics and Pneumatics-HERVEX*, Baile Govora, Romania, 7-9 November 2018, 77-85.

Mohammed, M., Oleiwi, J. K., Mohammed, A. M., & Mohamad Jawad, A. J. (2024). A review on the advancement of renewable natural fiber hybrid composites: Prospects, challenges, and industrial applications. *Journal of Renewable Materials, 12*(7), 1237–1290. DOI: 10.32604/jrm.2024.051201

Muthu, S. S. (2020). Introduction to sustainability and the textile supply chain and its environmental impact. In *Assessing the Environmental Impact of Textiles and the Clothing Supply Chain (The Textile Institute Book Series)* (pp. 1–32). Woodhead Publishing., DOI: 10.1016/B978-0-12-819783-7.00001-6

Neto, J., Queiroz, H., Aguiar, R., Lima, R., Cavalcanti, D., & Banea, M. D. (2022). A review of recent advances in hybrid natural fiber reinforced polymer composites. *Journal of Renewable Materials, 10*(3), 561–589. DOI: 10.32604/jrm.2022.017434

Nguyen, K. T., Navaratnam, S., Mendis, P., Zhang, K., Barnett, J., & Wang, H. (2020). Fire safety of composites in prefabricated buildings: From fibre reinforced polymer to textile reinforced concrete. Composites Part B: Engineering, 187, 107815. DOI: 10.1016/j.compositesb.2020.107815

Ogunleye, R. O., Rusnakova, S., Zaludek, M., & Emebu, S. (2022). The influence of ply stacking sequence on mechanical properties of carbon/epoxy composite laminates. Polymers, 14(24), 5566. DOI: 10.3390/polym14245566

Olanrewaju, O., Oladele, I. O., & Adelani, S. O. (2025). Recent Advances in Natural Fiber Reinforced Metal/Cement/Polymer Composites: An Overview of the Structure-Property Relationship for Engineering Applications. *Hybrid Advances*, 8, 100378. DOI: 10.1016/j.hybadv.2025.100378

Pavani, G., & Shashidhar, A. (2024). Mechanical Properties of Natural Composites: Hemp Fiber and Polypropylene Resin. *Journal of Physics: Conference Series, 2837*(1), 012002. DOI: 10.1088/1742-6596/2837/1/012002

Pednekar, P. P., Godiyal, S. C., Jadhav, K. R., & Kadam, V. J. (2017). Mesoporous silica nanoparticles: A promising multifunctional drug delivery system. In *Nanostructures for cancer therapy* (pp. 593–621). Elsevier., DOI: 10.1016/B978-0-323-46144-3.00023-4

Phiri, R., Rangappa, S. M., Siengchin, S., Oladijo, O. P., & Ozbakkaloglu, T. (2024). Advances in lightweight composite structures and manufacturing technologies: A comprehensive review. *Heliyon, 10*(21), e39661. DOI: 10.1016/j.heliyon.2024.e39661

Porwal, H., Soumya, G., Bhavana, P. H. S., Chandrashekhar, S., & Anand, A. V. (2023). Mechanical properties of green composites with hemp/sisal fiber and novolac resin. AIP Conference Proceedings, 2492, 040009. DOI: 10.1063/5.0113265

Pranathi, K., Ayub, S. M. A., & Paul, C. H. (2024). Composite Laminates with Epoxy Resin and E-Glass Fibers: Analysis of Mechanical Properties. *Journal of Physics: Conference Series, 2837*(1), 012008. DOI: 10.1088/1742-6596/2837/1/012008

Rana, S. S., Kumar, R., & Gupta, M. K. (2023). Investigation of mechanical and thermal behavior of sisal fibre reinforced polymer biocomposites. *Materials Today: Proceedings*. Advance online publication. DOI: 10.1016/j.matpr.2023.06.167

Rashid, A. B., Rayhan, A. M., Shaily, S. I., & Islam, S. M. (2024). An experimental study of physical, mechanical, and thermal properties of Rattan fiber reinforced hybrid epoxy resin laminated composite. *Results in Engineering*, *22*, 102053. DOI: 10.1016/j.rineng.2024.102053

Saba, N., Paridah, M. T., & Jawaid, M. (2015). Mechanical properties of kenaf fibre reinforced polymer composite: A review. *Construction and Building Materials, 76*, 87-96. DOI: 10.1016/j.conbuildmat.2014.11.043

Saha, M., Singh, H., Singh, M. K., Rangappa, S. M., & Siengchin, S. (2025). Advancements in natural fiber composites: market insights, surface modifications, advanced fabrication techniques and applications. *Sustainable Chemistry for Climate Action, 6*, 100081. DOI: 10.1016/j.scca.2025.100081

Salem, I. A. S., Rozyanty, A. R., Betar, B. O., Adam, T., Mohammed, M., & Mohammed, A. M. (2017). Study of the effect of surface treatment of kenaf fibre on mechanical properties of kenaf filled unsaturated polyester composite. *Journal of Physics: Conference Series, 908*, 012002. DOI: 10.1088/1742-6596/908/1/012002

Selvan, E. V., Ponshanmugakumar, A., Ramanan, N., & Naveen, E. (2021). Determination and investigation of mechanical behaviour on kenaf-sisal hybrid composite. *Materials Today: Proceedings, 46*, 3358-3362. DOI: 10.1016/j.matpr.2020.11.478

Senthilkumar, N., Deepanraj, B., Shaik, F., & Elsehly, E. M. (2024). Artificial Intelligence Based Investigation on Drilling of MWCNT-Seashell-Nylon66 Hybrid Polymer Composites by Fuzzy Inference System. In *2024 3rd International Conference for Innovation in Technology,* pp. 1-6. DOI: 10.1109/INOCON60754.2024.10511569

Skalková, P., Krmelová, V., Krmela, J., Ondrušová, D., Crkoň, A., & Benčíková, E. (2021). Composite materials with epoxy matrix and their properties. *IOP Conference Series. Materials Science and Engineering, 1199*(1), 012030. DOI: 10.1088/1757-899X/1199/1/012030

Solahuddin, B. A. (2022). A critical review on experimental investigation and finite element analysis on structural performance of kenaf fibre reinforced concrete. *Structures,* 35, 1030-1061). DOI: 10.1016/j.istruc.2021.11.056

Tang, H., Rogov, A. B., Soutis, C., & Yerokhin, A. (2023). Fabrication, interfacial and flexural properties of a polymer composite reinforced by γ-Al2O3/Al fibres. *Composites Part A: Applied Science and Manufacturing, 169*, 107502. DOI: 10.1016/j.compositesa.2023.107502

Thandavamoorthy, R., Alagarasan, J. K., Mohanavel, V., et. Al., (2024). Fabrication of kenaf fiber reinforced boron carbide fillers embedded epoxy matrix composite–An antimicrobial and structural analysis. *Journal of Materials Research and Technology, 33*, 2560-2567. DOI: 10.1016/j.jmrt.2024.09.236

Thirumurugan, A., Bhaskar, G. B., Poyyathappan, K., Karthik, R. J., Kumar, M. K., Somasundaram, G., & Venkatakrishnan, G. (2016). Investigations on aluminium wire mesh, banana fiber and glass fiber reinforced hybrid composites. *Indian J. Sci. Technol, 9*(42), 1-7. DOI: 10.17485/ijst/2016/v9i42/103306

Velmurugan, V., Kumar, D. D., & Thanikaikarasan, S. (2020). Experimental evaluation of mechanical properties of natural fibre reinforced polymer composites. *Materials Today: Proceedings, 33*, 3383-3388. DOI: 10.1016/j.matpr.2020.05.190

Villalón Fornés, I., Vaičiukynienė, D., Nizevičienė, D., Tamošaitis, G., & Pupeikis, D. (2022). The improvement of the thermal and acoustic insulation properties of phosphogypsum specimens by adding waste wood fibre. *Construction & Building Materials, 331*, 127341. DOI: 10.1016/j.conbuildmat.2022.127341

Yorseng, K., Rangappa, S. M., Pulikkalparambil, H., Siengchin, S., & Parameswaranpillai, J. (2020). Accelerated weathering studies of kenaf/sisal fiber fabric reinforced fully biobased hybrid bioepoxy composites for semi-structural applications: Morphology, thermo-mechanical, water absorption behavior and surface hydrophobicity. *Construction and Building Materials, 235*, 117464. DOI: 10.1016/j.conbuildmat.2019.117464

Zampaloni, M., Pourboghrat, F., Yankovich, S. A., Rodgers, B. N., Moore, J., Drzal, L. T., Mohanty, A.K., & Misra, M. J. (2007). Kenaf natural fiber reinforced polypropylene composites: A discussion on manufacturing problems and solutions. *Composites Part A: Applied Science and Manufacturing, 38*(6), 1569-1580. DOI: 10.1016/j.compositesa.2007.01.001

Zhu, D., Mobasher, B., & Rajan, S. D. (2011). Dynamic tensile testing of Kevlar 49 fabrics. *Journal of Materials in Civil Engineering, 23*(3). DOI: 10.1061/(ASCE)MT.1943-5533.0000156

Zuccarello, B., Bongiorno, F., & Militello, C. (2022). Basalt fiber hybridization effects on high-performance sisal-reinforced biocomposites. *Polymers, 14*(7), 1457. DOI: 10.3390/polym14071457

Chapter 13
Materials for Thin-Film Photovoltaics: Structure, Performance, Deposition Techniques, and Engineering Applications

Atef Raafat El Khatib
https://orcid.org/0000-0002-5357-7845
Lebanese International University, Lebanon

Hameed AlRashidi
Kuwait Fund for Arab Economic Development, Kuwait

ABSTRACT

The advancement of sustainable energy systems requires the development of high-performance photovoltaic materials that meet the demands of efficiency and scalability. Thin-film photovoltaic technologies have emerged as a promising solution. This chapter provides a review of second generation thin-film PV technologies, including amorphous silicon (a-Si), cadmium telluride (CdTe), copper indium selenide (CIS) and copper indium gallium selenide (CIGS), as well as emerging technologies including perovskites, dye-sensitized solar cells (DSSCs), and organic solar cells (OSCs). The chapter addresses material properties, device architectures, deposition processes, and the current performance trends. It also discusses challenges such as material scarcity, toxicity, and long-term stability. Case studies are provided on building-integrated photovoltaic (BIPV) systems for practical applications and performance outcomes. The chapter offers technical insights for researchers, engineers, and industry professionals engaged in developing or implementing thin-film photovoltaic technologies.

DOI: 10.4018/979-8-3373-9454-1.ch013

1. INTRODUCTION

The accelerating pace of climate change presents one of the most critical challenges of the 21st century. According to the Intergovernmental Panel on Climate Change (IPCC), the average global surface temperature in 2020 had already risen by 1.1 °C relative to pre-industrial levels. In the absence of effective mitigation strategies, projections indicate that global temperatures may reach or exceed 1.5 °C in the near term.

In this context, the transition to low-carbon and renewable energy sources has become a crucial need. Among the various renewable energy technologies, solar photovoltaics (PV) have emerged as a leading candidate due to their modularity and scalability applicability. PV systems are fundamentally materials-based technology, with energy conversion relying on the properties of semiconductors.

Over the past decades, significant research and industrial effort has been devoted to enhancing PV technologies in terms of efficiency, material usage, manufacturability, and integration. This progress has historically centered on crystalline silicon (c-Si) technologies, which continue to dominate commercial markets (Fazal & Rubaiee, 2023). However, new material systems and device architectures have been explored to address limitations in weight, flexibility, aesthetics, and manufacturing constraints. Thin-film photovoltaic technologies, have opened new pathways for PV deployment, particularly in building-integrated applications and low-weight environments.

This chapter addresses thin-film PV technologies, focusing on amorphous silicon (a-Si), cadmium telluride (CdTe), copper indium selenide and copper indium gallium selenide (CIS / CIGS), as well as emerging PV technologies including perovskites, organic solar cells (OSCs), and Dye-sensitized solar cells (DSSCs). Each technology is discussed in terms of properties, device structure, and deposition techniques. The chapter also presents real-world case studies for thin-film PV use in building-integrated photovoltaics (BIPV) applications.

By focusing on thin-film photovoltaic materials, this chapter addresses the scientific basis of second generation energy systems as well as the engineering challenges and solutions associated with their deployment. These include material scarcity, environmental toxicity, deposition scalability, stability under real-world conditions. As such, the chapter aligns closely with the broader theme of engineering solutions for modern challenges, offering both theoretical insight and applied knowledge relevant to materials scientists and engineers.

The objective of this chapter is to provide a comprehensive technical overview of thin-film photovoltaic technologies. It addresses both established technologies and emerging ones in terms of design, performance, and deposition. The scalability and suitability of these technologies in some applications is also discussed throughout the text. Ultimately, the chapter provides the reader with the necessary understanding of the current context of thin-film PV technologies.

2. OVERVIEW OF PHOTOVOLTAIC TECHNOLOGIES

2.1 Classification of PV Technologies

Photovoltaic technologies are commonly categorized based on their material composition and manufacturing approach into three broad generations: crystalline silicon (c-Si), thin-film (TFPV), and emerging or third-generation technologies.

Crystalline silicon (c-Si), comprising monocrystalline and polycrystalline silicon, is the most established and commercially dominant PV technology, accounting for the majority of global installations. These cells are manufactured from silicon wafers, typically 150–200 microns thick (B. Gupta et al., 2024), and offer high efficiencies with well-understood performance characteristics.

Thin-film technologies represent the second generation of PV and are characterized by the deposition of semiconductor materials in thin layers, usually on glass, metal, or plastic substrates. This class includes various technologies based on the materials used, yet Cadmium Telluride (CdTe), Copper Indium Gallium Selenide (CIGS/CIS), and Amorphous Silicon (a-Si) remain the more mature (IRENA, 2012). Gallium Arsenide (GaAs) Photovoltaic solar cells are based on thin-film deposition techniques, but they are generally not grouped with conventional thin-film PV technologies due to their unique material class, fabrication methods, and specialized applications. Thin-film materials allow for lower-cost manufacturing and lighter, flexible modules, due to their ability to efficiently absorb light and transform it into electric energy through a thin-film layer. However, this is typically at the expense of lower efficiencies compared to c-Si.

A third category, often labeled emerging or third-generation PV technologies, encompasses materials and concepts still undergoing rapid research and early-stage commercialization. These include perovskite solar cells, dye-sensitized solar cells (DSSC), organic photovoltaics (OPV) among others.

Many of these emerging, third-generation, technologies are structurally similar to thin-film PV, with active layers typically ranging from nanometers to a few microns thick. Therefore, from a structural perspective, these emerging PV technologies can be considered part of the broader thin-film category.

Figure 1 shows the categorization of PV Technologies that will be addressed in this chapter.

Figure 1. Categorization of PV technologies

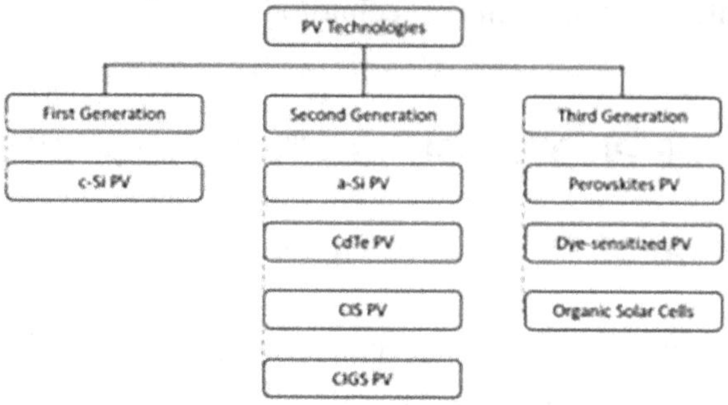

2.2. Working Principle and Key Properties

Photovoltaic (PV) cells convert sunlight into electricity by using the properties of semiconductors. When sunlight strikes the surface of a solar cell, photons from the light transfer their energy to electrons within the material. These electrons, previously bound in a lower-energy state, gain enough energy to jump into a higher-energy state (the conduction band), where they can move freely. The movement of these charge carriers, which are electrons and the corresponding "holes" they leave behind, creates an electric current when the cell is connected to an external circuit.

The minimum energy needed for this jump between energy states is known as the "bandgap" of the material. Materials with a lower bandgap, for example 1.1 eV for crystalline silicon, can absorb more of the sunlight spectrum, including low-energy photons like red and infrared. In contrast, materials with a higher bandgap, example 1.7 to 1.8 eV for amorphous silicon, are limited to absorbing higher-energy (shorter-wavelength) photons, which reduces their total solar energy harvesting potential.

Equally important is whether the bandgap is "direct" or "indirect". In direct bandgap materials, an electron can absorb a photon and directly transition across the bandgap without needing a change in momentum. This results in efficient light absorption, even in very thin layers. Indirect bandgap materials, like crystalline silicon, require both a photon and a phonon, which is a unit of lattice vibration, to facilitate this transition, making the process less probable and necessitating much thicker material layers to absorb sufficient light. This is why thin-film PV technologies typically use direct bandgap semiconductors. They offer high absorption efficiency in very thin layers, enabling lightweight or flexible designs.

It is important to note that bandgap value and whether it is direct or indirect describe the material's fundamental light absorption behavior, but do not solely determine overall efficiency. The real-world performance of a solar technology is influenced by a combination of factors, including the level of development in manufacturing processes, the degree of internal defects, long-term stability, and the cell architecture.

In addition to bandgap, another crucial property is the "absorption coefficient", that is a measure of how strongly a material absorbs light at a given wavelength. Even if a photon has enough energy to be absorbed, how quickly and effectively it gets absorbed depends on the absorption coefficient.

A high absorption coefficient means that photons are absorbed near the surface, allowing the active layer to be thin. In contrast, indirect bandgap materials like crystalline silicon have lower absorption coefficients near their bandgap, meaning that light penetrates deeper before being absorbed, and thus require much thicker layers to collect sufficient energy.

While solar panels are powered by light, they don't perform better with heat. As the temperature of a PV cell rises, its efficiency typically decreases. This behavior is captured by a metric known as the temperature coefficient, usually expressed as a percentage of power loss per degree Celsius above the standard test condition (25°C).

Table 1 summarizes the basic properties of crystalline silicon (c-Si) PV and thin-film PV. As can be noticed in the table, thin-film PV have direct bandgap and high absorption coefficients which allow them to generate electricity using a "thin film" of semi-conductor materials. Additionally, they often have lower temperature coefficients than c-Si PV, which makes them suitable for industrial deployment despite their lower efficiencies (Al Miaari & Ali, 2025)(Shah et al., 1999)(Taesoo & Abasifreke, 2016)(Bouanani et al., 2020). Figure 2 shows the absorption coefficient and bandgap values for visual comparison.

Table 1. Basic properties of c-Si and thin-film PV technologies

Material	Bandgap (eV)	Absorption Coefficient (cm^{-1})	Temperature Coefficient (%/°C)	Bandgap Type	Reference
c-Si	1.12	10^4	–0.45	Indirect bandgap	(Shah et al., 1999) (Al Miaari & Ali, 2025)
a-Si (Thin-Film)	1.75	1.7	–0.30	Direct-like behavior	(Al Miaari & Ali, 2025) (Taesoo & Abasifreke, 2016)
CIGS (Thin-Film)	1.1–1.2 (tunable)	$>10^5$	–0.26	Direct	(Taesoo & Abasifreke, 2016) (Bouanani et al., 2020)
CdTe (Thin-Film	1.46	1.1148×10^6	–0.25	Direct	(Al Miaari & Ali, 2025) (Taesoo & Abasifreke, 2016)

Figure 2. Basic bandgap and absorption coefficient of c-Si and thin-film PV technologies (Al Miaari & Ali, 2025)(Shah et al., 1999)(Taesoo & Abasifreke, 2016) (Bouanani et al., 2020)

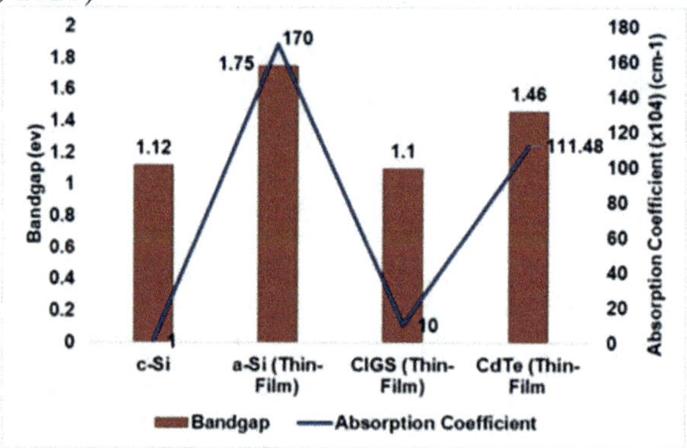

For emerging PV technologies, that are perovskites, SDDCs, and organic PV, published data exists, but values cannot be reflected as unified since they vary based on material composition, film thickness and morphology.

It is also noteworthy in the table that amorphous silicon, despite being an indirect bandgap material, it is often described as "direct-like" bandgap type. This is because in a-Si, the atomic structure is disordered, which causes the energy levels to overlap. This allows electrons to absorb photons and jump to the conduction band without the need of a phonon. However, it does not completely follow the same behavior as in direct bandgap materials

2.3. Why Thin-Film PV

The choice of photovoltaic material fundamentally influences how efficiently a solar cell performs, particularly in terms of light absorption, electrical conversion, and temperature behavior. As discussed earlier, properties such as bandgap type (direct vs. indirect), absorption coefficient, and temperature coefficient are central to this performance, besides other factors such as the carrier mobility, defect density and others. Thin-film photovoltaic technologies differ significantly from crystalline silicon (c-Si) in how they leverage these properties, making them more suited to certain applications that require flexible and light PV modules.

Crystalline silicon (c-Si), must be several hundred microns thick to absorb enough sunlight. This not only increases material usage but also limits design flexibility. In contrast, most thin-film materials are direct bandgap semiconductors with high absorption coefficients. They can absorb sunlight efficiently in layers as thin as 1–3 microns, enabling lightweight and flexible modules.

However, thin-film efficiencies, while improving, generally lag behind that of c-Si. Additionally, some technologies, for example CdTe and CIGS, are dependent on materials that are scarce or raise toxicity concerns.

The selection between thin-film PV or crystalline silicon is not a matter of superiority, but rather application-based. In applications where flexibility, weight, transparency, or aesthetic integration are prioritized, such as building-integrated photovoltaics (BIPV) (Al Miaari et al., 2023), thin-film technologies are advantageous. While, crystalline silicon remains a better choice in applications where high efficiency is required such as utility scale.

Table 2 represents a comprehensive comparison between c-Si PV and thin-film PV including emerging technologies.

Table 2. Basic comparison between c-Si PV and thin-film PV

Parameter	Crystalline Silicon (c-Si)	Thin-Film PV
Material Composition	Monocrystalline or polycrystalline silicon	Various materials, a-Si, CdTe, CIGS, perovskites, etc.
Cell Structure	Discrete wafer-based cells	Monolithically integrated or layered on substrates
Power Efficiency	Higher	Lower
Temperature Coefficient	Higher	Lower
Form Factor	Rigid, opaque, usually glass-laminated	Flexible or rigid, lightweight, semi-transparent options available
Transparency Options	Very limited, requires spacing between cells	Available, tunable transparency
Maturity of manufacturing processes	Mature, standardized, and cost-optimized	Less mature, some methods are still evolving
Aesthetic Integration	Limited	Adaptable architectural integration
Material Abundance	Silicon is abundant	Some materials such as Cd, In, Te, and Ga may pose availability or toxicity concerns
Main Use	High efficiency power generation with limited concern towards aesthetic appearance	Portable applications, BIPV or any application that requires light weight and flexibility

2.4. Market Share

Although the literature does not provide a clear market share analysis, various sources agree that c-Si PV highly dominate the market. Reported c-Si PV's share include 92% in 2014 (Weckend et al., 2016),, and 90% in 2023 (Fazal & Rubaiee, 2023). This can be referred to their proven efficiency and durability. In addition, c-Si PVs are highly available in the market which makes them affordable. Thin-film PV almost had the remaining shares in each of the above mentioned percentages. CdTe and CIGS record the highest share among thin-film PV technologies accounting for around 5% and 2% respectively in 2014 (Weckend et al., 2016). Although, it was forecasted that in 2020, the silicon-based PV market share will drop to 73.3% and the CdTe and CIGS shares will reach 5.2% each (Weckend et al., 2016), c-Si are still more dominant the market as reported earlier.

For emerging PV technologies, their current market share is almost negligible. However, research and development efforts are intensive on emerging PV technologies, especially perovskites as they have demonstrated rapidly improving conversion efficiencies. Around 80 companies around the globe are working on commercializing their manufacturing (IEA, 2024). Other emerging technologies remain suitable for niche applications only.

3. MATERIALS FOR THIN-FILM PV

Thin-film PV technologies can be realized through a wide range of material systems, deposition methods, and device structures. While these structures may differ, all photovoltaic devices share a few essential components.

At the core of any PV device are a light-absorbing layer to generate charge carriers, a charge separation mechanism, electrical contacts to extract the current, and a substrate or support layer that provides structural integrity. The choice of architecture reflects the electrical, optical, and chemical nature of the materials, as well as the intended manufacturing approach.

Scalability remains a defining challenge across all technologies. Some materials offer excellent efficiency at the laboratory scale but are difficult to mass produce due to instability, scarcity of raw elements, or complex fabrication requirements. Others trade high performance for simplicity and manufacturability.

The section below provides reviews the thin-film PV technologies starting with an overview, architecture, performance, and limitations.

3.1. Amorphous Silicon (a-Si) PV

Overview

In amorphous silicon (a-Si) PV, the lack of crystallinity in atoms results in a direct-like bandgap behavior and higher absorption coefficient which enables sunlight absorption within thin layers of around 1 μm (T. D. Lee & Ebong, 2017). In addition, a-Si PVs are valued for being of low-cost, and compatible with flexible substrates

To enhance its electronic properties, a-Si is commonly used in hydrogenated form (a-Si:H) (T. D. Lee & Ebong, 2017). Through this process, hydrogen atoms passivate dangling silicon bonds, reducing defect density and improving carrier transport.

Architecture

The standard a-Si:H solar cell is deposited in a p-i-n junction structure, which consists of a thin p-type layer, a thicker intrinsic a-Si:H absorber layer, and a thin n-type layer. The internal electric field formed across the p-i and i-n interfaces drives the separation and collection of photo-generated carriers.

In substrate configurations, this structure is deposited on a back contact (e.g., aluminum or silver) on metal, glass, or plastic substrates, and capped with a transparent conductive oxide (TCO) such as indium tin oxide (ITO) to allow light in and conduct electricity. Superstrate configurations reverse this order, often depositing on glass coated with TCO.

Typical layer thicknesses are 250–500 nm for the i-layer and around 20–30 nm for the doped layers (Carlson & Wronski, 1977). Figure 3 shows a generalized typical architecture of a-Si PV in p-i-n and n-i-p configuration.

Figure 3. Typical structure of a-Si PV

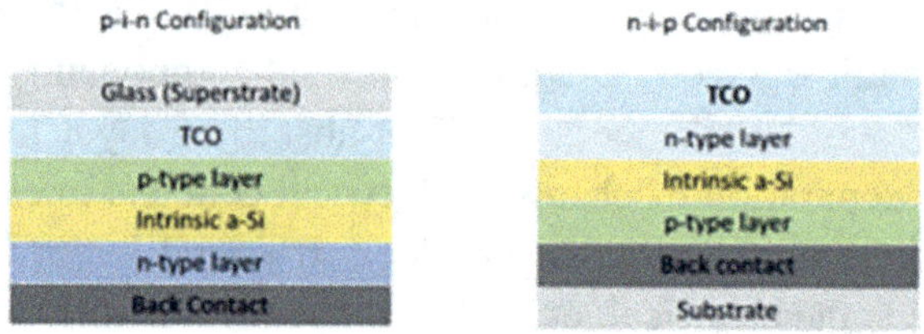

Performance

Amorphous silicon cells have historically lagged behind other PV technologies in terms of conversion efficiency. The initial device by Carlson and Wronski demonstrated 2.4% efficiency (Carlson & Wronski, 1976). A major milestone was the introduction of hydrogenated a-Si:H and light trapping designs by the mid-1980s, which further raised laboratory efficiencies (Yablonovitch & Cody, 1982).

The development of multi-junction architectures, particularly tandem and triple-junction cells, has been crucial to boosting stabilized efficiencies. (Guha et al., 1992) achieved around 11% efficiency using multi-layer structures, through further enhancement, the a-Si PV efficiency reached (*Best Research-Cell Efficiency Chart | Photovoltaic Research | NREL*, 2025).

Despite these improvements, performance remains constrained by material limitations such as lower carrier mobility, shorter diffusion lengths, and the wide bandgap, which limits infrared absorption. The International Renewable Energy Agency (Weckend et al., 2016) reported numbers from previous publications that the share of a-Si thin-film PV will phase-out before the year 2030 as more competitive thin-film technologies are becoming available.

Limitations

Amorphous silicon exhibits several intrinsic and extrinsic limitations. The most widely known is the Staebler-Wronski Effect (SWE), where prolonged light exposure introduces metastable defects in the a-Si:H layer, causing performance degradation that might reduce power output by 10–30% within the first 1000 hours (Fritzsche, 2001).

Carrier mobility in a-Si is substantially lower than in crystalline materials due to the absence of long-range atomic order. The disordered structure also introduces a high density of trap states, which reduce open-circuit voltage and fill factor (Juška, 1991)

3.2. Cadmium Telluride (CdTe) PV

Overview

Cadmium Telluride (CdTe) is the most scalable thin-film PV thanks to its high absorption coefficient, almost optimum bandgap, and cost-effective manufacturing. It can efficiently absorb incident sunlight within a thin layer of around 1 μm (Shah et al., 1999).

CdTe PV are widely studied for efficiency enhancement, such as through selenium alloying to enhance absorption in the near-infrared region (Sivasankar et al., 2025) and for applicability in various conditions (Alrashidi et al., 2022b) (Alrashidi et al., 2022a).

Architecture

CdTe solar cells are usually deposited in a superstrate configuration. Light enters through a transparent substrate which is usually glass. The substrate is coated with a transparent conductive oxide (TCO) to enhance electric conductivity while maintaining transparency.

The next layer, n-layer, is cadmium sulfide (CdS) or the oxygenated cadmium sulfide (CdS:O). This layer allows the formation of a heterojunction with the p-type CdTe absorber beneath (Sharma et al., 2022). The absorbing, p-type, is the CdTe or CdSeTe, after selenization.

Then, the back contact layer is made of either copper-doped graphite, or molybdenum, or gold. Encapsulation layer is then added for protection against moisture and ultraviolet degradation (Basol & McCandless, 2014).

The typical generalized architecture of CdTe PV cells is shown in Figure 4.

Figure 4. typical structure of CdTe PV

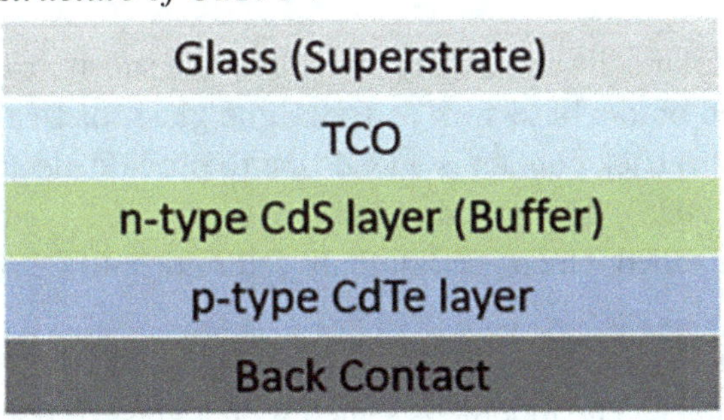

Performance

CdTe solar cells have experienced steady progress in efficiency since the first devices were demonstrated in the 1970s with values near 6% (Romeo et al., 2004). Subsequent improvements in deposition methods, interface quality, and doping strategies have elevated laboratory efficiencies above 20% (Kujovic et al., 2025).

Efforts to optimize the transparent front contact have also yielded performance enhancements. Replacing conventional SnO_2 with Cd_2SnO_4, and incorporating high-bandgap buffer layers such as Zn_2SnO_4 or modified CdS:O, has helped increase fill factors and overall device efficiency(Taesoo & Abasifreke, 2016).

Limitations

Despite its advantages, CdTe technology faces several material and process-related challenges. The most significant environmental concern is the use of cadmium which is a toxic heavy metal that requires strict safety measures during manufacturing, deployment, and end-of-life recycling.

In addition, the global supply of tellurium is limited, as it is primarily obtained as a byproduct of copper refining. The scarcity of tellurium hinders the scalability of CdTe PV technology if global demand continues to grow rapidly.

3.3. Copper Indium Selenide (CIS) and Copper Indium Gallium Selenide (CIGS) PV

Overview

Copper Indium Selenide (CIS) and Copper Indium Gallium Selenide (CIGS) are compound semiconductors used in thin-film photovoltaic (PV) technologies. Thanks to their direct bandgap and high absorption coefficients, they absorb light within 1–2 μm thick layers (Taesoo & Abasifreke, 2016).

Gallium is added to the CIS PV to become CIGS, which is the most common. This enhancement was made to enhance the bandgap and allowing tuning by varying the gallium-indium ratio. This resulted in more suitability for real-world applications.

Architecture

CIS and CIGS solar cells are usually fabricated in a substrate configuration. All layers are deposited onto a base such as soda-lime glass, or flexible polymer foil. A molybdenum (Mo) back contact is added first to provide electrical conductivity (Sivasankar et al., 2025). The absorber layer (CIS or CIGS) is then deposited. A buffer layer is then added which is commonly cadmium sulfide (CdS) (Sivasankar et al., 2025).

Transparent conductive oxide (TCO) is then deposited over the buffer layer to act as the front contact,. The module is sealed with a glass or polymer encapsulant for protection from moisture and other environmental factors that might induce (Kazmerski et al., 1976). Typical structure of CIGS PV is shown in Figure 5.

Figure 5. Typical structure of CIGS PV

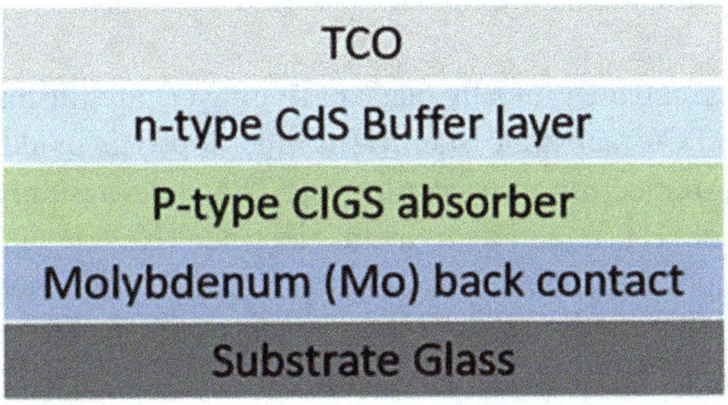

Performance

CIGS solar cells currently represent one of the most efficient thin-film PV technologies as laboratory-scale CIGS devices have surpassed 24% efficiency (Kumar & Giripunje, 2025).

Significant advancements have also focused on improving the cell's internal structure for enhanced environmental sustainability, this includes the replacement of cadmium sulfide (CdS) buffer layer with zinc-based materials like Zn(O,S). This change enhanced overall efficiency (N'guessan et al., 2023). Advancements in manufacturing and deposition techniques also supports more precise adjustment of material composition, leading to better performance.

Limitations

One limitation is that the manufacturing of CIGS requires careful control on the gallium-indium ratio. Also, the dependence on cadmium in the CdS buffer layer raises toxicity concerns, similar to CdTe cells (Genchi et al., 2020). Research is still ongoing for more environmental friendly buffer layers. Moreover, both indium and gallium are considered rare which affects the CIGS scalability and cost reduction.

3.4 Emerging PV Technologies

3.4.1 Perovskites

Overview

Perovskite solar cells (PSCs) are one of the most promising types of solar cells. Their technology has witnessed an unprecedented improvement in power conversion efficiency and cost-effectiveness.

Perovskites are made of materials with the general formula ABX_3. Where; A is a monovalent cation, B is a divalent metal (commonly lead is used), and X is a halide anion. These materials when crystallized demonstrated suitable properties for the

PV technologies such as direct bandgap and high absorption coefficient. They also benefit from their tunable nature to exhibit variable properties.

Architecture

Perovskite solar cell are typically made of five-layers architecture. A transparent conductive oxide (TCO) substrate, such as fluorine-doped tin oxide or indium-doped tin oxide. Then, An electron transport layer (ETL), a perovskite absorber layer, a hole transport layer (HTL), and a top metal electrode (Kegelmann et al., 2019).

They can be deposited either in an n–i–p structure or the inverted p–i–n structure. It is worth noting that these terms reflect the sequence of functional layers rather than doped junctions, but they are used for accessible description of the device. In both designs, the perovskite absorber is embedded between the HTL and the ETL. The configurations are represented in Figure 6.

Figure 6. Typical structure of Perovskite PV

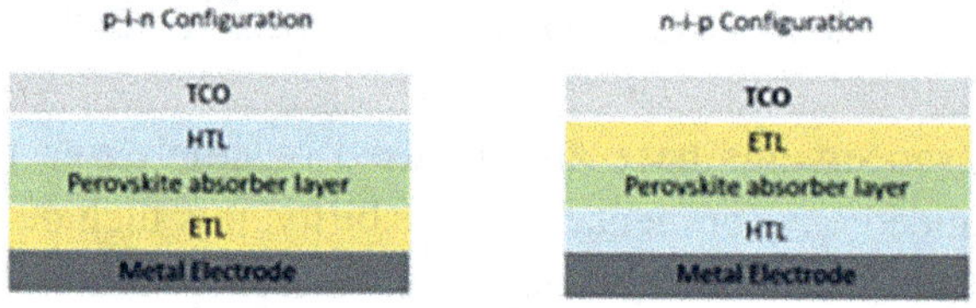

Performance

Perovskite solar cells have demonstrated a rapid increase in power conversion efficiency. From initial values around 3.8% in 2009, certified laboratory-scale efficiencies have exceeded 26% (Basumatary & Agarwal, 2022)(Jeong et al., 2021). This is thanks to the intensive research and development efforts in various aspects that material composition, precise crystallization control.

The ability to tune the bandgap through compositional engineering has also enabled their integration into highly efficient tandem devices, where wide-bandgap perovskites are paired with lower-bandgap absorbers like silicon (Ahn & Choi, 2024). Such tandem architectures have now achieved power conversion efficiencies above 33% (Min et al., 2021) (C. Yang et al., 2024).

Limitations

Perovskite materials are highly susceptible to degradation under in real-world environments as they are affected by moisture, heat, oxygen, and UV radiation (Miah et al., 2025). Encapsulation technologies can mitigate some degradation, but are still not sufficient (Afre & Pugliese, 2024). Figure 7 shows the recent advancements in efficiency and stability of Perovskites solar cells (Khatoon et al., 2023).

Scalability also presents notable challenges as fabrication methods are not still suitable for industrial-scale production (C. Yang et al., 2024) (Afre & Pugliese,

2024). In addition, environmental and health risks are associated with the use of lead which may cause physical damage or contamination risks.

Figure 7. Efficiency and stability of Perovskites PV cells (Khatoon et al., 2023)

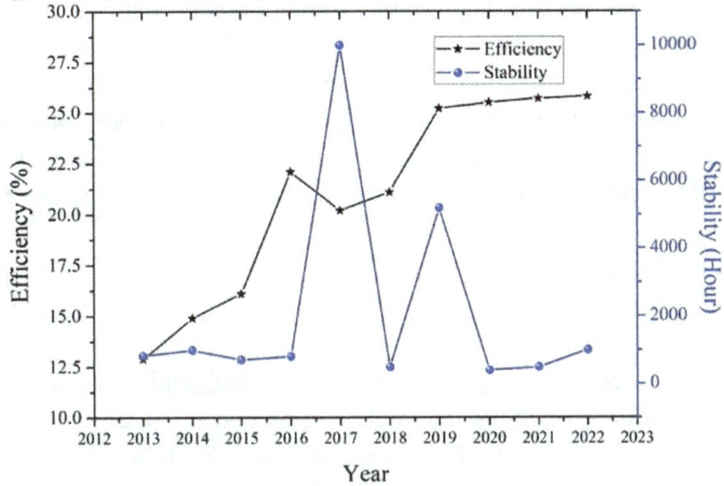

3.4.2 Dye-Sensitized Solar Cells (DSSC)

Overview

DSSCs use a photosensitizing dye to capture sunlight. The dye then inject electrons into a sponge-like, mesoporous wide-bandgap semiconductor which is commonly titanium dioxide (TiO_2). The electrons then travel towards a transparent conductive electrode where electricity is generated.

They offers several advantages including cost effectiveness, low-temperature deposition process, and applicability in diffuse and indoor lighting. DSSCs are also available in multiple colors which make them a good fit for applications that require aesthetic appearance (Hoseinnezhad & Gharanjig, 2018). They also do not involve any toxic materials (Hardin et al., 2012).

Architecture

DSSCs are typically made of a glass or plastic substrate coated with a TCO. On this substrate, a mesoporous film is deposited by soaking in a dye solution. The dye serves as the light-absorbing component. Electrons are injected into the network of of TiO_2. The photoanode is connected with a redox electrolyte contained between the photoanode and the counter electrode (Francis et al., 2021).

Performance

The power conversion efficiency of DSSCs reaches up to 13% only (*Best Research-Cell Efficiency Chart | Photovoltaic Research | NREL*, 2025). This value remains modest compared to other technologies but are competitive under low-light (early mornings and evenings) or diffuse lighting conditions, where DSSCs often outperform many conventional PV technologies (Francis et al., 2021).

Limitations

One major concern is the use of liquid electrolytes, which can evaporate or leak over time,. Encapsulation technologies and solid-state electrolytes are being investigated to overcome these concerns but often result in reduced efficiencies (Capitão et al., 2023). Another limitation lies in the long-term stability of the dye molecules.

3.4.3 Organic Solar Cells

Overview

Organic solar cells (OSCs) use carbon-based semiconducting materials, such as organic polymers or small organic molecules to convert sunlight into electricity. OSCs. Their key advantages lie in their mechanical flexibility, low material cost, lightweight nature, and compatibility with advanced processing techniques (IRENA, 2012).

These advantages make them highly promising for next-generation applications demanding adaptability and cost-efficiency, such as curved architectural elements and semi-transparent windows for building-integrated photovoltaics (BIPV).

Architecture

OSCs adopt a layered architecture that includes a transparent electrode, followed by a thin hole transport layer (HTL), then the active layer, and an electron transport layer (ETL) topped by a metal electrode. The active layer is an a blend of electron donor and electron acceptor materials that form a bulk heterojunction. (Kesinro et al., 2021). Typical architecture is shown in Figure 8.

Figure 8. Typical structure of organic solar cells

Performance

OSCs have relatively low power conversion efficiencies, yet recent material innovations have led to remarkable progress. The introduction of non-fullerene acceptors and optimized polymer donors has increased lab-scale efficiencies to around 21% (Ali et al., 2025). It is worth noting that, transparent OSCs have achieved average visible transmittances of 80%, which is close to full transparency, while maintaining a power conversion efficiencies of 10% (Ali et al., 2025).

Limitations

OSCs face significant challenges hindering their commercialization. The most critical is their stability and susceptibility to degradation from oxygen, moisture, heat, and UV radiation, which rapidly reduces their operational lifespan (Kesinro et al., 2021)(Al-Ali et al., 2025).

4. THIN-FILM DEPOSITION TECHNIQUES

The choice of deposition technique plays a central role in the development and commercialization of thin-film photovoltaic (PV) technologies. While each material has its own chemical and physical requirements, the selection of a deposition method is rarely based on material compatibility alone. Factors such as substrate constraints, processing temperature, scalability, roll-to-roll production potential, and cost also influence this decision.

Deposition refers to the process of forming thin layers of active or supporting materials on a substrate. These layers can be semiconducting, conducting, or insulating, and they must be applied with a high degree of control to ensure functionality and stability. Deposition techniques are commonly categorized in several ways, including vacuum-based vs. solution-based, or physical vs. chemical, or by the phase of the precursor (vapor-phase, solution-phase). Each categorization helps highlight a different aspect of how the material interacts with the substrate and the processing environment. This section categorizes deposition techniques as Vacuum-based, Solution-based, and Electrochemical.

The section reviews the most in-use deposition methods and clarifies their compatibility with specific thin-film PV technology Figure 9 shows the list of deposition methods that will be addressed. It is worth noting that there are other deposition techniques, yet only the ones relevant to the previously addressed PV technologies will be tackled.

Figure 9. Categorization of thin-film deposition techniques

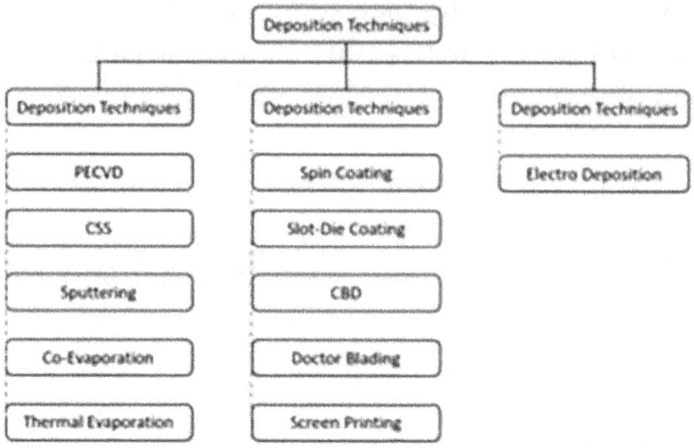

4.1. Vacuum-Based Deposition Techniques

Vacuum-based deposition techniques are a family of processes in which thin-film materials are deposited onto a substrate within a vacuum or low-pressure environment. The use of vacuum enables precise control over the atmosphere during deposition, reduces contamination from ambient gases like oxygen and moisture, and allows the use of high-purity source materials.

These techniques are suitable for deposition on large areas and complex device architectures. However, they are capital-intensive and energy-demanding. Additionally, despite some vacuum-based methods are convenient for low-temperature substrates, others require high process temperatures that limits their use on flexible or polymer-based surfaces.

4.1.1 Plasma-Enhanced Chemical Vapor Deposition (PECVD)

PECVD is a chemical process that involves the use of plasma to activate gaseous precursors. This enables the deposition of thin-film at relatively low substrate temperatures (around 250°C) (Crose et al., 2015)(C.-S. Yang et al., 2000).

The low operating temperature makes PECVD suitable for depositing hydrogenated amorphous silicon (a-Si:H) on flexible substrates such as glass and polymers (Kreiger et al., 2013). In addition, PECVD is a mature technology and industrially scalable process that can be used for large-area deposition. It also offers good control over film thickness and composition.

4.1.2 Close-Spaced Sublimation (CSS)

In CSS a solid source material and substrate are placed few millimeters apart inside a heated chamber. The heat causes the solid source to sublimate and condense uniformly as a thin-film on the substrate. This requires high temperature that are often in the range of 500 to 600 °C, and it is often used for the production of CdTe PVs (Ferekides et al., 1993).

CSS has the advantages of high deposition rates, material efficiency, scalability to large-area manufacturing, and high thin-film quality with minimal material waste. Additionaly, its setup is considered simple and of low cost. However, its high process temperature limits compatibility with flexible substrates (Doroody et al., 2021).

4.1.3 Sputtering

Sputtering is a physical deposition technique in which ions are accelerated toward a solid target material which causes atoms to be deposited as a thin film on a nearby substrate. It offers control over film thickness, composition, and uniformity.

Sputtering is a used for the deposition of buffer layers, transparent conductive oxides, and back contacts across several technologies (Kaim et al., 2021), most commonly in CIGS and CdTe (Machkih et al., 2024)(A. Gupta et al., 2006). It is also compatible with large-area which makes it suitable for scalable manufacturing lines. However, the technology is considered slow and may cause substrate damage.

4.1.4 Co-Evaporation

In co-evaporation multiple source materials are heated simultaneously in separate effusion cells to a temperature of around 400 to 600 °C (Cho et al., 2019). Their vapors then mix and condense to form a thin-film on the substrate.

Vapor rates can be tuned for control on thin-film composition, this makes it suitable for the fabrication of CIGS PV cells (Jung et al., 2010)(Sheu et al., 2016).

Co-evaporation produces dense, well-crystallized films, making it a go-to method in laboratory-scale and pilot-line production. However, the method is capital-intensive, requires careful process monitoring, and is generally slower than other deposition techniques, which limits its widespread industrial adoption compared to other alternatives (Dewi et al., 2024).

4.1.4 Thermal Evaporation

Thermal Evaporation is a technique where a solid source material is heated in a high-vacuum chamber until it vaporizes and condenses as a thin film onto a nearby, cooler substrate. The heating is typically achieved through resistive heating of a boat or filament, or by electron beam bombardment.

Operating under high vacuum prevents collisions with gas molecules, allowing the evaporated atoms to travel in a straight line to the substrate, enabling good directional control and relatively high purity films.

Thermal Evaporation is a versatile method primarily used for depositing metal electrodes (e.g., aluminum, silver, gold) that serve as front or back contacts in almost all types of thin-film solar cells, including a-Si, CdTe, CIGS, perovskites, and organic PV.

This process is characterized by a relatively simple setup, high material utilization, and compatibility with both rigid and flexible substrates, making it a common method in both research and industrial production lines (J. H. Lee et al., 2023)(Fadaam et al., 2019)(Ablekim et al., 2020).

4.2. Solution-Based Deposition Techniques

Solution-based deposition techniques involve the application of liquid-phase precursor materials, typically in the form of inks, suspensions, or chemical solutions, onto a substrate to form thin films after drying or annealing. These processes are carried out under ambient or low-pressure conditions and rely on relatively simple equipment setups, making them attractive for low-cost, scalable manufacturing. Thermal or chemical treatment is often required after the solution-based techniques to convert the precursor into a functional conducting film.

Solution-based deposition techniques are usually used for third generation PV technologies where lightweight, flexible, or semi-transparent devices are desired. However, they are limiated in achieving uniform films on large areas.

4.2.1 Spin Coating

Spin coating is a technique in which a liquid precursor is dropped onto the center of a substrate. The substrate spins at high speed allowing centrifugal force to spread the liquid into a uniform film. The film's thickness can be tuned by controlling the amount of dropped liquid offering good control over coating quality.

This technique is widely used in the fabrication of perovskite solar cells, organic solar cells, and dye-sensitized solar cells (Zheng et al., 2017). However, the technique involves material waste as some liquid is wasted during spinning, which limits its scalability for industrial manufacturing (Sahu et al., 2009).

4.2.2 Slot-Die Coating

Slot-die coating is a scalable technique in which a liquid precursor is continuously dispensed through a narrow slit onto a moving substrate. The coating head measures the flow and spreads the solution evenly, which offers uniform thin-film over large areas.

Slot-die is one of that most material-efficient techniques and is suitable for both rigid and flexible substrates, making it well-suited for roll-to-roll processing (Huang et al., 2017).

It is mainly used for perovskite and organic solar cells, where it offers a low-cost manufacturing (Patidar et al., 2020). The technique is yet to reach commercial scalability but it is currently widely used in pilot-scale production lines.

4.2.3 Chemical Bath Deposition (CBD)

CBD is a low-cost technique in which a substrate is immersed in a chemical solution. Under controlled temperature and pH conditions, the film gradually precipitates from the solution on the substrate surface. The process is relatively simple and requires at low temperatures (below 100 °C) which makes it accessible and cost-effective (Ezekoye et al., 2013). However, it does not offer control over film uniformity compared to other techniques. It also generates chemical waste that need careful management and disposal (De Guire et al., 2013).

The process is widely used to deposit buffer layers, particularly CdS, in CdTe and CIGS PVs (Munna et al., 2021)(Han et al., 2011)

4.2.4 Doctor Blading / Blade Coating

Doctor blading, also known as blade coating, is also a scalable technique in which a liquid, or paste, precursor solution or paste is spread across a substrate using flat blade. The film thickness can be controlled through the gap between the blade and the substrate, and the speed of the blade. It generates uniform films on large areas and is compatible with both rigid and flexible substrates, which makes it suitable for roll-to-roll processing (Hu et al., 2016).

It is commonly used for depositing DSSCs and is increasingly used in the scalable fabrication of absorber layers for perovskite and organic solar cells (Noorasid et al., 2022)(Wu et al., 2017).

4.2.5 Screen Printing

Screen printing is an industrially established solution-based deposition technique. It involves pushing a viscous ink, or paste, through the open areas of a fine mesh screen (stencil) onto the substrate, using a squeegee. It offers control over thin-film thickness and allows for selective deposition which makes it material-efficient (Clark, 1985). Yet, it can be considered as relatively slow.

Screen printing is a highly used for the industrial-scale fabrication of DSSCs, organic photovoltaics and perovskite solar cells (Raïssi et al., 2020)(Chen et al., 2022).

4.3. Electrochemical Deposition Techniques

Electrochemical deposition techniques uses electrical current to push dissolved precursor ions, that are dissolved in liquid electrolyte, into a substrate to form a thin-film. These methods are performed under ambient or mild conditions, which them attractive for low-cost, large-area fabrication.

The basic setup includes a working electrode, a counter electrode, and an electrolyte solution. When a electrical energy is applied, the ions are reduced and deposit onto the substrate surface. These techniques are valued for their simplicity, low energy requirements, and compatibility with large-area and flexible substrates.

Electrodeposition stands out as the most studied in the context of thin-film PV fabrication.

4.3.1 Electrodeposition

In electrodeposition, a thin film is formed on the substrate by applying an external voltage to drive the reduction of metal ions from a solution. The substrate acts as the cathode, attracting positively charged ions from the electrolyte, which are reduced and deposited as a solid layer. The process is typically carried out at room temperature and atmospheric pressure, using aqueous solutions containing the desired metal precursors.

In thin-film photovoltaics, electrodeposition is primarily used to fabricate metallic precursor layers for CIGS and, to a lesser extent, CdTe absorber materials, and also for depositing counter electrodes such as platinum in dye-sensitized solar cells (H. Lee et al., 2012). However, achieving precise stoichiometric control and uniform film quality can be challenging, and post-deposition treatment is essential for crystallinity and performance optimization.

4.4. Summary Tables

While this section has focused on the major deposition techniques relevant to the technologies addressed in the chapter, it is important to note that many other deposition methods exist. These include techniques such as electrophoretic deposition, electroless plating, spray pyrolysis, dip coating, and Langmuir-Blodgett assembly, and others.

Table 3 and Table 4 represent is a comparison among the addressed deposition categories and techniques respectively.

Table 3. Comparison of main features among Vacuum-Based, Solution-Based, and Electrochemical Deposition Techniques

Feature	Vacuum-Based Techniques	Solution-Based Techniques	Electrochemical Techniques
Typical Environment	Vacuum or low-pressure environment.	Ambient or low-pressure conditions.	Ambient or mild conditions.
Equipment & Cost	Capital-intensive, energy-demanding equipment.	Relatively simple equipment with lower cost.	Attractive for low-cost.
Control & Purity	Precise control, produces high-purity films.	Challenges in achieving uniform large-area films	Less control over film stoichiometry and uniformity.
Operating Temperature	Usually, high temperatures. Some compatible with low-temp substrates.	Low temperatures than vacuum-based.	Typically room temperature.
Advantages	Excellent control over layer quality,.	Low cost, high throughput, Roll-to-roll compatibility.	Simplicity, low energy requirements.
Limitations	High cost, energy-demanding. High temperature can restrict flexible use.	Challenges in uniformity and material stability.	Less control over film stoichiometry and uniformity. Requires Post-deposition treatment.

Table 4. Comparison among main Vacuum-based deposition techniques.

Feature	PECVD	(CSS)	Sputtering	Co-Evaporation	Thermal Evaporation
Typical Temperature Range	Relatively low temperatures (around 200–300 °C).	High temperatures (around 400–600 °C).	Room temperature or mildly heated.	High temperatures (around 400–600 °C)	Room temperature or mildly heated.
Key Advantages	Low-temperature, Precise composition control.	High deposition rates, excellent material utilization, low-cost.	High versatility Excellent adhesion, good uniformity over large areas.	Precise compositional control.	Relatively simple setup, low equipment cost.
Key Limitation	Use of hazardous precursor gases, Potential for plasma damage.	High process temperature limits compatibility with flexible substrates.	Slow, Potential for substrate damage	High cost, slow, requires high monitoring.	Radiant heat can affect substrate.
Primary PV Application(s)	Hydrogenated amorphous silicon (a-Si:H) absorber layers.	Cadmium Telluride (CdTe) absorber layers.	Commonly used for buffer layers, example in CIGS or CdTe.	Copper Indium Gallium Selenide (CIGS) absorber layers.	Metal electrodes for all thin-film PV technologies. Organic PV. Some perovskite layers.
Scalability	Scalable	Scalable.	Scalable	Laboratory scale	Scalable.

Table 5. Comparison among main solution-based deposition techniques.

Feature	Spin Coating	Slot-Die Coating	CBD	Doctor Blading / Blade Coating	Screen Printing
Typical Operating Conditions	Ambient atmosphere, room temperature.	Ambient atmosphere, room temperature	Low temperatures (below 100 °C), no vacuum required.	Ambient atmosphere, room temperature.	Ambient atmosphere, room temperature.
Key Advantage(s)	Simplicity, low cost. High film quality. Good control for lab scale.	Minimum material waste. Compatible with rigid and flexible substrates. High uniformity over large areas.	Low cost. High film quality.	Compatible with rigid and flexible roll-to-roll production. Material efficiency.	Precise patterning. Minimizes material waste.
Key Limitation(s)	Not suited for large-area production.	Requires precise control of process parameters.	Limited control over film uniformity and composition. Generates chemical waste requiring management.	Requires high control.	Resolution limitations for very fine features. Slow.

continued on following page

Table 5. Continued

Feature	Spin Coating	Slot-Die Coating	CBD	Doctor Blading / Blade Coating	Screen Printing
Primary PV Application(s)	Perovskites, Organic solar cells, Dye-Sensitized Solar Cells	Perovskite and Organic Solar Cells.	Buffer layers (particularly CdS) in CdTe and CIGS solar cells.	Dye-sensitized solar cells, Perovskites and Organic Solar Cells.	Dye-sensitized solar cells, Organic Solar Cells and perovskites.
Scalability	Lab-scale and prototyping.	Scalable.	Well-suited for industrial and research-scale fabrication.	Scalable	Scalable.

5. LIMITATIONS AND TRADE-OFFS OF THIN-FILM PV

As discussed earlier, thin-film photovoltaic technologies offer distinct advantages over conventional crystalline silicon, including lightweight design, mechanical flexibility, and lower material usage. However, these strengths are accompanied by trade-offs that affect performance, scalability, environmental impact, and long-term reliability.

A key challenge is the availability and scalability of raw materials. Thin-film technologies such as CIGS and CdTe rely on elements like indium, gallium, and tellurium, which are either rare, difficult to extract, or unevenly distributed geographically. For instance, tellurium and gallium are byproducts of copper and aluminum production, respectively, and their supply is constrained by the scale of those industries. In terms of scalability, it is estimated that a full global transition to CdTe PV by 2050 would require 737,000 metric tons of cadmium and 786,000 metric tons of Tellurium equivalent to 34 years of cadmium production and 1,500 years of tellurium production at current rates (Efaz et al., 2021). Similarly, CIGS technologies would require significant increases in gallium and indium production. In contrast, silicon remains far more abundant, being approximately 20,000 times more available than gallium and nearly 300 million times more available than tellurium (Jean et al., 2015).

Toxicity presents another layer of concern. Cadmium, used in CdTe, and lead, found in some emerging perovskite technologies, are both hazardous substances. While current encapsulation and recycling strategies limit exposure during a module's operational life, risks associated with mining, manufacturing, and disposal remain significant. These issues increase regulatory scrutiny and complicate waste management at the end-of-life stage. Indium and gallium, while less toxic, still pose environmental risks during extraction and are not easily recovered from end-of-life panels.

End-of-life recycling potential is increasingly viewed as an opportunity for raw material recovery and value creation. Figure 10 (Weckend et al., 2016) shows IRENA's estimates for the recoverable mass of common PV module components at end-of-life in (Ton). The largest share by mass is recyclable glass accounting for approximately 960,000 tons, followed by Polymer, aluminum, Silicon, and copper. These materials are readily recoverable using existing infrastructure and represent the majority of economic value in recycling streams. In contrast, thin-film PV elements such as cadmium, tellurium, indium, and gallium constitute a much smaller fraction of total mass and require more advanced recovery techniques.

Figure 10. Recovery Potential of materials used in PV production (Weckend et al., 2016)

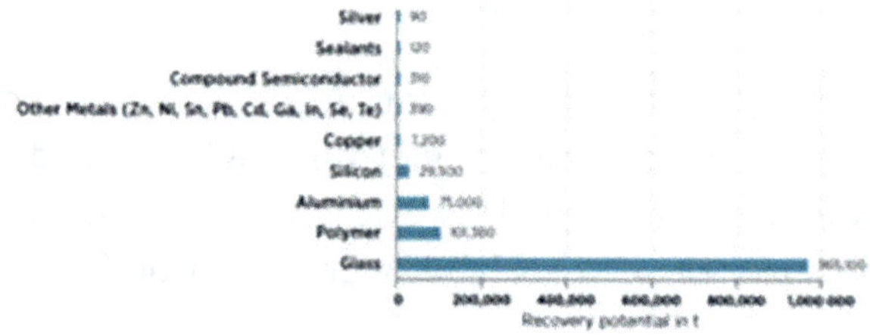

Efficiency and manufacturing limitations also shape the commercial viability of thin-film technologies. Laboratory-scale devices have achieved promising results, with perovskites reaching around 27%, and CIGS and CdTe modules around 23% efficiency potential (*Best Research-Cell Efficiency Chart | Photovoltaic Research | NREL*, 2025). However, it is crucial to note that these high efficiencies are typically achieved at the laboratory cell level, and commercial module efficiencies often lag behind due to challenges in degradation and maintaining uniformity across larger areas.

Despite these constraints, thin-film PV continues to contribute meaningfully to the global energy mix. While silicon remains the dominant material, thin-film modules now play a growing role in expanding access to solar energy and diversifying supply chains. The implementation of PV technologies to cover around 3% of global electricity demand helps save approximately 590 million tons of CO_2 yearly (Efaz et al., 2021).

6. THIN-FILM PV IN BUILDING INTEGRATED PHOTOVOLTAICS (BIPV): CASE STUDIES

Buildings are primary energy consumers. A report form IEA in 2018 stated that they account for 20–30% of total energy consumption in advanced countries such as Japan, Russia, India, USA, and China (IEA, 2018). Building-Integrated Photovoltaics (BIPV), is involves utilizing PV modules as part of the building structure, such as roofs, façades, and glazing, allowing them to serve as energy generators while being a conventional building materials (El Khatib et al., 2024).

Thin-film photovoltaics are particularly advantageous for BIPV applications thanks to their light weight, availability in flexible and semi-transparent form, and applicability for irregular or n curved geometries (Alrashidi et al., 2020).

This section below presents three real-world case studies on the use of thin-film for BIPV applications. For additional references, (Weerasinghe et al., 2021) conducted economic viability for 45 building in 12 countries.

6.1. Case Study 1: Helmholtz Zentrum Berlin (HZB) "Living Lab for BIPV" - Berlin, Germany (Helmholtz Zentrum Berlin, 2021)

– Building Type: Research laboratory building.
– Completion Year: June 2021
– System Size: 48.6 kWp
– Thin-Film PV Technology Used: CIGS (Copper Indium Gallium Selenide)
– Integration Type: Ventilated curtain-type façade on three sides (south, west, and north).
– Annual Energy Production: Approximately 30 MWh/year.
– Design:

The laboratory is two-story with a floor area of 62 x 16 m^2 and a height of around 13m. 360 frameless and homogeneously colored (blue) Avancis Skala 7003 modules (135 W each) occupy an area of around 380 m^2. The frameless design and concealed suspension system allowed for seamless combination with the building's metal curtain wall, achieving a sleek, uniform appearance. The architectural intent was to create a demonstration and living laboratory where BIPV is not just functional but also aesthetically integrated into the building's design.

The system is extensively monitored with 120 measuring points, serving as a "living lab" to validate simulation models and understand long-term degradation specific to CIGS in a façade application. Integration on the north side, a challenging orientation for PV, allows for critical analysis of performance under indirect and diffuse irradiation, leveraging CIGS's known advantage in such conditions. The energy produced is fully consumed directly within the research campus grid.

6.2. Case Study 2: Engie Fabricom Building - Zwijndrecht, Belgium (Solliance, 2019)

– Building Type: Existing commercial office/industrial building (retrofit).
– Completion Year: 2018.
– System Size: Around 20 m^2 (glass spandrels).
– Thin-Film PV Technology Used: Organic Photovoltaics (OPV).
– Integration Type: Retrofit of existing façade elements. Flexible OPV films were integrated into glass-glass BIPV modules for existing glass spandrels, and explicitly applied to Hoesch metal substrates to form solar-active steel façade elements.
– Total Energy Production (Initial): Generated 260 kWh shortly after installation (continuous monitoring planned).
– Design:

Extremely lightweight and flexible OPV films. This allowed for integration onto existing lightweight structures without requiring significant structural reinforcement, and their thin profile facilitated discreet integration matching the building's original aesthetics.

The OPV's known ability to perform well in low-light and high-temperature conditions is particularly advantageous for façade applications where modules may not always be optimally oriented or can experience significant heat build-up.

This project highlighted the unique potential of OPV for flexible, aesthetically matching retrofits. Continuous monitoring of ambient temperature, module temperature (especially for the steel façade elements to understand heat dissipation), and yield provides ongoing data to validate OPV performance in real-world BIPV scenarios.

6.3 Case Study 3: Copenhagen International School - Copenhagen, Denmark (AVENSTON, 2021)

– Building Type: Educational (International School).
– Completion Year: 2016.
– System Size: 700 kWp.
– Thin-Film PV Technology Used: amorphous silicon (a-Si)

– Integration Type: The building's entire 6,000 m^2 façade is composed of these specially designed photovoltaic glass panels, making the PV an integral part of the building's envelope and structural design.
– Annual Energy Production: 500,000 kWh per year.
– Percentage of Building's Energy Demand Met: The BIPV system covers 50% of the school's total annual electricity consumption.
– Design

Uniform green panels designed to interact with light, creating dynamic shades throughout the day. The choice of colored, semi-transparent BIPV glass was a central design decision from the earliest stages, emphasizing that thin-film allows for significant aesthetic flexibility and can be a defining architectural feature, not just a utilitarian add-on.

The design balances energy generation with a visual identity, showcasing how BIPV can be a central element of a building's character. The integration of semi-transparent thin-film PV into the façade also allows for controlled daylighting within the building, reducing the need for artificial lighting and contributing to occupant comfort.

7. SUMMARY AND RESEARCH GAPS

A common classification of PV technologies is by generation. crystalline silicon (c-Si) PV represent the first generation, thin-film PV are the second generation, and other emerging PV technologies are considered as the third generation. c-Si PVs dominate the market, thanks to their high efficiency, durability. However, they have limited design flexibility as they are typically rigid and opaque.

To address design and application concerns, thin-film PV technologies were developed to provide more adaptable as they are generally light, flexible and can be semi-transparent. However, they still lag behind c-Si in terms of efficiency and often rely on materials that raise toxicity and scarcity concerns. In response, other emerging thin-film based PV technologies are currently under development for performance and scalability. These often offer high lab-scale efficiencies but still face changes in their real-world performance stability.

The most mature second generation thin-film PV technologies include amorphous silicon (a-Si), cadmium telluride (CdTe), and copper indium gallium selenide (CIGS). a-Si is low-cost but has relatively lower efficiency and faces performance degradation under light exposure. CdTe offer high efficiency and proven commercial scalability, yet relies on toxic cadmium and scarce tellurium. CIGS, offers a tunable bandgap and relatively high efficiency, but suffers from material toxicity concerns as they depend on cadmium-based buffer layers.

Emerging photovoltaic technologies include perovskite solar cells, dye-sensitized solar cells (DSSCs), and organic solar cells (OSCs). Perovskites are the most promising as they have achieved remarkable lab-scale efficiencies. They still require research and development efforts for stability and scalable production. DSSCs offer moderate to low efficiencies, but they excel in niche applications that involve limited irradiance such as indoor lighting or early morning and early evenings. OSCs, are valued for their mechanical flexibility and ease of fabrication, but still face serious performance stability issues.

Thin-film PV technologies are deposited using various techniques, each vary in terms of cost, operating temperature, material compatibility, precision, throughput, and scalability. Vacuum-based techniques are commonly used for high-quality and precise films. Solution-based and electrochemical methods are of lower cost and usually suitable for emerging PV technologies.

A major concern in thin-film PV technologies is its dependence on scarce materials like indium, gallium, and tellurium, or materials that raise toxicity concerns such as cadmium and lead. While recycling efforts are used to recover valuable materials of the used PV modules, they are mainly able to recover glass, aluminum and polymers, rather than thin-film materials. These risks can be mitigated through intensive research on end-of-life management and on the scalability of emerging technologies.

The above overview highlights multiple areas for improvement that can be addressed through focused research efforts. These can be summarized as follows:

1- Identifying or developing alternative thin-film PV materials that are abundant and not toxic.
2- Advancing recycling and end-of-life management technologies that permits the recovery of thin-film materials and safe disposal.
3- Enhancing the long-term stability of the available PV technologies, especially the emerging PVs to enhance their scalability.
4- Bridging the gap between lab-scale and real-world efficiencies in emerging PV technologies.
5- Developing deposition methods to achieve ones that are simple, cost and energy-efficient, while being precise, and providing high throughput. This is key for thin-film PV scalability.
6- Improving inter-layer connections within PV cells to reduce energy losses. This can be achieved through addressing cell designs and exploring alternative buffer and interface layers.

REFERENCES

Ablekim, T., Duenow, J. N., Zheng, X., Moutinho, H., Moseley, J., Perkins, C. L., Johnston, S. W., O'Keefe, P., Colegrove, E., Albin, D. S., Reese, M. O., & Metzger, W. K. (2020). Thin-Film Solar Cells with 19% Efficiency by Thermal Evaporation of CdSe and CdTe. *ACS Energy Letters*, *5*(3), 892–896. DOI: 10.1021/acsenergylett.9b02836

Afre, R. A., & Pugliese, D. (2024). Perovskite Solar Cells: A Review of the Latest Advances in Materials, Fabrication Techniques, and Stability Enhancement Strategies. *Micromachines 2024, Vol. 15, Page 192, 15*(2), 192. DOI: 10.3390/mi15020192

Ahn, N., & Choi, M. (2024). Towards Long-Term Stable Perovskite Solar Cells: Degradation Mechanisms and Stabilization Techniques. *Advancement of Science*, *11*(4), 2306110. DOI: 10.1002/advs.202306110 PMID: 37997198

Al-Ali, S., Olabi, A. G., & Mahmoud, M. (2025). A review of solar photovoltaic technologies: developments, challenges, and future perspectives. *Energy Conversion and Management: X, 27*(February), 101057. DOI: 10.1016/j.ecmx.2025.101057

Al Miaari, A., & Ali, H. M. (2025). Recent advances on nano-enhanced phase change materials (NEPCMs) for photovoltaic thermal management and role of machine learning: A review of fundamentals, preparation, characterization, and thermo-physical properties. *Journal of Energy Storage*, *124*(April), 116544. DOI: 10.1016/j.est.2025.116544

Al Miaari, A., El Khatib, A., & Ali, H. M. (2023). Design and thermal performance of an innovative greenhouse. *Sustainable Energy Technologies and Assessments*, *57*, 103285. DOI: 10.1016/j.seta.2023.103285

Ali, A. O., Elgohr, A. T., El-Mahdy, M. H., Zohir, H. M., Emam, A. Z., Mostafa, M. G., Al-Razgan, M., Kasem, H. M., & Elhadidy, M. S. (2025). Advancements in photovoltaic technology: A comprehensive review of recent advances and future prospects. *Energy Conversion and Management: X, 26*(February), 100952. DOI: 10.1016/j.ecmx.2025.100952

Alrashidi, H., Ghosh, A., Issa, W., Sellami, N., Mallick, T. K., & Sundaram, S. (2020). Thermal performance of semitransparent CdTe BIPV window at temperate climate. *Solar Energy*, *195*, 536–543. DOI: 10.1016/j.solener.2019.11.084

Alrashidi, H., Issa, W., Sellami, N., Sundaram, S., & Mallick, T. (2022a). Thermal performance evaluation and energy saving potential of semi-transparent CdTe in Façade BIPV. *Solar Energy*, *232*, 84–91. DOI: 10.1016/j.solener.2021.12.037

Alrashidi, H., Issa, W., Sellami, N., Sundaram, S., & Mallick, T. (2022b). Thermal performance evaluation and energy saving potential of semi-transparent CdTe in Façade BIPV. *Solar Energy*, *232*, 84–91. DOI: 10.1016/j.solener.2021.12.037

AVENSTON. (2021). *Seven impressive BIPV systems in Europe | AVENSTON.* https://avenston.com/en/articles/7bipv/

Basol, B. M., & McCandless, B. (2014). Brief review of cadmium telluride-based photovoltaic technologies. *Journal of Photonics for Energy*, *4*(1), 040996. DOI: 10.1117/1.JPE.4.040996

Basumatary, P., & Agarwal, P. (2022). A short review on progress in perovskite solar cells. *Materials Research Bulletin*, *149*, 111700. DOI: 10.1016/j.materresbull.2021.111700

Best Research-Cell Efficiency Chart | Photovoltaic Research | NREL. (2025). https://www.nrel.gov/pv/cell-efficiency

Bouanani, B., Joti, A., Bachir Bouiadjra, F. S., & Kadid, A. (2020). Band gap and thickness optimization for improvement of CIGS/CIGS tandem solar cells using Silvaco software. *Optik (Stuttgart)*, *204*, 164217. Advance online publication. DOI: 10.1016/j.ijleo.2020.164217

Capitão, J., Martins, J., Emami, S., Ivanou, D., & Mendes, A. (2023). Fully glass frit encapsulated dye-sensitized solar cells: Challenges for hermetical sealing of electrolyte injection holes. *Solar Energy*, *249*, 476–484. DOI: 10.1016/j.solener.2022.12.001

Carlson, D. E., & Wronski, C. R. (1976). Amorphous silicon solar cell. *Applied Physics Letters*, *28*(11), 671–673. DOI: 10.1063/1.88617

Carlson, D. E., & Wronski, C. R. (1977). Solar cells using discharge-produced amorphous silicon. *Journal of Electronic Materials*, *6*(2), 95–106. DOI: 10.1007/BF02660377

Chen, C., Chen, J., Han, H., Chao, L., Hu, J., Niu, T., Dong, H., Yang, S., Xia, Y., Chen, Y., & Huang, W. (2022). Perovskite solar cells based on screen-printed thin films. *Nature*, *612*(7939), 266–271. DOI: 10.1038/s41586-022-05346-0 PMID: 36352221

Cho, M. Y., Kim, I. S., Park, C., Sohn, H., Lee, D., Koo, S. M., Cho, W. J., Kim, Y. N., Lee, D. W., & Oh, J. M. (2019). Multistage Degradation Mechanisms in Cu(In, Ga)Se2 Photovoltaic Modules Prepared by Co-Evaporation: Toward High Performances and Enhanced Stability. *ACS Applied Energy Materials*, *2*(10), 7171–7182. DOI: 10.1021/acsaem.9b01155

Clark, R. H. (1985). Screen Printing. In *Handbook of Printed Circuit Manufacturing* (pp. 216–244). Springer., DOI: 10.1007/978-94-011-7012-3_11

Crose, M., Sang-Il Kwon, J., Nayhouse, M., Ni, D., & Christofides, P. D. (2015). Multiscale modeling and operation of PECVD of thin film solar cells. *Chemical Engineering Science*, *136*, 50–61. DOI: 10.1016/j.ces.2015.02.027

De Guire, M. R., Bauermann, L. P., Parikh, H., & Bill, J. (2013). Chemical Bath Deposition. *Chemical Solution Deposition of Functional Oxide Thin Films*, *9783211993118*, 319–339. DOI: 10.1007/978-3-211-99311-8_14

Dewi, H. A., Erdenebileg, E., De Luca, D., Mhaisalkar, S. G., & Bruno, A. (2024). Accelerated MAPbI3 Co-evaporation: Productivity Gains without Compromising Performance. *ACS Energy Letters*, *4319–4322*(9), 4319–4322. Advance online publication. DOI: 10.1021/acsenergylett.4c01597

Doroody, C., Rahman, K. S., Rosly, H. N., Harif, M. N., Sopian, K., Abdullah, S. F., & Amin, N. (2021). A comprehensive comparative study of CdTe thin films grown on ultra-thin glass substrates by close-spaced sublimation and RF magnetron sputtering. *Materials Letters*, *293*, 129655. DOI: 10.1016/j.matlet.2021.129655

Efaz, E. T., Rhaman, M. M., Al Imam, S., Bashar, K. L., Kabir, F., Mourtaza, M. E., Sakib, S. N., & Mozahid, F. A. (2021). A review of primary technologies of thin-film solar cells. *Engineering Research Express*, *3*(3), 032001. DOI: 10.1088/2631-8695/ac2353

El Khatib, A., Al Rashidi, H., Al Miaari, A., & Hatamleh, Z. (2024). Thermal Performance Evaluation Techniques for Thin-Film Semi-Transparent PV Glazing Using Experimental Testing. *2024 6th International Symposium on Advanced Electrical and Communication Technologies, ISAECT 2024*. DOI: 10.1109/ISAECT64333.2024.10799601

Ezekoye, B. A., Offor, P. O., Ezekoye, V. A., & Ezema, F. I. (2013). Chemical Bath Deposition Technique of Thin Films: A Review. *International Journal of Scientific Research*, •••, 2.

Fadaam, S. A., Mustafa, M. H., Abd AlRazaK, A. H., & Shihab, A. A. (2019). Enhanced efficiency of CdTe Photovoltaic by thermal evaporation Vacuum. *Energy Procedia*, *157*, 635–643. DOI: 10.1016/j.egypro.2018.11.229

Fazal, M. A., & Rubaiee, S. (2023). Progress of PV cell technology: Feasibility of building materials, cost, performance, and stability. *Solar Energy*, *258*, 203–219. DOI: 10.1016/j.solener.2023.04.066

Ferekides, C., Britt, J., Ma, Y., & Killian, L. (1993). High efficiency CdTe solar cells by close spaced sublimation. *Conference Record of the IEEE Photovoltaic Specialists Conference, 389–393*, 389–393. Advance online publication. DOI: 10.1109/PVSC.1993.347151

Francis, O. I., Ikenna, A., Francis, O. I., & Ikenna, A. (2021). Review of Dye-Sensitized Solar Cell (DSSCs) Development. *Natural Science (Irvine, Calif.)*, *13*(12), 496–509. DOI: 10.4236/ns.2021.1312043

Fritzsche, H. (2001). Development in understanding and controlling the Staebler-Wronski effect in a-Si:H. *Annual Review of Materials Science, 31*(Volume 31, 2001), 47–79. https://doi.org/DOI: 10.1146/ANNUREV.MATSCI.31.1.47/CITE/REFWORKS

Genchi, G., Sinicropi, M. S., Lauria, G., Carocci, A., & Catalano, A. (2020). The Effects of Cadmium Toxicity. *International Journal of Environmental Research and Public Health 2020, Vol. 17, Page 3782, 17*(11), 3782. DOI: 10.3390/ijerph17113782

Guha, S., Yang, J., Banerjee, A., Glatfelter, T., & Xu, X. (1992). Advances in amorphous silicon alloy-based multijunction cells and modules. *AIP Conference Proceedings*, *268*(1), 64–71. DOI: 10.1063/1.42943

Gupta, A., Parikh, V., & Compaan, A. D. (2006). High efficiency ultra-thin sputtered CdTe solar cells. *Solar Energy Materials and Solar Cells*, *90*(15), 2263–2271. DOI: 10.1016/j.solmat.2006.02.029

Gupta, B., Min, K. H., Lee, Y., Tournet, J., Hoex, B., Jagadish, C., Tan, H. H., & Karuturi, S. (2024). From Rigid to Flexible: Progress, Challenges and Prospects of Thin c-Si Solar Energy Devices. *Advanced Energy Materials*, *14*(27), 2400743. DOI: 10.1002/aenm.202400743 PMID: 28296287

Han, J., Spanheimer, C., Haindl, G., Fu, G., Krishnakumar, V., Schaffner, J., Fan, C., Zhao, K., Klein, A., & Jaegermann, W. (2011). Optimized chemical bath deposited CdS layers for the improvement of CdTe solar cells. *Solar Energy Materials and Solar Cells*, *95*(3), 816–820. DOI: 10.1016/j.solmat.2010.10.027

Hardin, B. E., Snaith, H. J., & McGehee, M. D. (2012). The renaissance of dye-sensitized solar cells. *Nature Photonics*, *6*(3), 162–169. DOI: 10.1038/nphoton.2012.22

Helmholtz Zentrum Berlin. (2021). *Living lab for BIPV*. https://www.helmholtz-berlin.de/projects/pvcomb/forschen/living-lab-bipv/index_en.html

Hoseinnezhad, M., & Gharanjig, K. (2018). Synthesis and Application of an Organic Dye in Nanostructure Solar Cells Device. *International Journal of Chemical and Molecular Engineering, 12*(9), 438–441.

Hu, Q., Wu, H., Sun, J., Yan, D., Gao, Y., & Yang, J. (2016). Large-area perovskite nanowire arrays fabricated by large-scale roll-to-roll micro-gravure printing and doctor blading. *Nanoscale, 8*(9), 5350–5357. DOI: 10.1039/C5NR08277C PMID: 26883938

Huang, Y. C., Cha, H. C., Chen, C. Y., & Tsao, C. S. (2017). A universal roll-to-roll slot-die coating approach towards high-efficiency organic photovoltaics. *Progress in Photovoltaics: Research and Applications, 25*(11), 928–935. DOI: 10.1002/pip.2907

IEA. (2018). *Data & Statistics - IEA*. https://www.iea.org/data-and-statistics?country =WORLD&fuel=Energy supply&indicator=TPESbySource

IEA. (2024). *Trends in Photovoltaic Applications*. https://doi.org/https://doi.org/10 .69766/jnew6916

IRENA. (2012). *Renewable Energy Technologies: Cost Analysis Series* (Vol. 1, Issue 4/5).

Jean, J., Brown, P. R., Jaffe, R. L., Buonassisi, T., & Bulović, V. (2015). Pathways for solar photovoltaics. *Energy & Environmental Science, 8*(4), 1200–1219. DOI: 10.1039/C4EE04073B

Jeong, J., Kim, M., Seo, J., Lu, H., Ahlawat, P., Mishra, A., Yang, Y., Hope, M. A., Eickemeyer, F. T., Kim, M., Yoon, Y. J., Choi, I. W., Darwich, B. P., Choi, S. J., Jo, Y., Lee, J. H., Walker, B., Zakeeruddin, S. M., Emsley, L., … Kim, J. Y. (2021). Pseudo-halide anion engineering for α-FAPbI3 perovskite solar cells. *Nature 2021 592:7854, 592*(7854), 381–385. DOI: 10.1038/s41586-021-03406-5

Jung, S., Ahn, S. J., Yun, J. H., Gwak, J., Kim, D., & Yoon, K. (2010). Effects of Ga contents on properties of CIGS thin films and solar cells fabricated by co-evaporation technique. *Current Applied Physics, 10*(4), 990–996. DOI: 10.1016/j.cap.2009.11.082

Juška, G. (1991). Properties of free-carrier transport in a-Se and a-Si:H. *Journal of Non-Crystalline Solids, 137–138*(PART 1), 401–406. DOI: 10.1016/S0022-3093(05)80140-1

Kaim, P., Lukaszkowicz, K., Szindler, M., Szindler, M. M., Basiaga, M., & Hajduk, B. (2021). The Influence of Magnetron Sputtering Process Temperature on ZnO Thin-Film Properties. *Coatings 2021, Vol. 11, Page 1507, 11*(12), 1507. DOI: 10.3390/coatings11121507

Kazmerski, L. L., White, F. R., & Morgan, G. K. (1976). Thin-film CuInSe/sub 2// CdS heterojunction solar cells. *Appl. Phys. Lett.; (United States), 29:4*(4), 268–270. DOI: 10.1063/1.89041

Kegelmann, L., Tockhorn, P., Wolff, C. M., Márquez, J. A., Caicedo-Dávila, S., Korte, L., Unold, T., Lövenich, W., Neher, D., Rech, B., & Albrecht, S. (2019). Mixtures of Dopant-Free Spiro-OMeTAD and Water-Free PEDOT as a Passivating Hole Contact in Perovskite Solar Cells. *ACS Applied Materials & Interfaces, 11*(9), 9172–9181. DOI: 10.1021/acsami.9b01332 PMID: 30741517

Kesinro, R. O., Boyo, A. O., Akinyemi, M. L., Emetere, M. E., & Aizebeokhai, A. P. (2021). Progress on Organic Solar Cells: A Short Review. *IOP Conference Series. Earth and Environmental Science*, *665*(1), 012036. Advance online publication. DOI: 10.1088/1755-1315/665/1/012036

Khatoon, S., Kumar Yadav, S., Chakravorty, V., Singh, J., Bahadur Singh, R., Hasnain, M. S., & Hasnain, S. M. M. (2023). Perovskite solar cell's efficiency, stability and scalability: A review. *Materials Science for Energy Technologies*, *6*, 437–459. DOI: 10.1016/j.mset.2023.04.007

Kreiger, M. A., Shonnard, D. R., & Pearce, J. M. (2013). Life cycle analysis of silane recycling in amorphous silicon-based solar photovoltaic manufacturing. *Resources, Conservation and Recycling*, *70*, 44–49. DOI: 10.1016/j.resconrec.2012.10.002

Kujovic, L., Liu, X., Togay, M., Abbas, A., Law, A. M., Jones, L. O., Curson, K. M., Barth, K. L., Bowers, J. W., Walls, J. M., Oklobia, O., Lamb, D. A., Irvine, S. J. C., Zhang, W., Lee, C., Nagle, T., Lu, D., & Xiong, G. (2025). Development of ZnO Buffer Layers for As-Doped CdSeTe/CdTe Solar Cells with Efficiency Exceeding 20%. *Advanced Materials Technologies*, *2401364*(13), 2401364. Advance online publication. DOI: 10.1002/admt.202401364

Kumar, A., & Giripunje, S. M. (2025). A comparative numerical simulation study of CIGS solar cells with distinct back surface field layers for enhanced performance. *Journal of Physics and Chemistry of Solids*, *197*, 112436. DOI: 10.1016/j.jpcs.2024.112436

Lee, H., Yoon, H., Ji, C., Lee, D., Lee, J. H., Yun, J. H., & Kim, Y. (2012). Fabrication of CIGS films by electrodeposition method for photovoltaic cells. *Journal of Electronic Materials*, *41*(12), 3375–3381. DOI: 10.1007/s11664-012-2252-x

Lee, J. H., Kim, B. S., Park, J., Lee, J. W., & Kim, K. (2023). Opportunities and Challenges for Perovskite Solar Cells Based on Vacuum Thermal Evaporation. *Advanced Materials Technologies*, *8*(20), 2200928. DOI: 10.1002/admt.202200928

Lee, T. D., & Ebong, A. U. (2017). A review of thin film solar cell technologies and challenges. *Renewable & Sustainable Energy Reviews*, *70*(November), 1286–1297. DOI: 10.1016/j.rser.2016.12.028

Machkih, K., Oubaki, R., & Makha, M. (2024). A Review of CIGS Thin Film Semiconductor Deposition via Sputtering and Thermal Evaporation for Solar Cell Applications. *Coatings 2024, Vol. 14, Page 1088, 14*(9), 1088. DOI: 10.3390/coatings14091088

Miah, M. H., Rahman, M. B., Nur-E-Alam, M., Islam, M. A., Shahinuzzaman, M., Rahman, M. R., Ullah, M. H., & Khandaker, M. U. (2025). Key degradation mechanisms of perovskite solar cells and strategies for enhanced stability: Issues and prospects. *RSC Advances*, *15*(1), 628–654. DOI: 10.1039/D4RA07942F PMID: 39777157

Min, H., Lee, D. Y., Kim, J., Kim, G., Lee, K. S., Kim, J., Paik, M. J., Kim, Y. K., Kim, K. S., Kim, M. G., Shin, T. J., & Il Seok, S. (2021). Perovskite solar cells with atomically coherent interlayers on SnO2 electrodes. *Nature*, *598*(7881), 444–450. DOI: 10.1038/s41586-021-03964-8 PMID: 34671136

Munna, F. T., Selvanathan, V., Sobayel, K., Muhammad, G., Asim, N., Amin, N., Sopian, K., & Akhtaruzzaman, M. (2021). Diluted chemical bath deposition of CdZnS as prospective buffer layer in CIGS solar cell. *Ceramics International*, *47*(8), 11003–11009. DOI: 10.1016/j.ceramint.2020.12.222

N'guessan, A. I., Bouich, A., Touré, A., Soucasse, B. M., & Soro, D. (2023). Influence of Sulfur Content in Zn(O,S) Buffer Layer onto Copper Indium Gallium Sulfur-Based Solar Cells Through Surface Engineering at ZnO1−xSx/CIGS Interface. *Journal of the Minerals Metals & Materials Society*, *75*(10), 4332–4340. DOI: 10.1007/s11837-023-06018-8

Noorasid, N. S., Arith, F., Mustafa, A. N., Azam, M. A., Mahalingam, S., Chelvanathan, P., & Amin, N. (2022). Current advancement of flexible dye sensitized solar cell: A review. *Optik (Stuttgart)*, *254*, 168089. DOI: 10.1016/j.ijleo.2021.168089

Patidar, R., Burkitt, D., Hooper, K., Richards, D., & Watson, T. (2020). Slot-die coating of perovskite solar cells: An overview. *Materials Today. Communications*, *22*, 100808. DOI: 10.1016/j.mtcomm.2019.100808

Raïssi, M., Pellegrin, Y., Lefevre, F. X., Boujtita, M., Rousseau, D., Berthelot, T., & Odobel, F. (2020). Digital printing of efficient dye-sensitized solar cells (DSSCs). *Solar Energy*, *199*, 92–99. DOI: 10.1016/j.solener.2020.02.004

Romeo, N., Bosio, A., Canevari, V., & Podestà, A. (2004). Recent progress on CdTe/CdS thin film solar cells. *Solar Energy, 77*(6), 795–801. DOI: 10.1016/j.solener.2004.07.011

Sahu, N., Parija, B., & Panigrahi, S. (2009). Fundamental understanding and modeling of spin coating process: A review. *Indian Journal of Physics, 83*(4), 493–502. DOI: 10.1007/s12648-009-0009-z

Shah, A., Torres, P., Tscharner, R., Wyrsch, N., & Keppner, H. (1999). Photovoltaic Technology: The Case for Thin-Film Solar Cells. *Science, 285*(5428), 692–698. www.nsf.gov/sbe/srs/pubdata.htm. DOI: 10.1126/science.285.5428.692 PMID: 10426984

Sharma, I., Pawar, P. S., Kumar Yadav, R., Nandi, R., & Heo, J. (2022). Review on bandgap engineering in metal-chalcogenide absorber layer via grading: A trend in thin-film solar cells. *Solar Energy, 246*, 152–180. DOI: 10.1016/j.solener.2022.09.046

Sheu, H. H., Hsu, Y. T., Jian, S. Y., & Liang, S. C. (2016). The effect of Cu concentration in the photovoltaic efficiency of CIGS solar cells prepared by co-evaporation technique. *Vacuum, 131*, 278–284. DOI: 10.1016/j.vacuum.2016.07.008

Sivasankar, S. M., Amorim, C. de O., & Cunha, A. F. (2025). Progress in Thin-Film Photovoltaics: A Review of Key Strategies to Enhance the Efficiency of CIGS, CdTe, and CZTSSe Solar Cells. *Journal of Composites Science, 9*(3), 143. Advance online publication. DOI: 10.3390/jcs9030143

Solliance. (2019). *The key to a sustainable building façade - Solliance*. https://www.solliance.eu/2018/the-key-to-a-sustainable-building-facade/

Taesoo, D. L., & Abasifreke, U. E. (2016). A review of thin film solar cell technologies and challenges. *Renewable and Sustainable Energy Reviews, 70*(September 2015), 0–1. DOI: 10.1016/j.rser.2016.12.028

Weckend, S., Wade, A., & Heath, H. (2016). End-Of-Life Management Of Solar Panels. In *IRENA*. DOI: 10.1109/NRSC58893.2023.10152958

Weerasinghe, R. P. N. P., Yang, R. J., Wakefield, R., Too, E., Le, T., Corkish, R., Chen, S., & Wang, C. (2021). Economic viability of building integrated photovoltaics: A review of forty-five (45) non-domestic buildings in twelve (12) western countries. *Renewable & Sustainable Energy Reviews, 137*, 110622. DOI: 10.1016/j.rser.2020.110622

Wu, H., Zhang, C., Ding, K., Wang, L., Gao, Y., & Yang, J. (2017). Efficient planar heterojunction perovskite solar cells fabricated by in-situ thermal-annealing doctor blading in ambient condition. *Organic Electronics, 45*, 302–307. DOI: 10.1016/j.orgel.2017.03.017

Yablonovitch, E., & Cody, G. D. (1982). Intensity Enhancement in Textured Optical Sheets for Solar Cells. *IEEE Transactions on Electron Devices*, *29*(2), 300–305. DOI: 10.1109/T-ED.1982.20700

Yang, C., Hu, W., Liu, J., Han, C., Gao, Q., Mei, A., Zhou, Y., Guo, F., & Han, H. (2024). Achievements, challenges, and future prospects for industrialization of perovskite solar cells. *Light: Science & Applications 2024 13:1, 13*(1), 1–48. DOI: 10.1038/s41377-024-01461-x

Yang, C.-S., Smith, L. L., Arthur, C. B., & Parsons, G. N. (2000). Stability of low-temperature amorphous silicon thin film transistors formed on glass and transparent plastic substrates. *Journal of Vacuum Science & Technology. B, Microelectronics and Nanometer Structures : Processing, Measurement, and Phenomena : An Official Journal of the American Vacuum Society*, *18*(2), 683–689. DOI: 10.1116/1.591259

Zheng, J., Zhang, M., Lau, C. F. J., Deng, X., Kim, J., Ma, Q., Chen, C., Green, M. A., Huang, S., & Ho-Baillie, A. W. Y. (2017). Spin-coating free fabrication for highly efficient perovskite solar cells. *Solar Energy Materials and Solar Cells*, *168*, 165–171. DOI: 10.1016/j.solmat.2017.04.029

Chapter 14
Next-Generation Devices Enabled by RF Sputtered Wide Bandgap Semiconductors

Pramod Mandal
https://orcid.org/0000-0001-9797-0377
Faculty of Chemical Technology, University of Pardubice, Czech Republic

Anand Pandey
https://orcid.org/0000-0001-5195-2966
Sensor and Actuator Systems, Department of Physics, Chemistry and Biology, Linköping University, Sweden

ABSTRACT

RF-sputtered wide-bandgap (WBG) semiconductors demonstrate significant potential for next-generation electronics, optoelectronics, and power electronics devices due to their exceptional physical, thermal, and chemical stability. This review comprehensively explores their capabilities, the challenges associated with their synthesis and integration, as well as the critical process parameters that influence film quality. Additionally, it provides targeted optimization strategies and practical recommendations aimed at improving the structural and functional performance of RF-sputtered WBG semiconductor thin films, thereby enhancing their applicability in next-generation device technologies.

DOI: 10.4018/979-8-3373-9454-1.ch014

1. INTRODUCTION

Silicon (Si) is a well-known semiconductor since 1950 that has been utilized in most power electronics, electronics, and optoelectronic device applications (Akbar et al. 2025; Torky and Chow 2024). However, as device optimization and the manufacturing process progress, material properties become a crucial limiting factor for next-generation devices. Si, consisting of a low energy band gap (1.1 eV), low breakdown field (~0.3 MV/cm), high switching losses, and incapable of operating in harsh environments, leads to poorer performance (Ra, Han, and Zhao 2016). To date, several efforts have been invested in the more basic alternative of replacement of Si as a semiconductor material (Torky and Chow 2024). In that case, wide band-gap semiconductors have been emerging as the best alternative option of Si for next-generation devices, including power electronics, photodetectors, UV-sensors, UV-light emitting diodes (LEDs), laser diodes for extreme environments, etc., due to their wide bandgap (>3.3 eV), low conduction, low switching losses, large breakdown field, and operation at high temperature (Akbar et al. 2025; Chaudhary et al. 2023; Torky and Chow 2024). Wide bandgap semiconductors like; silicon carbide (SiC) (E_g =3.2 eV) (Varley, Shen, and Higashiwaki 2022), gallium nitride(GaN) (E_g=3.4) (Varley et al. 2022), aluminium nitride(AlN) (E_g=6.2 eV) (Cheng et al. 2024), Gallium oxide (E_g=4.9 eV) (Mandal et al. 2025), zinc oxide(ZnO) (E_g=3.37) (Fedorenko et al. 2024), and diamond (E_g=5.5 eV) (Lin et al. 2014), etc, have shown remarkable advancements in the performance for next generation device including high-power devices, solar-blind UV photodetector, UV-LEDs, and UV-sensor, etc (Suthar, Venkataramanaiah, and Suresh 2025; Torky and Chow 2024). We know that the fabrication method plays a substantial part in achieving the best performance of any semiconductor device. To date, the most commonly used method to fabricate wideband semiconductor thin film for devices available is pulse laser deposition (PLD) (Blumenschein, Paskova, and Muth 2019; Yu, Ou, and Wuu 2015), low-pressure chemical vapor deposition (LPCVD) (Mondal et al. 2020; Ranga et al. 2021), physical vapor deposition (PVD) radio frequency (RF) sputtering method (Mandal, Roy, and Singh 2022; Mandal, Singh, and Roy 2021; Mobtakeri et al. 2021), atomic layer deposition (ALD) (Ramachandran et al. 2014; Gu et al. 2022), molecular beam epitaxy (MBE) (Ngo et al. 2019), metal-organic chemical vapor deposition (MOCVD) (Park et al. 2019; Mi et al. 2014; Hernandez et al. 2021), sol-gel method (Shen et al. 2018; Cheah, Osman, and Poopalan 2020), *etc*. However, compared to the above method, the RF sputtering method is more favourable for fabricating the wideband semiconductor (Saikumar, Nehate, and Sundaram 2019). It is an eminent fact that the wide bandgap semiconductors act as an insulating material at room temperature due to the large energy bandgap until some highly conductive dopant (tin, aluminium, and gold) materials are doped for electrical conductivity (Mandal et al. 2024). The RF sputtering method can fabricate this material easily, as there is no need for a conductive material; in contrast, other methods need a conductive or chemically active material for fabrication. Also, it shows several advantages such as controllability in the quality of thin film properties as well as thickness by tailoring the working parameters such as sputtering power (W), deposition time (time), standoff distance between the substrate to target (cm), argon gas pressure (mbar), and post-deposition analysis (Mandal et al. 2025).

Thus, herein, we have provided a brief assessment of the PVD RF sputtered process for the deposition of various WBGs like SiC, GaN, AlN, Ga_2O_3, ZnO, and diamond, including their working principles, key parameters, and applications. We will also cover the fundamentals of existing WBGs, including their material properties and potential applications. Furthermore, we have also explored the thin film properties, such as structural, morphological, electrical, and optical features of RF-sputtered WBGs in detail, focusing on their relevance for next-generation device applications.

2. PVD RF SPUTTERING METHOD

In the 1960s, the term "physical vapor deposition" was first coined. Moreover, several technologies, such as gas chemistry, plasma, vacuum, including sputtering, thermal evaporation, and magnetic fields, have evolved to advance vacuum coating techniques (Baptista et al. 2018). Initially, the above method was complicated to understand. However, implementing the numerous modifications has improved the adhesion, structural properties, and composition of material deposition. Recently, PVD RF Sputtering is one of the best material deposition techniques for ceramic, oxide, metallic, and other compound materials, with controllable composition and thickness ranging from nanometres to millimetres (Baptista et al. 2018; Mattox 2010).

In this method, a sputtering target material and a pre-cleaned substrate are mounted inside a vacuum chamber, where a low pressure of approximately 10^{-6} to 10^{-8} mbar is preserved. Argon (Ar) gas is then supplied into the chamber, as a result, the electrons accelerate the neutral Ar atoms, ionizing them into Ar^+. These positively charged ions are directed towards the target (acting as the cathode), bombarding its surface. This collision causes atoms from the target material to dislodge and eject, subsequently depositing onto the substrate to form a thin film. The ejected atoms from the sputtering target gradually deposit onto the substrate, forming a thin film layer by layer (Mandal and Roy 2025; Baptista et al. 2018) as shown in Figure 1.

Figure 1. Schematic of the PVD RF sputtering method for thin film deposition.

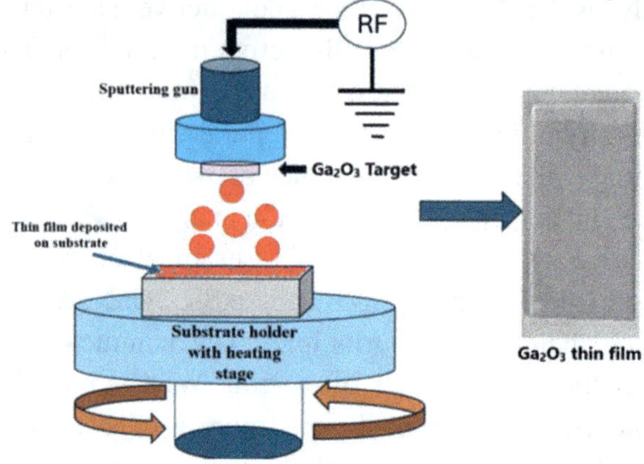

2.1. Working Parameters

The quality of the produced thin film material is very much dependent on the working parameters such as argon gas pressure (mbar), sputtering power (W), deposition time (minutes), substrate to target standoff distance, substrate temperature, and substrate orientation. These main working parameters are discussed below,

Figure 2. List of main working parameters of the RF sputtering method.

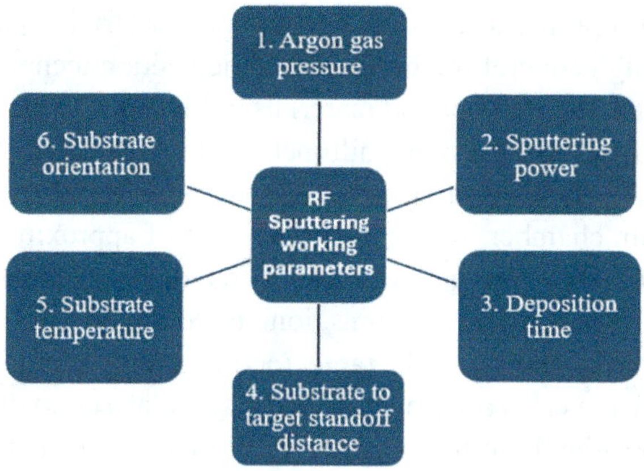

2.1.1. Argon Gas Pressure

In the RF sputtering method, Ar gas is usually used as an inert gas to generate plasma inside the chamber. Sometimes reactive gases like oxygen and nitrogen are used along with argon to create the reactive sputtering method for deposition. Ar gas pressure significantly impacts structure quality, composition (stoichiometry ratio), and thickness of thin films. The gas pressure should be kept reasonable during material deposition to achieve optimal surface roughness. As we know, lower Ar pressure might result in a rougher surface finish, and higher argon gas pressure results in a rougher surface roughness too (Chalana et al. 2019). Also, it can affect the structural properties like crystallinity and grain size of thin films. Low Ar gas pressure is very crucial to producing great crystallinity of thin films as well as larger grain size, whereas higher argon gas pressure tends to produce poor crystallinity (Chalana et al. 2019). Moreover, it affects the optical and electrical characteristics of thin films. Also, thin film fabricated at high Ar gas pressure tends to lower conductivity and reduce the transparency of materials like zinc oxide and titanium oxide, but increases the energy bandgap of the material (Acosta, Riech, and Martín-Tovar 2013; Nair et al. 2011). while lower argon pressure tends to have higher conductivity and produce an intrinsic bandgap.

2.1.2. Sputtering Power

The next essential working parameter in the RF sputtering method is sputtering power, it is a key parameter that plays a vital role during the deposition method. It has a great impact on the deposition rate, grain size, adhesion, surface morphology, crystallinity, and composition of thin films (Mandal, Singh, and Roy 2021). It is a well-known fact that material deposition at lower sputtering power tends to lower deposition rate, and poor adhesion of thin films, while higher sputtering power tends to higher deposition rate, denser and better adhered thin films. However, for achieving better crystallinity and stoichiometry of any thin film, a reasonable sputtering power is considered (Chelvanathan et al. 2017). Moreover, it has significant impacts on optical and electrical properties. Material deposited at a higher sputtering parameter results in higher conductivity and an upsurge in the bandgap (Astinchap, Moradian, and Gholami 2017), while lower sputtering tends to lower conductivity and the intrinsic energy bandgap.

2.1.3. Deposition Time

Deposition time also plays a vital role during deposition process; it is helpful to attain the desired thickness (Mandal et al. 2021). It also affects the growth of thin films, such as morphological, compositional, and structural properties. (El-Shaer et al. 2023). Lower deposition time during the deposition process tends to achieve a smoother surface finish but with lower thickness and stoichiometry. In comparison, a more extended time duration might be helpful to achieve the desired stoichiometry and excellent crystallinity with larger grain size and somewhat higher surface roughness. Moreover, deposition time can also impact the optical properties of a thin film, like transmittance, reflectance, energy bandgap, and electrical conductivity. The thin film produced at lower deposition time has higher transparency, while at higher deposition time, the thin film tends to increase optical absorptions and lower energy bandgap (bandgap narrowing) (Tondare et al. 2018). In the case of electrical conductivity, lower deposition time tends to lower electrical conductivity, while thin films deposited for a long duration have higher conductivity and carrier mobility.

2.1.4. Standoff Distance Between the Sputtering Targets and the Substrate

The critical parameter "the standoff distance between target and substrate" has a substantial impact on thin film quality, including optical and electrical properties. This parameter influences the energy, angular distribution, and scattering behaviour of sputtered atoms as they approach the substrate. At a short distance (~4 cm), the deposition rate of the thin film produced will be higher with non-uniformity and higher conductivity, but the thin film made at a longer distance (8-10 cm) has a uniform thin film, better crystallinity, good transparency and but lower electrical conductivity than the shorter distance (Lenis, Gómez, and Bolívar 2019).

2.1.5. Substrate Temperature

Likewise, the above parameters, substrate temperature, also impact the physical properties of RF sputtered thin films (Mandal et al. 2021). As the substrate temperature rises, adatom mobility on the surface improves, enabling better atomic arrangement and increasing crystallinity and grain size. This produces a dense film with smoother morphology and fewer defects. Higher temperatures decrease internal tension while increasing phase purity. Increased crystallinity improves transparency and allows for modification of the optical bandgap. Electrically, improved grain connectivity and reduced defect density result in greater carrier mobility and lower resistance. However, extremely high temperatures might cause film degradation or undesirable interdiffusion. Thus, adjusting substrate temperature is crucial for achieving the desired film performance in a wide range of applications (Wasa 2012).

2.1.6. Substrate Orientation

Substrate orientation also plays a critical role in achieving excellent surface morphology, structural, optical, and electrical properties in the RF sputtered method (Mandal et al. 2021). Different crystallographic orientations produce different surface energies and atomic arrangements, influencing the nucleation and growth mechanisms during film deposition. This results in different grain structures, surface morphologies, and favoured growth directions. Films deposited on well-aligned substrates display superior crystallinity and orientation matching, resulting in fewer defects and strain. Morphological characteristics like as grain size and texture are also highly reliant on substrate orientation. These structural changes directly impact optical qualities, including transmittance and bandgap, as well as electrical behaviour, such as carrier mobility and conductivity. Choosing an optimal substrate orientation is thus critical for the enhancement of thin film performance in electrical, optoelectronic, and sensing applications (Li et al. 2024).

Figure 3 shows how the deposition rate and surface smoothness of the thin film were affected by the above-mentioned effective RF sputtering parameters. Also effects of important parameters on device performance is discussed in the Table 1

Figure 3. Effect of RF sputtering parameters on (a) deposition rate and (b) surface smoothness of thin films.

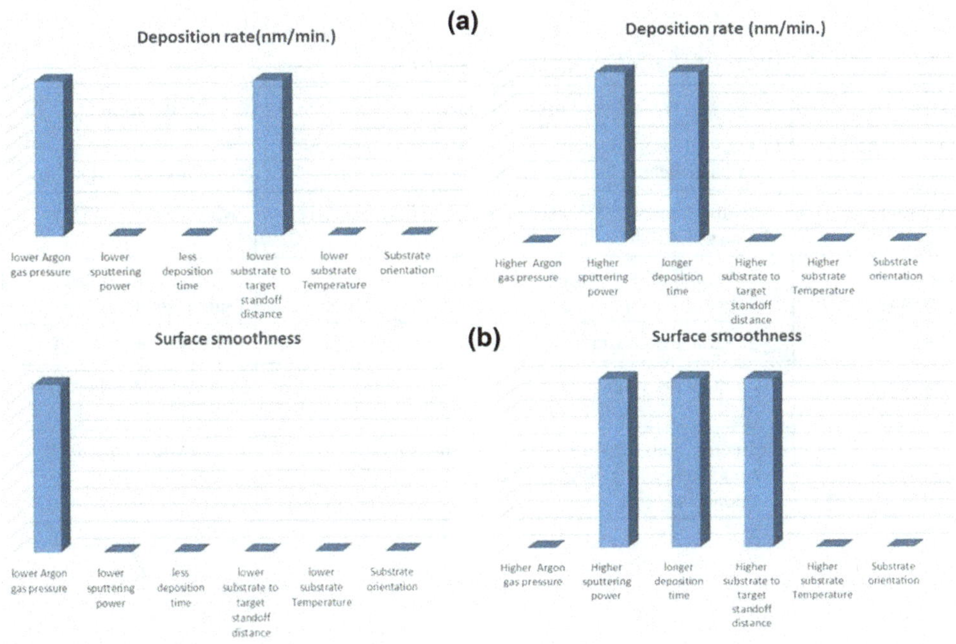

Table 1. Effects of RF sputtering parameters on device properties.

RF sputtering Parameters	Effect on Film Properties	Effect on Devices	Example of Device
Lower argon gas Pressure	Larger grain size, higher film density	Higher mobility → Lower Ron and better high-frequency performance	Power MOSFET, RF devices
Higher Substrate Temperature(C)	Better crystallinity, reduced defects	Higher breakdown voltage, improved thermal stability	HEMTs, Schottky diodes
Higher Sputtering Power(W)	Increased deposition rate, higher stress	Possible film cracking → reduced reliability	Thick oxide buffer layers
Reactive Gas Flow Ratio (Ar+O_2/N)	Controls stoichiometry and phase purity	Balanced carrier concentration, reduced leakage	Ga_2O_3 MOSFETs, ZnO TFTs

2.2. Application of the RF Sputtering Method

RF sputtering is an adaptable deposition method used extensively in industrial and research applications due to its ability to deposit insulating, semiconducting, and metallic materials with precise thickness, homogeneity, and adherence. The ability to tailor film properties like composition, thickness, and surface morphology makes RF sputtering a highly adaptable technique appropriate for various advanced applications. The broadly classified field where the RF sputtering method has been employed is discussed in Table 2.

Table 2. Application of the RF Sputtering Method

Area	Applications	References
Opto-electronics	LEDs, OLEDs, photovoltaics, Deep UV photodetectors (PDs), and electrochemical devices.	(Wu et al. 2024) (Mezdrogina et al. 2020)
electronics	Thin film transistor, UV-sensor, Resistive Random Access Memory (ReRAM), etc.	(Mathkari et al. 2025)
Power semiconductor	MOSFET, IGBTs, Schottky barrier diodes (SBDs), Rectifier	(Mourya et al. 2022) (Saikumar et al. 2023)
Optical	Eyewear lenses, mirrors, and decorative window glass	(Dobrikov et al. 2009)
Bio-medical	Coating on the surgical instrument	(Dragomir (Nicolescu) et al. 2023)
Industrial	Decorative metal coating (jewellery, watches, etc), Automotive coating (wheel, trims, and exhaust components)	(Vorobyova et al. 2023)

3. WIDE BANDGAP SEMICONDUCTOR

Recently, the wide bandgap semiconductors (WBGs), including silica carbide, gallium nitride, aluminium nitride, gallium oxide, zinc oxide, and diamond, have shown remarkable advancements in the performance for next-generation devices. WBGs consist of an energy bandgap (E_g) higher than >2 eV (Yoshikawa, Matsunami, and Nanishi 2007). These semiconductors are enabling next-generation electronic devices in medical, military, and social applications. Thus, we have explored the history, properties, and applications of these WGBs as discussed below.

3.1. Silica Carbide (SiC)

SiC is an evolving WBG for the next generation of device applications, particularly in battery electric vehicles (BEVs), as well as for optoelectronics devices such as LEDs and power devices such as MOSFETs (Sparavigna, 2018). SiC was first synthesized by Edward G. Acheson in 1891, and its name was termed "carborundum"(Sparavigna, 2018). It was first commercially employed in 1950 because of its wide bandgap properties (Wang et al. 2024). It is the solitary compound of group IV elements made up of silicon (Si) and carbon (C) with a low intrinsic carrier density of $5x10^{-9}$ cm^{-3} at ambient temperature and a wide bandgap (~3.2 eV), making it suitable for high-temperature electrical devices application (Wang et al. 2024). SiC exists in a variety of crystalline forms known as polytypes. Figure 4 displays the unit cell crystal structure of SiC. The most prevalent polytypes are cubic (3C-SiC) and hexagonal (4H-SiC and 6H-SiC), which are being used for various device applications (Wang et al. 2024).

If we compare the polytypes mentioned above, cubic is unstable at high temperatures, whereas hexagonal SiC is stable and widely employed in various device applications in recent years. The 4H-SiC and 6H-SiC exhibit a bandgap of 3.26 eV and 3.02 eV, electron mobility of 1150 and 100 $cm^2/V.s$, hole mobility of 120 and 100 $cm^2/V.s$, and breakdown field of 2.5-3 and 3 MV/cm, respectively, placing them in next-generation device applications groups (Wang et al. 2024). Since polytypes hexagonal 4H-SiC was first used as a wafer in 1994, it has dominated in the industrial and research work over 6H-SiC due to its superior electron mobility(Wang et al. 2024).

Figure 4. Unit cell of SiC structure (using Vesta software).

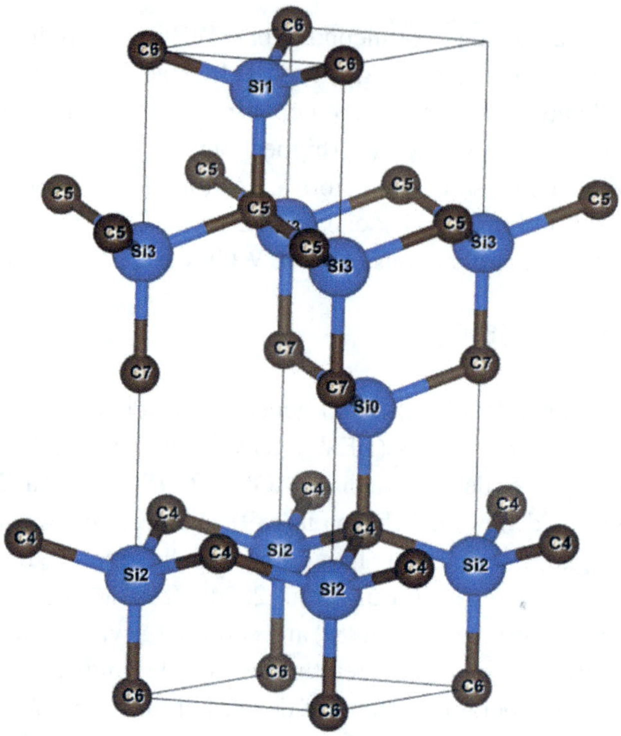

3.2. Gallium Nitride (GaN)

GaN comprises gallium and nitrogen atoms in its unit structure, as shown in Figure 5. GaN, which was first synthesized in 1930, became commercially available in the 1900s due to its application in the development of optoelectronics devices(blue and UV LEDs) (Sharma et al. 2023). It has a direct energy bandgap of ~3.4 eV, a breakdown field of 3.3 MV/cm, excellent thermal conductivity (2.5 Wcm/K), and an electron mobility (~2000 $cm^2/V.s$), making it highly efficient for extremely high-temperature environments (Meneghini et al. 2021). Due to these special characteristics, it has emerged as a widely used WBG for power semiconductor devices (MOSFETs and HEMTs), and optoelectronic devices (UV and blue LEDs) (Meneghini et al. 2021). Its potential is currently being investigated for next-generation power devices, including RF components for 5G, radar systems, and satellite communications, due to its high-frequency performance.

Figure 5. Unit cell structure of cubic GaN (using Vesta software).

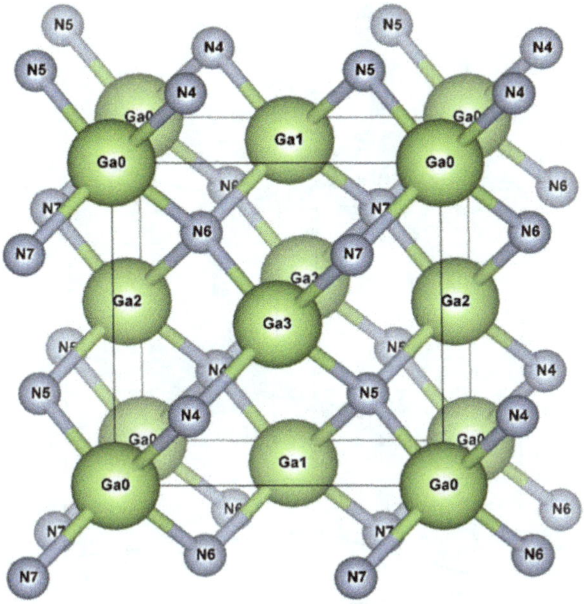

3.3. Aluminium Nitride (AlN)

AlN is an emerging WBG that has gained popularity recently due to its distinct physical, thermal, and electrical properties. It comes under a group of III-V compound semiconductors and consists of a wurtzite crystal structure, as shown in Figure 6 (Bian et al. 2015). AlN, which was first synthesized in the 19th century, remained mostly academic until technical advances in electronics and photonics revealed its promise. It is nowadays widely used in the fabrication and development of next-generation devices. AlN consists wide energy bandgap of ~6.2 eV, a large thermal conductivity (268-374 W/m·K), excellent electrical insulation, and a thermal expansion coefficient (Fu 2016; Rounds et al. 2018; Meneghini et al. 2021). It is increasingly being employed as a substrate and insulating layer in next-generation power electronics devices, particularly GaN and SiC (Galizia et al. 2024). Its compatibility with WBGs improves device efficiency and durability. Moreover, AlN thin films are critical components of bulk acoustic wave (BAW) and surface acoustic wave (SAW) filters, which require materials with high acoustic velocity but moderate dielectric loss (Galizia et al. 2024).

Figure 6. Unit cell Wurtzite structure of AlN (using Vesta software).

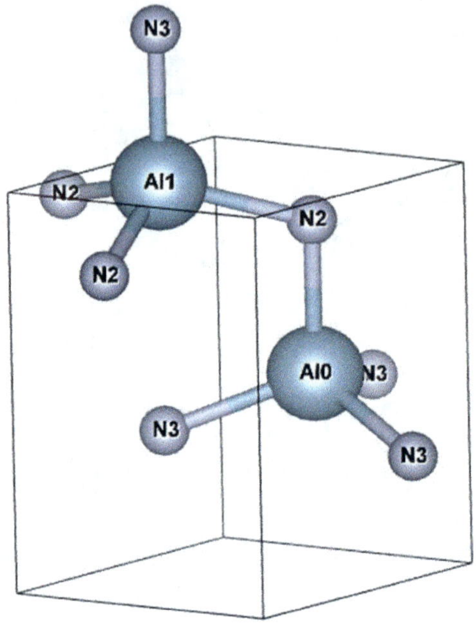

3.4. Gallium Oxide (Ga_2O_3)

Ga_2O_3 has been researched and developed for more than 70 years, although it has just recently gained popularity as a semiconductor material (Higashiwaki 2021; Busch 1989). It was discovered by Lecoq *et al.* (de Boisbaudran 1875) in 1875, but received attention for its application in lasers, sensors, optoelectronics, magneto-electronics, and photonic devices (Zhang et al. 2020; García-Carrión et al. 2020). It consists of n-type conductivity and belongs to the group (III) oxide materials (Litimein et al. 2009). It has mainly five polymorphs, i.e., α, β, γ, δ, and Ɛ-Ga_2O_3, which was discovered in 1952 by Roy *et al.* (Roy, Hill, and Osborn 1952). Amongst all the polymorphs, β-Ga_2O_3 is the most stable one as shown in Figure 7,

Figure 7. Unit cell structure of β-Ga_2O_3 (using Vesta software).

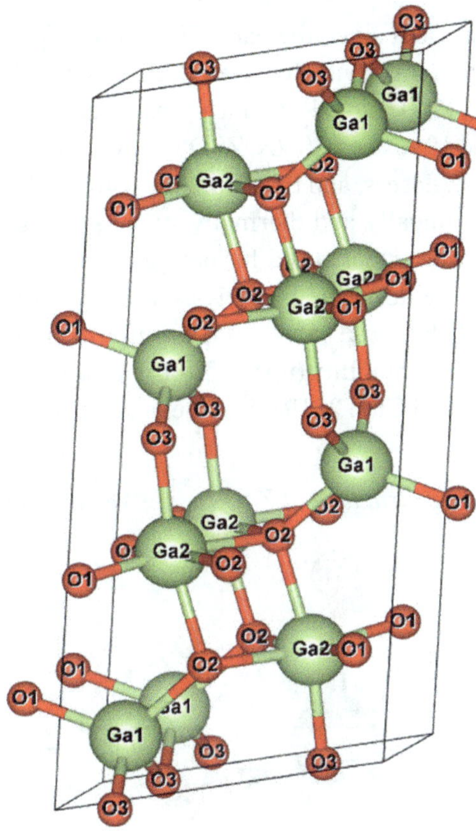

β-Ga_2O_3 has an E_g of ~4.6-4.9 eV, a melting point of 1795°C, a large electron mobility of 300 $cm^2V^{-1}S^{-1}$, a dielectric constant of ~10, a breakdown field of 8 MV/cm, a high Baliga Figure of merits of ~3214, and a poor thermal conductivity of ~0.11 $Wcm^{-1}K^{-1}$ (Liu et al. 2020; Mandal et al. 2024). β-Ga_2O_3 is a cost-effective candidate for next-generation power electronics due to its large efficiency, compact design, and ability to operate at high temperatures. Applications include electric automobiles, renewable energy systems, and aeronautical electronics. Furthermore, its profound UV transparency and excellent resistance make it perfect for use in UV photodetectors and transparent electronics.

3.5. Zinc Oxide (ZnO)

ZnO has been investigated since the early 20th century, primarily for its piezoelectric properties and as a pigment. However, due to its broad bandgap and strong exciton binding energy, ZnO gained popularity as a semiconductor in the 1970s (Mele et al. 2015). In the 1990s and 2000s, ZnO was widely used in optoelectronics like LEDs, solar cells, UV lasers, and thin-film transistors. The hexagonal wurtzite structure is the most chemically and thermally stable structure of ZnO, as shown in Figure 8, which has been widely used for next-generation devices such as solar cells, gas sensors, photodetectors, photo-catalysis, laser systems, and optoelectronic devices (UV-LEDs). ZnO has an E_g of ~3.37 eV, a high exciton binding energy (60 meV), strong n-type electron mobility (up to 155 $cm^2.V^{-1}.S^{-1}$), and outstanding visual transparency (Shin, Hur, and Choi 2019) (Davis et al. 2019) (Kermiche et al. 2022).

Figure 8. Unit cell wurtzite structure of ZnO (using Vesta software).

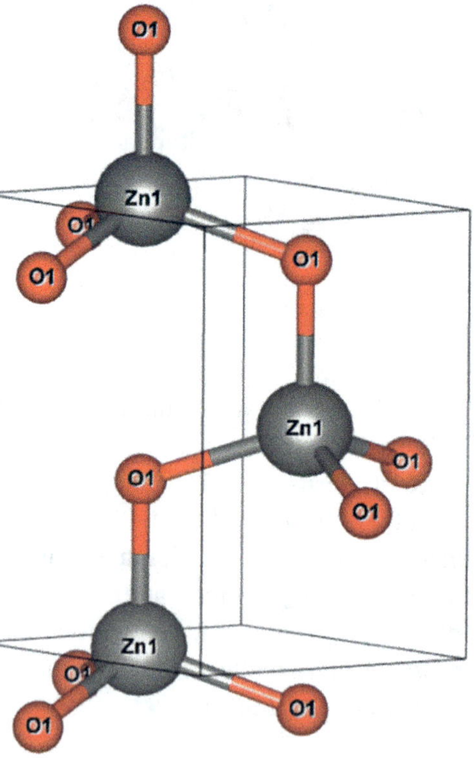

3.6. Diamond

Diamond, once treasured as a gemstone, has now appeared as a highly auspicious WBG for future power devices. Synthetic diamond manufacture, which has long been admired for its hardness and brightness, began in the 1950s, allowing for regulated creation in industrial and scientific applications. Recently, diamond has emerged as a viable WBG for future electronics that enable high-frequency operation and small power consumption. Diamond consists of an E_g of 5.5 eV, strong carrier mobility (4500 and 3800 cm^2/Vs for electrons and holes, respectively), thermal conductivity (2200 W/m.K), transparency, and a large breakdown field (>10 MV/cm) makes the potential for next-generation device for power electronics and electronics fields (Meneghini et al. 2021)(Umezawa 2018). The cubic diamond crystal structure is shown in Figure 9.

Figure 9. Unit cell cubic structure of diamond (using Vesta software).

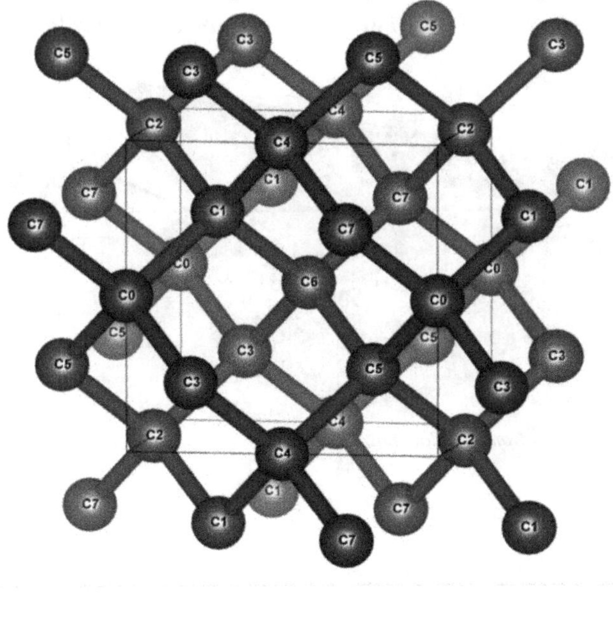

4. NEXT-GENERATION DEVICES

The important next-generation devices are solar-blind UV photodetectors, power devices (MOSFET and Schottky barrier diodes), and sensors. Basic details of these devices are discussed below.

4.1. Photodetector

Photodetectors (PDs) are devices that convert light into electrical signals, enabling the detection of various forms of electromagnetic radiation. The ultraviolet (UV) range spanning wavelengths from 200 to 280 nm is known as the "solar-blind" region. PDs operating within this range are termed solar-blind UV PDs (Arora & Kumar, 2020). These devices can detect extremely weak signals amidst strong background radiation from the Earth, thereby minimizing the chance of false readings. This makes them particularly valuable for applications requiring rapid identification of early warning signals. Key performance parameters of solar-blind photodetectors include dark current (I_{dark}), photocurrent (I_{photo}), external quantum efficiency (EQE), responsivity (R), photo-to-dark current ratio (PDCR), specific detectivity, signal-to-noise ratio (SNR), and response time. Figure 10 shows a Ga_2O_3-based Schottky contact metal-semiconductor-metal (MSM) structure photodetector (Mondal et al. 2020).

Figure 10. Schottky contact MSM structure solar-blind UV photodetector

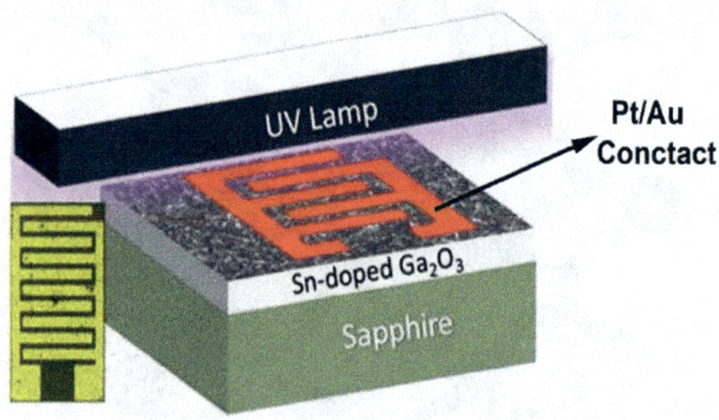

Adapted with permission from Mondal et al. (2020).

4.2. Power Electronic Device

Power electronics like MOSFETs and SBDs have played a significant part in the defence field, where reliability, efficiency, and high power are required. Figure 11(a) and (b) show the β-Ga_2O_3-based SBDs and a MOSFET device (Ji et al. 2024; Wang et al. 2020). These devices play a vital role in advanced propulsion systems used in drones, e-vehicles, and aircraft, where factors such as operational efficiency, effective thermal management, and weight reduction are essential. Key performance parameters for components like SBDs and MOSFETs include on-resistance, breakdown voltage, ideality factor, and Schottky barrier height. Notably, β-Ga_2O_3 has emerged as a promising material for these applications due to its exceptional versatility and capability to perform reliably in extreme environments, positioning it as a breakthrough candidate for future defence and high-power electronics technology.

*Figure 11. Pictorial representation of (**a**) β-Ga_2O_3-based SBD, and (**b**) β-Ga_2O_3-based MOSFET*

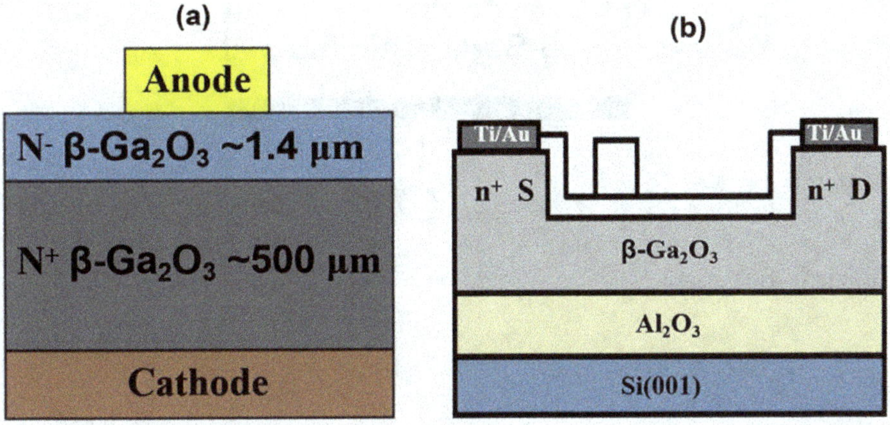

Source: Ji et al. (2024); Wang et al. (2020)

4.3. Sensor

Advances in material science and nanotechnology have substantially influenced the development of wide bandgap materials for UV sensors for next-generation device applications. WBG materials provide increased sensitivity, a larger spectrum response, and improved stability under harsh environmental conditions. In simple, sensors are a device that receives many types of signals, such as physical, chemical, or biological signals and turns them into electrical signal that provide improved sensitivity, downsizing, real-time data processing, and multifunctional capabilities for applications such as monitoring UV levels in the environment, and radiation detection for satellite protection, etc (Vetelino and Reghu 2017). The sensor consists of several performance characteristics like dynamic range, accuracy, saturation, calibration error, input/output signal, hysteresis, linearity, resolution, dead band, repeatability, output impedance, excitation, and dependability (Vetelino and Reghu 2017; Panda et al. 2024). Sensor based on Schottky barrier photodiodes, and nanowire-based sensors improve photosensitivity and speed. A typical WBGs-based SBD sensor side view is represented in Figure 12.

Figure 12. Typical schematic diagram of β-Ga_2O_3-based SBD sensor

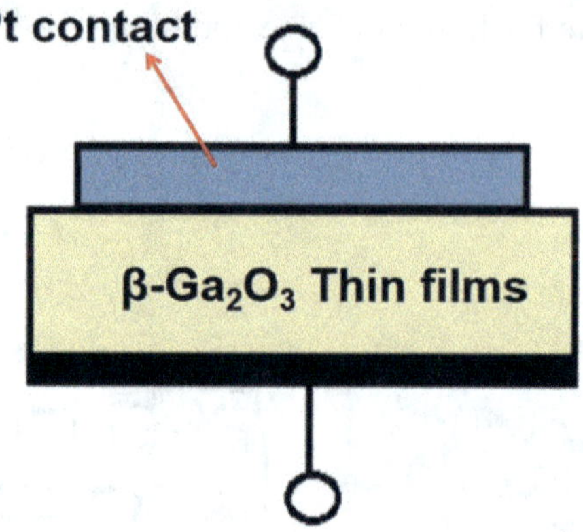

Source: Zhu et al. (2022).

5. NEXT-GENERATION DEVICES ENABLED BY RF SPUTTERED WBGs

Over the past decades, WBGs like GaN, SiC, Ga_2O_3, AlN, ZnO, and diamond, fabricated using the PVD RF sputtering method, have demonstrated significant advancements in the performance of high-power electronic devices, UV PDs, and sensors. In this section, we discuss the relevant morphological properties (e.g., surface roughness and grain structure analysed via scanning electron microscope(SEM) and atomic force microscopy(AFM), structural properties, optical characteristics, and the correlation between these properties and key device performance metrics pertinent to next-generation applications. Furthermore, we review and analyze several published studies focused on WBGs deposited via the RF sputtering method, highlighting their application in emerging device technologies.

5.1. Photodetector

5.1.1. SiC

Ferhati *et al.* (2022) utilized RF sputtering for SiC thin film fabrication with varying sputtering powers ranging from 60 W to 120 W. XRD, UV-Vis. Spectroscopy, EDS, SEM, and I-V measurements were employed to examine the materials' morphological, structural, and electrical properties device performance. Morphological analysis of all fabricated samples was evaluated through SEM, and the reported results indicated a dense and uniform surface film. Furthermore, XRD analysis results revealed that all the samples were amorphous. Amorphous-SiC alloy's optical behaviour may be modulated by the sputtering power, adjusting for favourable visible absorbance at high sputter power. This phenomenon is associated with the effect of RF power on SiC film structural characteristics and compositions. Measurements indicate that an amorphous-SiC PD with 60 W of RF power can detect UV light with a high responsivity (138 mA/W), high detectivity (7.8×10^{12} Jones), and low noise effects, while preserving solar blindness. The constructed device, on the other hand, exhibits extended photo response characteristics at high sputtering power, producing responsivity of 77 mA/W and 426 mA/W in the visible and UV regions, respectively.

5.1.2. Ga_2O_3

Chen *et al.* (2020) deposited a 492 nm β-Ga_2O_3 thin film on MgO substrate using sputtering power (100W) at different growth temperatures of 600-700 °C for a solar-blind UV PD. Upon the analysis, the sample annealed at 700 °C depicted the best morphological and structural properties. With annealing, the obtained sample at a lower temperature tends to have a weaker intensity of the XRD peak, but when the sample was annealed at a higher temperature, the XRD peak intensity was improved. Furthermore, the optical properties reported for the sample annealed at 700 °C show an average transmittance of 80% in the 400-760 nm region. Moreover, the energy bandgap measured through the Tauc's plot was obtained to be 4.91 eV. Therefore, based on optimal material properties, a device was fabricated (700 °C), and the device performance was measured using a current density of 70 μW/cm^2, illumination of 254 nm at 10 V bias. Upon analysis, an I_{dark} of 3.5 pA, an I_{photo} of 42 nA, a PDCR exceeding 10^4, and responsivity of 30 mA/W, with quick rise (0.07s) and decay times (0.06s) has been reported. While Liu *et al.* (2024) fabricated a Sn-doped Ga_2O_3-based solar-blind UV PD using the RF sputtering method on the sapphire substrate by varying sputtering power (0 to 30 W). The surface morphology designates that when the Sn concentration was increased, surface roughness was also increased. Also, when Sn concentration was increased on Ga_2O_3, it affected the intensity of the XRD peak. Moreover, the optical energy bandgap was observed to be decreased from 5.03 eV to 3.91 eV when the Sn concentration was increased. Furthermore, the MSM-structured UV photodetector device was fabricated, and its performance was then characterized. The I_{dark} obtained from a device fabricated at 0 to 30W was 2.46 nA,7.6 nA,11.48 nA, and 26.74 nA, respectively. Moreover, photocurrent under the illumination of 254 nm, 100 μW/ cm^2, under 10 V has been investigated for all samples. The I_{photo} was reported as 0.9 μA, 0.72 μA, 1.17 μA, and 2.72 μA, respectively, for the samples deposited at 0W, 10W, 20W, and 30W. The responsivity of these samples was reported to be increased from 0.67 A/W to 19.25 A/W when sputtering power was increased from 0 W to 30 W. The rise time and decay time reported for the samples fabricated at 30 W were 31 s and 22.83 s, respectively. The best result was obtained for the samples grown at 30 W. Further, Wang *et al.* (2019) fabricated Ga_2O_3 thin films on a sapphire substrate (0001) at parameters i.e., sputtering power (60W, 100W, and 150W), base chamber pressure of 5 x 10^{-4} Pa. From the analysis of morphological analysis(SEM), they concluded that sputtering power had significant effects on films thickness, growth rate and structural properties of thin films, when thin film fabricated at lower sputtering power(60W), the surface of thin film was reported to be rough, while at higher sputtering power(150W) surface of thin film reported to be uniform and lower roughness. Moreover, electrical characteristics on an MSM solar-blind PD was investigated and reported a high PDCR (>10), a decay time of 0.05 s, and a response time of 0.31 s, demonstrating the potential of PVD RF sputtered Ga_2O_3 thin films for fast-response UV detection.

5.1.3. ZnO

Mondal *et al.* (2021) observed a phosphorus (P)-doped ZnO-based UV PD using the RF sputtering method. They used a 100 W sputtering power, 32.4 SSCM Ar gas, and 1.8 SCCM O_2 for 4-hour deposition time of thin film on a PET substrate. Further, various characterization method was employed to study the structure, morphology, and electrical properties. The obtained thickness of the thin film was 445 nm with highly oriented to the c-plane axis. Moreover, upon FESEM and AFM analysis, the average grain diameter of the thin film was reported to be in the size 50-80 nm range with an average surface roughness value of 9.2 nm, respectively. Furthermore, a device was fabricated, where aluminium as an electrode on top of the thin film was deposited using a shadow mask for I-V characteristics. The P-doped ZnO film-based UV PD exhibits significantly higher UV photosensitivity, with responsivity and detectivity of 23.5 mA/W and 1.86×10^{13} Jones related to intrinsic ZnO's of 0.073 mA/W and 2.56×10^{10} Jones, respectively. Also, Rawal *et al.* (2024) had fabricated the n-type-iron (Fe)-doped-zinc oxide(n-$Zn_{1-x}Fe_xO$) based-solar-blind UV PD by the RF sputtering. Before device fabrication, undoped and iron 2.92% and 5.14% doped-$Zn_{1-x}Fe_xO$ film was investigated. The thin film fabricated was reported to be a uniform finish without any defects it. The crystallite size of the thin film was further calculated using the XRD data; it was reported for pristine and iron 2.92% and 5.14% doped-$Zn_{1-x}Fe_xO$ film as 13.2 nm, 16.2 nm, and 18.3 nm, respectively. Furthermore, a heterojunction device of n-iron-doped-$Zn_{1-x}Fe_xO$/p-Si was constructed, and its performance characteristics were investigated. UV radiations of 365 nm and 254 nm were employed to analyse the device's performance characteristics. After analysing the device performance parameters, it was reported that the PDs had recovery and response times of fewer than one second. The detectivity for n-ZnO/p-Si, n-$Zn_{0.97}Fe_{0.03}O$/p-Si, and n-$Zn_{0.95}Fe_{0.05}O$/p-Si devices were 0.321×10^{11}, 1.051×10^{11}, and 2.992×10^{11} Jones, respectively, at a fixed illumination intensity of 0.812 mW/cm^2 for a UVA source with a wavelength of 365 nm.

5.2. Power Electronic Device (MOSFET/SBDs)

5.2.1. Ga_2O_3

Yoon *et al.* (2022) reported a Sn-doped β-Ga_2O_3-based MOSFET device as a solar-blind phototransistor before that thin film was characterized with various characterization methods. The Sn-doped β-Ga_2O_3 thin film thickness was reported as 8 nm, which exhibited a polycrystalline nature. The maximum optical transmittance was observed as 88%, while the energy bandgap of Sn-doped β-Ga_2O_3 was reported as 5.77 eV. Moreover, device performance was carried out. At zero gate voltage, the fully depleted phototransistor demonstrated a high PDCR of ~10^4 and an I_{dark} of 29.3 pA. The gate-driver circuitry's simplicity and low standby power consumption are advantages of the typically off behaviour. β-Ga_2O_3's incredibly thin layer also raised the bandgap and produced a quantum confinement effect. In contrast to other β-Ga_2O_3-based phototransistors that have been described currently, this enabled the suggested device to accomplish detection in the deep UV region (~215 nm). While Yen *et al* (2023) fabricated β-Ga_2O_3/Si-based SBDs via TiN(111) buffer layer device by RF sputtering. The sputtering parameters, like 100 W sputtering power, 600 °C substrate temperature, 10 SCCM Ar gas flow, and 5 mtorr of working pressure evolved. Further thin film was analysed using different methods such as XRD, FE-SEM, AFM, etc. The thin film produced was uniform and defect-free. Furthermore, the device performance characteristics were investigated, according to literature, R_{on} of the as-deposited sample, samples annealed at (700 °C, 800 °C, 900 °C/1 min RTA annealed beta Ga_2O_3/TiN/Si), and Ga_2O_3/Si (RTA:800 C) SBDs were reported as 7.05×10^4, 5.38×10^5, 3.03×10^4, 2.69×10^6, and 6.04×10^7 m.Ω-cm^2, respectively. Furthermore, the V_{br} of the as-deposited sample, annealed samples (700 °C, 800 °C, and 900 °C/1 min RTA annealed β-Ga_2O_3/TiN/Si), and Ga_2O_3/Si (RTA:800 °C) SBDs were reported as 14V, 41V, 48V, 80V, and 9V, respectively.

5.2.2. ZnO

Rajan *et al.* (2016) fabricated a platinum (Pt)/ZnO Schottky diode by the RF sputtering method. The RF sputtering parameters utilized during the ZnO thin films deposition were a sputtering power of 100W, an Ar gas flow of 15 SCCM, and a distance between the sputtering target and substrate of 11 cm, respectively. The deposited ZnO thin film was characterized using various methods such as XRD, FESEM, AFM, and PL. The structural properties reported for ZnO thin film indicated a hexagonal wurtzite crystalline structure with a bandgap of 3.26 eV using both PL and UV-visible spectroscopy methods. Furthermore, a Pt/ZnO Schottky diode was constructed, and its performance characteristics were examined. According to (I-V) and (C-V) characterizations, the device performance parameters such as ideality factor, barrier height, and series resistance were reported as 2.25, 0.71 eV, and 95 Ω, respectively.

5.3. Sensor

5.3.1. SiC

Naderi *et al.* (2020) fabricated a porous SiC thin film-based high-performance UV detector for sensing applications. A 200 W sputtering power, 400 °C substrate temperature, and a 2-hour deposition time were employed, resulting in a film thickness of 1000 nm with an 8.33 nm/min deposition rate. The morphological analysis indicated porous SiC thin films; moreover, the energy bandgap obtained at different wavelengths by PL analysis was 2.57 eV at 482 nm, 2.67 eV at 463 nm, and 2.56 eV at 484 nm, respectively. Furthermore, a SiC-based MSM UV detector was fabricated using the electrode of a finger-shaped Nickel top surface. The device performance of the SiC-based UV detector reported a strong I_{photo} at 350 nm, 0.4 W/cm^2. This device's high sensitivity (54.51) can be attributed to its increased surface area and reduced crystal flaws, resulting in superior optical quality. This device's fast sensing speed (140 ms rise time, 110 ms recovery time) indicates optimal performance for a Schottky UV detector with an optimized SiC structure. Also, Kumar *et al.* (2020) fabricated a Pd/SiC-based sensor by the RF sputtering method, and the effect of the thickness of the thin film was analysed on device performance. During thin film deposition, the sputtering was kept at 6 cm from the substrate holder, and the sputtering chamber was evacuated to 10^{-6} Torr. The thickness of samples varied from 1090 nm, 1680 nm, and 2430 nm, with an amorphous structure. The AFM analysis suggested a dense and uniform all-over substrate with an average grain size of 7 nm, 10 nm, and 18 nm. Furthermore, device performance was measured, and the Pd/SiC-based sensor's performance was reported to decrease as SiC thickness increased. The Pd/SiC-based sensor was reported as highly sensitive to H_2 gas (2 bar) and has a rapid reaction response of 18 s at 250 °C.

5.3.2. Ga_2O_3

Sharma *et al.* (2025) fabricated Ga_2O_3-based UV sensor device by the RF sputtering at a sputtering power of 100W on the r-plane sapphire, a-plane, c-plane, and Si_3N_4 substrates at 800 °C and varying the O_2 gas flow from 0 to 15%, respectively. The thin films deposited at an O_2 gas flow of 10% showed the best crystallinity. Ga_2O_3 films fabricated on the r-plane sapphire and Si_3N_4 were reported to be amorphous or nanocrystalline thin films. Furthermore, the optical bandgaps in β-Ga_2O_3 films ranged from 5.22 to 5.36 eV, with smaller bandgaps found in nanocrystalline/amorphous Ga_2O_3. Further, an MSM-structured device was fabricated where Pt/Ti/Au was used as an electrode. On measuring the device performance characteristics of the fabricated device, it was revealed that when MSM photodetectors were fabricated on polycrystalline β-Ga_2O_3 films on a-plane sapphire, the UV-C PDCR was greater than 10^3. Similarly, MSM photodetectors were also able to attain PDCR of up to 10^3 on an amorphous Si_3N_4 substrate.

5.3.3. ZnO

Shin *et al.* (2019) fabricated a ZnO/PEDOT: PSS-based sensor using an RF magnetron sputtering method. The sputtering parameters utilized for ZnO thin films deposition on soda lime glass are sputtering parameters of 50 W, Ar gas flow of 8 sccm at 20 mTorr working pressure. Further, the thin films were characterized, and morphological analysis reported a dense thin film without any defects(cracks). Moreover, the structural analysis suggested the hexagonal wurtzite structure of ZnO with an optical energy bandgap of 3.25 eV. Furthermore, a device ZnO/PEDOT: PSS was constructed, and Ag(silver) was used as an electrode. Device performance characteristics were analysed further using 254 nm and 364 nm UV illumination and an applied bias of 5V. The I_{dark} of the device was reported as 37.969 nA, while the I_{photo} using 254 nm and 365 nm UV light illumination was 554 nA and 720.49 nA, respectively. Moreover, sensitivity properties were also measured; the sensitivity of the device was reported as 14.49 at 254 nm UV light illumination, while using 365 nm was 21.46, respectively.

Thus, a brief overview of WBGs-based next-generation devices has been presented. RF-sputtered WBGs like Ga_2O_3, SiC, and ZnO have emerged as promising candidates for future electronic, optoelectronic, and power electronic devices due to their superior physical and chemical characteristics. When fabricated as thin films via RF sputtering, these materials exhibit unique morphological, structural, and optical characteristics that significantly influence device performance. The morphological features of WBGs thin films-such as surface roughness, grain size, and optical parameters (e.g., transmittance, absorbance, and energy bandgap)-play critical roles in charge transport and interface quality. For example, RF-sputtered ZnO films typically show nanocrystalline or columnar structures with smooth surfaces, which minimize surface recombination, making them ideal for UV photodetectors. In contrast, diamond and SiC films often exhibit denser, more polycrystalline morphologies, which enhance thermal conductivity and mechanical durability, key factors for power MOSFETs and SBDs. Structurally, high crystalline quality and low defect density are essential. RF sputtering allows fine control over deposition conditions, facilitating the growth of textured or oriented films. For instance, highly c-axis-oriented AlN and GaN films exhibit enhanced piezoelectric and optical properties, improving UV detection sensitivity and sensor performance. Ga_2O_3, with its monoclinic phase, offers excellent breakdown strength and transparency, while diamond provides unmatched mechanical strength and chemical stability for operation in harsh environments. Optically, WBGs reveal better transmittance in the UV range and wide optical bandgaps, from approximately 3.2 eV for ZnO to ~5.5 eV for AlN and >5.5 eV for diamond, making them ideal for solar-blind UV detection and low-loss optical applications. These characteristics enable high responsivity and fast response times in UV photodetectors. Additionally, materials like SiC offer high electron mobility and thermal stability, allowing MOSFETs and SBDs to function at higher voltages and temperatures with lesser power losses. In device applications, performance metrics such as breakdown voltage, leakage current, responsivity, and sensitivity are strongly influenced by the film's structural integrity and interface quality. When optimized, RF-sputtered WBGs demonstrate superior performance, including low dark current in UV detectors, fast switching in power devices, and robust operation in chemically aggressive environments for sensors. Overall, the precise film control afforded by RF sputtering, combined with the intrinsic advantages of WBG materials, provides a strong platform for developing compact, high-efficiency, and durable devices for next-generation electronic and sensing technologies.

6. CHALLENGES ASSOCIATED WITH THE RF SPUTTERED WBGS FOR THE NEXT GENERATION DEVICE

Wide bandgap semiconductors (WBGs) such as ZnO, GaN, SiC, Ga_2O_3, AlN, and Diamond have, over recent decades, emerged as vital materials for next-generation electronic and optoelectronic devices due to their superior optical, electrical, thermal, and chemical properties. RF sputtering offers significant advantages for thin-film deposition owing to its scalability, homogeneity, and relatively low-temperature processing capabilities. However, several critical challenges persist, affecting the performance, reliability, and integration of RF-sputtered WBG-based devices. This section systematically discusses these challenges, their underlying mechanisms, current mitigation strategies, and their direct impact on device metrics.

6.1. Stoichiometry of Thin Film and Defect Management

Achieving precise stoichiometry in RF-sputtered WBG thin films is inherently challenging, especially in reactive sputtering environments. The issue stems from the differing sputtering yields and volatile nature of constituent elements (e.g., nitrogen in GaN, oxygen in ZnO, and Ga_2O_3), leading to compositional drift. This stoichiometry imbalance results in the formation of undesirable secondary phases, defects like oxygen or nitrogen vacancies, and impurity incorporation, which degrade film quality. For instance, nitrogen vacancies in GaN induce unintentional n-type conductivity, shifting device threshold voltages, and increasing leakage currents. Moreover, high-energy ion bombardment during sputtering induces intrinsic defects such as dislocations, point defects, and grain boundary flaws, which act as non-radiative recombination centers or scattering sites, thereby lowering carrier mobility and reducing breakdown voltages.

Mitigation strategies include:

- Co-sputtering method or target composition tuning to balance element loss.
- Reactive gas flow control and substrate biasing to stabilize plasma chemistry.
- Post-deposition annealing in controlled atmospheres to heal defects.

6.1.1 Impact on Device Metrics

Defects and stoichiometry drift lead to increased leakage currents, reduced breakdown voltages, and degraded optical emission efficiency in LEDs and power devices.

Observed trends:

Studies report that higher sputtering power increases deposition rate but exacerbates stoichiometric imbalance and defect generation, while substrate heating and low-pressure deposition can improve film density and reduce vacancy formation.

6.2. Epitaxy Thin Film and Crystallinity of Thin Film Growth

The performance of WBG devices is heavily dependent on the crystallinity of their active layers. RF sputtering typically yields polycrystalline or amorphous films due to limited adatom mobility on the substrate surface, making epitaxial growth on lattice-mismatched substrates difficult. Poor crystallinity introduces grain boundaries and dislocations, which impede charge transport and reduce thermal conductivity, thereby compromising device efficiency and lifetime.

Mitigation strategies include:

- Increasing substrate temperature during deposition to enhance surface diffusion.
- Using buffer layers or template layers to facilitate lattice matching.
- Applying post-deposition annealing to improve grain growth and reduce defects.

6.2.1 Impact on Device Metrics:

Low crystallinity increases defect density, reduces electron and hole mobilities, and limits thermal dissipation, affecting high-frequency and high-power device performance.

Observed trends:

Higher substrate temperatures and in-situ annealing lead to larger grain sizes and improved crystallinity, though at the expense of increased film stress. There is a trade-off between process temperature and film quality in thermally sensitive substrates.

6.3. Residual Stress and Adhesion Problem

The energetic particle flux inherent in RF sputtering induces intrinsic compressive or tensile stresses within the deposited films. These stresses, resulting from differences in atomic packing density, thermal expansion mismatches, and ion bombardment, can lead to film cracking, and delamination.

Mitigation strategies involve:

- Adjustment in the deposition parameters such as power density, working gas pressure, and substrate bias.

6.3.1 Impact on Device Metrics:

High residual stresses can compromise film integrity, reduce device yield, and cause premature mechanical or thermal failure, especially in high-power applications.

Observed trends:

Lower sputtering power and reduced gas pressure generally decrease intrinsic stress but may also affect deposition rate and uniformity. Optimizing deposition parameters remains crucial for balancing stress control and throughput.

6.4. Surface Roughness and Quality of the Thin Film

Surface morphology effectively affects the electrical, optical, and interfacial properties of WBG thin films. Rough surfaces tend to increase surface scattering, electrical properties instability, and higher leakage current in devices. Furthermore, poor interface quality between active layers and contacts or dielectric films can introduce additional charge trapping and recombination centers.

Mitigation strategies:

- Process parameter optimization (e.g., lower deposition rates, substrate heating).
- Post-deposition treatments (annealing).

6.4.1 Impact on Device Metrics

High surface roughness tend to degrades the electron and hole mobility, increases leakage current, and lowers dielectric breakdown strength, directly affecting device efficiency and reliability.

Observed trends:

Lower working pressures and moderate substrate temperatures generally improve surface smoothness. However, ultra-smooth surfaces (<1 nm RMS roughness) are difficult to achieve without post-deposition treatments.

7. CONCLUSIONS

In summary, this chapter focuses on RF-sputtered wide-bandgap semiconductors for future generation devices. The structure-property relationships of thin films deposited via the RF sputtering method, along with their device performance, have been reviewed. Further, the challenges associated with device integration into advanced technologies have also been discussed. RF sputtering is a powerful and versatile technique for depositing wide band gap semiconductor thin films with tailored properties essential for cutting-edge device applications. While these materials offer exceptional performance advantages, overcoming challenges related to stoichiometry control, film stress, and defect management remains critical. Continued advancements in processing methods and characterization techniques are imperative to fully harness the potential of RF-sputtered wide-band gap materials, facilitating the development of their widespread adoption in future power electronic and optoelectronic device technologies.

8. FUTURE SCOPE

Wide bandgap semiconductors have sparked widespread interest in next-generation device applications because of their excellent optical, electrical, thermal, and chemical stability compared to traditional semiconductors (Si). As the requirement for high-frequency, high-power, and energy-efficient electronic devices increases, these materials are projected to play an important role in next-generation technology. RF sputtering, a versatile physical vapor deposition technology, provides a scalable and cost-effective method for producing high-quality thin films of these WBG semiconductors while maintaining perfect control over film composition and microstructure. RF-sputtered SiC films are gaining attention for applications in power electronics and harsh-environment sensors due to their excellent thermal stability, high breakdown field, and radiation hardness. Ga_2O_3, with a very large bandgap and a high critical electric field, is considered a potential candidate for future power devices and solar-blind UV photodetectors. Its ability to be deposited at comparatively low temperatures via RF sputtering enables integration on various substrates, including flexible and temperature-sensitive platforms. Meanwhile, ZnO exhibits promising optoelectronic and piezoelectric properties, making it suitable for transparent electronics and piezoelectric sensors. In the future, RF-sputtered WBG semiconductors may play a significant role in the growth of compact, efficient, and long-lasting devices in industries like aerospace electronics, renewable energy systems, electric vehicles, and wearable technology. Continued breakthroughs in sputtering process optimization and material engineering will improve the performance and commercial viability of next-generation device architectures.

ACKNOWLEDGEMENT

The authors sincerely thank the University of Pardubice, Czech Republic, for their valuable support throughout the writing process and acknowledge the assistance of an AI tool for enhancing the manuscript's language.

REFERENCES

Acosta, M., Riech, I., & Martín-Tovar, E. (2013). Effects of the argon pressure on the optical band gap of zinc oxide thin films grown by nonreactive RF sputtering. *Advances in Condensed Matter Physics*, *2013*, 970976. Advance online publication. DOI: 10.1155/2013/970976

Akbar, G., Di Fatta, A., Rizzo, G., Ala, G., Romano, P., & Imburgia, A. (2025). Comprehensive review of wide-bandgap (WBG) devices: SiC MOSFET and its failure modes affecting reliability. *Physchem*, *5*(1), 10. Advance online publication. DOI: 10.3390/physchem5010010

Astinchap, B., Moradian, R., & Gholami, K. (2017). Effect of sputtering power on optical properties of prepared TiO_2 thin films by thermal oxidation of sputtered Ti layers. *Materials Science in Semiconductor Processing*, *63*, 169–175. DOI: 10.1016/j.mssp.2017.02.007

Baptista, A., Silva, F., Porteiro, J., Míguez, J., & Pinto, G. (2018). Sputtering physical vapour deposition (PVD) coatings: A critical review on process improvement and market trend demands. *Coatings*, *8*(11), 402. Advance online publication. DOI: 10.3390/coatings8110402

Bian, Y., Liu, M., Ke, G., Chen, Y., DiBattista, J., Chan, E., & Yang, Y. (2015). Aluminum nitride thin film growth and applications for heat dissipation. *Surface and Coatings Technology*, *267*, 65–69. DOI: 10.1016/j.surfcoat.2014.11.060

Blumenschein, N., Paskova, T., & Muth, J. F. (2019). Effect of growth pressure on PLD-deposited gallium oxide thin films for deep-UV photodetectors. *Physica Status Solidi (A). Physica Status Solidi. A, Applications and Materials Science*, *216*(20), 1900098. Advance online publication. DOI: 10.1002/pssa.201900098

Busch, G. (1989). Early history of the physics and chemistry of semiconductors—From doubts to fact in a hundred years. *European Journal of Physics*, *10*(4), 254–264. DOI: 10.1088/0143-0807/10/4/002

Chalana, S. R., Kavitha, V. S., Srinivasan, P., Ahamed, S. R., Tiwari, G., & Pillai, V. P. M. (2019). Influence of argon pressure on the crystalline phase formation of RF sputtered zinc sulfide thin films. *Materials Today: Proceedings*, *29*, 917–922. DOI: 10.1016/j.matpr.2020.05.347

Chaudhary, O. S., Denaï, M., Refaat, S. S., & Pissanidis, G. (2023). Technology and applications of wide bandgap semiconductor materials: Current state and future trends. *Energies*, *16*(18), 6689. Advance online publication. DOI: 10.3390/en16186689

Cheah, L. B., Osman, R. A. M., & Poopalan, P. (2020). Ga_2O_3 thin films by sol-gel method its optical properties. *AIP Conference Proceedings*, *2203*, 020001. Advance online publication. DOI: 10.1063/1.5142120

Chelvanathan, P., Shahahmadi, S. A., Arith, F., Sobayel, K., Akhtaruzzaman, M., Sopian, K., Alharbi, F. H., Tabet, N., & Amin, N. (2017). Effects of RF magnetron sputtering deposition process parameters on the properties of molybdenum thin films. *Thin Solid Films*, *638*, 213–219. DOI: 10.1016/j.tsf.2017.07.057

Chen, X., Mi, W., Wu, J., Yang, Z., Zhang, K., Zhao, J., Luan, C., & Wei, Y. L. (2020). A solar-blind photodetector based on β-Ga_2O_3 film deposited on MgO (100) substrates by RF magnetron sputtering. *Vacuum*, *180*, 109632. Advance online publication. DOI: 10.1016/j.vacuum.2020.109632

Cheng, Z., Wang, X., Gao, J., Wang, M., Wang, A., Bo, H., Guo, Z., Zou, W., & Ma, X. (2024). Deposition and structural investigation of uniform AlN(100) films at wafer scale through RF magnetron sputtering. *Ceramics International*, *50*(16), 28601–28608. DOI: 10.1016/j.ceramint.2024.05.170

Davis, K., Yarbrough, R., Froeschle, M., White, J., & Rathnayake, H. (2019). Band gap engineered zinc oxide nanostructures via a sol-gel synthesis of solvent driven shape-controlled crystal growth. *RSC Advances*, *9*(26), 14638–14648. DOI: 10.1039/C9RA02091H PMID: 35516315

de Boisbaudran, L. (1875). On the chemical and spectroscopic characters of a new metal (gallium). *The London, Edinburgh and Dublin Philosophical Magazine and Journal of Science*, *50*(332), 414–416. DOI: 10.1080/14786447508641312

Dobrikov, G. H., Rassovska, M. M., Andreev, N. M., Boyadzhiev, S. I., Gesheva, K. A., Ivanova, T. M., Sharlandjiev, P. S., & Nazarowa, D. I. (2009). Development of transparent heat mirrors based on metal oxide thin film structures. *Thin Solid Films*, *518*(4), 1091–1094. DOI: 10.1016/j.tsf.2009.07.203

Dragomir (Nicolescu), L., Antoniac, A., Manescu (Paltanea), V., Robu, A., Dinu, M., Pana, I., Cotrut, C. M., Kamel, E., Antoniac, I., Rau, J. V., & Vladescu (Dragomir), A. (. (2023). Preparation and characterization of hydroxyapatite coating by magnetron sputtering on Mg–Zn–Ag alloys for orthopaedic trauma implants. *Ceramics International*, *49*(16), 26274–26288. DOI: 10.1016/j.ceramint.2023.05.116

El-Shaer, A., Ezzat, S., Habib, M. A., Alduaij, O. K., Meaz, T. M., & El-Attar, S. A. (2023). Influence of deposition time on structural, morphological, and optical properties of CdS thin films grown by low-cost chemical bath deposition. *Crystals*, *13*(5), 788. Advance online publication. DOI: 10.3390/cryst13050788

Fedorenko, A. V., Bozhko, K. M., Kachur, N. V., Kosiakovskiy, A. V., & Maslov, V. P. (2024). Optical and electrical properties of zinc oxide nanofilms deposited using the sol-gel method. *Semiconductor Physics. Quantum Electronics and Optoelectronics*, *27*(1), 117–123. DOI: 10.15407/spqeo27.01.117

Ferhati, H., Djeffal, F., Bendjerad, A., Foughali, L., Benhaya, A., & Saidi, A. (2022). Highly-detective tunable band-selective photodetector based on RF sputtered amorphous SiC thin-film: Effect of sputtering power. *Journal of Alloys and Compounds*, *907*, 164464. Advance online publication. DOI: 10.1016/j.jallcom.2022.164464

Fu, X. (2016). Aluminum nitride wide band-gap semiconductor and its basic characteristics. In *Proceedings of the 2016 2nd International Conference on Education, Management, Information and Mechanical Engineering (EMIM 2016)* (pp. 555–558). https://doi.org/DOI: 10.2991/emim-16.2016.117

Galizia, B., Fiorenza, P., Schilirò, E., Pecz, B., Foragassy, Z., Greco, G., Saggio, M., Cascino, S., Lo Nigro, R., & Roccaforte, F. (2024). Towards aluminum oxide/aluminum nitride insulating stacks on 4H–SiC by atomic layer deposition. *Materials Science in Semiconductor Processing*, *174*, 108244. Advance online publication. DOI: 10.1016/j.mssp.2024.108244

García-Carrión, M., Ramírez-Castellanos, J., Nogales, E., Méndez, B., You, C. C., Karazhanov, S., & Marstein, E. S. (2020). Hybrid solar cells with β- and γ-gallium oxide nanoparticles. *Materials Letters*, *261*, 127088. Advance online publication. DOI: 10.1016/j.matlet.2019.127088

Gu, L., Ma, H. P., Shen, Y., Zhang, J., Chen, W. J., Yang, R. Y., Wu, F., Yang, L., Zeng, Y. X., Wang, X. R., Zhu, J. T., & Zhang, Q. C. (2022). Temperature-dependent oxygen annealing effect on the properties of Ga_2O_3 thin film deposited by atomic layer deposition. *Journal of Alloys and Compounds*, *925*, 166727. Advance online publication. DOI: 10.1016/j.jallcom.2022.166727

Hernandez, A., Islam, M. M., Saddatkia, P., Codding, C., Dulal, P., Agarwal, S., Janover, A., Novak, S., Huang, M., Dang, T., Snure, M., & Selim, F. A. (2021). MOCVD growth and characterization of conductive homoepitaxial Si-doped Ga_2O_3. *Results in Physics*, *25*, 104167. Advance online publication. DOI: 10.1016/j.rinp.2021.104167

Higashiwaki, M. (2021). β-Gallium oxide devices: Progress and outlook. *Physica Status Solidi. Rapid Research Letters*, *15*(11), 2100357. Advance online publication. DOI: 10.1002/pssr.202100357

Ji, X., Wang, J., Qi, S., Liang, Y., Hu, S., Zheng, H., Zhang, S., Yue, J., Qi, X., Li, S., Liu, Z., Shu, L., Tang, W., & Li, P. (2024). Improvement of Ga_2O_3 vertical Schottky barrier diode by constructing NiO/Ga_2O_3 heterojunction. *Journal of Semiconductors*, *45*(4), 042503. Advance online publication. DOI: 10.1088/1674-4926/45/4/042503

Kermiche, F., Taabouche, A., Bouabellou, A., Hanini, F., & Bouachiba, Y. (2022). Behavior study of ZnO thin films grown by PLD for several applications. *Crystallography Reports*, *67*(7), 1239–1245. DOI: 10.1134/S1063774522070069

Kumar, M., Kumar, A., Gautam, Y. K., Chandra, R., Goyat, M. S., Tewari, B. S., & Tewari, R. K. (2020). Influence of SiC thin films thickness on the electrical properties of Pd/SiC thin films for hydrogen gas sensor. *Vacuum*, *182*, 109750. Advance online publication. DOI: 10.1016/j.vacuum.2020.109750

Lenis, J. A., Gómez, M. A., & Bolívar, F. J. (2019). Effect of deposition temperature and target-substrate distance on the structure, phases, mechanical and tribological properties of multi-layer HA-Ag coatings obtained by RF magnetron sputtering. *Surface and Coatings Technology*, *378*, 124936. Advance online publication. DOI: 10.1016/j.surfcoat.2019.124936

Li, K., Tang, R., Zhu, C., & Chen, T. (2024). Critical review on crystal orientation engineering of antimony chalcogenide thin film for solar cell applications. *Advancement of Science*, *11*(2), 2304963. Advance online publication. DOI: 10.1002/advs.202304963 PMID: 37939308

Lin, C. R., Wei, D. H., BenDao, M. K., Chen, W. E., & Liu, T. Y. (2014). Development of high-performance UV detector using nanocrystalline diamond thin film. *International Journal of Photoenergy*, *2014*, 492152. Advance online publication. DOI: 10.1155/2014/492152

Litimein, F., Rached, D., Khenata, R., & Baltache, H. (2009). FPLAPW study of the structural, electronic, and optical properties of Ga_2O_3: Monoclinic and hexagonal phases. *Journal of Alloys and Compounds*, *488*(1), 148–156. DOI: 10.1016/j.jallcom.2009.08.092

Liu, X., Liu, Q., Zhao, B., Ren, Y., Tao, B. W., & Zhang, W. L. (2020). Comparison of β-Ga_2O_3 thin films grown on r-plane and c-plane sapphire substrates. *Vacuum*, *178*, 109435. Advance online publication. DOI: 10.1016/j.vacuum.2020.109435

Liu, Y., Huang, R., Lin, T., Dang, J., Huang, H., Shi, J., & Chen, S. (2024). Preparation of Sn-doped Ga_2O_3 thin films and MSM ultraviolet detectors using magnetron co-sputtering. *Materials (Basel)*, *17*(13), 3227. Advance online publication. DOI: 10.3390/ma17133227 PMID: 38998311

Mandal, P., Kumar, S., Pandey, A., Katariya, L., Mondal, A., & Bag, A. (2024). Unveiling structural and optical properties of Sn-doped β-Ga_2O_3: A correlation of experimental and theoretical observations. *Materials Science and Engineering B*, *302*, 117266. Advance online publication. DOI: 10.1016/j.mseb.2024.117266

Mandal, P., Mondal, A., Pandey, A., Romeo, S., & Bag, A. (2025). Investigating role of annealing in shaping morphological, structural, tribological, and optical characteristics of gallium oxide (Ga_2O_3): Insights from DFT analysis. *Journal of Alloys and Compounds*, *1011*, 178367. Advance online publication. DOI: 10.1016/j.jallcom.2024.178367

Mandal, P., & Roy, S. (2025). Ultra-wide bandgap β-gallium oxide (Ga_2O_3) advancing next-generation defense technologies. DOI: 10.4018/979-8-3373-0933-0.ch011.

Mandal, P., Roy, S., & Singh, U. P. (2022). Investigation on the optical and electrical performance of aluminium doped gallium oxide thin films. *Optical and Quantum Electronics*, *54*(8), 515. Advance online publication. DOI: 10.1007/s11082-022-03851-0

Mandal, P., Singh, U. P., & Roy, S. (2021a). A review on the effects of PVD RF sputtering parameters on rare earth oxide thin films and their applications. *IOP Conference Series. Materials Science and Engineering*, *1166*(1), 012022. Advance online publication. DOI: 10.1088/1757-899X/1166/1/012022

Mandal, P., Singh, U. P., & Roy, S. (2021b). Optical performance of europium-doped β gallium oxide PVD thin films. *Journal of Materials Science Materials in Electronics*, *32*(4), 3958–3965. DOI: 10.1007/s10854-020-05137-1

Mathkari, R., Liehr, M., Ravindra, P., Pareis, R., Beckmann, K., Tokranova, N., Schujman, S., Saraf, I., Van der Straten, O., Gong, N., Ando, T., & Cady, N. (2025). Investigation of the effect of oxygen partial pressure during reactive sputtering of tantalum oxide resistive random access memory switching layer. *Materials Science in Semiconductor Processing*, *186*, 109060. Advance online publication. DOI: 10.1016/j.mssp.2024.109060

Mattox, D. M. (2010). *Handbook of physical vapor deposition (PVD) processing* (2nd ed.). Elsevier.

Mele, P., Endo, T., Arisawa, S., Li, C., & Tsuchiya, T. (Eds.). (2015). *Oxide thin films, multilayers, and nanocomposites*. Springer. DOI: 10.1007/978-3-319-14478-8

Meneghini, M., De Santi, C., Abid, I., Buffolo, M., Cioni, M., Khadar, R. A., Nela, L., Zagni, N., Chini, A., Medjdoub, F., Meneghesso, G., Verzellesi, G., Zanoni, E., & Matioli, E. (2021). GaN-based power devices: Physics, reliability, and perspectives. *Journal of Applied Physics*, *130*(18), 181101. Advance online publication. DOI: 10.1063/5.0061354

Mezdrogina, M. M., Vinogradov, A. Y., Kozhanova, Y. V., & Borsuk, E. A. (2020). LED structures based on ZnO films obtained by RF magnetron sputtering for the UV spectral range. *Technical Physics*, *65*(3), 434–439. DOI: 10.1134/S1063784220030159

Mi, W., Du, X., Luan, C., Xiao, H., & Ma, J. (2014). Electrical and optical characterizations of β-Ga_2O_3:Sn films deposited on MgO (110) substrate by MOCVD. *RSC Advances*, *4*(58), 30579–30583. DOI: 10.1039/C4RA02479F

Mobtakeri, S., Akaltun, Y., Özer, A., Kılıç, M., Tüzemen, E. Ş., & Gür, E. (2021). Gallium oxide films deposition by RF magnetron sputtering; a detailed analysis on the effects of deposition pressure and sputtering power and annealing. *Ceramics International*, *47*(2), 1721–1727. DOI: 10.1016/j.ceramint.2020.08.289

Mondal, A., Yadav, M. K., Shringi, S., & Bag, A. (2020). Extremely low dark current and detection range extension of Ga_2O_3 UV photodetector using Sn alloyed nanostructures. *Nanotechnology*, *31*(29), 295201. Advance online publication. DOI: 10.1088/1361-6528/ab82d4 PMID: 32208373

Mondal, S., Ghosh, S., & Basak, D. (2021). Extraordinarily high ultraviolet photodetection by defect tuned phosphorus doped ZnO thin film on flexible substrate. *Materials Research Bulletin*, *144*, 111490. Advance online publication. DOI: 10.1016/j.materresbull.2021.111490

Mourya, S. K., Malik, G., Alisha, , Kumar, B., & Chandra, R. (2022). The role of non-homogeneous barrier on the electrical performance of 15R–SiC Schottky diodes grown by in-situ RF sputtering. *Materials Science in Semiconductor Processing*, *149*, 106855. Advance online publication. DOI: 10.1016/j.mssp.2022.106855

Naderi, N., & Moghaddam, M. (2020). Ultra-sensitive UV sensors based on porous silicon carbide thin films on silicon substrate. *Ceramics International*, *46*(9), 13821–13826. DOI: 10.1016/j.ceramint.2020.02.173

Nair, P. B., Justinvictor, V. B., Daniel, G. P., Joy, K., Ramakrishnan, V., & Thomas, P. V. (2011). Effect of RF power and sputtering pressure on the structural and optical properties of TiO_2 thin films prepared by RF magnetron sputtering. *Applied Surface Science*, *257*(24), 10869–10875. DOI: 10.1016/j.apsusc.2011.07.125

Ngo, T. S., Le, D. D., Song, J. H., & Hong, S. K. (2019). Growth and characterization of gallium oxide films grown with nitrogen by plasma-assisted molecular-beam epitaxy. *Thin Solid Films*, *682*, 93–98. DOI: 10.1016/j.tsf.2019.05.029

Panda, S., Mehlawat, S., Dhariwal, N., Sanger, A., & Kumar, A. (2024). Comprehensive review on gas sensors: Unveiling recent developments and addressing challenges. *Materials Science and Engineering B*, *308*, 117616. Advance online publication. DOI: 10.1016/j.mseb.2024.117616

Park, J. H., McClintock, R., Jaud, A., Dehzangi, A., & Razeghi, M. (2019). MOCVD grown β-Ga_2O_3 metal-oxide-semiconductor field effect transistors on sapphire. *Applied Physics Express*, *12*(9), 091001. Advance online publication. DOI: 10.7567/1882-0786/ab3b2a

Ra, S., Han, L., & Zhao, H. (2016). Synthesis of wide bandgap Ga_2O_3. *Semiconductors*, *50*(4), 1002–1009. DOI: 10.1134/S1063782624602152

Rajan, L., Periasamy, C., & Sahula, V. (2016). Comprehensive study on electrical and hydrogen gas sensing characteristics of Pt/ZnO nanocrystalline thin film-based Schottky diodes grown on n-Si substrate using RF sputtering. *IEEE Transactions on Nanotechnology*, *15*(2), 201–208. DOI: 10.1109/TNANO.2015.2513102

Ramachandran, R. K., Dendooven, J., Botterman, J., Sree, S. P., Poelman, D., Martens, J. A., Poelman, H., & Detavernier, C. (2014). Plasma enhanced atomic layer deposition of Ga_2O_3 thin films. *Journal of Materials Chemistry. A, Materials for Energy and Sustainability*, *2*(45), 19232–19238. DOI: 10.1039/C4TA05007J

Ranga, P., Bhattacharyya, A., Whittaker-Brooks, L., Scarpulla, M. A., & Krishnamoorthy, S. (2021). N-type doping of low-pressure chemical vapor deposition grown β-Ga_2O_3 thin films using solid-source germanium. *Journal of Vacuum Science & Technology. A, Vacuum, Surfaces, and Films*, *39*(3), 030404. Advance online publication. DOI: 10.1116/6.0001004

Rawal, I., Goyal, P. K., & Dahiya, S. (2024). Highly sensitive self-powered solar-blind ultraviolet photodetectors based on n-$Zn_{1-X}Fe_XO$/p-Si heterojunctions. *Sensors and Actuators. A, Physical*, *371*, 115268. Advance online publication. DOI: 10.1016/j.sna.2024.115268

Rounds, R., Sarkar, B., Klump, A., Hartmann, C., Nagashima, T., Kirste, R., Franke, A., Bickermann, M., Kumagai, Y., Sitar, Z., & Collazo, R. (2018). Thermal conductivity of single-crystalline AlN. DOI: 10.7567/apex.11.071001

Roy, R., Hill, V. G., & Osborn, E. F. (1952). Polymorphism of Ga_2O_3 and the system Ga_2O_3–H_2O. *Journal of the American Chemical Society*, *74*(3), 719–722. DOI: 10.1021/ja01123a039

Saikumar, A. K., Nehate, S. D., & Sundaram, K. B. (2019). Review—RF sputtered films of Ga_2O_3. *ECS Journal of Solid State Science and Technology : JSS*, *8*(7), Q3064–Q3078. DOI: 10.1149/2.0141907jss

Saikumar, A. K., Sundaresh, S., Nehate, S. D., Phelps, J., Abdolvand, R., & Sundaram, K. B. (2023). Work function estimation of gallium-doped zinc oxide using transparent gate electrode MOSFET. *ECS Journal of Solid State Science and Technology : JSS*, *12*(3), 033010. Advance online publication. DOI: 10.1149/2162-8777/acbede

Sharma, R. C., Nandal, R., Tanwar, N., Yadav, R., Bhardwaj, J., & Verma, A. (2023). Gallium arsenide and gallium nitride semiconductors for power and optoelectronics devices applications. *Journal of Physics: Conference Series*, *2426*(1), 012008. Advance online publication. DOI: 10.1088/1742-6596/2426/1/012008

Sharma, S., Nagib, H., Le, P. Y., Allen, M. W., Holland, A. S., Partridge, J. G., & Tran, H. N. (2025). Structural, surface, electrical and UVC sensing properties of high temperature RF sputtered gallium oxide thin films. *Materials Today Electronics*, *11*, 100139. Advance online publication. DOI: 10.1016/j.mtelec.2025.100139

Shen, H., Yin, Y., Tian, K., Baskaran, K., Duan, L., Zhao, X., & Tiwari, A. (2018). Growth and characterization of β-Ga_2O_3 thin films by sol-gel method for fast-response solar-blind ultraviolet photodetectors. *Journal of Alloys and Compounds*, *766*, 601–608. DOI: 10.1016/j.jallcom.2018.06.313

Shin, S. G., Hur, J., & Choi, H. W. (2019). A study of ZnO/PEDOT:PSS based UV sensors with RF sputter. *Japanese Journal of Applied Physics, 58*(SD). *Japanese Journal of Applied Physics*, *SD0804*(SD), SDDE09. Advance online publication. DOI: 10.7567/1347-4065/ab0ff3

Sparavigna, A. C. (2018). The invention of carborundum, the synthetic silicon carbide. [HAL Id: hal-01686436; likely preprint or report.]

Suthar, A. N., Venkataramanaiah, J., & Suresh, Y. (2025). Conventional, wide-bandgap, and hybrid power converters: A comprehensive review. *Renewable & Sustainable Energy Reviews*, *213*, 115419. Advance online publication. DOI: 10.1016/j.rser.2025.115419

Tondare, R. S., Shivaraj, B. W., Narasimhamurthy, H. N., Krishna, M., & Subramanyam, T. K. (2018). Effect of deposition time on structural, electrical and optical properties of aluminium doped ZnO thin films by RF magnetron sputtering. *Materials Today: Proceedings*, *5*(1), 2710–2715. DOI: 10.1016/j.matpr.2018.01.052

Torky, M., & Chow, T. P. (2024). Determination of the effective critical breakdown field for Si, wide, and extreme bandgap semiconductor superjunction devices. *AIP Advances*, *14*(5), 055001. Advance online publication. DOI: 10.1063/5.0211696

Umezawa, H. (2018). Recent advances in diamond power semiconductor devices. *Materials Science in Semiconductor Processing*, *78*, 147–156. DOI: 10.1016/j.mssp.2018.01.007

Varley, J. B., Shen, B., & Higashiwaki, M. (2022). Wide bandgap semiconductor materials and devices. *Journal of Applied Physics*, *131*(23), 230401. Advance online publication. DOI: 10.1063/5.0100601

Vetelino, J. F., & Reghu, A. (2017). *Introduction to sensors* (2nd ed.). CRC Press., DOI: 10.1201/9781315218274

Vorobyova, M., Biffoli, F., Giurlani, W., Martinuzzi, S. M., Linser, M., Caneschi, A., & Innocenti, M. (2023). PVD for decorative applications: A review. *Materials (Basel)*, *16*(14), 4919. Advance online publication. DOI: 10.3390/ma16144919 PMID: 37512195

Wang, H., Lai, P., Islam, M. Z., Hasan, A. S. M. K., Di Mauro, A., Anika, N. E., Russell, R., Feng, Z., Chen, K., Faruque, A., White, T., Chen, Z., & Mantooth, H. A. (2024). A review of silicon carbide CMOS technology for harsh environments. *Materials Science in Semiconductor Processing*, *178*, 108422. Advance online publication. DOI: 10.1016/j.mssp.2024.108422

Wang, J., Ye, L., Wang, X., Zhang, H., Li, L., Kong, C., & Li, W. (2019). High transmittance β-Ga_2O_3 thin films deposited by magnetron sputtering and post-annealing for solar-blind ultraviolet photodetector. *Journal of Alloys and Compounds*, *803*, 9–15. DOI: 10.1016/j.jallcom.2019.06.224

Wang, Y. B., Xu, W. H., You, T. G., Mu, F. W., Hu, H. D., Liu, Y., Huang, H., Suga, T., Han, G. Q., Ou, X., & Hao, Y. (2020). β-Ga_2O_3 MOSFETs on the Si substrate fabricated by the ion-cutting process. *Science China. Physics, Mechanics & Astronomy*, *63*(7), 277311. Advance online publication. DOI: 10.1007/s11433-020-1533-0

Wasa, K. (2012). *Sputtering systems* (2nd ed.). Elsevier. DOI: 10.1016/B978-1-4377-3483-6.00003-6

Wu, C., Zhu, Z., Li, J., Tu, C., Lv, P., & Wang, Y. (2024). Fabrication of h-BN solar-blind ultraviolet detectors by RF magnetron sputtering. *Next Materials, 3*, 100075. Advance online publication. DOI: 10.1016/j.nxmate.2023.100075

Yen, C. C., Singh, A. K., Wu, P. W., Chou, H. Y., & Wuu, D. S. (2023). Interface engineering in epitaxial growth of sputtered β-Ga_2O_3 films on Si substrates via TiN (111) buffer layer for Schottky barrier diodes. *Materials Today. Advances, 17*, 100348. Advance online publication. DOI: 10.1016/j.mtadv.2023.100348

Yoon, Y., Hwang, W. S., & Shin, M. (2022). Solar-blind ultrathin Sn-doped polycrystalline Ga_2O_3 UV phototransistor for normally off operation. *Advanced Photonics Research, 3*(5), 2100316. Advance online publication. DOI: 10.1002/adpr.202100316

Yoshikawa, A., Matsunami, H., & Nanishi, Y. (2007). Development and applications of wide bandgap semiconductors. In *Wide bandgap semiconductors: Fundamental properties and modern photonic and electronic devices* (pp. 1–24). Springer., DOI: 10.1007/978-3-540-47235-3_1

Yu, F.-P., Ou, S.-L., & Wuu, D.-S. (2015). Pulsed laser deposition of gallium oxide films for high performance solar-blind photodetectors. *Optical Materials Express, 5*(5), 1240–1249. DOI: 10.1364/OME.5.001240

Zhang, X., Wang, L., Wang, X., Chen, Y., Shao, Q., Wu, G., Wang, X., Lin, T., Shen, H., Wang, J., Meng, X., & Chu, J. (2020). High-performance β-Ga_2O_3 thickness dependent solar blind photodetector. *Optics Express, 28*(3), 4169–4177. DOI: 10.1364/OE.385470 PMID: 32122074

Zhu, J., Xu, Z., Ha, S., Li, D., Zhang, K., Zhang, H., & Feng, J. (2022). Gallium oxide for gas sensor applications: A comprehensive review. *Materials (Basel), 15*(20), 7339. Advance online publication. DOI: 10.3390/ma15207339 PMID: 36295403

Compilation of References

Abbas, A.-G. N., Aziz, F. N. A. A., Abdan, K., Mohd Nasir, N. A., & Norizan, M. N. (2022). Kenaf Fibre Reinforced Cementitious Composites. *Fibers (Basel, Switzerland), 10*(1), 3. Advance online publication. DOI: 10.3390/fib10010003

Abbood, I. S., Odaa, S., Hasan, K. F., & Jasim, M. A. (2021). Properties evaluation of fiber reinforced polymers and their constituent materials used in structures – A review. *Materials Today: Proceedings, 43*(Part 2), 1003–1008. DOI: 10.1016/j.matpr.2020.07.636

Abdellaoui, H., Raji, M., Bouhfid, R., & Qaiss, A. el K. (2019). Investigation of the deformation behavior of epoxy-based composite materials. In Failure Analysis in Biocomposites, Fibre-Reinforced Composites and Hybrid Composites (Woodhead Publishing Series in Composites Science and Engineering), pp. (chapter 2). Woodhead Publishing, 29-49. DOI: 10.1016/B978-0-08-102293-1.00002-4

Abdullatef, M. S., Alzubaidi, F. N., Al-Tamimi, A., & Mahmood, Y. A. (2023). Fatigue Life Estimation of High Strength 2090-T83 Aluminum Alloy under Pure Torsion Loading using Various Machine Learning Techniques. *Fluid Dyn. Mater. Process., 19*(8), 2083–2107. DOI: 10.32604/fdmp.2023.027266

Ablekim, T., Duenow, J. N., Zheng, X., Moutinho, H., Moseley, J., Perkins, C. L., Johnston, S. W., O'Keefe, P., Colegrove, E., Albin, D. S., Reese, M. O., & Metzger, W. K. (2020). Thin-Film Solar Cells with 19% Efficiency by Thermal Evaporation of CdSe and CdTe. *ACS Energy Letters, 5*(3), 892–896. DOI: 10.1021/acsenergylett.9b02836

Acosta, M., Riech, I., & Martín-Tovar, E. (2013). Effects of the argon pressure on the optical band gap of zinc oxide thin films grown by nonreactive RF sputtering. *Advances in Condensed Matter Physics, 2013*, 970976. Advance online publication. DOI: 10.1155/2013/970976

Acquesta, A., & Monetta, T. (2020). As-Built EBM and DMLS Ti-6Al-4V Parts: Topography–Corrosion Resistance Relationship in a Simulated Body Fluid. *Metals*, *10*(8), 1015. DOI: 10.3390/met10081015

Afre, R. A., & Pugliese, D. (2024). Perovskite Solar Cells: A Review of the Latest Advances in Materials, Fabrication Techniques, and Stability Enhancement Strategies. *Micromachines 2024, Vol. 15, Page 192, 15*(2), 192. DOI: 10.3390/mi15020192

Agueda, J. R. H. S., Chen, Q., Maalihan, R. D., Ren, J., Da Silva, Í. G. M., Dugos, N. P., Caldona, E. B., & Advincula, R. C. (2021). 3D printing of biomedically relevant polymer materials and biocompatibility. *MRS Communications*, *11*(2), 197–212. DOI: 10.1557/s43579-021-00038-8

Ahangar, P., Cooke, M. E., Weber, M. H., & Rosenzweig, D. H. (2019). Current Biomedical Applications of 3D Printing and Additive Manufacturing. *Applied Sciences (Basel, Switzerland)*, *9*(8), 1713. DOI: 10.3390/app9081713

Ahmad, B., van der Veen, S. O., Fitzpatrick, M. E., & Guo, H. (2018). Measurement and modelling of residual stress in wire-feed additively manufactured titanium. *Materials Science and Technology*, *34*(18), 2250–2259. DOI: 10.1080/02670836.2018.1528747

Ahmad, M. N., Ishak, M. R., Mohammad Taha, M., Mustapha, F., Leman, Z., Anak Lukista, D. D., Irianto, , & Ghazali, I. (2022). Application of Taguchi Method to Optimize the Parameter of Fused Deposition Modeling (FDM) Using Oil Palm Fiber Reinforced Thermoplastic Composites. *Polymers*, *14*(11), 2140. DOI: 10.3390/polym14112140

Ahn, N., & Choi, M. (2024). Towards Long-Term Stable Perovskite Solar Cells: Degradation Mechanisms and Stabilization Techniques. *Advancement of Science*, *11*(4), 2306110. DOI: 10.1002/advs.202306110 PMID: 37997198

Akbar, G., Di Fatta, A., Rizzo, G., Ala, G., Romano, P., & Imburgia, A. (2025). Comprehensive review of wide-bandgap (WBG) devices: SiC MOSFET and its failure modes affecting reliability. *Physchem*, *5*(1), 10. Advance online publication. DOI: 10.3390/physchem5010010

Akgün, M. (2023). Measurement and optimization of cutting forces, surface roughness and temperature in turning of AZ91 Mg alloy. *Sadhana*, *48*(2), 60. DOI: 10.1007/s12046-023-02113-3

Akram, J., Pal, D., & Stucker, B. (2019). Establishing flow stress and elongation relationships as a function of microstructural features of ti6al4v alloy processed using slm. *Designs*, *3*(2), 1–11. DOI: 10.3390/designs3020021

Al Miaari, A., & Ali, H. M. (2025). Recent advances on nano-enhanced phase change materials (NEPCMs) for photovoltaic thermal management and role of machine learning: A review of fundamentals, preparation, characterization, and thermo-physical properties. *Journal of Energy Storage*, *124*(April), 116544. DOI: 10.1016/j.est.2025.116544

Al Miaari, A., El Khatib, A., & Ali, H. M. (2023). Design and thermal performance of an innovative greenhouse. *Sustainable Energy Technologies and Assessments*, *57*, 103285. DOI: 10.1016/j.seta.2023.103285

Al-Ali, S., Olabi, A. G., & Mahmoud, M. (2025). A review of solar photovoltaic technologies: developments, challenges, and future perspectives. *Energy Conversion and Management: X, 27*(February), 101057. DOI: 10.1016/j.ecmx.2025.101057

Al-Azad, N., Asril, M. F. M., & Shah, M. K. M. (2021). A review on development of natural fibre composites for construction applications. *Journal of Materials Science and Chemical Engineering*, *9*(7), 1–9. DOI: 10.4236/msce.2021.97001

Alcaraz, J. Y.II, Foqué, W., Sharma, A., & Tjahjowidodo, T. (2024). Indirect porosity detection and root-cause identification in WAAM. *Journal of Intelligent Manufacturing*, *35*(4), 1607–1628. DOI: 10.1007/s10845-023-02128-x

Alemu, D., Getachew, E., & Mondal, A. K. (2023). Study on the Physicochemical Properties of Chitosan and their Applications in the Biomedical Sector. *International Journal of Polymer Science*, *2023*, 1–13. DOI: 10.1155/2023/5025341

Alexandre-Franco, M. F., Kouider, R., Al-Karany, R. K., Cuerda-Correa, E. M., & Al-Kassir, A. (2024). Recent Advances in polymer science and fabrication processes for enhanced microfluidic applications: An Overview. *Micromachines*, *15*(9), 1137. DOI: 10.3390/mi15091137

Ali, A. O., Elgohr, A. T., El-Mahdy, M. H., Zohir, H. M., Emam, A. Z., Mostafa, M. G., Al-Razgan, M., Kasem, H. M., & Elhadidy, M. S. (2025). Advancements in photovoltaic technology: A comprehensive review of recent advances and future prospects. *Energy Conversion and Management: X, 26*(February), 100952. DOI: 10.1016/j.ecmx.2025.100952

Ali, M. A., Mufti, N. A., Sana, M., Tlija, M., Farooq, M. U., & Haber, R. (2024). Enhancing high-speed EDM performance of hybrid aluminium matrix composite by genetic algorithm integrated neural network optimization. *Journal of Materials Research and Technology*, *31*, 4113–4127. DOI: 10.1016/j.jmrt.2024.07.077

Ally, S., Spelt, J. K., & Papini, M. (2012). Prediction of machined surface evolution in the abrasive jet micro-machining of metals. *Wear, 292–293*, 89–99. DOI: 10.1016/j.wear.2012.05.029

Almer, J. D., & Winholtz, R. A. (2008). *X-Ray Stress Analysis.*, DOI: 10.1007/978-0-387-30877-7_28

Al-Mhanna, M. A.-M. H., Al-Ethari, H., & Jasim, T. A. (2023). Influence of Warm Squeezing on Some Properties of AA7075 Alloy Prepared by Stir Casting. *2023 14th International Conference on Mechanical and Aerospace Engineering (ICMAE)*, 60–63. DOI: 10.1109/ICMAE59650.2023.10424449

Alrashidi, H., Ghosh, A., Issa, W., Sellami, N., Mallick, T. K., & Sundaram, S. (2020). Thermal performance of semitransparent CdTe BIPV window at temperate climate. *Solar Energy, 195*, 536–543. DOI: 10.1016/j.solener.2019.11.084

Alrashidi, H., Issa, W., Sellami, N., Sundaram, S., & Mallick, T. (2022a). Thermal performance evaluation and energy saving potential of semi-transparent CdTe in Façade BIPV. *Solar Energy, 232*, 84–91. DOI: 10.1016/j.solener.2021.12.037

Altin Karataş, M., & Biberci, M. A. (2023). Statistical Analysis of WEDM Machining Parameters of Ti-6Al-4V Alloy Using Taguchi-Based Grey Relational Analysis and Artificial Neural Network. *Experimental Techniques, 47*(4), 851–870. DOI: 10.1007/s40799-022-00601-5

Anjum, M. N. (2023). eNext: An IoT and AI driven solution to the plugged-ear pandemic. *IEEE Internet of Things Journal, 10*(13), 11940–11941. DOI: 10.1109/JIOT.2023.3244800

Anoop Johny, & Thiagarajan, C. (2022). Investigation, analysis, optimization of reciprocating wire-cut electrical discharge machining of titanium alloy with molybdenum electrode. *Journal of Ceramic Processing Research, 23*(4), 443–458.

Arefin, A. M. E., Khatri, N. R., Kulkarni, N., & Egan, P. F. (2021). Polymer 3D Printing Review: Materials, process, and design Strategies for medical applications. *Polymers, 13*(9), 1499. DOI: 10.3390/polym13091499

Arunnath, A., Madhu, S., & Tufa, M. (2022). Experimental Investigation and Optimization of Material Removal Rate and Tool Wear in the Machining of Aluminum-Boron Carbide (Al-B4C) Nanocomposite Using EDM Process. *Advances in Materials Science and Engineering, 2022*, 1–11. DOI: 10.1155/2022/4254024

Ashby, M. F. (2006). The properties of foams and lattices. *Philosophical Transactions of the Royal Society A: Mathematical, Physical and Engineering Sciences, 364*(1838), 15-30.

Ashok Sonawane, S., & Sitaram Wangikar, S. (2022). Multi-superiority response optimization of WEDM for titanium grade-5 using PCA based TOPSIS theory. *Materials Today: Proceedings*, *62*, 841–847. DOI: 10.1016/j.matpr.2022.04.044

Aslam, S., Aslam, S., Wang, T., Feng, D., & Zhang, S. (2025). A novel and secure machine learning-based Hyperledger blockchain for IoT healthcare. *IEEE Internet of Things Journal*, *12*(15), 31126–31140. Advance online publication. DOI: 10.1109/JIOT.2025.3574471

Assad, H., Assad, A., & Kumar, A. (2023). Recent developments in 3D Bio-Printing and its biomedical applications. *Pharmaceutics*, *15*(1), 255. DOI: 10.3390/pharmaceutics15010255

Astinchap, B., Moradian, R., & Gholami, K. (2017). Effect of sputtering power on optical properties of prepared TiO_2 thin films by thermal oxidation of sputtered Ti layers. *Materials Science in Semiconductor Processing*, *63*, 169–175. DOI: 10.1016/j.mssp.2017.02.007

ASTM D3039, Standard Test Method for Tensile Properties of Polymer Matrix Composite Materials.

Ataee, A., Li, Y., & Wen, C. (2019). A comparative study on the nanoindentation behavior, wear resistance and in vitro biocompatibility of SLM manufactured CP–Ti and EBM manufactured Ti64 gyroid scaffolds. *Acta Biomaterialia*, *97*, 587–596. DOI: 10.1016/j.actbio.2019.08.008

Aufa, A. N., Hassan, M. Z., & Ismail, Z. (2022). Recent advances in Ti-6Al-4V additively manufactured by selective laser melting for biomedical implants: Prospect development. *Journal of Alloys and Compounds*, *896*, 163072. DOI: 10.1016/j.jallcom.2021.163072

AVENSTON. (2021). *Seven impressive BIPV systems in Europe | AVENSTON*. https://avenston.com/en/articles/7bipv/

Awais, M., Khan, W., Akram, T., & Nam, Y. (2025). Energy-efficient discrete cosine transform architecture using reversible logic for IoT-enabled consumer electronics. [or access via IEEE Xplore]. *IEEE Access : Practical Innovations, Open Solutions*, *13*, 12292–12307. DOI: 10.1109/ACCESS.2025.3528261

Babu, B. S., & Kumar, K. (2021). *Nanomaterials and Nanocomposites: Characterization, Processing, and Applications*. CRC Press. Retrieved from https://books.google.co.in/books?id=RxwfEAAAQBAJ

Bagherpour, E., Pardis, N., Reihanian, M., & Ebrahimi, R. (2019). An overview on severe plastic deformation: Research status, techniques classification, microstructure evolution, and applications. *International Journal of Advanced Manufacturing Technology, 100*(5-8), 1647–1694. DOI: 10.1007/s00170-018-2652-z

Bai, Y., & Williams, C. B. (2015). An exploration of binder jetting of copper. *Rapid Prototyping Journal, 21*(2), 177–185. DOI: 10.1108/RPJ-12-2014-0180

Balaji, V. R., Kalvinathan, V., Dheepanchakkravarthy, A., & Muthuvel, P. (2021). IoT enabled smart irrigation system. In *2021 International Conference on Advancements in Electrical, Electronics, Communication, Computing and Automation (ICAECA)* (pp. 1–6).

Balla, V. K., Bodhak, S., Bose, S., & Bandyopadhyay, A. (2010). Porous tantalum structures for bone implants: Fabrication, mechanical and in vitro biological properties. *Acta Biomaterialia, 6*(8), 3349–3359. DOI: 10.1016/j.actbio.2010.01.046

Bandyopadhyay, A., Espana, F., Balla, V. K., Bose, S., Ohgami, Y., & Davies, N. M. (2010). Influence of porosity on mechanical properties and in vivo response of Ti6Al4V implants. *Acta Biomaterialia, 6*(4), 1640–1648. DOI: 10.1016/j.actbio.2009.11.011

Baptista, A., Silva, F., Porteiro, J., Míguez, J., & Pinto, G. (2018). Sputtering physical vapour deposition (PVD) coatings: A critical review on process improvement and market trend demands. *Coatings, 8*(11), 402. Advance online publication. DOI: 10.3390/coatings8110402

Baragetti, S., Cavalleri, S., & Tordini, F. (2011). Fatigue behavior of notched Ti-6Al-4V in air and corrosive environment. *Procedia Engineering, 10*, 2435–2440. DOI: 10.1016/j.proeng.2011.04.401

Barath Kumar, M. D., & Manikandan, M. (2022). Assessment of process, parameters, residual stress mitigation, post treatments and finite element analysis simulations of wire arc additive manufacturing technique. *Metals and Materials International, 28*(1), 54–111. DOI: 10.1007/s12540-021-01015-5

Baskaran, J. (2025). Metallic 3D-Printed Scaffolds with Functional Coatings: Advancing Tissue Regeneration through Bioactive and Antibacterial Surfaces. In *Innovations and Applications of Advanced Biomaterials in Healthcare and Engineering* (pp. 253-276): IGI Global Scientific Publishing.

Baskaran, J., Muthukannan, D., Shukla, R., & Gokuldoss, P. K. (2024). Manufacturability and deformation studies on a novel metallic lattice structure fabricated by Selective Laser Melting. *Vacuum, 222*, 113065. DOI: 10.1016/j.vacuum.2024.113065

Basol, B. M., & McCandless, B. (2014). Brief review of cadmium telluride-based photovoltaic technologies. *Journal of Photonics for Energy*, *4*(1), 040996. DOI: 10.1117/1.JPE.4.040996

Basu, B., Gowtham, N. H., Xiao, Y., Kalidindi, S. R., & Leong, K. W. (2022). Biomaterialomics: Data science-driven pathways to develop fourth-generation biomaterials. *Acta Biomaterialia*, *143*, 1–25. DOI: 10.1016/j.actbio.2022.02.027

Basumatary, P., & Agarwal, P. (2022). A short review on progress in perovskite solar cells. *Materials Research Bulletin*, *149*, 111700. DOI: 10.1016/j.materresbull.2021.111700

Beaman, J. J., Bourell, D. L., Seepersad, C. C., & Kovar, D. (2020). Additive manufacturing review: Early past to current practice. *Journal of Manufacturing Science and Engineering*, *142*(11), 110812. Advance online publication. DOI: 10.1115/1.4048193

Behjat, A., Sanaei, S., Mosallanejad, M. H., Atapour, M., Sheikholeslam, M., Saboori, A., & Iuliano, L. (2024). A novel titanium alloy for load-bearing biomedical implants: Evaluating the antibacterial and biocompatibility of Ti536 produced via electron beam powder bed fusion additive manufacturing process. *Biomaterials Advances*, *163*, 213928. DOI: 10.1016/j.bioadv.2024.213928

Benafan, O., Noebe, R., Padula, I. I. S.II, Brown, D., Vogel, S., & Vaidyanathan, R. (2014). Thermomechanical cycling of a NiTi shape memory alloy-macroscopic response and microstructural evolution. *International Journal of Plasticity*, *56*, 99–118. DOI: 10.1016/j.ijplas.2014.01.006

Benedetti, M., & Fontanari, V. (2004). The effect of bimodal and lamellar microstructures of Ti–6Al–4V on the behaviour of fatigue cracks emanating from edge-notches. *Fatigue & Fracture of Engineering Materials & Structures*, *27*(11), 1073–1089. DOI: 10.1111/j.1460-2695.2004.00825.x

Besekar, N., & Bhattacharyya, B. (2022). Experimental investigation and characterization of NiTinol shape memory alloy during wire electrochemical machining. *Journal of Manufacturing Processes*, *81*, 346–361. DOI: 10.1016/j.jmapro.2022.07.019

Best Research-Cell Efficiency Chart | Photovoltaic Research | NREL. (2025). https://www.nrel.gov/pv/cell-efficiency

Bezek, L. B., & Williams, C. B. (2023). Process-structure-property effects of ultraviolet curing in multi-material jetting additive manufacturing. *Additive Manufacturing*, *73*, 103640. Advance online publication. DOI: 10.1016/j.addma.2023.103640

Bharathi, P., Sampath Kumar, T., & Anbuchezhiyan, G. (2025). Multioptimization analysis of machining characteristics on spark electrical discharge machining of Al/SiC and Al/SiC/B 4 C composites. *Engineering Research Express*, *7*(1), 015518. DOI: 10.1088/2631-8695/ada723

Bhardwaj, A. R., Meshram, P. D., Chakule, R. R., & Gupta, P. (2025). Processing of aluminum metal matrix composite with titanium carbide reinforcement. *Materials and Manufacturing Processes*, *40*(9), 1–15. DOI: 10.1080/10426914.2025.2491413

Bhat, A., & Kakandikar, G. (2019). Manufacture of silicon carbide reinforced aluminium 6061 metal matrix composites for enhanced sliding wear properties. *Manufacturing Review*, *6*, 24. DOI: 10.1051/mfreview/2019021

Bhattacharya, S., Mishra, A., Singh, T., Bandyopadhyay, S., Kumar, S., Dey, G. K., & Kain, V. (2019). Simulation of residual stress using thermomechanical analysis on wire EDM cut surface. *Materials Today: Proceedings*, *19*, 462–467. DOI: 10.1016/j.matpr.2019.07.636

Bhowmick, S., Prajapati, G., Biswas, N., & Majumdar, G. (2024). Fuzzy Modelling and Grey Relation Analysis of Electric Discharge Machining Performed on Inconel 718 Superalloy. *2024 4th International Conference on Computer, Communication, Control & Information Technology (C3IT)*, 1–6. DOI: 10.1109/C3IT60531.2024.10829413

Bhuiyan, M. S., Mutoh, Y., & McEvily, A. J. (2012). The influence of mechanical surface treatments on fatigue behavior of extruded AZ61 magnesium alloy. *Materials Science and Engineering A*, *549*, 69–75. DOI: 10.1016/j.msea.2012.04.007

Bhushan, R. K. (2020). Impact of nose radius and machining parameters on surface roughness, tool wear and tool life during turning of AA7075/SiC composites for green manufacturing. *Mechanics of Advanced Materials and Modern Processes*, *6*(1), 1. DOI: 10.1186/s40759-020-00045-7

Bian, Y., Liu, M., Ke, G., Chen, Y., DiBattista, J., Chan, E., & Yang, Y. (2015). Aluminum nitride thin film growth and applications for heat dissipation. *Surface and Coatings Technology*, *267*, 65–69. DOI: 10.1016/j.surfcoat.2014.11.060

Bikas, H., Stavropoulos, P., & Chryssolouris, G. (2016). Additive manufacturing methods and modelling approaches: A critical review. *International Journal of Advanced Manufacturing Technology*, *83*(1-4), 389–405. DOI: 10.1007/s00170-015-7576-2

Blaiszik, B., Jones, A., Sottos, N. R., & White, S. R. (2014). Microencapsulation of gallium–indium (Ga–In) liquid metal for self-healing applications. *Journal of Microencapsulation, 31*(4), 350–354. DOI: 10.3109/02652048.2013.858790

Blumenschein, N., Paskova, T., & Muth, J. F. (2019). Effect of growth pressure on PLD-deposited gallium oxide thin films for deep-UV photodetectors. *Physica Status Solidi (A). Physica Status Solidi. A, Applications and Materials Science, 216*(20), 1900098. Advance online publication. DOI: 10.1002/pssa.201900098

Bouanani, B., Joti, A., Bachir Bouiadjra, F. S., & Kadid, A. (2020). Band gap and thickness optimization for improvement of CIGS/CIGS tandem solar cells using Silvaco software. *Optik (Stuttgart), 204*, 164217. Advance online publication. DOI: 10.1016/j.ijleo.2020.164217

Bouchaala, K. (2023). Review of Aluminium Lithium Alloy Manufacturing Procedures, Mechanical Characteristics and Precipitations for Usage in Aerospace Applications. *Journal of Astrophysics & Aerospace Technology, 11*, 1. DOI: 10.37421/2329-6542.2023.11.249

Boyer, R. R. (1996). An overview on the use of titanium in the aerospace industry. *Materials Science and Engineering A, 213*(1-2), 103–114. DOI: 10.1016/0921-5093(96)10233-1

Breuning, C., Pistor, J., Markl, M., & Körner, C. (2022). Basic Mechanism of Surface Topography Evolution in Electron Beam Based Additive Manufacturing. *Materials (Basel), 15*(14), 4754. DOI: 10.3390/ma15144754

Busch, G. (1989). Early history of the physics and chemistry of semiconductors—From doubts to fact in a hundred years. *European Journal of Physics, 10*(4), 254–264. DOI: 10.1088/0143-0807/10/4/002

Butler, K. T., Davies, D. W., Cartwright, H., Isayev, O., & Walsh, A. (2018). Machine learning for molecular and materials science. *Nature, 559*(7715), 547–555. DOI: 10.1038/s41586-018-0337-2

Capitão, J., Martins, J., Emami, S., Ivanou, D., & Mendes, A. (2023). Fully glass frit encapsulated dye-sensitized solar cells: Challenges for hermetical sealing of electrolyte injection holes. *Solar Energy, 249*, 476–484. DOI: 10.1016/j.solener.2022.12.001

Carlson, D. E., & Wronski, C. R. (1976). Amorphous silicon solar cell. *Applied Physics Letters, 28*(11), 671–673. DOI: 10.1063/1.88617

Carlson, D. E., & Wronski, C. R. (1977). Solar cells using discharge-produced amorphous silicon. *Journal of Electronic Materials, 6*(2), 95–106. DOI: 10.1007/BF02660377

Cavaliere, P., & De Marco, P. P. (2007). Superplastic behaviour of friction stir processed AZ91 magnesium alloy produced by high pressure die cast. *Journal of Materials Processing Technology*, *184*(1–3), 77–83. DOI: 10.1016/j.jmatprotec.2006.11.005

Chalana, S. R., Kavitha, V. S., Srinivasan, P., Ahamed, S. R., Tiwari, G., & Pillai, V. P. M. (2019). Influence of argon pressure on the crystalline phase formation of RF sputtered zinc sulfide thin films. *Materials Today: Proceedings*, *29*, 917–922. DOI: 10.1016/j.matpr.2020.05.347

Chamos, A. N., Pantelakis, S. G., Haidemenopoulos, G. N., & Kamouts, E. (2008). Tensile and fatigue behaviour of wrought magnesium alloys AZ31 and AZ61. *Fatigue & Fracture of Engineering Materials & Structures*, *31*(9), 812–821. DOI: 10.1111/j.1460-2695.2008.01267.x

Chaudhary, O. S., Denaï, M., Refaat, S. S., & Pissanidis, G. (2023). Technology and applications of wide bandgap semiconductor materials: Current state and future trends. *Energies*, *16*(18), 6689. Advance online publication. DOI: 10.3390/en16186689

Cheah, L. B., Osman, R. A. M., & Poopalan, P. (2020). Ga_2O_3 thin films by sol-gel method its optical properties. *AIP Conference Proceedings*, *2203*, 020001. Advance online publication. DOI: 10.1063/1.5142120

Chelvanathan, P., Shahahmadi, S. A., Arith, F., Sobayel, K., Akhtaruzzaman, M., Sopian, K., Alharbi, F. H., Tabet, N., & Amin, N. (2017). Effects of RF magnetron sputtering deposition process parameters on the properties of molybdenum thin films. *Thin Solid Films*, *638*, 213–219. DOI: 10.1016/j.tsf.2017.07.057

Chen, C., Chen, J., Han, H., Chao, L., Hu, J., Niu, T., Dong, H., Yang, S., Xia, Y., Chen, Y., & Huang, W. (2022). Perovskite solar cells based on screen-printed thin films. *Nature*, *612*(7939), 266–271. DOI: 10.1038/s41586-022-05346-0 PMID: 36352221

Chen, C., Wang, X., Wang, Y., Yang, D., Yao, F., Zhang, W., Wang, B., Sewvandi, G. A., Yang, D., & Hu, D. (2020). Additive manufacturing of piezoelectric materials. *Advanced Functional Materials*, *30*(52), 2005141. DOI: 10.1002/adfm.202005141

Chen, F., Miao, X., Tang, Y., & Yin, S. (2017). A review on recent advances in machining methods based on abrasive jet polishing (AJP). *International Journal of Advanced Manufacturing Technology*, *90*(1-4), 785–799. DOI: 10.1007/s00170-016-9405-7

Cheng, X. Y., Li, S. J., Murr, L. E., Zhang, Z. B., Hao, Y. L., Yang, R., Medina, F., & Wicker, R. B. (2012). Compression deformation behavior of Ti–6Al–4V alloy with cellular structures fabricated by electron beam melting. *Journal of the Mechanical Behavior of Biomedical Materials, 16*, 153–162. DOI: 10.1016/j.jmbbm.2012.10.005

Cheng, X., Shen, Z., & Zhang, Y. (2024). Bioinspired 3D flexible devices and functional systems. *National Science Review, 11*(3), nwad314. DOI: 10.1093/nsr/nwad314

Cheng, Z., Wang, X., Gao, J., Wang, M., Wang, A., Bo, H., Guo, Z., Zou, W., & Ma, X. (2024). Deposition and structural investigation of uniform AlN(100) films at wafer scale through RF magnetron sputtering. *Ceramics International, 50*(16), 28601–28608. DOI: 10.1016/j.ceramint.2024.05.170

Chen, H., Liu, Y., Balabani, S., Hirayama, R., & Huang, J. (2023). Machine learning in predicting printable biomaterial formulations for direct ink writing. *Research, 6*, 0197. Advance online publication. DOI: 10.34133/research.0197

Chen, J., Chen, R., Liao, H., He, Y., Chen, P., Tian, Y., Zhang, B., & Zhang, Z. (2024a). Improving Joint Performance of Friction Stir Welded 2195-O Al-Li Alloy by Post-Weld Treatment and Rolling Deformation. *Journal of Materials Research and Technology, 29*, 5048–5059. DOI: 10.1016/j.jmrt.2024.03.009

Chen, J.-P., Gu, L., & He, G.-J. (2020). A review on conventional and nonconventional machining of SiC particle-reinforced aluminium matrix composites. *Advances in Manufacturing, 8*(3), 279–315. DOI: 10.1007/s40436-020-00313-2

Chen, S., Gao, H., Zhang, Y., Wu, Q., Gao, Z., & Zhou, X. (2022). Review on residual stresses in metal additive manufacturing: Formation mechanisms, parameter dependencies, prediction and control approaches. *Journal of Materials Research and Technology, 17*, 2950–2974. DOI: 10.1016/j.jmrt.2022.02.054

Chen, W.-M., Xie, Y. M., Imbalzano, G., Shen, J., Xu, S., Lee, S.-J., & Lee, P. V. S. (2016). Lattice Ti structures with low rigidity but compatible mechanical strength: Design of implant materials for trabecular bone. *International Journal of Precision Engineering and Manufacturing, 17*(6), 793–799. DOI: 10.1007/s12541-016-0097-6

Chen, X., Mi, W., Wu, J., Yang, Z., Zhang, K., Zhao, J., Luan, C., & Wei, Y. L. (2020). A solar-blind photodetector based on β-Ga_2O_3 film deposited on MgO (100) substrates by RF magnetron sputtering. *Vacuum, 180*, 109632. Advance online publication. DOI: 10.1016/j.vacuum.2020.109632

Chen, X., Shah, K., Dong, S., Peterson, L., Callagon La Plante, E., & Sant, G. (2020). Elucidating the corrosion-related degradation mechanisms of a Ti-6Al-4V dental implant. *Dental Materials, 36*(3), 431–441. DOI: 10.1016/j.dental.2020.01.008

Chen, Y., Pan, X., Peng, J., & Wang, F. (2022). Effect of Coupling Damage on Fatigue Behavior of 2198-T8 Aluminum-Lithium Alloy Thin Sheet for Fuselage. *Fatigue & Fracture of Engineering Materials & Structures*, *45*(3), 754–769. DOI: 10.1111/ffe.13631

Chen, Y., Yu, J., Ge, X., Sun, Y., Sun, L., Zhou, W., & Zhao, G. (2024b). Cryogenic Deformation Behavior, Constitutive Modeling and Microstructure Evolution of Solution-Treated 2195 Al-Li Alloy at High Strain Rates. *Journal of Materials Research and Technology*, *31*, 2915–2929. DOI: 10.1016/j.jmrt.2024.07.013

Chen, Z., Yuan, L., Pan, Z., Zhu, H., Ma, N., Ding, D., & Li, H. (2025). A comprehensive review and future perspectives of simulation approaches in wire arc additive manufacturing (WAAM). *International Journal of Extreme Manufacturing*, *7*(2), 022016. DOI: 10.1088/2631-7990/ada099

Chia, H. N., & Wu, B. M. (2015). Recent advances in 3D printing of biomaterials. *Journal of Biological Engineering*, *9*(1), 4. Advance online publication. DOI: 10.1186/s13036-015-0001-4

Childerhouse, T., M'Saoubi, R., Franca, L., Weston, N., Crawforth, P., & Jackson, M. (2022). The influence of machining induced surface integrity and residual stress on the fatigue performance of Ti-6Al-4V following polycrystalline diamond and coated cemented carbide milling. *International Journal of Fatigue*, *163*, 107054. DOI: 10.1016/j.ijfatigue.2022.107054

Chinna, R., Sharma, P., & Fernandes, F. (2025). Machinability and surface integrity analysis of Ti-17 alloy using WEDC for advanced aero-engine application. *Heliyon*, *11*(1), e41461. DOI: 10.1016/j.heliyon.2024.e41461

Chmielewska, A., Jahadakbar, A., Wysocki, B., Elahinia, M., Święszkowski, W., & Dean, D. (2022). Chemical polishing of additively manufactured, porous, nickel–titanium skeletal fixation plates. *3D Printing. Additive Manufacturing*, *9*(4), 269–277.

Cho, M. Y., Kim, I. S., Park, C., Sohn, H., Lee, D., Koo, S. M., Cho, W. J., Kim, Y. N., Lee, D. W., & Oh, J. M. (2019). Multistage Degradation Mechanisms in Cu(In, Ga)Se2 Photovoltaic Modules Prepared by Co-Evaporation: Toward High Performances and Enhanced Stability. *ACS Applied Energy Materials*, *2*(10), 7171–7182. DOI: 10.1021/acsaem.9b01155

Chophla, E., Sharda, S., Madan, S., & Chopra, A. (2024). IoT-driven healthcare: A new era of connected care. In *2024 2nd International Conference on Advances in Computation, Communication and Information Technology (ICAICCIT)* (pp. 585–592). DOI: 10.1109/ICAICCIT64383.2024.10912296

Choudhary, O. P., Jaiswal, S., & Das, S. (2021). IoT enabled condition monitoring of low voltage motors using fuzzy inference system. In *2021 IEEE 5th International Conference on Condition Assessment Techniques in Electrical Systems (CATCON)* (pp. 132–133). DOI: 10.1109/CATCON52335.2021.9670486

Cho, Y., Beak, J. W., Sagong, M., Ahn, S., Nam, J. S., & Kim, I. (2025). Electrospinning and nanofiber technology: Fundamentals, innovations, and applications. *Advanced Materials, 37*(28), 2570190. Advance online publication. DOI: 10.1002/adma.202570190

Chua, Z. Y., Ahn, I. H., & Moon, S. K. (2017). Process monitoring and inspection systems in metal additive manufacturing: Status and applications. *International Journal of Precision Engineering and Manufacturing-Green Technology, 4*(2), 235–245. DOI: 10.1007/s40684-017-0029-7

Chudinova, E. A., Surmeneva, M. A., Timin, A. S., Karpov, T. E., Wittmar, A., Ulbricht, M., Ivanova, A., Loza, K., Prymak, O., Koptyug, A., Epple, M., & Surmenev, R. A. (2019). Adhesion, proliferation, and osteogenic differentiation of human mesenchymal stem cells on additively manufactured Ti6Al4V alloy scaffolds modified with calcium phosphate nanoparticles. *Colloids and Surfaces. B, Biointerfaces, 176*, 130–139. DOI: 10.1016/j.colsurfb.2018.12.047

Clark, R. H. (1985). Screen Printing. In *Handbook of Printed Circuit Manufacturing* (pp. 216–244). Springer., DOI: 10.1007/978-94-011-7012-3_11

Colegrove, P. A., Coules, H. E., Fairman, J., Martina, F., Kashoob, T., Mamash, H., & Cozzolino, L. D. (2013). Microstructure and residual stress improvement in wire and arc additively manufactured parts through high-pressure rolling. *Journal of Materials Processing Technology, 213*(10), 1782–1791. DOI: 10.1016/j.jmatprotec.2013.04.012

Collins, J. A., & Tovey, F. M. (1972). Fretting fatigue mechanisms and the effect of direction of fretting motion on fatigue strength. *Testing and Evaluation*, 420–424.

Connecticut Resources Recovery Authority. Projects. Bridgeport Project/Southwest Division History (2013), Retrieved June 8, 2016, from http://www.crra.org/pages/proj_bridgeport_history.htm

Couper, M. J., Neeson, A. E., & Griffiths, J. R. (1990). Casting defects and the fatigue behaviour of an aluminium casting alloy. *Fatigue & Fracture of Engineering Materials & Structures, 13*(3), 213–227. DOI: 10.1111/j.1460-2695.1990.tb00594.x

Crose, M., Sang-Il Kwon, J., Nayhouse, M., Ni, D., & Christofides, P. D. (2015). Multiscale modeling and operation of PECVD of thin film solar cells. *Chemical Engineering Science*, *136*, 50–61. DOI: 10.1016/j.ces.2015.02.027

Crupi, V., Epasto, G., Guglielmino, E., & Squillace, A. (2017). Influence of microstructure [alpha+beta and beta] on very high cycle fatigue behaviour of Ti-6Al-4V alloy. *International Journal of Fatigue*, *95*, 64–67. DOI: 10.1016/j.ijfatigue.2016.10.002

Cruz, A., Villena, M. L., Castro, I. M., Cervantes, J. P., Ondra, S. A., Quijano, R., & Alipio, M. (2022). Vitaband: IoT-driven health monitoring system for home quarantine COVID-19 patient using LoRaWAN. In *2022 IEEE International Conference on Internet of Things and Intelligence Systems (IoTaIS)* (pp. 351–357). IEEE. DOI: 10.1109/IoTaIS56727.2022.9975992

Curtis, B. (2018). *Regulatory considerations for medical polymers*. Foster Corporation. https://www.fostercomp.com/regulatory-considerations-for-medical-polymers/

Cvijovic-Alagic, I., Gubeljak, N., Rakin, M., Cvijovic, Z., & Geric, K. (2014). Microstructural morphology effects on fracture resistance and crack tip strain distribution in Ti–6Al–4V alloy for orthopedic implants. *Materials & Design*, *53*, 870–880. DOI: 10.1016/j.matdes.2013.07.097

Dabees, S., Mirzaei, S., Kaspar, P., Holcman, V., & Sobola, D. (2022). Characterization and Evaluation of Engineered Coating Techniques for Different Cutting Tools—Review. *Materials (Basel)*, *15*(16), 5633. DOI: 10.3390/ma15165633

Danylenko, Y. A. (2018). Characteristics and classification of innovation and innovation process. Science and innovation, 14(3), 14-26.

Das, S., & Joshi, S. N. (2020). Estimation of wire strength based on residual stresses induced during wire electric discharge machining. *Journal of Manufacturing Processes*, *53*, 406–419. DOI: 10.1016/j.jmapro.2020.03.015

Davis, K., Yarbrough, R., Froeschle, M., White, J., & Rathnayake, H. (2019). Band gap engineered zinc oxide nanostructures via a sol-gel synthesis of solvent driven shape-controlled crystal growth. *RSC Advances*, *9*(26), 14638–14648. DOI: 10.1039/C9RA02091H PMID: 35516315

Dawood, A., Marti, B. M., Sauret-Jackson, V., & Darwood, A. (2015). 3D printing in dentistry. *British Dental Journal*, *219*(11), 521–529. DOI: 10.1038/sj.bdj.2015.914

de Boisbaudran, L. (1875). On the chemical and spectroscopic characters of a new metal (gallium). *The London, Edinburgh and Dublin Philosophical Magazine and Journal of Science*, *50*(332), 414–416. DOI: 10.1080/14786447508641312

De Guire, M. R., Bauermann, L. P., Parikh, H., & Bill, J. (2013). Chemical Bath Deposition. *Chemical Solution Deposition of Functional Oxide Thin Films, 9783211993118*, 319–339. DOI: 10.1007/978-3-211-99311-8_14

De Souza, A. A., De Cássia Braga Granja, R., De Melo, E. C. C., De Oliveira, W. F., De Albuquerque, P. B. S., Correia, M. T. D. S., & Carneiro-Da-Cunha, M. D. G. (2022). Biodegradable synthetic polymers in biomedical application: A review. *Brazilian Journal of Health Review*, *5*(6), 24628–24644. DOI: 10.34119/bjhrv5n6-217

Deepa Pandey. (2025). Healthcare Additive Manufacturing Market.

Deepanraj, B., Saravanan, A. M., Senthilkumar, N., Afzal, A. A., & Afzal, A. R. (2025). Drilling characteristics optimization of polymer composite fortified with eggshells using Box–Behnken design and zebra optimization algorithm. *Results in Engineering*, *25*, 104102. DOI: 10.1016/j.rineng.2025.104102

Delai, O. Y. (2011). *Research on microstructure evolution and dynamic recrystallization behavior of titanium alloy TB6 and TA15 during β forging*. Nanjing University of Aeronautics and Astronautics.

Delgado, P., Cuesta, I. I., Alegre, J. M., & D'az, A. (2016). State of the art of Deep Rolling. *Precision Engineering*, *46*, 1–10. DOI: 10.1016/j.precisioneng.2016.05.001

Deng, J., Zhang, X., Niu, P., Liu, L., & Wang, J. (2006). Wear of ceramic nozzles by dry sand blasting. *Tribology International*, *39*(3), 274–280. DOI: 10.1016/j.triboint.2004.07.026

Deschamps, A., Decreus, B., Geuser, F. D., Dorin, T., & Weyland, M. (2013). The influence of precipitation on plastic deformation of Al–Cu–Li alloys. *Acta Materialia*, *61*(11), 4011–4021. DOI: 10.1016/j.actamat.2013.03.015

Dewi, H. A., Erdenebileg, E., De Luca, D., Mhaisalkar, S. G., & Bruno, A. (2024). Accelerated MAPbI3 Co-evaporation: Productivity Gains without Compromising Performance. *ACS Energy Letters*, *4319–4322*(9), 4319–4322. Advance online publication. DOI: 10.1021/acsenergylett.4c01597

Dhaliwal, K. (2018). Biodegradable Polymers and their Role in Drug Delivery Systems. *Biomedical Journal of Scientific & Technical Research*, *11*(1). Advance online publication. DOI: 10.26717/BJSTR.2018.11.002056

Dhinakaran, V., Ajith, J., Fahmidha, A. F. Y., Jagadeesha, T., Sathish, T., & Stalin, B. (2020). Wire Arc Additive Manufacturing (WAAM) process of nickel based superalloys–A review. *Materials Today: Proceedings*, *21*, 920–925. DOI: 10.1016/j.matpr.2019.08.159

Dhinakarraj, C. K., Senthilkumar, N., Palanikumar, K., & Deepanraj, B. (2025). Machinability evaluation of magnesium composite using response surface methodology and nature-inspired metaheuristic algorithms. [IJIDeM]. *International Journal on Interactive Design and Manufacturing*, *19*(4), 2607–2627. DOI: 10.1007/s12008-024-01866-3

Di Prima, M., Coburn, J., Hwang, D., Kelly, J., Khairuzzaman, A., & Ricles, L. (2016). Additively manufactured medical products – the FDA perspective. 3D Printing in Medicine, 2(1), 1. https://doi.org/DOI: 10.1186/s41205-016-0005-9

Ding, D., Zhao, R., Lu, Q., Pan, Z., Li, H., Wang, K., & He, K. (2021). A shape control strategy for wire arc additive manufacturing of thin-walled aluminium structures with sharp corners. *Journal of Manufacturing Processes*, *64*, 253–264. DOI: 10.1016/j.jmapro.2021.01.029

Dinovitzer, M., Chen, X., Laliberte, J., Huang, X., & Frei, H. (2019). Effect of wire and arc additive manufacturing (WAAM) process parameters on bead geometry and microstructure. *Additive Manufacturing*, *26*, 138–146. DOI: 10.1016/j.addma.2018.12.013

Distler, T., Polley, C., Shi, F., Schneidereit, D., Ashton, M. D., Friedrich, O., Kolb, J. F., Hardy, J. G., Detsch, R., Seitz, H., & Boccaccini, A. R. (2021). Electrically conductive and 3D-Printable oxidized Alginate-Gelatin Polypyrrole:PSS hydrogels for tissue engineering. *Advanced Healthcare Materials*, *10*(9), 2001876. Advance online publication. DOI: 10.1002/adhm.202001876

Divakar, S., Sardar, S., Sah, S., & Das, D. (2025). A state-of-the-art review on SiC and MWCNTs reinforced hybrid metal matrix composites: Processing, properties, and applications. *Hybrid Advances*, *10*, 100454. DOI: 10.1016/j.hybadv.2025.100454

Dobrikov, G. H., Rassovska, M. M., Andreev, N. M., Boyadzhiev, S. I., Gesheva, K. A., Ivanova, T. M., Sharlandjiev, P. S., & Nazarowa, D. I. (2009). Development of transparent heat mirrors based on metal oxide thin film structures. *Thin Solid Films*, *518*(4), 1091–1094. DOI: 10.1016/j.tsf.2009.07.203

Dolabella, S., Borzì, A., Dommann, A., & Neels, A. (2022). Lattice Strain and Defects Analysis in Nanostructured Semiconductor Materials and Devices by High-Resolution X-Ray Diffraction: Theoretical and Practical Aspects. *Small Methods*, *6*(2), 2100932. Advance online publication. DOI: 10.1002/smtd.202100932

Dolinar, D., Gorenšek, M., Avsec, K., Šetina Batič, B., Hočevar, M., Godec, M., Žužek, B., Debeljak, M., Jenko, M., Grant, J. T., & Kocjančič, B. (2023). Mechanisms of Premature Fracture in Modular Neck Stems Made of CoCrMo/Ti6Al4V and Ti6Al4V/Ti6Al4V Alloy. *Coatings*, *13*(7), 1255. DOI: 10.3390/coatings13071255

Dongming, G. (2024). High-Performance Manufacturing. *International Journal of Extreme Manufacturing*, *6*(6), 060201. DOI: 10.1088/2631-7990/ad7426

Doroody, C., Rahman, K. S., Rosly, H. N., Harif, M. N., Sopian, K., Abdullah, S. F., & Amin, N. (2021). A comprehensive comparative study of CdTe thin films grown on ultra-thin glass substrates by close-spaced sublimation and RF magnetron sputtering. *Materials Letters*, *293*, 129655. DOI: 10.1016/j.matlet.2021.129655

Dragomir (Nicolescu), L., Antoniac, A., Manescu (Paltanea), V., Robu, A., Dinu, M., Pana, I., Cotrut, C. M., Kamel, E., Antoniac, I., Rau, J. V., & Vladescu (Dragomir), A. (. (2023). Preparation and characterization of hydroxyapatite coating by magnetron sputtering on Mg–Zn–Ag alloys for orthopaedic trauma implants. *Ceramics International*, *49*(16), 26274–26288. DOI: 10.1016/j.ceramint.2023.05.116

Dry-ice blasting (2018). Retrieved November 12, 2019, from http://yourmoveinfo.blogspot.com/2018/06/dry-ice-blasting.html

Dutta, B., & Froes, F. H. (2017). The Additive Manufacturing (AM) of titanium alloys. *Metal Powder Report*, *72*(2), 96–106. DOI: 10.1016/j.mprp.2016.12.062

Dutta, S. D., Hexiu, J., Patel, D. K., Ganguly, K., & Lim, K. (2020). 3D-printed bioactive and biodegradable hydrogel scaffolds of alginate/gelatin/cellulose nanocrystals for tissue engineering. *International Journal of Biological Macromolecules*, *167*, 644–658. DOI: 10.1016/j.ijbiomac.2020.12.011

Dutta, S., & Narala, S. K. R. (2021). Effect of tool nose radius in turning of novel Mg alloy. *Materials Today: Proceedings*, *38*, 2675–2679. DOI: 10.1016/j.matpr.2020.08.334

Dutta, S., Sarma, D. K., Vora, J., Chaudhari, R., Bhowmik, A., Samal, P., & Khanna, S. (2025). A State-of-the-Art review on Micro-Machining of nitinol shape memory alloys and optimization of process variables considering the future trends of research. *Journal of Manufacturing Materials Processing*, *9*(6), 183. DOI: 10.3390/jmmp9060183

Du, Y., Liu, Z., Shen, M., Wu, J., Zhang, K., Jiang, C., & Zhing, J. (2025). An Efficient Parametric Modeling, Evaluation and Optimization Strategy for Aerodynamic Configuration Design of EVTOL Aircraft. *Aerosp. Res. Commun.*, *3*, 14986. DOI: 10.3389/arc.2025.14986

Dwivedi, P., Maheshwari, S., & Siddiquee, A. N. (2021). Issues and Requirements for Aluminum Alloys Used in Aircraft Components: State of the Art. *Russian Journal of Non-Ferrous Metals*, *62*(2), 212–225. DOI: 10.3103/S1067821221020048

Efaz, E. T., Rhaman, M. M., Al Imam, S., Bashar, K. L., Kabir, F., Mourtaza, M. E., Sakib, S. N., & Mozahid, F. A. (2021). A review of primary technologies of thin-film solar cells. *Engineering Research Express*, *3*(3), 032001. DOI: 10.1088/2631-8695/ac2353

El Hawary, O., Boccarusso, L., Ansell, M. P., Durante, M., & Pinto, F. (2023). An overview of natural fiber composites for marine applications. *Journal of Marine Science and Engineering*, *11*(5), 1076. DOI: 10.3390/jmse11051076

El Khatib, A., Al Rashidi, H., Al Miaari, A., & Hatamleh, Z. (2024). Thermal Performance Evaluation Techniques for Thin-Film Semi-Transparent PV Glazing Using Experimental Testing. *2024 6th International Symposium on Advanced Electrical and Communication Technologies, ISAECT 2024*. DOI: 10.1109/ISAECT64333.2024.10799601

Elfaleh, I., Abbassi, F., Habibi, M., Ahmad, F., Guedri, M., Nasri, M., & Garnier, C. (2023). A comprehensive review of natural fibers and their composites: An eco-friendly alternative to conventional materials. *Results in Engineering*, *19*, 101271. DOI: 10.1016/j.rineng.2023.101271

El-Shaer, A., Ezzat, S., Habib, M. A., Alduaij, O. K., Meaz, T. M., & El-Attar, S. A. (2023). Influence of deposition time on structural, morphological, and optical properties of CdS thin films grown by low-cost chemical bath deposition. *Crystals*, *13*(5), 788. Advance online publication. DOI: 10.3390/cryst13050788

Estrin, Y., & Vinogradov, A. (2010). Fatigue behaviour of light alloys with ultrafine grain structure produced by severe plastic deformation: An overview. *International Journal of Fatigue*, *32*(6), 898–907. DOI: 10.1016/j.ijfatigue.2009.06.022

Etesami, S. A., Fotovvati, B., & Asadi, E. (2022). Heat treatment of Ti-6Al-4V alloy manufactured by laser-based powder-bed fusion: Process, microstructures, and mechanical properties correlations. *Journal of Alloys and Compounds*, *895*, 162618. DOI: 10.1016/j.jallcom.2021.162618

Everton, S. K., Hirsch, M., Stravroulakis, P., Leach, R. K., & Clare, A. T. (2016). Review of in-situ process monitoring and in-situ metrology for metal additive manufacturing. *Materials & Design*, *95*, 431–445. DOI: 10.1016/j.matdes.2016.01.099

Ezekoye, B. A., Offor, P. O., Ezekoye, V. A., & Ezema, F. I. (2013). Chemical Bath Deposition Technique of Thin Films: A Review. *International Journal of Scientific Research*, •••, 2.

Fadaam, S. A., Mustafa, M. H., Abd AlRazaK, A. H., & Shihab, A. A. (2019). Enhanced efficiency of CdTe Photovoltaic by thermal evaporation Vacuum. *Energy Procedia*, *157*, 635–643. DOI: 10.1016/j.egypro.2018.11.229

Fadag, H. A., Mall, S., & Jain, V. K. (2008). A Finite Element Analysis of Fretting Fatigue Crack Growth Behavior in Ti–6Al–4V. *Engineering Fracture Mechanics*, *75*(6), 1384–1399. DOI: 10.1016/j.engfracmech.2007.07.003

Fan, D., Staufer, U., & Accardo, A. (2019). Engineered 3D polymer and hydrogel microenvironments for cell culture applications. *Bioengineering (Basel, Switzerland)*, *6*(4), 113. DOI: 10.3390/bioengineering6040113

Farjaminejad, S., Farjaminejad, R., Hasani, M., Garcia-Godoy, F., Abdouss, M., Marya, A., Harsoputranto, A., & Jamilian, A. (2024). Advances and Challenges in Polymer-Based Scaffolds for Bone Tissue Engineering: A path towards Personalized Regenerative Medicine. *Polymers*, *16*(23), 3303. DOI: 10.3390/polym16233303

Farsadi, T., Ahmadi, M., Sahin, M., Khodaparast, H. H., Kayran, A., & Friswell, M. I. (2024). High Aspect Ratio Composite Wings: Geometrically Nonlinear Aeroelasticity, Multi-Disciplinary Design Optimization, Manufacturing, and Experimental Testing. *Aerospace (Basel, Switzerland)*, *11*(3), 193. DOI: 10.3390/aerospace11030193

Fazal, M. A., & Rubaiee, S. (2023). Progress of PV cell technology: Feasibility of building materials, cost, performance, and stability. *Solar Energy*, *258*, 203–219. DOI: 10.1016/j.solener.2023.04.066

Fedorenko, A. V., Bozhko, K. M., Kachur, N. V., Kosiakovskiy, A. V., & Maslov, V. P. (2024). Optical and electrical properties of zinc oxide nanofilms deposited using the sol-gel method. *Semiconductor Physics. Quantum Electronics and Optoelectronics*, *27*(1), 117–123. DOI: 10.15407/spqeo27.01.117

Ferekides, C., Britt, J., Ma, Y., & Killian, L. (1993). High efficiency CdTe solar cells by close spaced sublimation. *Conference Record of the IEEE Photovoltaic Specialists Conference*, *389–393*, 389–393. Advance online publication. DOI: 10.1109/PVSC.1993.347151

Ferhati, H., Djeffal, F., Bendjerad, A., Foughali, L., Benhaya, A., & Saidi, A. (2022). Highly-detective tunable band-selective photodetector based on RF sputtered amorphous SiC thin-film: Effect of sputtering power. *Journal of Alloys and Compounds*, *907*, 164464. Advance online publication. DOI: 10.1016/j.jallcom.2022.164464

Fernández-Zabalza, A., Rodríguez-Díaz, Á., Veiga, F., Suárez, A., Uralde, V., Ballesteros, T., & Alfaro, J. R. (2025). AI-driven predictive modeling of homogeneous bead geometry for WAAM processes. *International Journal of Advanced Manufacturing Technology*. Advance online publication. DOI: 10.1007/s00170-025-16061-1

Fiore, V., & Valenza, A. (2013). Epoxy resins as a matrix material in advanced fibre-reinforced polymer (FRP) composites. In Advanced Fibre-Reinforced Polymer (FRP) Composites for Structural Applications (pp. 88–121). Woodhead Publishing. DOI: 10.1533/9780857098641.1.88

Fitzpatrick, M., Fry, A., Holdway, P., Kandil, F. A., Shackleton, J., & Suominen, L. (2002). *NPL Good Practice Guide no. 52: determination of residual stresses by x-ray diffraction.*

Frame, L. D., Rambarran, I., Sala, K., & Sanders, C. (2020). Impacts of Machining and Heat Treating Practices on Residual Stresses in Alpha-Beta Titanium Alloys. *Journal of Materials Engineering and Performance*, *29*(6), 3626–3637. DOI: 10.1007/s11665-020-04843-5

Francis, O. I., Ikenna, A., Francis, O. I., & Ikenna, A. (2021). Review of Dye-Sensitized Solar Cell (DSSCs) Development. *Natural Science (Irvine, Calif.)*, *13*(12), 496–509. DOI: 10.4236/ns.2021.1312043

Frazier, W. E. (2014). Metal Additive Manufacturing: A Review. *Journal of Materials Engineering and Performance*, *23*(6), 1917–1928. DOI: 10.1007/s11665-014-0958-z

Freiburg, D., Aßmuth, R., Garcia Carballo, R., Biermann, D., Henneberg, J., & Merklein, M. (2019). Adaption of tool surface for sheet-bulk metal forming by means of pressurized air wet abrasive jet machining. *Production Engineering*, *13*(1), 71–77. DOI: 10.1007/s11740-018-0858-4

Friderikos, O., Sagris, D., David, C. N., & Korlos, A. (2020). Simulation of Adiabatic Shear Bands in Orthogonal Machining of Ti6Al4V Using a Rigid-Viscoplastic Finite Element Analysis. *Metals*, *10*(3), 338. DOI: 10.3390/met10030338

Fritzsche, H. (2001). Development in understanding and controlling the Staebler-Wronski effect in a-Si:H. *Annual Review of Materials Science, 31*(Volume 31, 2001), 47–79. https://doi.org/DOI: 10.1146/ANNUREV.MATSCI.31.1.47/CITE/REFWORKS

Frunz verde, D., Muntean, S., M rginean, G., Campian, V., Marşavina, L., Terzi, R., & Şerban, V. (2010). Failure analysis of a Francis turbine runner. *IOP Conference Series. Earth and Environmental Science*, *12*(1), 12115. DOI: 10.1088/1755-1315/12/1/012115

Fuller, C. B., Mahoney, M. W., Bingel, W. H., Calabrese, M., & London, B. (2007). Tensile and fatigue properties of friction stir processed NiAl bronze. *Materials Science Forum*, *539*, 3751–3756. DOI: 10.4028/www.scientific.net/MSF.539-543.3751

Fu, X. (2016). Aluminum nitride wide band-gap semiconductor and its basic characteristics. In *Proceedings of the 2016 2nd International Conference on Education, Management, Information and Mechanical Engineering (EMIM 2016)* (pp. 555–558). https://doi.org/DOI: 10.2991/emim-16.2016.117

Gadge, M., Lohar, G., & Chinchanikar, S. (2022). A review on micro-blasting as surface treatment technique for improved cutting tool performance. *Materials Today: Proceedings*, *64*, 725–730. DOI: 10.1016/j.matpr.2022.05.196

Gaikwad, M. U., Krishnamoorthy, A., & Jatti, V. S. (2021). Investigating the influence of electrical discharge machining process parameter on fatigue strength during machining of titanium grade-2. *Materials Today: Proceedings*, *46*, 8951–8957. DOI: 10.1016/j.matpr.2021.05.367

Galizia, B., Fiorenza, P., Schilirò, E., Pecz, B., Foragassy, Z., Greco, G., Saggio, M., Cascino, S., Lo Nigro, R., & Roccaforte, F. (2024). Towards aluminum oxide/ aluminum nitride insulating stacks on 4H–SiC by atomic layer deposition. *Materials Science in Semiconductor Processing*, *174*, 108244. Advance online publication. DOI: 10.1016/j.mssp.2024.108244

Gancarz, T., Dobosz, A., Bogno, A.-A., Cempura, G., Schell, N., Chulist, R., & Henein, H. (2021). Characterization of Rapidly Solidified Al-Mg-Sc Alloys with Li Addition. *Materials Characterization*, *178*, 111290. DOI: 10.1016/j.matchar.2021.111290

Gangil, N., Siddiquee, A. N., & Maheshwari, S. (2020). *Composite Fabrication on Age-Hardened Alloy using Friction Stir Processing*. CRC Press. DOI: 10.1201/9781003003038

Gangwar, A., Kumar, V., Yaylacı, M., & Panda, S. K. (2024). Computational modelling and mechanical characteristics of polymeric hybrid composite materials: An extensive review. *Archives of Computational Methods in Engineering*, *31*, 3901–3921. DOI: 10.1007/s11831-024-10097-4

Gao, C., Wang, C., Jin, H., Wang, Z., Li, Z., Shi, C., Leng, Y., Yang, F., Liu, H., & Wang, J. (2018a). Additive manufacturing technique-designed metallic porous implants for clinical application in orthopedics. *RSC Advances*, *8*(44), 25210–25227. DOI: 10.1039/C8RA04815K

Gao, K., Zhang, Y., Yi, J., Dong, F., & Chen, P. (2024). Overview of Surface Modification Techniques for Titanium Alloys in Modern Material Science: A Comprehensive Analysis. *Coatings*, *14*(1), 148. DOI: 10.3390/coatings14010148

Gao, W., Bi, Y., Li, X., Dong, A., Wang, J., & Yang, X. (2024). Variational Method-based Trajectory Optimization for Hybrid Airships. *Aerospace (Basel, Switzerland)*, *11*(4), 250. DOI: 10.3390/aerospace11040250

Gao, Z., Yang, S., Meng, X., Zhentao, W., & Peng, Z. (2021). Study on fatigue crack growth of electron beam selective melting of titanium alloy. *Materials Research Express*, *8*(9), 96521. DOI: 10.1088/2053-1591/ac2444

Garbacz, H., Motyka, M., Ziaja, W., Lewandowska, M., Sieniawski, J., & Topolski, K. (2011). High cycle fatigue strength of hydrostatically extruded nanocrystalline CP titanium. *Proceedings of the 12th World Conference on Titanium*, 1137–1140.

García-Carrión, M., Ramírez-Castellanos, J., Nogales, E., Méndez, B., You, C. C., Karazhanov, S., & Marstein, E. S. (2020). Hybrid solar cells with β- and γ-gallium oxide nanoparticles. *Materials Letters*, *261*, 127088. Advance online publication. DOI: 10.1016/j.matlet.2019.127088

Gebremichael, G., Mekonone, S. T., & Baye, T. M.. (2025). Mechanical and microstructural characterization of sisal fiber-reinforced polyester laminate composites for improved durability in automotive applications. *Journal of Materials Science: Composites*, *6*(1), 1. Advance online publication. DOI: 10.1186/s42252-024-00064-4

Ge, H., Zhang, X., & Liu, Y. (2024). Numerical and Experimental Study on the Jet Behavior of Ultrafine Dry Powder Based on a Supersonic Nozzle. *ACS Omega*, *9*(46), 46574–46587. DOI: 10.1021/acsomega.4c08542

Ge, J., Lin, Y., Qi, H., Li, Y., Li, X., Li, C., Li, Z., & Xu, K. (2024). The impact of ultrasonic-induced jet morphology on polishing efficiency. *International Journal of Mechanical Sciences*, *284*, 109764. DOI: 10.1016/j.ijmecsci.2024.109764

Genchi, G., Sinicropi, M. S., Lauria, G., Carocci, A., & Catalano, A. (2020). The Effects of Cadmium Toxicity. *International Journal of Environmental Research and Public Health 2020, Vol. 17, Page 3782*, *17*(11), 3782. DOI: 10.3390/ijerph17113782

Geng, R., Cheng, Y., Chao, L., Wei, Z., & Ma, N. (2023). Microstructure and solute concentration analysis of epitaxial growth during wire and arc additive manufacturing of aluminum alloy. *Crystals*, *13*(5), 776. DOI: 10.3390/cryst13050776

Gherissi, A., & Einasri, I. (2019). Experimental and numerical study on jet abrasive machining of aluminum 2024 – T3. *Acta Technica Napocensis, Series: Applied Mathematics. Mechanical Engineering (New York, N.Y.)*, *62*(1), 149–156.

Ghimire, R., & Raji, A. (2024). Use of Artificial Intelligence in Design, Development, Additive Manufacturing, and Certification of Multifunctional Composites for Aircraft, Drones, and Spacecraft. *Applied Sciences (Basel, Switzerland), 14*(3), 1187. DOI: 10.3390/app14031187

Ghobeity, A., Krajac, T., Burzynski, T., Papini, M., & Spelt, J.K. Surface evolution models in abrasive jet micromachining. *Wear, 264*(3-4), 185–198. https://doi.org/ DOI: 10.1016/j.wear.2007.02.020

Ghonem, H. (2010). Microstructure and fatigue crack growth mechanisms in high temperature titanium alloys. *International Journal of Fatigue, 32*(9), 1448–1460. DOI: 10.1016/j.ijfatigue.2010.02.001

Gibson, I., Rosen, D. W., & Stucker, B. (2015). *Additive manufacturing technologies* (2nd ed.). Springer. DOI: 10.1007/978-1-4939-2113-3

Gibson, I., Rosen, D., Stucker, B., & Khorasani, M. (2021a). *Additive Manufacturing Technologies* (Vol. 17). Springer International Publishing., DOI: 10.1007/978-3-030-56127-7

Giladi, R., & Menachi, E. (2024). Validating Aircraft Noise Models: Aviation Environmental Design Tool at Heathrow. *Journal of Air Transport Management, 116*, 102557. DOI: 10.1016/j.jairtraman.2024.102557

Gilbert, J. L., Mali, S., Urban, R. M., Silverton, C. D., & Jacobs, J. J. (2012). In vivo oxide-induced stress corrosion cracking of Ti-6Al-4V in a neck–stem modular taper: Emergent behavior in a new mechanism of in vivo corrosion. *Journal of Biomedical Materials Research. Part B, Applied Biomaterials, 100B*(2), 584–594. DOI: 10.1002/jbm.b.31943

Gill, S. S., Arora, H., Jidesh, , & Sheth, V. (2017). On the development of Antenna feed array for space applications by additive manufacturing technique. *Additive Manufacturing, 17*, 39–46. DOI: 10.1016/j.addma.2017.06.010

Giummarra, C., Thomas, B., & Rioja, R. J. (2007). New aluminum lithium alloys for aerospace applications. *Proceedings of the Light Metals Technology Conference*, 313.

Golden, P. J., & Grandt, A. F.Jr. (2004). Fracture Mechanics Based Fretting Fatigue Life Predictions in Ti–6Al–4V. *Engineering Fracture Mechanics, 71*(15), 2229–2243. DOI: 10.1016/j.engfracmech.2003.10.005

Goyal, V., Yadav, A., & Mukherjee, R. (2021). Thermographic camera-based E-IoT enabled architecture for smart poultry farm. In *2021 5th International Conference on Electrical, Electronics, Communication, Computer Technologies and Optimization Techniques (ICEECCOT)* (pp. 772–775).

Gradeen, A. G. (2012). *Cryogenic abrasive jet machining of polydimethylsiloxane and polytetrafluoroethylene at different temperatures*. Ryerson University, Toronto., DOI: 10.32920/ryerson.14652027

Gradeen, A. G., Spelt, J. K., & Papini, M. (2012). Cryogenic abrasive jet machining of polydimtylsiloxane at different temperatures. *Wear*, *274-275*, 335–344. DOI: 10.1016/j.wear.2011.09.013

Grigoriev, S., Sotova, C., Vereschaka, A., Uglov, V., & Cherenda, N. (2023). Modifying Coatings for Medical Implants Made of Titanium Alloys. *Metals*, *13*(4), 718. DOI: 10.3390/met13040718

Guest, D. E. (2018). Aluminum alloys in the aerospace industry. Retrieved October 28, 2019, from https://www.thomasnet.com/insights/aluminum-alloys-in-the-aerospace-industry/

Guha, S., Yang, J., Banerjee, A., Glatfelter, T., & Xu, X. (1992). Advances in amorphous silicon alloy-based multijunction cells and modules. *AIP Conference Proceedings*, *268*(1), 64–71. DOI: 10.1063/1.42943

Gu, L., Ma, H. P., Shen, Y., Zhang, J., Chen, W. J., Yang, R. Y., Wu, F., Yang, L., Zeng, Y. X., Wang, X. R., Zhu, J. T., & Zhang, Q. C. (2022). Temperature-dependent oxygen annealing effect on the properties of Ga_2O_3 thin film deposited by atomic layer deposition. *Journal of Alloys and Compounds*, *925*, 166727. Advance online publication. DOI: 10.1016/j.jallcom.2022.166727

Guo, F., Duan, S., Matsuda, K., Liu, X., & Zou, Y. (2025). The Microstructure and Corrosion Behavior Analysis of Different Aging Al-Zn-Mg-Cu Alloys Manufactured by Additive Friction Stir Deposition. *Materials Characterization*, *225*, 115187. DOI: 10.1016/j.matchar.2025.115187

Gupta, K., & Davim, J. P. (2020). *High-Speed Machining*. Academic Press. Retrieved from https://books.google.co.in/books?id=fZ_NDwAAQBAJ

Gupta, A., Parikh, V., & Compaan, A. D. (2006). High efficiency ultra-thin sputtered CdTe solar cells. *Solar Energy Materials and Solar Cells*, *90*(15), 2263–2271. DOI: 10.1016/j.solmat.2006.02.029

Gupta, B., Min, K. H., Lee, Y., Tournet, J., Hoex, B., Jagadish, C., Tan, H. H., & Karuturi, S. (2024). From Rigid to Flexible: Progress, Challenges and Prospects of Thin c-Si Solar Energy Devices. *Advanced Energy Materials*, *14*(27), 2400743. DOI: 10.1002/aenm.202400743 PMID: 28296287

Gupta, K. (2024). Developments in Abrasive Water Jet Machining of Nonmetals. *Jurnal Poli-Teknologi*, *23*(2), 73–82. DOI: 10.32722/pt.v23i2.6910

Guru, P. R. F., Khan, M. D., Panigrahi, S. K., & Ram, G. D. J. (2015). Enhancing strength, ductility and machinability of a Al-Si cast alloy by friction stir processing. *Journal of Manufacturing Processes, 18*, 67–74. DOI: 10.1016/j.jmapro.2015.01.005

Haglin, J. M., Eltorai, A. E. M., Gil, J. A., Marcaccio, S. E., Botero-Hincapie, J., & Daniels, A. H. (2016a). Patient-Specific Orthopaedic Implants. *Orthopaedic Surgery*, *8*(4), 417–424. DOI: 10.1111/os.12282

Hajjioui, E. A., Bouchaâla, K., Faqir, M., & Essadiqi, E. (2023). A Review of Manufacturing Processes, Mechanical Properties and Precipitations for Aluminum Lithium Alloys used in Aeronautic Applications. *Heliyon*, *9*(3), e12565. DOI: 10.1016/j.heliyon.2022.e12565

Halder, T., Barot, H., Kumar, B., Kaushik, V., Patel, H., Bhut, H., Saha, B., Poddar, S., & Acharya, N. (2024). An Insight into Biodegradable Polymers and their Biomedical Applications for Wound Healing. *Current Pharmaceutical Design*, *30*(31), 2425–2444. DOI: 10.2174/0113816128295935240425101509

Halloran, J. W. (1999). Freeform fabrication of ceramics. *British Ceramic Transactions*, *98*(6), 299–303. DOI: 10.1179/096797899680633

Hamad, K., Kaseem, M., Yang, H. W., Deri, F., & Ko, Y. G. (2015). Properties and medical applications of polylactic acid: A review. *Express Polymer Letters*, *9*(5), 435–455. DOI: 10.3144/expresspolymlett.2015.42

Hamrani, A., Bouarab, F. Z., Agarwal, A., Ju, K., & Akbarzadeh, H. (2023). Advancements and applications of multiple wire processes in additive manufacturing: A comprehensive systematic review. *Virtual and Physical Prototyping*, *18*(1), e2273303. DOI: 10.1080/17452759.2023.2273303

Hanawa, T. (2019). Titanium–Tissue Interface Reaction and Its Control With Surface Treatment. *Frontiers in Bioengineering and Biotechnology*, *7*, 170. Advance online publication. DOI: 10.3389/fbioe.2019.00170

Handbook, M. (1989). *Machining* (9th ed., Vol. 16). ASM International.

Hang, R., Yao, X., Bai, L., & Hang, R. (2025). Evolving Biomaterials Design from Trial and Error to Intelligent Innovation. *Acta Biomaterialia*, *197*, 29–47. Advance online publication. DOI: 10.1016/j.actbio.2025.03.013

Han, J., Spanheimer, C., Haindl, G., Fu, G., Krishnakumar, V., Schaffner, J., Fan, C., Zhao, K., Klein, A., & Jaegermann, W. (2011). Optimized chemical bath deposited CdS layers for the improvement of CdTe solar cells. *Solar Energy Materials and Solar Cells*, *95*(3), 816–820. DOI: 10.1016/j.solmat.2010.10.027

Hanush, S. S., & Manjaiah, M. (2022). Topology Optimization of Aerospace Part to Enhance the Performance by Additive Manufacturing Process. *Materials Today: Proceedings*, *62*, 7373–7378. DOI: 10.1016/j.matpr.2022.02.074

Hao, L. C., Sapuan, S. M., Hassan, M. R., & Sheltami, R. M. (2018). Natural fiber reinforced vinyl polymer composites. In Sapuan, S. M. (Ed.), *Natural fibre reinforced vinyl ester and vinyl polymer composites* (pp. 27–70). Woodhead Publishing., DOI: 10.1016/B978-0-08-102160-6.00002-0

Hardin, B. E., Snaith, H. J., & McGehee, M. D. (2012). The renaissance of dye-sensitized solar cells. *Nature Photonics*, *6*(3), 162–169. DOI: 10.1038/nphoton.2012.22

Hariharan, V., & Rajeshkumar, G. (2024). Investigation of mechanical properties of sisal fiber and sugar palm fiber reinforced hybrid composites. Qeios, January, 1–9. https://www.qeios.com/read/JT0ETV

Haus"ol, T., H"oppel, H. W., & G"oken, M. (2010). Tailoring materials properties of UFG aluminium alloys by accumulative roll bonded sandwich-like sheets. *Journal of Materials Science, 45*, 4733–4738.

Haydar, H. J., & Al-Deen, J. (1973). AbidAli, A. K., & Mahmoud, A. A. (2021). Improved performance of Ti6Al4V alloy in Biomedical applications - Review. *Journal of Physics: Conference Series, 012146*(1). DOI: 10.1088/1742-6596/1973/1/012146

He, B., Wang, Y., Wang, C., Ren, Y., & Yang, G. (2024). Effect of gradient decrease heat-input combined with inter-layer cooling on microstructure and mechanical properties of WAAM-Ti-6Al-2Zr-1Mo-1V alloy. *Materials Today. Communications*, *39*, 108628. DOI: 10.1016/j.mtcomm.2024.108628

Hegde, A., Shetty, R., Nayak, R., Shetty, S., & Shetty, S. V. (2024). Optimizing the Die-Sink EDM Machinability of AISI 316L Using Ti-6Al-4V-SiCp Electrodes: A Computational Approach. *Journal of Manufacturing and Materials Processing*, *8*(5), 202. DOI: 10.3390/jmmp8050202

Heinl, P., Körner, C., & Singer, R. F. (2008). Selective Electron Beam Melting of Cellular Titanium: Mechanical Properties. *Advanced Engineering Materials*, *10*(9), 882–888. DOI: 10.1002/adem.200800137

Heinz, A., Haszler, A., Keidel, C., Moldenhauer, S., Benedictus, R., & Miller, W. S. (2000). Recent development in aluminium alloys for aerospace applications. *Materials Science and Engineering A*, *280*(1), 102–107. DOI: 10.1016/S0921-5093(99)00674-7

Helmholtz Zentrum Berlin. (2021). *Living lab for BIPV.* https://www.helmholtz-berlin.de/projects/pvcomb/forschen/living-lab-bipv/index_en.html

Hernandez, A., Islam, M. M., Saddatkia, P., Codding, C., Dulal, P., Agarwal, S., Janover, A., Novak, S., Huang, M., Dang, T., Snure, M., & Selim, F. A. (2021). MOCVD growth and characterization of conductive homoepitaxial Si-doped Ga_2O_3. *Results in Physics*, *25*, 104167. Advance online publication. DOI: 10.1016/j.rinp.2021.104167

Higashiwaki, M. (2021). β-Gallium oxide devices: Progress and outlook. *Physica Status Solidi. Rapid Research Letters*, *15*(11), 2100357. Advance online publication. DOI: 10.1002/pssr.202100357

Holzknecht, E. (2005). Shot blasting goes high tech. Case studies document how new systems reduced manufacturing costs for end users. *Metal Finishing*, *103*(7-8), 25–27. DOI: 10.1016/S0026-0576(05)80460-X

Hoseinnezhad, M., & Gharanjig, K. (2018). Synthesis and Application of an Organic Dye in Nanostructure Solar Cells Device. *International Journal of Chemical and Molecular Engineering*, *12*(9), 438–441.

Hosseini, A., Shahverdi, M., Sourani, Z., Habibi, R., Ahangari, Z., Shirian, S., & Zafari, F. (2023). A Review on Synthesis, Characterization and Applications of Polycaprolactone as a Novel Drug Delivery System and Tissue Engineering. *Journal of Inflammatory Diseases*, *27*(3). Advance online publication. DOI: 10.69107/jid-150907

Huang, C., Zhao, Y., Xin, S., Zhou, W., Li, Q., & Zeng, W. (2017). Effect of microstructure on tensile properties of Ti–5Al–5Mo–5V–3Cr–1Zr alloy. *Journal of Alloys and Compounds*, *693*, 582–591. DOI: 10.1016/j.jallcom.2016.09.233

Huang, J., & Rodrigue, D. (2022). Stiffness behavior of sisal fiber reinforced foam concrete under flexural loading. *Journal of Natural Fibers*, *19*(15), 12251–12267. DOI: 10.1080/15440478.2022.2054896

Huang, R., Riddle, M., Graziano, D., Warren, J., Das, S., Nimbalkar, S., Cresko, J., & Masanet, E. (2016). Energy and emissions saving potential of additive manufacturing: The case of lightweight aircraft components. *Journal of Cleaner Production*, *135*, 1559–1570. DOI: 10.1016/j.jclepro.2015.04.109

Huang, Y. C., Cha, H. C., Chen, C. Y., & Tsao, C. S. (2017). A universal roll-to-roll slot-die coating approach towards high-efficiency organic photovoltaics. *Progress in Photovoltaics: Research and Applications*, *25*(11), 928–935. DOI: 10.1002/pip.2907

Hua, S., De Matos, M. B. C., Metselaar, J. M., & Storm, G. (2018). Current Trends and Challenges in the Clinical translation of nanoparticulate Nanomedicines: Pathways for translational development and Commercialization. *Frontiers in Pharmacology*, *9*, 790. Advance online publication. DOI: 10.3389/fphar.2018.00790

Hunter, A. C., & Moghimi, S. M. (2016). Smart polymers in drug delivery: A biological perspective. *Polymer Chemistry*, *8*(1), 41–51. DOI: 10.1039/C6PY00676K

Hu, Q., Wu, H., Sun, J., Yan, D., Gao, Y., & Yang, J. (2016). Large-area perovskite nanowire arrays fabricated by large-scale roll-to-roll micro-gravure printing and doctor blading. *Nanoscale*, *8*(9), 5350–5357. DOI: 10.1039/C5NR08277C PMID: 26883938

Hu, T., Fang, J., Shen, Y., Li, M., Wang, B., Xu, Z., & Hu, W. (2024). Advances of naturally derived biomedical polymers in tissue engineering. *Frontiers in Chemistry*, *12*, 1469183. Advance online publication. DOI: 10.3389/fchem.2024.1469183

Hu, Y. (2025). Characteristics of abrasive jet erosion manufacturing technology: A comparative review. *International Journal of Advanced Manufacturing Technology*, *138*(9-10), 1–51. DOI: 10.1007/s00170-025-15698-2

Hu, Z., & Zhou, R. (2024). Review on some important research progresses in Biodegradable Plastics/Polymers. *Recent Progress in Materials*, *06*(02), 1–19. DOI: 10.21926/rpm.2402015

Ichim, M., Mure an, E. I., & Codău, E. (2024). Natural-fiber-reinforced polymer composites for furniture applications. *Polymers*, *16*(22), 3113. DOI: 10.3390/polym16223113 PMID: 39599204

IEA. (2018). *Data & Statistics - IEA*. https://www.iea.org/data-and-statistics?country=WORLD&fuel=Energy supply&indicator=TPESbySource

IEA. (2024). *Trends in Photovoltaic Applications*. https://doi.org/https://doi.org/10.69766/jnew6916

Ilangaweera, L. S., Naramada, S. E. M. H., Sajith, N. T. M., Aashiq, M. N. M., & Hinas, M. N. A. (2024). IOT based smart plant growing and caring platform for urban environments. In *2024 9th International Conference on Information Technology Research (ICITR)* (pp. 1–6). DOI: 10.1109/ICITR64794.2024.10857784

IRENA. (2012). *Renewable Energy Technologies: Cost Analysis Series* (Vol. 1, Issue 4/5).

Isupov, M. G., & Lukin, L. L. (2008). Characteristics of gas–abrasive jets. *Russian Engineering Research*, *28*(6), 566–570. DOI: 10.3103/S1068798X08060117

Jafar, R. J. M., Spelt, J. K., & Papini, M. (2013). Surface roughness and erosion rate of abrasive jet micro-machined channels: Experiments and analytical model. *Wear, 303*(1-2), 138–145. DOI: 10.1016/j.wear.2013.03.005

Jagadeesh, B., Duraiselvam, M., & Prashanth, K. (2023). Deformation behavior of metallic lattice structures with symmetrical gradients of porosity manufactured by metal additive manufacturing. *Vacuum, 211*, 111955. DOI: 10.1016/j.vacuum.2023.111955

Jagadeesh, B., Duraiselvam, M., & Prashanth, K. (2024). Deformation and energy absorption studies on FBCC and FBCCz lattice structures with symmetrical density gradients produced by L-PBF of Ti-6Al-4V alloy. *Materials Today: Proceedings*. Advance online publication. DOI: 10.1016/j.matpr.2024.02.008

Jagdale, M., Ambhore, N., Chaudhari, R., Kulkarni, A., & Abdullah, M. (2025). Experimental investigation of process parameters in Wire-EDM of Ti-6Al-4 V. *Scientific Reports, 15*(1), 5652. DOI: 10.1038/s41598-025-90486-2

Jain, V. K. (2013). *Abrasive jet micromachining*. Retrieved November 1, 2019, from https://home.iitk.ac.in/~vkjain/L-4-AJMM_ME698D-21-03-13.pdf

Jamalpur, B., Peddinti, D. R., Pawar, S. A., Zagade, A. D., Podduturi, M. R., & Thakur, J. (2025). IoT-driven data analytics: Transforming digital marketing campaigns with real-time consumer insights. In *2025 International Conference on Pervasive Computational Technologies (ICPCT)* (pp. 875–879). DOI: 10.1109/ICPCT64145.2025.10940271

Jana, S., Mishra, R. S., Baumann, J. B., & Grant, G. (2010). Effect of friction stir processing on fatigue behavior of an investment cast Al–7Si–0.6 Mg alloy. *Acta Materialia, 58*(3), 989–1003. DOI: 10.1016/j.actamat.2009.10.015

Jang, T.-S., Kim, D., Han, G., Yoon, C.-B., & Jung, H.-D. (2020a). Powder based additive manufacturing for biomedical application of titanium and its alloys: A review. *Biomedical Engineering Letters, 10*(4), 505–516. DOI: 10.1007/s13534-020-00177-2

Jaseem, J. I., Immanuel, R. J., Rao, P. N., Khan, F., Sahoo, B. N., Panigrahi, S. K., & Kamaraj, M. (2018). Synergetic effect of cryorolling and post-roll aging on simultaneous increase in wear resistance and mechanical properties of an Al-Cu alloy. *Journal of Tribology, 140*(November), 1–11. DOI: 10.1115/1.4040162

Jean, J., Brown, P. R., Jaffe, R. L., Buonassisi, T., & Bulović, V. (2015). Pathways for solar photovoltaics. *Energy & Environmental Science, 8*(4), 1200–1219. DOI: 10.1039/C4EE04073B

Jeong, J., Kim, M., Seo, J., Lu, H., Ahlawat, P., Mishra, A., Yang, Y., Hope, M. A., Eickemeyer, F. T., Kim, M., Yoon, Y. J., Choi, I. W., Darwich, B. P., Choi, S. J., Jo, Y., Lee, J. H., Walker, B., Zakeeruddin, S. M., Emsley, L., … Kim, J. Y. (2021). Pseudo-halide anion engineering for α-FAPbI3 perovskite solar cells. *Nature 2021 592:7854, 592*(7854), 381–385. DOI: 10.1038/s41586-021-03406-5

Jiang, L., Zhang, Z., Hu, H., He, X., Fu, H., & Xie, J. (2023). A Rapid and Effective Method for Alloy Materials Design via Sample Data Transfer Machine Learning. *npj Computational Materials, 9*(1), 26. DOI: 10.1038/s41524-023-00979-9

Jiang, M., Fang, H., & Tian, H. (2025). Latest advancements and trends in biomedical polymers for disease prevention, diagnosis, treatment, and clinical application. *Journal of Controlled Release, 380*, 138–174. DOI: 10.1016/j.jconrel.2025.01.063

Jia, Z., Dou, H., Liu, Y., Wang, Z., Zhang, H., Yue, T., & Mu, L. (2025). Study on Ductile Fracture Model considering Pre-Strain and its Application during Incremental Forming for 2091 Aluminum *Alloy. Engineering Fracture Mechanics, 322*, 111177. DOI: 10.1016/j.engfracmech.2025.111177

Jing, J., Lai, H., & Sun, L. (2024). Research on AI and IoT-driven monitoring in military physical fitness training. In *2024 6th International Conference on Frontier Technologies of Information and Computer (ICFTIC)* (pp. 1668–1672). https://doi.org/DOI: 10.1109/ICFTIC64248.2024.10912941

Ji, X., Wang, J., Qi, S., Liang, Y., Hu, S., Zheng, H., Zhang, S., Yue, J., Qi, X., Li, S., Liu, Z., Shu, L., Tang, W., & Li, P. (2024). Improvement of Ga_2O_3 vertical Schottky barrier diode by constructing NiO/Ga_2O_3 heterojunction. *Journal of Semiconductors, 45*(4), 042503. Advance online publication. DOI: 10.1088/1674-4926/45/4/042503

Johny, A., & Thiagarajan, C. (2020). Investigation of surface integrity and it's optimization on pure titanium using molybdenum wire by reciprocated travelling WEDM – A review. *Materials Today: Proceedings, 33*, 2581–2584. DOI: 10.1016/j.matpr.2019.12.251

Ju, J., Zhou, Y., Wang, K., Liu, Y., Li, J., Kang, M., & Wang, J. (2020). Tribological investigation of additive manufacturing medical Ti6Al4V alloys against Al2O3 ceramic balls in artificial saliva. *Journal of the Mechanical Behavior of Biomedical Materials, 104*, 103602. DOI: 10.1016/j.jmbbm.2019.103602

Jujjavarapu, S. E., Chandrasekhar, K., Naik, S., Toppo, A. L., & Thakur, V. (2020). Sources of Natural Polymers from Plants with Green Nanoparticles. In *CRC Press eBooks* (pp. 81–102). https://doi.org/DOI: 10.1201/9781351045155-4

Jung, S., Ahn, S. J., Yun, J. H., Gwak, J., Kim, D., & Yoon, K. (2010). Effects of Ga contents on properties of CIGS thin films and solar cells fabricated by co-evaporation technique. *Current Applied Physics*, *10*(4), 990–996. DOI: 10.1016/j.cap.2009.11.082

Juška, G. (1991). Properties of free-carrier transport in a-Se and a-Si:H. *Journal of Non-Crystalline Solids*, *137–138*(PART 1), 401–406. DOI: 10.1016/S0022-3093(05)80140-1

Kaim, P., Lukaszkowicz, K., Szindler, M., Szindler, M. M., Basiaga, M., & Hajduk, B. (2021). The Influence of Magnetron Sputtering Process Temperature on ZnO Thin-Film Properties. *Coatings 2021, Vol. 11, Page 1507, 11*(12), 1507. DOI: 10.3390/coatings11121507

Kalirajan, C., Dukle, A., Nathanael, A. J., Oh, T., & Manivasagam, G. (2021). A Critical review on Polymeric Biomaterials for Biomedical applications. *Polymers*, *13*(17), 3015. DOI: 10.3390/polym13173015

Kannan, G. R., Sivakumar, M., Jagadeesh, B., & Balaji, N. (2024). Powder Bed Fusion Process–State of Art. In *Advances in Additive Manufacturing* (pp. 89-107). DOI: 10.1002/9781394238316.ch5

Kanti, P. K., Kumar, H. G. P., Wanatasanappan, V. V., Kumar, A., & Regasa, M. B. (2025). Graphene's Frontier in Aerospace: Current Applications, Challenges, and Future Directions for Space Engineering. *Nanoscale Advances*, *7*(12), 3603–3618. DOI: 10.1039/D4NA00934G

Kara, A., Distler, T., Polley, C., Schneidereit, D., Seitz, H., Friedrich, O., Tihminlioglu, F., & Boccaccini, A. R. (2022). 3D printed gelatin/decellularized bone composite scaffolds for bone tissue engineering: Fabrication, characterization and cytocompatibility study. *Materials Today. Bio*, *15*, 100309. DOI: 10.1016/j.mtbio.2022.100309

Karimah, A., Ridho, M. R., Munawar, S. S., Adi, D. S., Ismadi, R. D., Damayanti, R., Subiyanto, B., Fatriasari, W., & Fudholi, A. (2023). A review on natural fibers for development of eco-friendly bio-composite: Characteristics and utilizations. *Journal of Materials Research and Technology*, *13*, 2442–2458. DOI: 10.1016/j.jmrt.2021.06.014

Kaur, N., Malik, P. K., Krishna, G., Kumar, M., Aluvala, S., & Singh, P. (2025). An IoT-driven workers health monitoring system utilizing LoRa technology for real-time vital signs collection, tracking and cloud-based data logging at industrial entry points. In *2025 International Conference on Intelligent Control, Computing and Communications (IC3)* (pp. 423–429). DOI: 10.1109/IC363308.2025.10956658

Kavimani, V., Gopal, P. M., Keerthiveettil Ramakrishnan, S., Giri, J., Alarifi, A., & Sathish, T. (2024). Predictive modelling and optimization of WEDM parameter for Mg–Li alloy using ANN integrated CRITIC-WASPAS approach. *Heliyon*, *10*(15), e35194. DOI: 10.1016/j.heliyon.2024.e35194

Kazmerski, L. L., White, F. R., & Morgan, G. K. (1976). Thin-film CuInSe/sub 2// CdS heterojunction solar cells. *Appl. Phys. Lett.; (United States), 29:4*(4), 268–270. DOI: 10.1063/1.89041

Kegelmann, L., Tockhorn, P., Wolff, C. M., Márquez, J. A., Caicedo-Dávila, S., Korte, L., Unold, T., Lövenich, W., Neher, D., Rech, B., & Albrecht, S. (2019). Mixtures of Dopant-Free Spiro-OMeTAD and Water-Free PEDOT as a Passivating Hole Contact in Perovskite Solar Cells. *ACS Applied Materials & Interfaces*, *11*(9), 9172–9181. DOI: 10.1021/acsami.9b01332 PMID: 30741517

Ke, J.-H., Tsai, F.-C., Hung, J.-C., & Yan, B.-H. (2012). Characteristics study of flexible magnetic abrasive in abrasive jet machining. *Procedia CIRP*, *1*, 679–680. DOI: 10.1016/j.procir.2012.05.025

Kermiche, F., Taabouche, A., Bouabellou, A., Hanini, F., & Bouachiba, Y. (2022). Behavior study of ZnO thin films grown by PLD for several applications. *Crystallography Reports*, *67*(7), 1239–1245. DOI: 10.1134/S1063774522070069

Kesinro, R. O., Boyo, A. O., Akinyemi, M. L., Emetere, M. E., & Aizebeokhai, A. P. (2021). Progress on Organic Solar Cells: A Short Review. *IOP Conference Series. Earth and Environmental Science*, *665*(1), 012036. Advance online publication. DOI: 10.1088/1755-1315/665/1/012036

Kessel, F., Friess, M., Rauh, C., & Wagner, A. (2025). Tailored Directional Porosity in Ceramic Matrix Composites (CMCs) for Hypersonic Applications. *CEAS Space Journal*, *17*(4), 623–633. DOI: 10.1007/s12567-024-00566-8

Khadse, C., & Hajare, S. (2025). Sensor based smart wearable devices for women: A review. In *2025 International Conference on Machine Learning and Autonomous Systems (ICMLAS)* (pp. 1933–1939). DOI: 10.1109/ICMLAS64557.2025.10968950

Khalil, H. A., Tehrani, M., Davoudpour, Y., Bhat, A., Jawaid, M., & Hassan, A. (2012). Natural fiber reinforced poly(vinyl chloride) composites: A review. *Journal of Reinforced Plastics and Composites*, *32*(5), 330–356. DOI: 10.1177/0731684412458553

Khan, N. Z., Siddiquee, A. N., & Khan, Z. A. (2017). *Friction stir welding: dissimilar aluminum alloys*.

Khan, F. (2018, July). M., Karthik, G. M., Panigrahi, S. K., & Ram, G. D. J. (2019). Friction stir processing of QE22 magnesium alloy to achieve ultra fi ne- grained microstructure with enhanced room temperature ductility and texture weakening. *Materials Characterization, 147*, 365–378. DOI: 10.1016/j.matchar.2018.11.020

Khan, F., Hossain, N., Mim, J. J., Rahman, S. M. M., Iqbal, M. J., Billah, M., & Chowdhury, M. A. (2025). Advances of composite materials in automobile applications – A review. *Journal of Engineering Research, 13*(2), 1001–1023. DOI: 10.1016/j.jer.2024.02.017

Khan, F., & Rokkala, U. (2023). Development of high strength and corrosion resistance Mg-Zn-Dy/HA-Ag composite for temporary implant applications. *Materials Letters, 347*, 134604. Advance online publication. DOI: 10.1016/j.matlet.2023.134604

Khan, H. M., Karabulut, Y., Kitay, O., Kaynak, Y., & Jawahir, I. S. (2021). Influence of the post-processing operations on surface integrity of metal components produced by laser powder bed fusion additive manufacturing: A review. *Machining Science and Technology, 25*(1), 118–176. DOI: 10.1080/10910344.2020.1855649

Khatoon, S., Kumar Yadav, S., Chakravorty, V., Singh, J., Bahadur Singh, R., Hasnain, M. S., & Hasnain, S. M. M. (2023). Perovskite solar cell's efficiency, stability and scalability: A review. *Materials Science for Energy Technologies, 6*, 437–459. DOI: 10.1016/j.mset.2023.04.007

Khoo, Z. X., An, J., Chua, C. K., Shen, Y. F., Kuo, C. N., & Liu, Y. (2018). Effect of heat treatment on repetitively scanned SLM NiTi shape memory alloy. *Materials (Basel), 12*(1), 77. DOI: 10.3390/ma12010077

Khorasani, A., Gibson, I., Veetil, J. K., & Ghasemi, A. H. (2020). A review of technological improvements in laser-based powder bed fusion of metal printers. *International Journal of Advanced Manufacturing Technology, 108*(1–2), 191–209. DOI: 10.1007/s00170-020-05361-3

Khosrozadeh, B. (2024). Surface integrity of Ti-6Al-4V alloy in hybrid electrical discharge machining processes. *Proceedings of the Institution of Mechanical Engineers. Part E, Journal of Process Mechanical Engineering, 238*(5), 2258–2267. DOI: 10.1177/09544089231160464

Kilian, K. A., & Moghe, P. V. (2016). High throughput strategies for the design, discovery, and analysis of biomaterials. *Acta Biomaterialia, 34*, v–vi. DOI: 10.1016/j.actbio.2016.03.019

Kim, J.-H., Kim, B.-S., Im, D., Park, J. H., Moon, D.-J., & Park, Y.-J. (2024). Optimization of High-Speed Angular Contact Ball Bearing for Aircraft Gearbox utilizing and Evolutionary Multi.Objective Algorithm, NSGA-III. *Advances in Mechanical Engineering*, *16*(9), 1–14. DOI: 10.1177/16878132241272197

Kim, W. J., Hong, S. I., Kim, Y. S., Min, S. H., Jeong, H. T., & Lee, J. D. (2003). Texture development and its effect on mechanical properties of an AZ61 Mg alloy fabricated by equal channel angular pressing. *Acta Materialia*, *51*(11), 3293–3307. DOI: 10.1016/S1359-6454(03)00161-7

Kishawy, H. A., & Hosseini, A. (2018). *Machining Difficult-to-Cut Materials: Basic Principles and Challenges*. Springer International Publishing. Retrieved from https://books.google.co.in/books?id=vYtoDwAAQBAJ

Kociubczyk, A. I., Grau, J. E., Sarutti, J. L., Méndez, C. M., Gregorutti, R. W., & Ares, A. E. (2023). Solidification Analysis, Microhardness, and Corrosion Resistance of a Cast Hip Prosthesis in ASTM F745 Austenitic Stainless Steel. *Journal of Materials Engineering and Performance*, *32*(3), 1065–1075. DOI: 10.1007/s11665-022-07164-x

Kong, H., & Liu, C. (2018). Analysis of the effect of fretting wear on key structural components of TB6 titanium alloy with actual failure cases. *IOP Conference Series. Materials Science and Engineering*, *439*, 22011. DOI: 10.1088/1757-899X/439/2/022011

Kordonski, W. I., Shorey, A. B., & Tricard, M. (2006). Magnetorheological jet (MR JET™) finishing technology. *Transactions of the ASME, 128,* 20-26. DOI: 10.1115/1.2140802

Kordonski, W., & Shorey, A. (2007). Magnetorheological (MR) jet finishing technology. *Journal of Intelligent Material Systems and Structures*, *18*(12), 1127–1130. DOI: 10.1177/1045389X07083139

Kreiger, M. A., Shonnard, D. R., & Pearce, J. M. (2013). Life cycle analysis of silane recycling in amorphous silicon-based solar photovoltaic manufacturing. *Resources, Conservation and Recycling*, *70*, 44–49. DOI: 10.1016/j.resconrec.2012.10.002

Kridli, G. T., El-Gizawy, A. S., & Lederich, R. (1998). Development of Process Maps for Superplastic Forming of Weldalite™ 049. *Materials Science and Engineering A*, *244*(2), 224–232. DOI: 10.1016/S0921-5093(97)00545-5

Kujovic, L., Liu, X., Togay, M., Abbas, A., Law, A. M., Jones, L. O., Curson, K. M., Barth, K. L., Bowers, J. W., Walls, J. M., Oklobia, O., Lamb, D. A., Irvine, S. J. C., Zhang, W., Lee, C., Nagle, T., Lu, D., & Xiong, G. (2025). Development of ZnO Buffer Layers for As-Doped CdSeTe/CdTe Solar Cells with Efficiency Exceeding 20%. *Advanced Materials Technologies*, *2401364*(13), 2401364. Advance online publication. DOI: 10.1002/admt.202401364

Kumar, S. (2019). Course Name: Unconventional Machining Processes. Retrieved November, 8, from http://docplayer.net/137171374-Course-name-unconventional-machining-processes-semester-v-course-code-ame507.html

Kumar, A. F., Khan, M., Panigrahi, S. K., & Chaudhari, G. P. (2021). Microstructural evolution and corrosion behaviour of friction stir-processed QE22 magnesium alloy. *Corrosion Reviews*, *39*(4), 351–360. DOI: 10.1515/corrrev-2020-0083

Kumar, A., & Giripunje, S. M. (2025). A comparative numerical simulation study of CIGS solar cells with distinct back surface field layers for enhanced performance. *Journal of Physics and Chemistry of Solids*, *197*, 112436. DOI: 10.1016/j.jpcs.2024.112436

Kumar, A., Kumar, V., & Kumar, J. (2014). Surface integrity and material transfer investigation of pure titanium for rough cut surface after wire electro discharge machining. *Proceedings of the Institution of Mechanical Engineers. Part B, Journal of Engineering Manufacture*, *228*(8), 880–901. DOI: 10.1177/0954405413513013

Kumar, M., Kumar, A., Gautam, Y. K., Chandra, R., Goyat, M. S., Tewari, B. S., & Tewari, R. K. (2020). Influence of SiC thin films thickness on the electrical properties of Pd/SiC thin films for hydrogen gas sensor. *Vacuum*, *182*, 109750. Advance online publication. DOI: 10.1016/j.vacuum.2020.109750

Kumar, R., & Neeraj. (2015). Minimization of material removal rate in abrasive jet machining of tempered glass using Taguchi method. *International Journal of Recent Research in Civil and Mechanical Engineering*, *2*(1), 253–260.

Kundan, D. K., & Prasad, E. K. (2015). An experimental work on multi-roller burnishing process on aluminum alloy. [IJAPME]. *International Journal of Advances in Production and Mechanical Engineering*, *1*(4), 24–28.

Kurowiak, J., Klekiel, T., & Będziński, R. (2023). Biodegradable Polymers in Biomedical Applications: A Review-Developments, Perspectives and Future Challenges. *International Journal of Molecular Sciences*, *24*(23), 16952. DOI: 10.3390/ijms242316952

Kwon, D.-K., & Lee, J.-H. (2022). Performance Improvement of Micro-Abrasive Jet Blasting Process for Al 6061. *Processes (Basel, Switzerland), 10*(11), 2247. DOI: 10.3390/pr10112247

Kyratsis, P., Tzotzis, A., & Davim, J. P. (2023). *3D FEA Simulations in Machining*. Springer International Publishing. Retrieved from https://books.google.co.in/books?id=p9mzEAAAQBAJ

la Monaca, A., Murray, J. W., Liao, Z., Speidel, A., Robles-Linares, J. A., Axinte, D. A., Hardy, M. C., & Clare, A. T. (2021). Surface integrity in metal machining - Part II: Functional performance. *International Journal of Machine Tools & Manufacture, 164*, 103718. DOI: 10.1016/j.ijmachtools.2021.103718

Lalegani Dezaki, M., Serjouei, A., Zolfagharian, A., Fotouhi, M., Moradi, M., Ariffin, M. K. A., & Bodaghi, M. (2022). A review on additive/subtractive hybrid manufacturing of directed energy deposition (DED) process. *Advanced Powder Materials, 1*(4), 100054. DOI: 10.1016/j.apmate.2022.100054

Lanning, D. B., Haritos, G. K., Nicholas, T., & Maxwell, D. C. (2001). Low-cycle fatigue/high-cycle fatigue interactions in notched Ti–6Al–4V. *Fatigue & Fracture of Engineering Materials & Structures, 24*(9), 565–577. DOI: 10.1046/j.1460-2695.2001.00411.x

Lee, H., Yoon, H., Ji, C., Lee, D., Lee, J. H., Yun, J. H., & Kim, Y. (2012). Fabrication of CIGS films by electrodeposition method for photovoltaic cells. *Journal of Electronic Materials, 41*(12), 3375–3381. DOI: 10.1007/s11664-012-2252-x

Lee, J. H., Kim, B. S., Park, J., Lee, J. W., & Kim, K. (2023). Opportunities and Challenges for Perovskite Solar Cells Based on Vacuum Thermal Evaporation. *Advanced Materials Technologies, 8*(20), 2200928. DOI: 10.1002/admt.202200928

Lee, K. Y., & Mooney, D. J. (2011). Alginate: Properties and biomedical applications. *Progress in Polymer Science, 37*(1), 106–126. DOI: 10.1016/j.progpolymsci.2011.06.003

Lee, T. D., & Ebong, A. U. (2017). A review of thin film solar cell technologies and challenges. *Renewable & Sustainable Energy Reviews, 70*(November), 1286–1297. DOI: 10.1016/j.rser.2016.12.028

Leng, J.-X., Shen, Y., Huang, W., An, K., Zhou, C.-C., & Wang, Z.-G. (2024). Multi-Fidelity Data Mining-based Design Optimization Framework for Scramjet-based Aircraft. *Energy, 311*, 133448. DOI: 10.1016/j.energy.2024.133448

Lenis, J. A., Gómez, M. A., & Bolívar, F. J. (2019). Effect of deposition temperature and target-substrate distance on the structure, phases, mechanical and tribological properties of multi-layer HA-Ag coatings obtained by RF magnetron sputtering. *Surface and Coatings Technology*, *378*, 124936. Advance online publication. DOI: 10.1016/j.surfcoat.2019.124936

Letov, N., Velivela, P. T., Sun, S., & Zhao, Y. F. (2021). Challenges and Opportunities in Geometric Modeling of Complex Bio-Inspired Three-Dimensional Objects Designed for Additive Manufacturing. *Journal of Mechanical Design*, *143*(12), 121705. Advance online publication. DOI: 10.1115/1.4051720

Le, V. T., Mai, D. S., Doan, T. K., & Paris, H. (2021). Wire and arc additive manufacturing of 308L stainless steel components: Optimization of processing parameters and material properties. *Engineering Science and Technology, an International Journal*, *24*(4), 1015–1026. DOI: 10.1016/j.jestch.2021.01.009

Leverant, G. R., Langer, B. S., Yuen, A., & Hopkins, S. W. (1979). Surface residual stresses, surface topography and the fatigue behavior of Ti-6AI-4V. *Metallurgical Transactions. A, Physical Metallurgy and Materials Science*, *10*(2), 251–257. DOI: 10.1007/BF02817635

Leyens, C., & Peters, M. (2006). *Titanium and Titanium Alloys: Fundamentals and Applications*. Wiley. Retrieved from https://books.google.co.in/books?id=okSA1N1xAxgC

Liang, R., Ji, Y., Wang, S., & Liu, S. (2016). Effect of microstructure on fracture toughness and fatigue crack growth behavior of Ti17 alloy. *Metals*, *6*(8), 186. Advance online publication. DOI: 10.3390/met6080186

Liang, S., Xu, F., Wu, J., & Deng, X. (2023). Editorial: Biomedical applications of water-soluble synthetic polymers. *Frontiers in Bioengineering and Biotechnology*, *11*, 1235931. Advance online publication. DOI: 10.3389/fbioe.2023.1235931

Libretexts. (2024). *30.1: Chain-Growth polymers*. Chemistry LibreTexts. https://chem.libretexts.org/@go/page/474086?pdf

Li, F., Chen, C., & Chen, X. (2024). Tremendous advances, multifaceted challenges and feasible future prospects of biodegradable medical polymer materials. *RSC Advances*, *14*(44), 32267–32283. DOI: 10.1039/D4RA00075G

Li, H., Li, X., Li, Y., Gao, G., Wen, K., Li, Z., Zhang, Y., & Xiong, B. (2023). Exploration of Alloying Elements of High Specific Modulus Al-Li Alloy based on Machine Learning. *Materials (Basel)*, *17*(1), 92. DOI: 10.3390/ma17010092

Li, J. I., Huang, J. I., Liu, D. Y., Chen, Y. I., Zhang, X. H., & Ma, P. C. (2019). Distribution and evolution of aging precipitates in Al–Cu–Li alloy with high Li concentration. *Transactions of Nonferrous Metals Society of China*, *29*(1), 15–24. DOI: 10.1016/S1003-6326(18)64910-6

Li, J., Li, Q., Deng, Y., Zhang, W., & Yin, H. (2025). Micro-Abrasive Air Jet Machining Technology for Fabrication of Helical Grooves on Bovine Bone. *Micromachines*, *16*(2), 149. DOI: 10.3390/mi16020149

Li, K., Tang, R., Zhu, C., & Chen, T. (2024). Critical review on crystal orientation engineering of antimony chalcogenide thin film for solar cell applications. *Advancement of Science*, *11*(2), 2304963. Advance online publication. DOI: 10.1002/advs.202304963 PMID: 37939308

Li, M., Contreras-Almengor, O., Ordoño, J., Aguilar Vega, C., Zapata Martínez, R., Echeverry-Rendón, M., Díaz-Lantada, A., & Molina-Aldareguia, J. (2025). The microstructure and mechanical properties of nitinol manufactured by LPBF: Differences between Ni-rich and Ti-rich compositions. *Virtual and Physical Prototyping*, *20*(1), e2476006. DOI: 10.1080/17452759.2025.2476006

Lin, C. R., Wei, D. H., BenDao, M. K., Chen, W. E., & Liu, T. Y. (2014). Development of high-performance UV detector using nanocrystalline diamond thin film. *International Journal of Photoenergy*, *2014*, 492152. Advance online publication. DOI: 10.1155/2014/492152

Lin, S. (2024). Advanced Materials and Their Impact on Aircraft Performance. *J. Aeronaut. Aerospace Eng.*, *13*(3), 355.

Lin, S., Ye, B., Peng, C., Ran, M., Jie, D., Du, Y., & Yin, D. (2025). Enhancing interfacial bonding and mechanical properties of CF/EP composites through low-temperature jet plasma treatment. *International Journal of Adhesion and Adhesives*, *141*, 104067. DOI: 10.1016/j.ijadhadh.2025.104067

Lin, Y.-C., Chen, Y.-F., Wang, A.-C., & Sei, W.-E. (2012). Machining performance on hybrid process of abrasive jet machining and electrical discharge machining. *Transactions of Nonferrous Metals Society of China*, *22*, 775–780. DOI: 10.1016/S1003-6326(12)61803-2

Lin, Y.-C., Hung, J.-C., Lee, H.-M., Wang, A.-C., & Fan, S.-F. (2018). Machining performances of electrical discharge machining combined with abrasive jet machining. *Procedia CIRP*, *68*, 162–167. DOI: 10.1016/j.procir.2017.12.040

Li, S., Lin, H., Zhang, T., Sui, J., & Wang, C. (2021). High-Speed Machining of Malleable Cast Iron by Various Cutting Tools Coated by Physical Vapor Deposition. *Chinese Journal of Mechanical Engineering*, *34*(1), 46. DOI: 10.1186/s10033-021-00561-8

Li, S.-S., Yue, X., Li, Q.-Y., Peng, H.-L., Dong, B.-X., Liu, T.-S., Yang, H.-Y., Fan, J., Shu, S.-L., Qiu, F., & Jiang, Q.-C. (2023). Development and Applications of Aluminum Alloys for Aerospace Industry. *Journal of Materials Research and Technology*, *27*, 944–983. DOI: 10.1016/j.jmrt.2023.09.274

Litimein, F., Rached, D., Khenata, R., & Baltache, H. (2009). FPLAPW study of the structural, electronic, and optical properties of Ga_2O_3: Monoclinic and hexagonal phases. *Journal of Alloys and Compounds*, *488*(1), 148–156. DOI: 10.1016/j.jallcom.2009.08.092

Liu, J., & Kulak, M. (2000). A New Paradigm in the Design of Aluminum Alloys for Aerospace Applications. *Mater. Sci. Forum, 331-337*, 127-142. doi: DOI: 10.4028/www.scientific.net/MSF.331.337.127

Liu, Y., Briones, J., Zhou, R., & Magotra, N. (2017). Study of secure boot with a FPGA-based IoT device. In *2017 IEEE 60th International Midwest Symposium on Circuits and Systems (MWSCAS)* (pp. 1053–1056). DOI: 10.1109/MWSCAS.2017.8053108

Liu, C., Zhang, X., Zhao, L., Liu, F., Chen, X., Yao, Y., & Li, J. (2019). Signal quality assessment and lightweight QRS detection for wearable ECG SmartVest system. *IEEE Internet of Things Journal*, *6*(2), 1363–1374. DOI: 10.1109/JIOT.2018.2844090

Liu, G., Huang, C., Su, R., Özel, T., Liu, Y., & Xu, L. (2019). 3D FEM simulation of the turning process of stainless steel 17-4PH with differently texturized cutting tools. *International Journal of Mechanical Sciences*, *155*, 417–429. DOI: 10.1016/j.ijmecsci.2019.03.016

Liu, J. F., & Guo, Y. B. (2016). Residual Stress Modeling in Electric Discharge Machining (EDM) by Incorporating Massive Random Discharges. *Procedia CIRP*, *45*, 299–302. DOI: 10.1016/j.procir.2016.02.060

Liu, R., Xu, S., Xu, C., Sun, K., Ju, X., Yang, X., Pan, Y., Li, J., Ren, G., & Wang, L. (2025). Study on the effect of ultrasonic rolling on the microstructure and fatigue properties of A03600 aluminum alloy. *Journal of Alloys and Compounds*, *1010*, 178124. https://doi.org/https://doi.org/10.1016/j.jallcom.2024.178124. DOI: 10.1016/j.jallcom.2024.178124

Liu, S., & Shin, Y. C. (2019a). Additive manufacturing of Ti6Al4V alloy: A review. *Materials & Design*, *164*, 107552. DOI: 10.1016/j.matdes.2018.107552

Liu, X., Liu, Q., Zhao, B., Ren, Y., Tao, B. W., & Zhang, W. L. (2020). Comparison of β-Ga_2O_3 thin films grown on r-plane and c-plane sapphire substrates. *Vacuum, 178*, 109435. Advance online publication. DOI: 10.1016/j.vacuum.2020.109435

Liu, X., Zheng, C., Luo, X., Wang, X., & Jiang, H. (2019). Recent advances of collagen-based biomaterials: Multi-hierarchical structure, modification and biomedical applications. *Materials Science and Engineering C, 99*, 1509–1522. DOI: 10.1016/j.msec.2019.02.070

Liu, Y., Huang, R., Lin, T., Dang, J., Huang, H., Shi, J., & Chen, S. (2024). Preparation of Sn-doped Ga_2O_3 thin films and MSM ultraviolet detectors using magnetron co-sputtering. *Materials (Basel), 17*(13), 3227. Advance online publication. DOI: 10.3390/ma17133227 PMID: 38998311

Liu, Y., Zhang, S., Zhang, J., Yao, K., Luo, M., & Liu, Y. (2024). Towards the Design of Future Aircraft: A Critical Review on the Tools and Methodologies. *Aerosp. Res. Commun., 2*, 13096. DOI: 10.3389/arc.2024.13096

Li, X., Wang, L., & Lu, Y. (2025). Three-Dimensional Disordered Alloy Metamaterials: A New Platform of Structure-Function Integration. *Materials Futures, 4*(1), 012001. DOI: 10.1088/2752-5724/ada172

Li, Y., Annan, C., Su, J., Li, Y., Zhang, Y., Li, Z., & Shi, Y. (2025). An overview on ceramic multi-material additive manufacturing: Progress and challenges. *International Journal of Extreme Manufacturing, 7*(4), 042005. DOI: 10.1088/2631-7990/adbc75

Li, Y., Lin, J., & Zhao, Z. (2024). Phase field study on the grain size dependent electrical resistivity-strain response in superelastic nanocrystalline NiTi alloys. *Journal of Materials Research and Technology, 32*, 4092–4105. DOI: 10.1016/j.jmrt.2024.08.215

Li, Y., Mai, Y.-W., & Ye, L. (2000). Sisal fibre and its composites: A review of recent developments. *Composites Science and Technology, 60*(11), 2037–2055. DOI: 10.1016/S0266-3538(00)00101-9

Lone, N. F., Ali, N., Qazi, A. M., Bajaj, D., & Siddiquee, A. N. (2024). New Developments in Bend Forming of Welded Sheet Metals. In *Comprehensive Materials Processing (Second Edition)* (Vol. 12, pp. 27–40). DOI: 10.1016/B978-0-323-96020-5.00034-0

Lopes, S. I., Pinho, P., Marques, P., Abreu, C., Carvalho, N. B., & Ferreira, J. (2021). Contactless smart screening in nursing homes: An IoT-enabled solution for the COVID-19 era. In *2021 17th International Conference on Wireless and Mobile Computing, Networking and Communications (WiMob)* (pp. 145–150).

Lopes, R. C., Nossa, T., Lustri, W. R., Lombardo, G., Errea, M. I., & Trovatti, E. (2024). Renewable polymers in biomedical applications: From the bench to the market. *Journal of Renewable Materials*, *12*(4), 643–666. DOI: 10.32604/jrm.2024.048957

Lores, A., Azurmendi, N., Agote, I., & Zuza, E. (2019). A review on recent developments in binder jetting metal additive manufacturing: Materials and process characteristics. *Powder Metallurgy*, *62*(5), 267–296. DOI: 10.1080/00325899.2019.1669299

Lu, B., Li, D., & Tian, X. (2015). Development trends in additive manufacturing and 3D printing. *Engineering (Beijing)*, *1*(1), 85–89. DOI: 10.15302/J-ENG-2015012

Lu, H., Shen, Q., Chen, X., Wen, M., & Jayalakshmi, S. (2024). Dual wire arc additive manufacturing of compositionally graded Alx-Co-Cr-Fe-Ni high-entropy alloy. *Journal of Materials Research and Technology*, *29*, 4052–4062. DOI: 10.1016/j.jmrt.2024.02.089

Lutjering, G. (1999). Property optimization through microstructural control in titanium and aluminum alloys. *Materials Science and Engineering A*, *263*(2), 117–126. DOI: 10.1016/S0921-5093(98)01169-1

Lütjering, G., & Williams, J. C. (2003). *Titanium Matrix Composites.*, DOI: 10.1007/978-3-540-71398-2_9

Lv, K., Zhu, C., Zheng, J., Wang, X., & Chen, B. (2019). Precipitation of T_1 phase in 2198 Al-Li Alloy Studied by Atomic-Resolution HAADF-STEM. *Journal of Materials Research*, *34*(20), 3535–3544. DOI: 10.1557/jmr.2019.136

Machado, L., & Lagoudas, D. (2008). Thermomechanical constitutive modeling of SMAs. In *Shape Memory Alloys: Modeling and Engineering Applications* (pp. 121-187): Springer. DOI: 10.1007/978-0-387-47685-8_3

Machkih, K., Oubaki, R., & Makha, M. (2024). A Review of CIGS Thin Film Semiconductor Deposition via Sputtering and Thermal Evaporation for Solar Cell Applications. *Coatings 2024, Vol. 14, Page 1088*, *14*(9), 1088. DOI: 10.3390/coatings14091088

Madhu, S., & Balasubramanian, M. (2015). A review on abrasive jet machining process parameters. *Applied Mechanics and Materials*, *766-767*, 629–634. . DOI: 10.4028/www.scientific.net/AMM.766-767.629

Madunuri, R., Chitta, S., Bonam, V. S. M., Vangoor, V. K. R., Yellepeddi, S. M., & Ravi, C. S. (2024). IoT-driven smart healthcare systems for remote patient monitoring and management. In *2024 Asian Conference on Intelligent Technologies (ACOIT)* (pp. 1–7). DOI: 10.1109/ACOIT62457.2024.10939717

Mahamood, R. M., Akinlabi, S. A., Shatalov, M., Murashkin, E. V., & Akinlabi, E. T. (2019). Additive Manufacturing / 3D Printing Technology: A Review. *Annals of Dunarea de Jos University of Galati Fascicle XII Welding Equipment and Technology*, *30*, 51–58. DOI: 10.35219/awet.2019.07

Maharubin, S., Hu, Y., Sooriyaarachchi, D., Cong, W., & Tan, G. Z. (2019). Laser engineered net shaping of antimicrobial and biocompatible titanium-silver alloys. *Materials Science and Engineering C*, *105*, 110059. DOI: 10.1016/j.msec.2019.110059

Mahjoub, R., Mohamad Yatim, J., Mohd Sam, A. R., & Hashemi, S. H. (2014). Tensile properties of kenaf fiber due to various conditions of chemical fiber surface modifications. *Construction & Building Materials*, *55*, 103–113. DOI: 10.1016/j.conbuildmat.2014.01.036

Mahmood, M. A., Chioibasu, D., Ur Rehman, A., Mihai, S., & Popescu, A. C. (2022). Post-Processing Techniques to Enhance the Quality of Metallic Parts Produced by Additive Manufacturing. *Metals*, *12*(1), 77. DOI: 10.3390/met12010077

Mahmud, A., Ding, D., Kiani, A., & Hasan, M. (2020). Corporate Social Responsibility Programs and Community Perceptions of Societal Progress in Bangladesh: A Multimethod Approach. *SAGE Open*, *10*(2), 2158244020924046. Advance online publication. DOI: 10.1177/2158244020924046

Mahmud, M. Z. A., Rabbi, S. F., Islam, M. D., & Hossain, N. (2025). Synthesis and applications of natural fiber-reinforced epoxy composites: A comprehensive review. *SPE Polymers*, *6*(1), e10161. DOI: 10.1002/pls2.10161

Mahrous, M. A., Abdelghany, M. A., Bonney, C., Farag, H., & Jasiuk, I. M. (2024). Irradiation-induced microstrain and dislocation density in additively manufactured 316H stainless steel. *Journal of Materials Research and Technology*, *33*, 8306–8320. DOI: 10.1016/j.jmrt.2024.11.060

Majzoobi, G. H., Azadikhah, K., & Nemati, J. (2009). The effects of deep rolling and shot peening on fretting fatigue resistance of Aluminum-7075-T6. *Materials Science and Engineering A*, *516*(1–2), 235–247. DOI: 10.1016/j.msea.2009.03.020

Mandal, P., & Roy, S. (2025). Ultra-wide bandgap β-gallium oxide (Ga_2O_3) advancing next-generation defense technologies. DOI: 10.4018/979-8-3373-0933-0.ch011.

Mandal, P., Kumar, S., Pandey, A., Katariya, L., Mondal, A., & Bag, A. (2024). Unveiling structural and optical properties of Sn-doped β-Ga_2O_3: A correlation of experimental and theoretical observations. *Materials Science and Engineering B*, *302*, 117266. Advance online publication. DOI: 10.1016/j.mseb.2024.117266

Mandal, P., Mondal, A., Pandey, A., Romeo, S., & Bag, A. (2025). Investigating role of annealing in shaping morphological, structural, tribological, and optical characteristics of gallium oxide (Ga_2O_3): Insights from DFT analysis. *Journal of Alloys and Compounds*, *1011*, 178367. Advance online publication. DOI: 10.1016/j.jallcom.2024.178367

Mandal, P., Roy, S., & Singh, U. P. (2022). Investigation on the optical and electrical performance of aluminium doped gallium oxide thin films. *Optical and Quantum Electronics*, *54*(8), 515. Advance online publication. DOI: 10.1007/s11082-022-03851-0

Mandal, P., Singh, U. P., & Roy, S. (2021a). A review on the effects of PVD RF sputtering parameters on rare earth oxide thin films and their applications. *IOP Conference Series. Materials Science and Engineering*, *1166*(1), 012022. Advance online publication. DOI: 10.1088/1757-899X/1166/1/012022

Mandal, P., Singh, U. P., & Roy, S. (2021b). Optical performance of europium-doped β gallium oxide PVD thin films. *Journal of Materials Science Materials in Electronics*, *32*(4), 3958–3965. DOI: 10.1007/s10854-020-05137-1

Mandal, S., Sadeghianjahromi, A., & Wang, C.-C. (2022). Experimental and numerical investigations on molten metal atomization techniques – A critical review. *Advanced Powder Technology*, *33*(11), 103809. DOI: 10.1016/j.apt.2022.103809

Mani, K., Singh, K. K., & Litoriya, R. (2024). IoT-driven revolution: Transforming elderly healthcare through dynamic heart health monitoring and analysis. In *2024 International Conference on Advances in Computing Research on Science Engineering and Technology (ACROSET)* (pp. 1–8). DOI: 10.1109/ACROSET62108.2024.10743704

Manoj, I., & Narendranath, S. (2020). Variation and artificial neural network prediction of profile areas during slant type taper profiling of triangle at different machining parameters on Hastelloy X by wire electric discharge machining. *Proceedings of the Institution of Mechanical Engineers. Part E, Journal of Process Mechanical Engineering*, *234*(6), 673–683. DOI: 10.1177/0954408920938614

Mansor, M. S. M., Raja, S., Yusof, F., Muhamad, M. R., Manurung, Y. H., Adenan, M. S., & Ren, J. (2024). Integrated approach to Wire Arc Additive Manufacturing (WAAM) optimization: Harnessing the synergy of process parameters and deposition strategies. *Journal of Materials Research and Technology*, *30*, 2478–2499. DOI: 10.1016/j.jmrt.2024.03.170

Marattukalam, J. J., Singh, A. K., Datta, S., Das, M., Balla, V. K., Bontha, S., & Kalpathy, S. K. (2015). Microstructure and corrosion behavior of laser processed NiTi alloy. *Materials Science and Engineering C*, *57*, 309–313. DOI: 10.1016/j.msec.2015.07.067

Marin, E., & Lanzutti, A. (2023). Biomedical Applications of Titanium Alloys: A Comprehensive Review. *Materials (Basel)*, *17*(1), 114. DOI: 10.3390/ma17010114

Markopoulos, A. P. (2012). *Finite Element Method in Machining Processes*. Springer London. Retrieved from https://books.google.co.in/books?id=FM8_dRux4sgC

Martina, F., Ding, J., Williams, S., Caballero, A., Pardal, G., & Quintino, L. (2019). Tandem metal inert gas process for high productivity wire arc additive manufacturing in stainless steel. *Additive Manufacturing*, *25*, 545–550. DOI: 10.1016/j.addma.2018.11.022

Martina, F., Roy, M. J., Szost, B. A., Terzi, S., Colegrove, P. A., Williams, S. W., Withers, P. J., Meyer, J., & Hofmann, M. (2016). Residual stress of as-deposited and rolled wire+arc additive manufacturing Ti–6Al–4V components. *Materials Science and Technology*, *32*(14), 1439–1448. DOI: 10.1080/02670836.2016.1142704

Mathias, L. E. T., Pinotti, V. E., Batistão, B. F., Rojas-Arias, N., Figueira, G., Andreoli, A. F., & Gargarella, P. (2024). Metal powder as feedstock for laser-based additive manufacturing: From production to powder modification. *Journal of Materials Research*, *39*(1), 19–47. DOI: 10.1557/s43578-023-01271-8

Mathkari, R., Liehr, M., Ravindra, P., Pareis, R., Beckmann, K., Tokranova, N., Schujman, S., Saraf, I., Van der Straten, O., Gong, N., Ando, T., & Cady, N. (2025). Investigation of the effect of oxygen partial pressure during reactive sputtering of tantalum oxide resistive random access memory switching layer. *Materials Science in Semiconductor Processing*, *186*, 109060. Advance online publication. DOI: 10.1016/j.mssp.2024.109060

Matsuno, H. (2001). Biocompatibility and osteogenesis of refractory metal implants, titanium, hafnium, niobium, tantalum and rhenium. *Biomaterials*, *22*(11), 1253–1262. DOI: 10.1016/S0142-9612(00)00275-1

Mattox, D. M. (2010). *Handbook of physical vapor deposition (PVD) processing* (2nd ed.). Elsevier.

Ma, Y., Li, J., Lei, J., Tang, Z., Liu, Y., & Yang, R. (2010). Influences of Microstructure on Fatigue Crack Propagating Path and Crack Growth Rates in TC4ELI Alloy. *Chin Shu Hsueh Pao*, *46*(9), 1086–1092. DOI: 10.3724/SP.J.1037.2010.00155

Ma, Z. Y. (2008). Friction Stir Processing Technology: A Review. *Metallurgical and Materials Transactions. A, Physical Metallurgy and Materials Science*, *39*(3), 642–658. DOI: 10.1007/s11661-007-9459-0

Mazza, E., & Ehret, A. E. (2015). Mechanical biocompatibility of highly deformable biomedical materials. *Journal of the Mechanical Behavior of Biomedical Materials. Journal of the Mechanical Behavior of Biomedical Materials*, *48*, 100–124. DOI: 10.1016/j.jmbbm.2015.03.023

Md, F. K., & Panigrahi, S. K.F Khan MD. (2018). Achieving excellent superplasticity in an ultrafine-grained QE22 alloy at both high strain rate and low-temperature regimes. *Journal of Alloys and Compounds*, *747*, 71–82. DOI: 10.1016/j.jallcom.2018.02.294

Meenia, S., & Khan, F. (2016). Particle refinement and fine-grain formation leading to enhanced mechanical behaviour in a hypo-eutectic Al-Si alloy subjected to multi-pass friction stir processing. *Materials Characterization*, *113*, 134–143. DOI: 10.1016/j.matchar.2016.01.011

Mehrotra, D., Kumar, S., Mehrotra, P., Khanna, R., Khanna, V., Eggbeer, D., & Evans, P. (2021). Patient specific total temporomandibular joint reconstruction: A review of biomaterial, designs, fabrication and outcomes. *Journal of Oral Biology and Craniofacial Research*, *11*(2), 334–343. DOI: 10.1016/j.jobcr.2021.02.014

Melentiev, R., & Fang, F. (2018). Review. Recent advances and challenges of abrasive jet machining. *CIRP Journal of Manufacturing Science and Technology*, *22*, 1–20. DOI: 10.1016/j.cirpj.2018.06.001

Mele, P., Endo, T., Arisawa, S., Li, C., & Tsuchiya, T. (Eds.). (2015). *Oxide thin films, multilayers, and nanocomposites*. Springer. DOI: 10.1007/978-3-319-14478-8

Memon, A. R., Li, J., Egger, J., & Chen, X. (2021). A review on patient-specific facial and cranial implant design using Artificial Intelligence (AI) techniques. *Expert Review of Medical Devices*, *18*(10), 985–994. DOI: 10.1080/17434440.2021.1969914

Meneghini, M., De Santi, C., Abid, I., Buffolo, M., Cioni, M., Khadar, R. A., Nela, L., Zagni, N., Chini, A., Medjdoub, F., Meneghesso, G., Verzellesi, G., Zanoni, E., & Matioli, E. (2021). GaN-based power devices: Physics, reliability, and perspectives. *Journal of Applied Physics*, *130*(18), 181101. Advance online publication. DOI: 10.1063/5.0061354

Mercelis, P., & Kruth, J. (2006). Residual stresses in selective laser sintering and selective laser melting. *Rapid Prototyping Journal*, *12*(5), 254–265. DOI: 10.1108/13552540610707013

Meteyer, S., Xu, X., Perry, N., & Zhao, Y. F. (2014). Energy and Material Flow Analysis of Binder-jetting Additive Manufacturing Processes. *Procedia CIRP*, *15*, 19–25. DOI: 10.1016/j.procir.2014.06.030

Meyer, P., Cans, Y., Ferton, D., & Reboul, M. (1987). The Metallurgy of Industrial Al-Li Alloys. *Journal de Physique. Colloque*, *48*(C3), 131–138. DOI: 10.1051/jphyscol:1987316

Mezdrogina, M. M., Vinogradov, A. Y., Kozhanova, Y. V., & Borsuk, E. A. (2020). LED structures based on ZnO films obtained by RF magnetron sputtering for the UV spectral range. *Technical Physics*, *65*(3), 434–439. DOI: 10.1134/S1063784220030159

Miah, M. H., Rahman, M. B., Nur-E-Alam, M., Islam, M. A., Shahinuzzaman, M., Rahman, M. R., Ullah, M. H., & Khandaker, M. U. (2025). Key degradation mechanisms of perovskite solar cells and strategies for enhanced stability: Issues and prospects. *RSC Advances*, *15*(1), 628–654. DOI: 10.1039/D4RA07942F PMID: 39777157

Middleton, B. (2016). Composites: Manufacture and application. In Design and Manufacture of Plastic Components for Multifunctionality: Structural Composites, Injection Molding, and 3D Printing (Chapter 3, pp. 53–101).

Min, H., Lee, D. Y., Kim, J., Kim, G., Lee, K. S., Kim, J., Paik, M. J., Kim, Y. K., Kim, K. S., Kim, M. G., Shin, T. J., & Il Seok, S. (2021). Perovskite solar cells with atomically coherent interlayers on SnO2 electrodes. *Nature*, *598*(7881), 444–450. DOI: 10.1038/s41586-021-03964-8 PMID: 34671136

Mishra, D. M. (2025). A Review of Manufacturing Processes for Aluminum-Lithium Alloys and Their Applications. *Premier Journal of Engineering*, *3*, 100007. DOI: 10.70389/PJE.100007

Mishra, S. F., Khan, M. D., & Panigrahi, S. K. (2022). A crystal plasticity based approach to establish role of grain size and crystallographic texture in the Tension–Compression yield asymmetry and strain hardening behavior of a Magnesium–Silver–Rare Earth alloy. *Journal of Magnesium and Alloys*, *10*(9), 2546–2562. DOI: 10.1016/j.jma.2021.08.021

Mistry, M., Palkhade, S., Patel, V., Shah, D., Vyas, K., & Shah, S. (2018). A review on abrasive jet machining. *International Journal of Scientific Research in Science, Engineering and Technology*, *4*(1), 571–574. DOI: 10.1016/0924-0136(93)90005-Q

Mi, W., Du, X., Luan, C., Xiao, H., & Ma, J. (2014). Electrical and optical characterizations of β-Ga_2O_3:Sn films deposited on MgO (110) substrate by MOCVD. *RSC Advances*, *4*(58), 30579–30583. DOI: 10.1039/C4RA02479F

Mndlovu, H., Kumar, P., Du Toit, L. C., & Choonara, Y. E. (2024). A review of biomaterial degradation assessment approaches employed in the biomedical field. *NPJ Materials Degradation*, *8*(1), 66. Advance online publication. DOI: 10.1038/s41529-024-00487-1

Mobtakeri, S., Akaltun, Y., Özer, A., Kılıç, M., Tüzemen, E. Ş., & Gür, E. (2021). Gallium oxide films deposition by RF magnetron sputtering; a detailed analysis on the effects of deposition pressure and sputtering power and annealing. *Ceramics International*, *47*(2), 1721–1727. DOI: 10.1016/j.ceramint.2020.08.289

Moenck, K., Koch, J., Rath, J.-E., Büsch, L., Gierecker, J., Kähler, F., Kalscheuer, F., Masuhr, C., Kipping, J., Prünte, P., Schoepflin, D., Eschen, H., Wulff, L.A., Rodeck, R., Wende, G., Gomse, M., & Schüppstuhl, T. (2025). Industry 5.0 in Aircraft Production and MRO: Challenges and Opportunities. *CEAS Aeronaut. J.* doi: DOI: 10.1007/s13272-025-00832-3

Mohammed, A., & Abdullah, A. (2018) Scanning Electron Microscopy (SEM): A Review. *Proceedings of 2018 International Conference on Hydraulics and Pneumatics-HERVEX*, Baile Govora, Romania, 7-9 November 2018, 77-85.

Mohammed, M., Oleiwi, J. K., Mohammed, A. M., & Mohamad Jawad, A. J. (2024). A review on the advancement of renewable natural fiber hybrid composites: Prospects, challenges, and industrial applications. *Journal of Renewable Materials*, *12*(7), 1237–1290. DOI: 10.32604/jrm.2024.051201

Mondal, A., Yadav, M. K., Shringi, S., & Bag, A. (2020). Extremely low dark current and detection range extension of Ga_2O_3 UV photodetector using Sn alloyed nanostructures. *Nanotechnology*, *31*(29), 295201. Advance online publication. DOI: 10.1088/1361-6528/ab82d4 PMID: 32208373

Mondal, S., Ghosh, S., & Basak, D. (2021). Extraordinarily high ultraviolet photodetection by defect tuned phosphorus doped ZnO thin film on flexible substrate. *Materials Research Bulletin*, *144*, 111490. Advance online publication. DOI: 10.1016/j.materresbull.2021.111490

Morrison, R. J., Kashlan, K. N., Flanangan, C. L., Wright, J. K., Green, G. E., Hollister, S. J., & Weatherwax, K. J. (2015). Regulatory Considerations in the Design and Manufacturing of Implantable 3D-Printed Medical Devices. *Clinical and Translational Science*, *8*(5), 594–600. DOI: 10.1111/cts.12315

Moshiri, A., Zarei-Hanzaki, A., Anousheh, A. S., Abedi, H. R., Sohn, S. S., Yang, J., & Berto, F. (2021). On the fatigue and dwell-fatigue behavior of a low-density steel and the correlated microstructure origin of damage mechanism. *Journal of Materials Research and Technology*, *15*, 6136–6154. DOI: 10.1016/j.jmrt.2021.10.135

Mourya, S. K., Malik, G., Alisha, , Kumar, B., & Chandra, R. (2022). The role of non-homogeneous barrier on the electrical performance of 15R–SiC Schottky diodes grown by in-situ RF sputtering. *Materials Science in Semiconductor Processing*, *149*, 106855. Advance online publication. DOI: 10.1016/j.mssp.2022.106855

Mu, H., He, F., Yuan, L., Hatamian, H., Commins, P., & Pan, Z. (2024). Online distortion simulation using generative machine learning models: A step toward digital twin of metallic additive manufacturing. *Journal of Industrial Information Integration*, *38*, 100563. DOI: 10.1016/j.jii.2024.100563

Mujahid, S., Boostani, A. F., Paudel, Y., Oppedal, A., Paliwal, B., Rhee, H., & El Kadiri, H. (2024). Examination of Novel Titanium-639 Alloy as a Means of Balancing Strength and Ductility through Molybdenum Addition Rather than Prolonged Aging Heat Treatment. *Metals*, *14*(7), 828. DOI: 10.3390/met14070828

Mumtaz, K. A., & Hopkinson, N. (2010). Selective Laser Melting of thin wall parts using pulse shaping. *Journal of Materials Processing Technology*, *210*(2), 279–287. DOI: 10.1016/j.jmatprotec.2009.09.011

Munna, F. T., Selvanathan, V., Sobayel, K., Muhammad, G., Asim, N., Amin, N., Sopian, K., & Akhtaruzzaman, M. (2021). Diluted chemical bath deposition of CdZnS as prospective buffer layer in CIGS solar cell. *Ceramics International*, *47*(8), 11003–11009. DOI: 10.1016/j.ceramint.2020.12.222

Muthu, S. S. (2020). Introduction to sustainability and the textile supply chain and its environmental impact. In *Assessing the Environmental Impact of Textiles and the Clothing Supply Chain (The Textile Institute Book Series)* (pp. 1–32). Woodhead Publishing., DOI: 10.1016/B978-0-12-819783-7.00001-6

N'guessan, A. I., Bouich, A., Touré, A., Soucasse, B. M., & Soro, D. (2023). Influence of Sulfur Content in Zn(O,S) Buffer Layer onto Copper Indium Gallium Sulfur-Based Solar Cells Through Surface Engineering at ZnO1−xSx/CIGS Interface. *Journal of the Minerals Metals & Materials Society*, *75*(10), 4332–4340. DOI: 10.1007/s11837-023-06018-8

Naderi, N., & Moghaddam, M. (2020). Ultra-sensitive UV sensors based on porous silicon carbide thin films on silicon substrate. *Ceramics International*, *46*(9), 13821–13826. DOI: 10.1016/j.ceramint.2020.02.173

Nair, P. B., Justinvictor, V. B., Daniel, G. P., Joy, K., Ramakrishnan, V., & Thomas, P. V. (2011). Effect of RF power and sputtering pressure on the structural and optical properties of TiO_2 thin films prepared by RF magnetron sputtering. *Applied Surface Science, 257*(24), 10869–10875. DOI: 10.1016/j.apsusc.2011.07.125

Nakai, M., Niinomi, M., Komine, K., Liu, H., Morisada, Y., & Fujii, H. (2019). High-cycle fatigue properties of an easily hot-workable (α+ β)-type titanium alloy butt joint prepared by friction stir welding below β transus temperature. *Materials Science and Engineering A, 742*, 553–563. DOI: 10.1016/j.msea.2018.10.103

Nakamura, H., Nakamura, M., Osuga, N., & Miyazawa, H. (2006). Application of an air-abrasive cutting apparatusin the pediatric dental field: Cutting using chitin-chitosan grains. *Pediatric Dental Journal: International Journal of Japanese Society of Pediatric Dentistry, 16*(1), 57–66. DOI: 10.1016/S0917-2394(06)70066-0

Namdar, M., & Jahromi, S. A. J. (2015). Influence of ECAP on the fatigue behavior of age-hardnable 2xxx aluminum alloy. *International Journal of Minerals Metallurgy and Materials, 3*(3), 285–291. DOI: 10.1007/s12613-015-1072-4

Nanda, H. S., Yang, L., Hu, J., Mao, H., & Jiang, S. (2022). Editorial: Biodegradable Polymers for Biomedical Applications. *Frontiers in Materials, 9*, 944755. Advance online publication. DOI: 10.3389/fmats.2022.944755

Nayak, R. K., Pradhan, M. K., & Sahoo, A. K. (2022). *Machining of Nanocomposites*. CRC Press LLC. Retrieved from https://books.google.co.in/books?id=ZEW9zgEACAAJ

Neto, J., Queiroz, H., Aguiar, R., Lima, R., Cavalcanti, D., & Banea, M. D. (2022). A review of recent advances in hybrid natural fiber reinforced polymer composites. *Journal of Renewable Materials, 10*(3), 561–589. DOI: 10.32604/jrm.2022.017434

Neve, S. L. (2019). *Fire Behaviour of Magnesium Alloys.*

Ngo, T. D., Kashani, A., Imbalzano, G., Nguyen, K. T. Q., & Hui, D. (2018). Additive manufacturing (3D printing): A review of materials, methods, applications and challenges. *Composites. Part B, Engineering, 143*, 172–196. DOI: 10.1016/j.compositesb.2018.02.012

Ngo, T. S., Le, D. D., Song, J. H., & Hong, S. K. (2019). Growth and characterization of gallium oxide films grown with nitrogen by plasma-assisted molecular-beam epitaxy. *Thin Solid Films, 682*, 93–98. DOI: 10.1016/j.tsf.2019.05.029

Nguyen, K. T., Navaratnam, S., Mendis, P., Zhang, K., Barnett, J., & Wang, H. (2020). Fire safety of composites in prefabricated buildings: From fibre reinforced polymer to textile reinforced concrete. Composites Part B: Engineering, 187, 107815. DOI: 10.1016/j.compositesb.2020.107815

Nguyen, V. T., Minh, P. S., Uyen, T. M. T., Do, T. T., Ngoc, H. V. T., Le, M. T., & Tien Nguyen, V. T. (2023). WAAM technique: Process parameters affecting the mechanical properties and microstructures of low-carbon steel. *Metals*, *13*(5), 873. DOI: 10.3390/met13050873

Ng, W. L., An, J., & Chua, C. K. (2024). Process, Material, and Regulatory Considerations for 3D Printed Medical Devices and Tissue Constructs. *Engineering (Beijing)*, *36*, 146–166. DOI: 10.1016/j.eng.2024.01.028

NIE, J. F. (2012). Precipitation and Hardening in Magnesium Alloys. *Metallurgical and Materials Transactions. A, Physical Metallurgy and Materials Science*, *43A*, 3891–3939.

Nie, X., He, W., Li, Q. P., Long, N., & Chai, Y. (2013). Experimental Investigation on Microstructure and Mechanical Properties of TC17 Titanium Alloy Treated by Laser Shock Peening with Different Laser Fluence. *Journal of Laser Applications*, *25*(4), 042001–1–042001–042006.

Nied, H. F., Taylor, C. A., & Delorenzi, H. G. (1990). Three-dimensional finite element simulation of thermoforming. *Polymer Engineering and Science*, *30*(20), 1314–1322. DOI: 10.1002/pen.760302009

Nie, L., Zhan, Y., Liu, H., & Tang, C. (2014). Novel β-type Zr–Mo–Ti alloys for biological hard tissue replacements. *Materials & Design*, *53*, 8–12. DOI: 10.1016/j.matdes.2013.07.008

Nie, X., He, W., Zhou, L., Li, Q., & Wang, X. (2014). Experimental Investigation of Laser Shock Peening on TC6 Titanium Alloy to Improve High Cycle Fatigue Performance. *Materials Science and Engineering A*, *594*, 161–167. DOI: 10.1016/j.msea.2013.11.073

Ni, J., Ling, H., Zhang, S., Wang, Z., Peng, Z., Benyshek, C., Zan, R., Miri, A. K., Li, Z., Zhang, X., Lee, J., Lee, K.-J., Kim, H.-J., Tebon, P., Hoffman, T., Dokmeci, M. R., Ashammakhi, N., Li, X., & Khademhosseini, A. (2019). Three-dimensional printing of metals for biomedical applications. *Materials Today. Bio*, *3*, 100024. DOI: 10.1016/j.mtbio.2019.100024

Nikkhoo, A. & Esmaeili, A. (2023). Effect of Different Aero-Structural Optimization in the Commercial Airplane. J. Comput. Appl. Mech., 54(2), 268-284. doi: 10.22059/jcamesh.2023.355499.809

Niu, F., Bi, W., Liu, Z., Sun, X., Ma, G., & Wu, D. (2024). 304-Ni60B (NiCrBSi) composite components using WAAM-LC integrated hybrid manufacturing. *International Journal of Advanced Manufacturing Technology*, *132*(1-2), 463–474. DOI: 10.1007/s00170-024-13376-3

Noorasid, N. S., Arith, F., Mustafa, A. N., Azam, M. A., Mahalingam, S., Chelvanathan, P., & Amin, N. (2022). Current advancement of flexible dye sensitized solar cell: A review. *Optik (Stuttgart)*, *254*, 168089. DOI: 10.1016/j.ijleo.2021.168089

Ogunleye, R. O., Rusnakova, S., Zaludek, M., & Emebu, S. (2022). The influence of ply stacking sequence on mechanical properties of carbon/epoxy composite laminates. Polymers, 14(24), 5566. DOI: 10.3390/polym14245566

Okokpujie, I. P., Chima, P. C., & Tartibu, L. K. (2022). Experimental and 3D-Deform Finite Element Analysis on Tool Wear during Turning of Al-Si-Mg Alloy. *Lubricants (Basel, Switzerland)*, *10*(12), 341. DOI: 10.3390/lubricants10120341

Okokpujie, I. P., Tartibu, L. K., Chima, P. C., & Ogundipe, A. T. (2023). A finite element based investigation of tool wear rate via machining of Al6061 alloys using deform-3D. *Materials Today: Proceedings*. Advance online publication. DOI: 10.1016/j.matpr.2023.05.566

Olanrewaju, O., Oladele, I. O., & Adelani, S. O. (2025). Recent Advances in Natural Fiber Reinforced Metal/Cement/Polymer Composites: An Overview of the Structure-Property Relationship for Engineering Applications. *Hybrid Advances*, 8, 100378. DOI: 10.1016/j.hybadv.2025.100378

Oleff, A., Küster, B., Stonis, M., & Overmeyer, L. (2021). Process monitoring for material extrusion additive manufacturing: A state-of-the-art review. *Progress in Additive Manufacturing*, *6*(4), 705–730. DOI: 10.1007/s40964-021-00192-4

Oleksy, M., Dynarowicz, K., & Aebisher, D. (2023). Advances in Biodegradable Polymers and Biomaterials for Medical Applications - A Review. *Molecules (Basel, Switzerland)*, *28*(17), 6213. DOI: 10.3390/molecules28176213

Oliveira, C., Sousa, D., Teixeira, J. A., Ferreira-Santos, P., & Botelho, C. M. (2023). Polymeric biomaterials for wound healing. *Frontiers in Bioengineering and Biotechnology*, *11*, 1136077. Advance online publication. DOI: 10.3389/fbioe.2023.1136077

Omiyale, B. O., Ogedengbe, I. I., Olugbade, T. O., Rasheed, A. A., Ogbeyemi, A., & Farayibi, P. K. (2025). Cold metal transfer WAAM of aluminum alloys: Influence of processing parameters. *International Journal of Advanced Manufacturing Technology*, *136*(5), 1967–1987. DOI: 10.1007/s00170-024-14989-4

Outeiro, J. (2018). Residual Stresses in Machining Operations. In CIRP Encyclopedia of Production Engineering (pp. 1–13). Berlin, Heidelberg: Springer Berlin Heidelberg. DOI: 10.1007/978-3-642-35950-7_16811-1

Ou, W., Knapp, G. L., Mukherjee, T., Wei, Y., & DebRoy, T. (2021). An improved heat transfer and fluid flow model of wire-arc additive manufacturing. *International Journal of Heat and Mass Transfer*, *167*, 120835. DOI: 10.1016/j.ijheatmasstransfer.2020.120835

Pai, A., Shenoy Baloor, S., & Kini, C. R. (2023). Taguchi Analysis of the Deformation Characteristics of Single-Point Cutting Tool with Micro-Tool Coatings During Orthogonal Machining. *Journal of The Institution of Engineers (India): Series D*, *104*(2), 811–820. DOI: 10.1007/s40033-022-00437-8

Panda, S., Mehlawat, S., Dhariwal, N., Sanger, A., & Kumar, A. (2024). Comprehensive review on gas sensors: Unveiling recent developments and addressing challenges. *Materials Science and Engineering B*, *308*, 117616. Advance online publication. DOI: 10.1016/j.mseb.2024.117616

Pang, B., Zhang, Y., Li, J., Wang, X., Chang, M., & Bai, J. (2024). Data-Driven Surrogate Model for Aerodynamic Design using Separable Shape Tensor Method. *Chinese Journal of Aeronautics*, *37*(9), 41–58. DOI: 10.1016/j.cja.2024.03.014

Pan, Y., Yu, M., Xu, C., Zhang, J., & Geng, L. (2024). High-performance 2319 aluminum alloy via CMT-WAAM: Microstructure, porosity, and mechanical properties. *Metals*, *14*(7), 14. DOI: 10.3390/met14070797

Papageorgiou, G. Z. (2018). Thinking Green: Sustainable Polymers from Renewable Resources. *Polymers*, *10*(9), 952. DOI: 10.3390/polym10090952

Pappalardo, D., Mathisen, T., & Finne-Wistrand, A. (2019). Biocompatibility of Resorbable Polymers: A Historical perspective and Framework for the future. *Biomacromolecules*, *20*(4), 1465–1477. DOI: 10.1021/acs.biomac.9b00159

Parihar, R. S., Sahu, R. K., & Srinivasu, G. (2017a). Finite Element Analysis of Cutting Forces Generated in Turning Process using Deform 3D Software. *Materials Today: Proceedings*, *4*(8), 8432–8438. DOI: 10.1016/j.matpr.2017.07.188

Park, D.-S., Cho, M.-W., Lee, H., & Cho, W.-S. (2004). Micro-grooving of glass using micro-abrasive jet machining. *Journal of Materials Processing Technology*, *146*(2), 234–240. DOI: 10.1016/j.jmatprotec.2003.11.013

Park, J. H., McClintock, R., Jaud, A., Dehzangi, A., & Razeghi, M. (2019). MOCVD grown β-Ga_2O_3 metal-oxide-semiconductor field effect transistors on sapphire. *Applied Physics Express*, *12*(9), 091001. Advance online publication. DOI: 10.7567/1882-0786/ab3b2a

Patel, P. A. (2011). Effective use and maintenance of shot-blasting machines. Retrieved November 3, 2019, from https://www.pshotblast.com/articles/effective-use-maintenance-page-1.pdf

Patidar, R., Burkitt, D., Hooper, K., Richards, D., & Watson, T. (2020). Slot-die coating of perovskite solar cells: An overview. *Materials Today. Communications*, *22*, 100808. DOI: 10.1016/j.mtcomm.2019.100808

Paul, A. A., Nguyen, Q. H., & Shen, W. (2024). 3D printed magnetostrictive polymer composites (MPCs) for wireless stress sensing. *Materials & Design*, *247*, 113402. DOI: 10.1016/j.matdes.2024.113402

Pavani, G., & Shashidhar, A. (2024). Mechanical Properties of Natural Composites: Hemp Fiber and Polypropylene Resin. *Journal of Physics: Conference Series*, *2837*(1), 012002. DOI: 10.1088/1742-6596/2837/1/012002

Pednekar, P. P., Godiyal, S. C., Jadhav, K. R., & Kadam, V. J. (2017). Mesoporous silica nanoparticles: A promising multifunctional drug delivery system. In *Nanostructures for cancer therapy* (pp. 593–621). Elsevier., DOI: 10.1016/B978-0-323-46144-3.00023-4

Pei, E., Bernard, A., Gu, D., Klahn, C., Monzón, M., Petersen, M., & Sun, T. (Eds.). (2023). *Springer Handbook of Additive Manufacturing*. Springer International Publishing., DOI: 10.1007/978-3-031-20752-5

Peng, X., Kong, L., Fuh, J. Y. H., & Wang, H. (2021). A Review of Post-Processing Technologies in Additive Manufacturing. *Journal of Manufacturing and Materials Processing*, *5*(2), 38. DOI: 10.3390/jmmp5020038

Periayah, M. H., Halim, A. S., & Saad, A. Z. M. (2016). Chitosan: A promising marine polysaccharide for biomedical research. *Pharmacognosy Reviews/Bioinformatics Trends. Pharmacognosy Reviews*, *10*(19), 39. DOI: 10.4103/0973-7847.176545

Perumal, G., Senthilkumar, N., Satheshkumar, D., & Anandhakumar, S. (2025). Comparative analysis of solution treatment with short and long-term aging on mechanical, tribological, and corrosion potential of Ti-6Al-4V alloy. *Multiscale and Multidisciplinary Modeling. Experiments and Design*, *8*(3), 159. DOI: 10.1007/s41939-025-00759-6

Pesode, P., & Barve, S. (2022a). Additive manufacturing of metallic biomaterials and its biocompatibility. *Materials Today: Proceedings*. Advance online publication. DOI: 10.1016/j.matpr.2022.11.248

Peters, A. B., Zhang, D., Chen, S., Ott, C., Oses, C., Curtarolo, S., McCue, I., Pollock, T. M., & Prameela, S. E. (2024). Materials Design for Hypersonics. *Nature Communications*, *15*(1), 3328. DOI: 10.1038/s41467-024-46753-3

Phiri, R., Rangappa, S. M., Siengchin, S., Oladijo, O. P., & Ozbakkaloglu, T. (2024). Advances in lightweight composite structures and manufacturing technologies: A comprehensive review. *Heliyon, 10*(21), e39661. DOI: 10.1016/j.heliyon.2024.e39661

Pollock, L., & Wild, G. (2024a). An Examination of High-Speed Aircraft - Part 1: Past, Present, and Future. *Transp. Eng.*, *18*, 100290. DOI: 10.1016/j.treng.2024.100290

Pollock, L., & Wild, G. (2024b). An Examination of High-Speed Aircraft – Part 2: Safety and Reliability. *Transp. Eng.*, *18*, 100269. DOI: 10.1016/j.treng.2024.100269

Polmear, I. J. (1989). *Light alloys: metallurgy of the light metals* (2nd ed.). Edward Arnold.

Popov, V. V.Jr, Muller-Kamskii, G., Katz-Demyanetz, A., Kovalevsky, A., Usov, S., Trofimcow, D., Dzhenzhera, G., & Koptyug, A. (2019). Additive manufacturing to veterinary practice: Recovery of bony defects after the osteosarcoma resection in canines. *Biomedical Engineering Letters*, *9*(1), 97–108. DOI: 10.1007/s13534-018-00092-7

Popov, V. V.Jr, Muller-Kamskii, G., Kovalevsky, A., Dzhenzhera, G., Strokin, E., Kolomiets, A., & Ramon, J. (2018). Design and 3D-printing of titanium bone implants: Brief review of approach and clinical cases. *Biomedical Engineering Letters*, *8*(4), 337–344. DOI: 10.1007/s13534-018-0080-5

Porwal, H., Soumya, G., Bhavana, P. H. S., Chandrashekhar, S., & Anand, A. V. (2023). Mechanical properties of green composites with hemp/sisal fiber and novolac resin. AIP Conference Proceedings, 2492, 040009. DOI: 10.1063/5.0113265

Potzies, C., & Kainer, K. U. (2004). Fatigue of Magnesium Alloys. *Advanced Engineering Materials*, *6*(5), 281–289. DOI: 10.1002/adem.200400021

Pradhan, S., & Dhupal, D. (2022). An integrated approach of simulation, modeling and computer-aided design of hot abrasive jet machining setup. *Journal of Advanced Manufacturing Systems, 21*(03), 427–472. DOI: 10.1142/S0219686722500123

Pradhan, S., & Dhupal, D. (2023). Machining of aluminium nitride ceramic with a hAJM process using SiC abrasives: Experiments and simulation. *Journal of Manufacturing Processes, 95*, 68–90. DOI: 10.1016/j.jmapro.2023.04.011

Pragadish, N., Kaliappan, S., Subramanian, M., Natrayan, L., Satish Prakash, K., Subbiah, R., Kumar, T., & Ch, A. (2023). Optimization of cardanol oil dielectric-activated EDM process parameters in machining of silicon steel. *Biomass Conversion and Biorefinery, 13*(15), 14087–14096. DOI: 10.1007/s13399-021-02268-1

Pramanik, A. (2021). *Machining and Tribology: Processes, Surfaces, Coolants, and Modeling*. Elsevier. Retrieved from https://books.google.co.in/books?id=RksiEAAAQBAJ

Pramanik, A., Basak, A. K., Littlefair, G., Debnath, S., Prakash, C., Singh, M. A., Marla, D., & Singh, R. K. (2020). Methods and variables in Electrical discharge machining of titanium alloy – A review. *Heliyon, 6*(12), e05554. DOI: 10.1016/j.heliyon.2020.e05554

Pranathi, K., Ayub, S. M. A., & Paul, C. H. (2024). Composite Laminates with Epoxy Resin and E-Glass Fibers: Analysis of Mechanical Properties. *Journal of Physics: Conference Series, 2837*(1), 012008. DOI: 10.1088/1742-6596/2837/1/012008

Prasad, N. E., Srivatsan, T. S., Wanhill, R., Malakondaiah, G., & Kutumbarao, V. V. (2014). Fatigue Behavior of Aluminum–Lithium Alloys. In *Aluminum–Lithium Alloys* (pp. 341–349). Processing, Properties, and Applications. DOI: 10.1016/B978-0-12-401698-9.00011-2

Prasad, S. R., Rishendra, D., Krishna, N. V., Dinesh, S., & Hamraj, S. (2025). Machinability characteristics of hot air abrasive jet machining. *Journal of Nonlinear Analysis and Optimization, 16*(1).

Prathyusha, G., Krishnan, M. B. M., & Deepak, S. P. K. (2024). Integration of IoT-enabled UAVs in smart agriculture: Types, applications, trends, challenges. In *2024 7th International Conference on Circuit Power and Computing Technologies (ICCPCT)* (pp. 1124–1130).

Precision, H. Abrasive Jet Machining (2012). Retrieved October 31, 2019, from-https://web.archive.org/web/20120304025100/http://www.texasairsonics.com/Applications.html

Priya, S. S., Al-Fatlawy, M. H., Khare, N., Mahalakshmi, V., & Ganesh, S. S. (2023). Machine and deep learning classifications for IoT-enabled healthcare devices. In *2023 4th International Conference on Computation, Automation and Knowledge Management (ICCAKM)* (pp. 1–7). DOI: 10.1109/ICCAKM58659.2023.10449617

Puertas-Bartolomé, M., Mora-Boza, A., & García-Fernández, L. (2021). Emerging Biofabrication Techniques: A review on natural polymers for biomedical applications. *Polymers*, *13*(8), 1209. DOI: 10.3390/polym13081209

Purohit, A., Jain, S., Nema, P., Vishwakarma, H., & Jain, P. K. (2022). Intelligent or smart polymers: Advance in novel drug delivery. *Journal of Drug Delivery and Therapeutics*, *12*(5), 208–216. DOI: 10.22270/jddt.v12i5.5578

Qi, H., Qin, S., Cheng, Z., Teng, Q., Hong, T., & Xie, Y. (2021). Towards understanding performance enhancing mechanism of micro-holes on K9 glasses using ultrasonic vibration-assisted abrasive slurry jet. *Journal of Manufacturing Processes*, *64*, 585–593. DOI: 10.1016/j.jmapro.2021.01.048

Radhakrishnan, S., Nagarajan, S., Belaid, H., Farha, C., Iatsunskyi, I., Coy, E., Soussan, L., Huon, V., Bares, J., Belkacemi, K., Teyssier, C., Balme, S., Miele, P., Cornu, D., Kalkura, N., Cavaillès, V., & Bechelany, M. (2020). Fabrication of 3D printed antimicrobial polycaprolactone scaffolds for tissue engineering applications. *Materials Science and Engineering C*, *118*, 111525. DOI: 10.1016/j.msec.2020.111525

Radhika, C., Shanmugam, R., Ramoni, M., & Gnanavel, B. K. (2024). A Review on Additive Manufacturing for Aerospace Application. *Materials Research Express*, *11*(2), 022001. DOI: 10.1088/2053-1591/ad21ad

Rafi, K., Zhonghong Liu, A., Di Prima, M., Bates, P., & Seifi, M. (2022). Regulatory and standards development in medical additive manufacturing. *MRS Bulletin*, *47*(1), 98–105. DOI: 10.1557/s43577-022-00266-w

Raïssi, M., Pellegrin, Y., Lefevre, F. X., Boujtita, M., Rousseau, D., Berthelot, T., & Odobel, F. (2020). Digital printing of efficient dye-sensitized solar cells (DSSCs). *Solar Energy*, *199*, 92–99. DOI: 10.1016/j.solener.2020.02.004

Rajaa, J. M., Manee, M. J., Mohammed, M. A., Mohammed, M. A., & Sharif, S. F. A. (2024). IoT-driven physical training and fitness management in colleges. In *2024 International Conference on Emerging Research in Computational Science (ICERCS)* (pp. 1–6). DOI: 10.1109/ICERCS63125.2024.10895309

Rajan, L., Periasamy, C., & Sahula, V. (2016). Comprehensive study on electrical and hydrogen gas sensing characteristics of Pt/ZnO nanocrystalline thin film-based Schottky diodes grown on n-Si substrate using RF sputtering. *IEEE Transactions on Nanotechnology*, *15*(2), 201–208. DOI: 10.1109/TNANO.2015.2513102

Raj, P., Biju, P. L., Deepanraj, B., & Senthilkumar, N. (2023). Optimizing the machining conditions in turning hybrid aluminium nanocomposites adopting teaching–learning based optimization and MOORA technique. [IJIDeM]. *International Journal on Interactive Design and Manufacturing*. Advance online publication. DOI: 10.1007/s12008-023-01450-1

Ramachandran, K., Boopalan, V., Bear, J. C., & Subramani, R. (2022). Multi-walled carbon nanotubes (MWCNTs)-reinforced ceramic nanocomposites for aerospace applications: A review. *Journal of Materials Science*, *57*(6), 3923–3953. DOI: 10.1007/s10853-021-06760-x

Ramachandran, R. K., Dendooven, J., Botterman, J., Sree, S. P., Poelman, D., Martens, J. A., Poelman, H., & Detavernier, C. (2014). Plasma enhanced atomic layer deposition of Ga_2O_3 thin films. *Journal of Materials Chemistry. A, Materials for Energy and Sustainability*, *2*(45), 19232–19238. DOI: 10.1039/C4TA05007J

Rambabu, P., Prasad, N. E., Kutumbarao, V. V., & Wanhill, R. J. H. (2017). Aluminium Alloys for Aerospace Applications. In Eswara Prasad, N., & Wanhill, R. J. H. (Eds.), *Aerospace Materials and Material Technologies* (Vol. 1, pp. 29–52). Springer., DOI: 10.1007/978-981-10-2134-3_2

Rana, S. S., Kumar, R., & Gupta, M. K. (2023). Investigation of mechanical and thermal behavior of sisal fibre reinforced polymer biocomposites. *Materials Today: Proceedings*. Advance online publication. DOI: 10.1016/j.matpr.2023.06.167

Ranga, P., Bhattacharyya, A., Whittaker-Brooks, L., Scarpulla, M. A., & Krishnamoorthy, S. (2021). N-type doping of low-pressure chemical vapor deposition grown β-Ga_2O_3 thin films using solid-source germanium. *Journal of Vacuum Science & Technology. A, Vacuum, Surfaces, and Films*, *39*(3), 030404. Advance online publication. DOI: 10.1116/6.0001004

Ranjan, P., & Hiremath, S. S. (2023). Finite element modeling and experimental validation of turning process using bio-mimicked structured tool for outcomes relevant to industry. *Machining Science and Technology*, *27*(3), 209–246. DOI: 10.1080/10910344.2023.2215304

Rao, M. S., & Padal, K. T. B. (2022). New Developments of Aluminum Alloys for Future Generation Applications – A Review. *AIP Conference Proceedings*, *2648*, 030011. DOI: 10.1063/5.0114421

Ra, S., Han, L., & Zhao, H. (2016). Synthesis of wide bandgap Ga_2O_3. *Semiconductors, 50*(4), 1002–1009. DOI: 10.1134/S1063782624602152

Rashid, A. B., Rayhan, A. M., Shaily, S. I., & Islam, S. M. (2024). An experimental study of physical, mechanical, and thermal properties of Rattan fiber reinforced hybrid epoxy resin laminated composite. *Results in Engineering*, *22*, 102053. DOI: 10.1016/j.rineng.2024.102053

Ravi Kumar, N. V., Blandin, J. J., Suéry, M., & Grosjean, E. (2003). Effect of alloying elements on the ignition resistance of magnesium alloys. *Scripta Materialia, 49*(3), 225–230. DOI: 10.1016/S1359-6462(03)00263-X

Rawal, I., Goyal, P. K., & Dahiya, S. (2024). Highly sensitive self-powered solar-blind ultraviolet photodetectors based on n-$Zn_{1-x}Fe_xO$/p-Si heterojunctions. *Sensors and Actuators. A, Physical*, *371*, 115268. Advance online publication. DOI: 10.1016/j.sna.2024.115268

Reddy, G. K., & Esanakula, J. R. (2023). Optimization of Operating Parameters of Wire EDM Using Design of Experiments Criteria. *2023 International Conference on Self Sustainable Artificial Intelligence Systems (ICSSAS)*, 1504–1509. DOI: 10.1109/ICSSAS57918.2023.10331679

Reddy, M. S. B., Ponnamma, D., Choudhary, R., & Sadasivuni, K. K. (2021). A comparative review of natural and Synthetic biopolymer composite scaffolds. *Polymers, 13*(7), 1105. DOI: 10.3390/polym13071105

Redwan, K., Ahmed, M., Turjo, K. I. H., Munna, M. M., Hossain, S. S. A., & Al Sohan, M. F. A. (2024). Smart IoT-driven shopping assistance system for cognitive impairment patients: Ensuring safe and independent shopping experiences. In *2024 10th International Conference on Computing, Engineering and Design (ICCED)* (pp. 1–6).

Richard O'Leary, R. S. P. P. G. H. (2016). An Investigation into the Recycling of Ti-6Al-4V Powder Used Within SLM to Improve Sustainability. In Impact. *The Journal of Innovation Impact*, *8*(2), 377–388.

Rioja, R. J., & Liu, J. (2012). The Evolution of Al-Li Base Products for Aerospace and Space Applications. *Metallurgical and Materials Transactions. A, Physical Metallurgy and Materials Science*, *43*(9), 3325–3337. DOI: 10.1007/s11661-012-1155-z

Rodinò, S., & Maletta, C. (2024). Design Considerations and Applications of Shape Memory Alloy-based Actuation in Morphing Structures: A Systematic Review. *Prog. Eng. Sci.*, *1*(4), 100021. DOI: 10.1016/j.pes.2024.100021

Rodrigues, T. A., Duarte, V., Miranda, R. M., Santos, T. G., & Oliveira, J. P. (2019). Current status and perspectives on wire and arc additive manufacturing (WAAM). *Materials (Basel)*, *12*(7), 1121. DOI: 10.3390/ma12071121

Rodríguez-González, P., Ruiz-Navas, E. M., & Gordo, E. (2023). Wire arc additive manufacturing (WAAM) for aluminum-lithium alloys: A review. *Materials (Basel)*, *16*(4), 16. DOI: 10.3390/ma16041375

Romeo, N., Bosio, A., Canevari, V., & Podestà, A. (2004). Recent progress on CdTe/CdS thin film solar cells. *Solar Energy*, *77*(6), 795–801. DOI: 10.1016/j.solener.2004.07.011

Ronald, B. S., Raman, R., Vekariya, D., Baskar, V. V., Rahul, J., & Srinivasan, C. (2024). SVM-enhanced dynamic audiology screening with IoT-driven monitoring for personalized hearing health management. In *2024 2nd International Conference on Self Sustainable Artificial Intelligence Systems (ICSSAS)* (pp. 1028–1033). DOI: 10.1109/ICSSAS64001.2024.10760912

Roos, S., & Rännar, L.-E. (2024). Surface roughness reduction in electron beam powder bed fusion additive manufacturing of super duplex stainless steel 2507: Investigating optimisation techniques and face orientation-dependent irregularities. *Progress in Additive Manufacturing*. Advance online publication. DOI: 10.1007/s40964-024-00763-1

Roseti, L., Parisi, V., Petretta, M., Cavallo, C., Desando, G., Bartolotti, I., & Grigolo, B. (2017). Scaffolds for Bone Tissue Engineering: State of the art and new perspectives. *Materials Science and Engineering C*, *78*, 1246–1262. DOI: 10.1016/j.msec.2017.05.017

Rounds, R., Sarkar, B., Klump, A., Hartmann, C., Nagashima, T., Kirste, R., Franke, A., Bickermann, M., Kumagai, Y., Sitar, Z., & Collazo, R. (2018). Thermal conductivity of single-crystalline AlN. DOI: 10.7567/apex.11.071001

Roy, M. (2025). Degradation of C_f-SiC Composites under Hypersonic Conditions. *Wear*, *576-577*, 206141. DOI: 10.1016/j.wear.2025.206141

Roy, R. K. (2010). *A Primer on the Taguchi Method* (2nd ed.). Society of Manufacturing Engineers., Retrieved from https://books.google.co.in/books?id=k5VBsRZfzQsC

Roy, R., Hill, V. G., & Osborn, E. F. (1952). Polymorphism of Ga_2O_3 and the system Ga_2O_3–H_2O. *Journal of the American Chemical Society*, *74*(3), 719–722. DOI: 10.1021/ja01123a039

Rozvany, G. (2000). *The SIMP method in topology optimization-theoretical background, advantages and new applications.* Paper presented at the 8th Symposium on Multidisciplinary Analysis and Optimization. DOI: 10.2514/6.2000-4738

Rumi, R. I., Pavel, M. I., Islam, E., Shakir, M. B., & Hossain, M. A. (2019). IoT enabled prescription reading smart medicine dispenser implementing maximally stable extremal regions and OCR. In *2019 Third International Conference on I-SMAC (IoT in Social, Mobile, Analytics and Cloud) (I-SMAC)* (pp. 134–138). DOI: 10.1109/I-SMAC47947.2019.9032709

Runge, J. M. (2018). The metallurgy of anodizing aluminum: Connecting science to practice. In *The Metallurgy of Anodizing Aluminum: Connecting Science to Practice*. Springer International Publishing., DOI: 10.1007/978-3-319-72177-4

Saba, N., Paridah, M. T., & Jawaid, M. (2015). Mechanical properties of kenaf fibre reinforced polymer composite: A review. *Construction and Building Materials, 76*, 87-96. DOI: 10.1016/j.conbuildmat.2014.11.043

Saboori, A., Aversa, A., Marchese, G., Biamino, S., Lombardi, M., & Fino, P. (2019). Application of Directed Energy Deposition-Based Additive Manufacturing in Repair. *Applied Sciences (Basel, Switzerland)*, *9*(16), 3316. DOI: 10.3390/app9163316

Sağbaş, B., & Gavcar, B. (2022). Additive Manufacturing Of Titanium Alloys For Biomedical Applications. *Uluborlu Mesleki Bilimler Dergisi.*, *5*(1), 54–74.

Saha, M., Singh, H., Singh, M. K., Rangappa, S. M., & Siengchin, S. (2025). Advancements in natural fiber composites: market insights, surface modifications, advanced fabrication techniques and applications. *Sustainable Chemistry for Climate Action, 6*, 100081. DOI: 10.1016/j.scca.2025.100081

Sahoo, A. K., Kumar, R., & Panda, A. (2024). *Sustainable Hard Machining: Implementation and Assessment*. CRC Press. Retrieved from https://books.google.co.in/books?id=OaALEQAAQBAJ

Sahoo, B. N., & Khan, F. (2018). Microstructural modification and its effect on strengthening mechanism and yield asymmetry of in-situ TiC-TiB2/ AZ91 magnesium matrix composite. *Materials Science and Engineering A*, *724*(March), 269–282. DOI: 10.1016/j.msea.2018.03.060

Sahu, N., Parija, B., & Panigrahi, S. (2009). Fundamental understanding and modeling of spin coating process: A review. *Indian Journal of Physics*, *83*(4), 493–502. DOI: 10.1007/s12648-009-0009-z

Sahu, S. B. B. J., & Mohapatra, S. R. (2015). Design of AJM, analysis of process parameter and their effect on MRR. *International Journal of Analytical. Experimental and Finite Element Analysis*, *2*(1), 24–28.

Saikumar, A. K., Nehate, S. D., & Sundaram, K. B. (2019). Review—RF sputtered films of Ga_2O_3. *ECS Journal of Solid State Science and Technology : JSS*, *8*(7), Q3064–Q3078. DOI: 10.1149/2.0141907jss

Saikumar, A. K., Sundaresh, S., Nehate, S. D., Phelps, J., Abdolvand, R., & Sundaram, K. B. (2023). Work function estimation of gallium-doped zinc oxide using transparent gate electrode MOSFET. *ECS Journal of Solid State Science and Technology : JSS*, *12*(3), 033010. Advance online publication. DOI: 10.1149/2162-8777/acbede

Salem, I. A. S., Rozyanty, A. R., Betar, B. O., Adam, T., Mohammed, M., & Mohammed, A. M. (2017). Study of the effect of surface treatment of kenaf fibre on mechanical properties of kenaf filled unsaturated polyester composite. *Journal of Physics: Conference Series*, *908*, 012002. DOI: 10.1088/1742-6596/908/1/012002

Salmi, M. (2021). Additive Manufacturing Processes in Medical Applications. *Materials (Basel)*, *14*(1), 191. DOI: 10.3390/ma14010191

Salvati, E., & Korsunsky, A. M. (2020). Micro-scale measurement & FEM modelling of residual stresses in AA6082-T6 Al alloy generated by wire EDM cutting. *Journal of Materials Processing Technology*, *275*, 116373. DOI: 10.1016/j.jmatprotec.2019.116373

Sames, W. J., List, F., Pannala, S., Dehoff, R. R., & Babu, S. S. (2016). The metallurgy and processing science of metal additive manufacturing. *J. International Materials Reviews*, *61*(5), 315–360. DOI: 10.1080/09506608.2015.1116649

Samir, A., Ashour, F. H., Hakim, A. A., & Bassyouni, M. (2022). Recent advances in biodegradable polymers for sustainable applications. *NPJ Materials Degradation*, *6*(1), 68. Advance online publication. DOI: 10.1038/s41529-022-00277-7

Sanaei, N., & Fatemi, A. (2021). Defects in additive manufactured metals and their effect on fatigue performance: A state-of-the-art review. *Progress in Materials Science*, *117*, 100724. DOI: 10.1016/j.pmatsci.2020.100724

Sanchez-Rexach, E., Johnston, T. G., Jehanno, C., Sardon, H., & Nelson, A. (2020). Sustainable Materials and Chemical Processes for Additive Manufacturing. *Chemistry of Materials*, *32*(17), 7105–7119. DOI: 10.1021/acs.chemmater.0c02008

Sankar, M. R. (2017). Introduction to abrasive machining and finishing processes. Retrieved November, 15, from https://nptel.ac.in/content/storage2/nptel_data3/html/mhrd/ict/text/112103250/lec14.pdf

Saran, R. N., Mali, H. S., & Malik, A. (2025). Surface texturing of axisymmetric thin-walled cylindrical components: A review. *Surface Engineering*, *41*(5), 532–563. DOI: 10.1177/02670844251351765

Saravanan, S. K., Parvathy, P. R., & Meenakshi, R., K, P., GaneshBabu, T. R., & Meenakshi, B. (2025). Predictive analytics for ear health management using IoT and ensemble learning. In *2025 International Conference on Intelligent Control, Computing and Communications (IC3)* (pp. 319–324). DOI: 10.1109/IC363308.2025.10957573

Saravanan, K., Francis, J. X., Sudeshkumar, M. P., Maridurai, T., Suyamburajan, V., & Jayaseelan, V. (2022). Optimization of SiC abrasive parameters on machining of Ti-6Al-4V alloy in AJM using Taguchi-Grey relational method. *Silicon*, *14*, 997–1004. DOI: 10.1007/s12633-020-00918-z

Sarmast, A., Schubnell, J., Preußner, J., Hinterstein, M., & Carl, E. (2023). Residual stress analysis in industrial parts: A comprehensive comparison of XRD methods. *Journal of Materials Science*, *58*(44), 16905–16929. DOI: 10.1007/s10853-023-09069-z

Saroia, J., Yanen, W., Wei, Q., Zhang, K., Lu, T., & Zhang, B. (2018). A review on biocompatibility nature of hydrogels with 3D printing techniques, tissue engineering application and its future prospective. *Bio-Design and Manufacturing*, *1*(4), 265–279. DOI: 10.1007/s42242-018-0029-7

Sasikumar, C., & Oyyaravelu, R. (2024). Mechanical properties and microstructure of SS 316 L created by WAAM based on GMAW. *Materials Today. Communications*, *38*, 107807. DOI: 10.1016/j.mtcomm.2023.107807

Satchanska, G., Davidova, S., & Petrov, P. D. (2024). Natural and synthetic polymers for biomedical and environmental applications. *Polymers*, *16*(8), 1159. DOI: 10.3390/polym16081159

Satyanarayana, K., Venu Gopal, A., & Bangaru Babu, P. (2013, November 15). Finite Element Simulation of Cutting Forces in Turning Ti6Al4V Using DEFORM 3D. *Volume 2A. Advances in Manufacturing*. Advance online publication. DOI: 10.1115/IMECE2013-62868

Schubnell, J., Carl, E., Widerspan, V., & Collmann, M. (2023). Determination of Loading and Residual Stresses on Offshore Jacket Structures by X-ray Diffraction. *Journal of Marine Science and Engineering*, *11*(7), 1304. DOI: 10.3390/jmse11071304

Seah, I., Ong, C., Liu, Z., & Su, X. (2022). Polymeric biomaterials in the treatment of posterior segment diseases. *Frontiers in Medicine, 9*, 949543. Advance online publication. DOI: 10.3389/fmed.2022.949543

Selvan, E. V., Ponshanmugakumar, A., Ramanan, N., & Naveen, E. (2021). Determination and investigation of mechanical behaviour on kenaf-sisal hybrid composite. *Materials Today: Proceedings, 46*, 3358-3362. DOI: 10.1016/j.matpr.2020.11.478

Semenova, I. P., Polyakov, A. V., Polyakova, V. V., Huang, Y., Valiev, R. Z., & Langdon, T. G. (2016). High-Cycle Fatigue Behavior of an Ultrafine-Grained Ti–6Al–4V Alloy Processed by ECAP and Extrusion. *Advanced Engineering Materials, 18*(12), 2057–2062. DOI: 10.1002/adem.201500630

Sengottaiyan, S., Shaisundaram, V. S., Shameer Basha, M., Deepanraj, B., & Senthilkumar, N. (2025). Effect of duplex and austenite stainless steel fillers on mechanical, microstructural, and corrosion characteristics of dissimilar welds using Cold Metal Transfer welding. *Results in Engineering, 26*, 104818. DOI: 10.1016/j.rineng.2025.104818

Senthilkumar, N. (2024). Machining Studies on AISI 4140 Steel as Automotive Axle and Drive Shafts in Automobile via Numerical Simulation. *SAE Technical Paper*, 2024-01-5232. DOI: 10.4271/2024-01-5232

Senthilkumar, N., Deepanraj, B., Shaik, F., & Elsehly, E. M. (2024). Artificial Intelligence Based Investigation on Drilling of MWCNT-Seashell-Nylon66 Hybrid Polymer Composites by Fuzzy Inference System. In *2024 3rd International Conference for Innovation in Technology,* pp. 1-6. DOI: 10.1109/INOCON60754.2024.10511569

Senthilkumar, N., Perumal, G., Palanikumar, K., Deepanraj, B., & Basha, M. S. (2025). Wear and mechanical characteristics of coated Ti-6Al-4 V alloy with an alumina-13 wt.% titania -40 wt.% yitrium stabilized zirconia ceramic composite via plasma spraying technique. *Results in Engineering, 25*, 103777. DOI: 10.1016/j.rineng.2024.103777

Serani, A., Diez, M., & Quagliarella, D. (2024). Aerodynamic Shape Optimization in Transonic Conditions through Parametric Model Embedding. *Aerospace Science and Technology, 155*, 109611. DOI: 10.1016/j.ast.2024.109611

Shah, A., Torres, P., Tscharner, R., Wyrsch, N., & Keppner, H. (1999). Photovoltaic Technology: The Case for Thin-Film Solar Cells. *Science, 285*(5428), 692–698. www.nsf.gov/sbe/srs/pubdata.htm. DOI: 10.1126/science.285.5428.692 PMID: 10426984

Shanthar, R., Chen, K., & Abeykoon, C. (2023). Powder-Based Additive Manufacturing: A Critical Review of Materials, Methods, Opportunities, and Challenges. *Advanced Engineering Materials*, *25*(19), 2300375. Advance online publication. DOI: 10.1002/adem.202300375

Shapiro, A. A., Borgonia, J. P., Chen, Q. N., Dillon, R. P., McEnerney, B., Polit-Casillas, R., & Soloway, L. (2016). Additive Manufacturing for Aerospace Flight Applications. *Journal of Spacecraft and Rockets*, *53*(5), 952–959. DOI: 10.2514/1.A33544

Sharabian, E., Leary, M., Fraser, D., & Gulizia, S. (2022). Electron beam powder bed fusion of copper components: A review of mechanical properties and research opportunities. *International Journal of Advanced Manufacturing Technology*, *122*(2), 513–532. DOI: 10.1007/s00170-022-09922-6

Sharma, K., & Pathak, A. K. (2025). Transforming radiology with IoT-enabled smart wireless solutions in healthcare infrastructure. In *2025 3rd International Conference on Disruptive Technologies (ICDT)* (pp. 806–809).

Sharma, I., Pawar, P. S., Kumar Yadav, R., Nandi, R., & Heo, J. (2022). Review on bandgap engineering in metal-chalcogenide absorber layer via grading: A trend in thin-film solar cells. *Solar Energy*, *246*, 152–180. DOI: 10.1016/j.solener.2022.09.046

Sharman, A. R. C., Hughes, J. I., & Ridgway, K. (2006). An analysis of the residual stresses generated in Inconel 718TM when turning. *Journal of Materials Processing Technology*, *173*(3), 359–367. DOI: 10.1016/j.jmatprotec.2005.12.007

Sharma, R. C., Nandal, R., Tanwar, N., Yadav, R., Bhardwaj, J., & Verma, A. (2023). Gallium arsenide and gallium nitride semiconductors for power and optoelectronics devices applications. *Journal of Physics: Conference Series*, *2426*(1), 012008. Advance online publication. DOI: 10.1088/1742-6596/2426/1/012008

Sharma, S., Nagib, H., Le, P. Y., Allen, M. W., Holland, A. S., Partridge, J. G., & Tran, H. N. (2025). Structural, surface, electrical and UVC sensing properties of high temperature RF sputtered gallium oxide thin films. *Materials Today Electronics*, *11*, 100139. Advance online publication. DOI: 10.1016/j.mtelec.2025.100139

Shen, H., Lin, J., Zhou, Z., & Liu, B. (2022). Effect of induction heat treatment on residual stress distribution of components fabricated by wire arc additive manufacturing. *Journal of Manufacturing Processes*, *75*, 331–345. DOI: 10.1016/j.jmapro.2022.01.018

Shen, H., Yin, Y., Tian, K., Baskaran, K., Duan, L., Zhao, X., & Tiwari, A. (2018). Growth and characterization of β-Ga_2O_3 thin films by sol-gel method for fast-response solar-blind ultraviolet photodetectors. *Journal of Alloys and Compounds*, *766*, 601–608. DOI: 10.1016/j.jallcom.2018.06.313

Shereen, Y., Shahid, M. F., & Tanveer, M. H. (2025). Self-aware AI-based anomaly detection for AI IoT enabled smart healthcare system. In *2025 4th International Conference on Computing and Information Technology (ICCIT)* (pp. 406–413). https://doi.org/ (or access via IEEE Xplore: https://ieeexplore.ieee.org/document/10989421)DOI: 10.1109/ICCIT.2025.10989421

Shetty, R., Pai, R. B., Rao, S. S., & Nayak, R. (2009). Taguchi's technique in machining of metal matrix composites. *Journal of the Brazilian Society of Mechanical Sciences and Engineering*, *31*(1). Advance online publication. DOI: 10.1590/S1678-58782009000100003

Sheu, H. H., Hsu, Y. T., Jian, S. Y., & Liang, S. C. (2016). The effect of Cu concentration in the photovoltaic efficiency of CIGS solar cells prepared by co-evaporation technique. *Vacuum*, *131*, 278–284. DOI: 10.1016/j.vacuum.2016.07.008

Shie, M., Chang, W., Wei, L., Huang, Y., Chen, C., Shih, C., Chen, Y., & Shen, Y. (2017). 3D Printing of Cytocompatible Water-Based Light-Cured Polyurethane with Hyaluronic Acid for Cartilage Tissue Engineering Applications. *Materials (Basel)*, *10*(2), 136. DOI: 10.3390/ma10020136

Shi, L., Fang, Y., Dai, Q., Huang, W., & Wang, X. (2017). Surface texturing on SiC by multiphase jet machining with microdiamond abrasives. *Materials and Manufacturing Processes*, *33*(13), 1–7. DOI: 10.1007/s00170-015-8272-y

Shin, S. G., Hur, J., & Choi, H. W. (2019). A study of ZnO/PEDOT:PSS based UV sensors with RF sputter. *Japanese Journal of Applied Physics, 58*(SD). *Japanese Journal of Applied Physics*, *SD0804*(SD), SDDE09. Advance online publication. DOI: 10.7567/1347-4065/ab0ff3

Shyha, I., & Huo, D. (2021). *Advances in Machining of Composite Materials: Conventional and Non-conventional Processes*. Springer International Publishing. Retrieved from https://books.google.co.in/books?id=aVw0EAAAQBAJ

Siddharthan, B., Rajiev, R., Saravanan, S., & Naveen, T. K. (2020). Effect of Silicon Carbide in Mechanical Properties of Aluminium Alloy Based Metal Matrix Composites. *IOP Conference Series. Materials Science and Engineering*, *764*(1), 012040. DOI: 10.1088/1757-899X/764/1/012040

Simchi, A., Petzoldt, F., Hartwig, T., Hein, S. B., Barthel, B., & Reineke, L. (2022). Binder Jet 3d Printing of Biomedical Grade Ti-6al-4v Alloy: The Role of Particle Size, Microstructural Features, and Mechanical Properties. SSRN Electronic Journal. https://doi.org/DOI: 10.2139/ssrn.4127679

Simchi, A., Petzoldt, F., Hartwig, T., Hein, S. B., Barthel, B., & Reineke, L. (2023). Microstructural development during additive manufacturing of biomedical grade Ti-6Al-4V alloy by three-dimensional binder jetting: Material aspects and mechanical properties. *International Journal of Advanced Manufacturing Technology*, *127*(3–4), 1541–1558. DOI: 10.1007/s00170-023-11661-1

Simöes, S. (2024). High-Performance Advanced Composites in Multifunctional Material Design: State of the Art, Challenges, and Future Directions. *Materials (Basel)*, *17*(23), 5997. DOI: 10.3390/ma17235997

Singh, A. B., Khandelwal, C., & Dangayach, G. S. (2024). Revolutionizing healthcare materials: Innovations in processing, advancements, and challenges for enhanced medical device integration and performance. Journal of Micromanufacturing. https://doi.org/DOI: 10.1177/25165984241256234

Singh, A., Kapil, S., & Das, M. (2020). A comprehensive review of the methods and mechanisms for powder feedstock handling in directed energy deposition. *Additive Manufacturing*, *35*, 101388. DOI: 10.1016/j.addma.2020.101388

Singhal, V., Shelly, D., Saxena, A., Gupta, R., Verma, V. K., & Jain, A. (2025). Study of the Influence of Nanoparticle Reinforcement on the Mechanical and Tribological Performance of Aluminum Matrix Composites—A Review. *Lubricants (Basel, Switzerland)*, *13*(2), 93. DOI: 10.3390/lubricants13020093

Singh, G., Mehta, A., & Vasudev, H. (2024). Sustainability of additive manufacturing: A comprehensive review. *Progress in Additive Manufacturing*, *9*(6), 2249–2272. DOI: 10.1007/s40964-024-00579-z

Singh, S. R., & Khanna, P. (2021). Wire arc additive manufacturing (WAAM): A new process to shape engineering materials. *Materials Today: Proceedings*, *44*, 118–128. DOI: 10.1016/j.matpr.2020.08.030

Sing, S. L. (2022). Perspectives on Additive Manufacturing Enabled BetaTitanium Alloys for Biomedical Applications. *International Journal of Bioprinting*, *8*(1), 478. DOI: 10.18063/ijb.v8i1.478

Sing, S. L., An, J., Yeong, W. Y., & Wiria, F. E. (2016). Laser and electron-beam powder-bed additive manufacturing of metallic implants: A review on processes, materials and designs. *Journal of Orthopaedic Research, 34*(3), 369–385. DOI: 10.1002/jor.23075

Sinha, M. K., Setti, D., Ghosh, S., & Rao, P. V. (2016). An alternate method for optimisation of minimum quantity lubrication parameters in surface grinding. *International Journal of Machining and Machinability of Materials, 18*(5-6), 586–605. DOI: 10.1504/IJMMM.2016.078983

Sivasankar, S. M., Amorim, C. de O., & Cunha, A. F. (2025). Progress in Thin-Film Photovoltaics: A Review of Key Strategies to Enhance the Efficiency of CIGS, CdTe, and CZTSSe Solar Cells. *Journal of Composites Science, 9*(3), 143. Advance online publication. DOI: 10.3390/jcs9030143

Skalková, P., Krmelová, V., Krmela, J., Ondrušová, D., Crkoň, A., & Benčíková, E. (2021). Composite materials with epoxy matrix and their properties. *IOP Conference Series. Materials Science and Engineering, 1199*(1), 012030. DOI: 10.1088/1757-899X/1199/1/012030

Slătineanu, L. (2013). Abrasive jet finishing. In M. Korzynski (Ed.), *Nonconventional finishing technologies* (pp. 61-82). Warsaw: Polish Scientific Publishers PWN. ISBN-10 : 8301172649

Slătineanu, L., Nagîţ, G., Dodun, O., Coteaţă, M., Chinesta, F., Gonçalves-Coelho, A., Pamies Teixeira J., San Juan, M., Santo, L., & Santos, F. (2004). *Non-traditional manufacturing processes*. Chişinău, Republic of Moldova: Tehnica Info Publishing House. ISBN (Print)9975-63-164-9

Slikkerveer, P. J., Bouten, P. C. P., Veld, F. H., & Scholten, H. (1998). Erosion and damage by sharp particles. *Wear, 217*(2), 237–250. DOI: 10.1016/S0043-1648(98)00187-2

Solahuddin, B. A. (2022). A critical review on experimental investigation and finite element analysis on structural performance of kenaf fibre reinforced concrete. *Structures,* 35, 1030-1061). DOI: 10.1016/j.istruc.2021.11.056

Solliance. (2019). *The key to a sustainable building façade - Solliance*. https://www.solliance.eu/2018/the-key-to-a-sustainable-building-facade/

Song, W., Wang, P., Wan, D., Qian, G., Correia, J., & Berto, F. (2021). Fatigue crack growth behavior of Ni-Cr-Mo-V steel welded joints considering strength mismatch effect. *International Journal of Fatigue, 151*, 106389. DOI: 10.1016/j.ijfatigue.2021.106389

Song, Y., Ma, Y., Chen, H., Chen, H., Zhang, T., & Gao, Z. (2021). The effects of tensile and compressive dwells on creep-fatigue behavior and fracture mechanism in welded joint of P92 steel. *Materials Science and Engineering A*, *813*, 141129. DOI: 10.1016/j.msea.2021.141129

Soni, H., S, N., & M R, R. (2018). Experimental Investigation on Effects of Wire Electro Discharge Machining of Ti50Ni45Co5 Shape Memory Alloys. *Silicon*, *10*(6), 2483–2490. DOI: 10.1007/s12633-018-9780-9

Soni, R., Verma, R., Garg, R. K., & Sharma, V. (2024). A Critical Review of Recent Advances in the Aerospace Materials. *Materials Today: Proceedings*, *113*, 180–184. DOI: 10.1016/j.matpr.2023.08.108

Soong, S. Z., Lai, W. L., & Kay Lup, A. N. (2023). Atomization of metal and alloy powders: Processes, parameters, and properties. *AIChE Journal. American Institute of Chemical Engineers*, *69*(11), e18217. Advance online publication. DOI: 10.1002/aic.18217

Sparavigna, A. C. (2018). The invention of carborundum, the synthetic silicon carbide. [HAL Id: hal-01686436; likely preprint or report.]

Sridhar, N., Aezhisai Vallavi, M. S., & Mugilan, T. (2023). An integrated approach of FEM analysis using DEFORM 3D and experimental investigation of forces developed in Al-Si7Mg. *Materials Today: Proceedings*, *80*, 888–895. DOI: 10.1016/j.matpr.2022.11.324

Srinivasa Rao, P., Ramji, K., & Satyanarayana, B. (2016). Effect of wire EDM conditions on generation of residual stresses in machining of aluminum 2014 T6 alloy. *Alexandria Engineering Journal*, *55*(2), 1077–1084. DOI: 10.1016/j.aej.2016.03.014

Srinivasan, D., Senthilkumar, N., Ganesh, M., & Perumal, G. (2024). *Characterization studies on nano aluminium composite reinforced with montmorillonite nanoclay and titanium carbide*. Multiscale and Multidisciplinary Modeling, Experiments and Design., DOI: 10.1007/s41939-024-00374-x

Srivastava, M., Jayakumar, V., Udayan, Y., Sathishkumar, M., Muthu, S. M., Gautam, P., & Nag, A. (2024). Additive Manufacturing of Titanium alloy for Aerospace Applications: Insights into the Process, Microstructure, and Mechanical Properties. *Applied Materials Today*, *41*, 102481. DOI: 10.1016/j.apmt.2024.102481

Srivastava, M., Rathee, S., Patel, V., Kumar, A., & Koppad, P. G. (2022). A review of various materials for additive manufacturing: Recent trends and processing issues. *Journal of Materials Research and Technology*, *21*, 2612–2641. DOI: 10.1016/j.jmrt.2022.10.015

Starke, E. A., & Staley, J. T. (2011). 24 - Application of modern aluminium alloys to aircraft. In Lumley, R. (Ed.), *Fundamentals of Aluminium Metallurgy: Production, Processing and Applications* (pp. 747–783). Woodhead Publishing., DOI: 10.1533/9780857090256.3.747

Staroselsky, A., Voytovych, D., & Acharya, R. (2024). Prediction of Ni-based alloy microstructure in wire arc additive manufacturing from cellular automata model. *Computational Materials Science*, *233*, 112721. Advance online publication. DOI: 10.1016/j.commatsci.2023.112721

Stoliar, I. A., Shepelevich, V. G., Wendler, E., & Tashlykova-Bushkevich, I. I. (2021). Effect of Lithium on the Structural-Phase State of Rapidly Solidified Al-Mg-Li Alloy during Heat Treatment. *Journal of Surface Investigation. X-ray, Synchrotron and Neutron Techniques*, *15*(4), 752–758. DOI: 10.1134/S1027451021040194

Su, C., Chen, X., Gao, C., & Wang, Y. (2019). Effect of heat input on microstructure and mechanical properties of Al-Mg alloys fabricated by WAAM. *Applied Surface Science*, *486*, 431–440. DOI: 10.1016/j.apsusc.2019.04.255

Sun, H., Zou, B., Quan, T., Wang, X., Ma, X., Xing, H., Wei, S., & Huang, C. (2025). Multi-material ceramic hybrid additive manufacturing based on vat photopolymerization and material extrusion compound process. *Additive Manufacturing*, *97*, 104627. DOI: 10.1016/j.addma.2024.104627

Sun, Y. J., Yang, R., Xie, L., Li, Y. B., Wang, S. L., Li, H. X., Wang, W. R., & Zhang, J. S. (2022). Interfacial Bonding and Corrosion Behaviors of HVOF-Sprayed Fe-based Amorphous Coating on 8090 Al-Li Alloy. *Surface and Coatings Technology*, *436*, 128316. DOI: 10.1016/j.surfcoat.2022.128316

Sun, Y., Aindow, M., & Hebert, R. J. (2018). Comparison of virgin Ti-6Al-4V powders for additive manufacturing. *Additive Manufacturing*, *21*, 544–555. DOI: 10.1016/j.addma.2018.02.011

Suresh, S., Natarajan, E., Markandan, K., Sudhagar, S., Anbuchezhiyan, G., & Elayaraja, R. (2025). Strategic enhancement of joint strength in dissimilar Al/Mg alloys: A workable approach for aerospace application. *Discover Materials*, *5*(1), 23. DOI: 10.1007/s43939-025-00194-x

Suthar, A. N., Venkataramanaiah, J., & Suresh, Y. (2025). Conventional, wide-bandgap, and hybrid power converters: A comprehensive review. *Renewable & Sustainable Energy Reviews*, *213*, 115419. Advance online publication. DOI: 10.1016/j.rser.2025.115419

Suthar, S. B., Chithra, R., Abhinaya, G. K., & Harikumar, H. (2023). Smart waste segregation and collection system with IoT-enabled monitoring and analytics. In *2023 International Conference on Recent Advances in Electrical, Electronics, Ubiquitous Communication, and Computational Intelligence (RAEEUCCI)* (pp. 1–6). DOI: 10.1109/RAEEUCCI57140.2023.10134423

Svetlizky, D., Das, M., Zheng, B., Vyatskikh, A. L., Bose, S., Bandyopadhyay, A., Schoenung, J. M., Lavernia, E. J., & Eliaz, N. (2021). Directed energy deposition (DED) additive manufacturing: Physical characteristics, defects, challenges and applications. *Materials Today*, *49*, 271–295. DOI: 10.1016/j.mattod.2021.03.020

Tabatabaeian, A., Ghasemi, A. R., Shokrieh, M. M., Marzbanrad, B., Baraheni, M., & Fotouhi, M. (2022). Residual Stress in Engineering Materials: A Review. *Advanced Engineering Materials*, *24*(3), 2100786. Advance online publication. DOI: 10.1002/adem.202100786

Takeuchi, E., Furuya, Y., Nagashima, N., & Matsuoka, S. (2008). The effect of frequency on the giga-cycle fatigue properties of a Ti-6Al-4V alloy. *Fatigue & Fracture of Engineering Materials & Structures*, *31*(7), 599–605. DOI: 10.1111/j.1460-2695.2008.01257.x

Takizawa, T., Nakayama, N., Haniu, H., Aoki, K., Okamoto, M., Nomura, H., Tanaka, M., Sobajima, A., Yoshida, K., Kamanaka, T., Ajima, K., Oishi, A., Kuroda, C., Ishida, H., Okano, S., Kobayashi, S., Kato, H., & Saito, N. (2018). Titanium Fiber Plates for Bone Tissue Repair. *Advanced Materials*, *30*(4), 1703608. Advance online publication. DOI: 10.1002/adma.201703608

Taleghani, A. S., & Torabi, F. (2024). Recent Developments in Aerodynamics. *Frontiers of Mechanical Engineering*, *10*, 1537383. DOI: 10.3389/fmech.2024.1537383

Tamayo, J. A., Riascos, M., Vargas, C. A., & Baena, L. M. (2021). Additive manufacturing of Ti6Al4V alloy via electron beam melting for the development of implants for the biomedical industry. *Heliyon*, *7*(5), e06892. DOI: 10.1016/j.heliyon.2021.e06892

Tamizharasan, T., & Senthil Kumar, N. (2012). Optimization of cutting inserts geometry using DEFORM-3D: Numerical simulation and experimental validation. *International Journal of Simulation Modelling*, *11*(2), 65–76. DOI: 10.2507/IJSIMM11(2)1.200

Tanaka, M., Sato, K., Kitakami, E., Kobayashi, S., Hoshiba, T., & Fukushima, K. (2014). Design of biocompatible and biodegradable polymers based on intermediate water concept. *Polymer Journal*, *47*(2), 114–121. DOI: 10.1038/pj.2014.129

Tanarro, Á., Mallor, F., Offermans, N., Peplinski, A., Vinuesa, R., & Schlatter, P. (2020). Enabling Adaptive Mesh Refinement for Spectral-Element Simulations of Turbulence Around Wing Sections. *Flow, Turbulence and Combustion*, *105*(2), 415–436. DOI: 10.1007/s10494-020-00152-y

Tang, H., Rogov, A. B., Soutis, C., & Yerokhin, A. (2023). Fabrication, interfacial and flexural properties of a polymer composite reinforced by γ-Al2O3/Al fibres. *Composites Part A: Applied Science and Manufacturing, 169*, 107502. DOI: 10.1016/j.compositesa.2023.107502

Tang, H. P., Qian, M., Liu, N., Zhang, X. Z., Yang, G. Y., & Wang, J. (2015). Effect of Powder Reuse Times on Additive Manufacturing of Ti-6Al-4V by Selective Electron Beam Melting. *Journal of the Minerals Metals & Materials Society*, *67*(3), 555–563. DOI: 10.1007/s11837-015-1300-4

Tan, X. P., Tan, Y. J., Chow, C. S. L., Tor, S. B., & Yeong, W. Y. (2017). Metallic powder-bed based 3D printing of cellular scaffolds for orthopaedic implants: A state-of-the-art review on manufacturing, topological design, mechanical properties and biocompatibility. *Materials Science and Engineering C*, *76*, 1328–1343. DOI: 10.1016/j.msec.2017.02.094

Tareq, S., Rahman, T., Poudel, B., Chung, H., & Kwon, P. (2024). Heat treatment protocol for additively manufactured nitinol shape memory alloys in biomedical applications. *Materials Science and Engineering A*, *897*, 146274. DOI: 10.1016/j.msea.2024.146274

Taşdemir, A., & Nohut, S. (2021). An overview of wire arc additive manufacturing (WAAM) in shipbuilding industry. *Ships and Offshore Structures*, *16*(7), 797–814. DOI: 10.1080/17445302.2020.1786232

Teja, N. S., & Reddy, G. V. (2017). Experimental investigation of MRR and surface finish using abrasive jet machining. *International Journal of Scientific and Engineering Research*, *8*(7), 622–629. https://www.ijtimes.com/index.php/ijtimes/article/view/2256

Tej, Y. V., Potharaju, V. S., Jakkidi, P. R., Sharma, A., & Reddy, N. S. (2024). Deep learning techniques for image recognition in IoT-enabled surveillance systems. In *2024 Asian Conference on Intelligent Technologies (ACOIT)* (pp. 1–8).

Teng, Z., Wu, H., Huang, Z., & Starke, P. (2021). Effect of mean stress in very high cycle fretting fatigue of a bearing steel. *International Journal of Fatigue*, *149*, 106262. DOI: 10.1016/j.ijfatigue.2021.106262

Thandavamoorthy, R., Alagarasan, J. K., Mohanavel, V., et. Al., (2024). Fabrication of kenaf fiber reinforced boron carbide fillers embedded epoxy matrix composite–An antimicrobial and structural analysis. *Journal of Materials Research and Technology, 33*, 2560-2567. DOI: 10.1016/j.jmrt.2024.09.236

Thangam, S., Joshith, G., Santhosh, S., Harish, V., & Kumari, J. J. J. (2024). Smart steps to sporting success: IoT-driven footwear innovations. In *2024 15th International Conference on Computing Communication and Networking Technologies (ICCCNT)* (pp. 1–6).

Thangaraj, M., Ponmalar, P. P., & Anuradha, S. (2015). Internet of Things (IoT) enabled smart autonomous hospital management system - A real world health care use case with the technology drivers. In *2015 IEEE International Conference on Computational Intelligence and Computing Research (ICCIC)* (pp. 1–8). DOI: 10.1109/ICCIC.2015.7435678

The Master of Blasters (2012). Retrieved November 3, 2019, from http://www.raga.at/download/en.pdf.

Thirumalvalavan, S., Perumal, G., Senthilkumar, N., & Selvarasu, S. (2025). Enhancing Tribological Characteristics of Titanium Grade-5 Alloy through HVOF Thermal-Sprayed WC-Co Nano Coatings by TOPSIS and Golden Jack Optimization Algorithm. *Recent Patents on Nanotechnology, 19*(4), 544–567. DOI: 10.2174/0118722105306841240808092616

Thirumurugan, A., Bhaskar, G. B., Poyyathappan, K., Karthik, R. J., Kumar, M. K., Somasundaram, G., & Venkatakrishnan, G. (2016). Investigations on aluminium wire mesh, banana fiber and glass fiber reinforced hybrid composites. *Indian J. Sci. Technol, 9*(42), 1-7. DOI: 10.17485/ijst/2016/v9i42/103306

Thompson, S. M., Bian, L., Shamsaei, N., & Yadollahi, A. (2015). An overview of Direct Laser Deposition for additive manufacturing; Part I: Transport phenomena, modeling and diagnostics. *Additive Manufacturing, 8*, 36–62. DOI: 10.1016/j.addma.2015.07.001

Todd, R. H., Allen, D. K., & Alting, L. (1994). *Manufacturing processes. Reference guide*. Industrial Press Inc.

Tondare, R. S., Shivaraj, B. W., Narasimhamurthy, H. N., Krishna, M., & Subramanyam, T. K. (2018). Effect of deposition time on structural, electrical and optical properties of aluminium doped ZnO thin films by RF magnetron sputtering. *Materials Today: Proceedings, 5*(1), 2710–2715. DOI: 10.1016/j.matpr.2018.01.052

Tong, J., Bowen, C. R., Persson, J., & Plummer, A. (2017). Mechanical properties of titanium-based Ti–6Al–4V alloys manufactured by powder bed additive manufacture. *Materials Science and Technology*, *33*(2), 138–148. DOI: 10.1080/02670836.2016.1172787

Tôn, L.-P., Lê, L.-S., & Nguyen, M.-S. (2021). Micraspis: A computer-aided proposal toward programming and architecting smart IoT wearables. *IEEE Access : Practical Innovations, Open Solutions*, *9*, 105393–105408. DOI: 10.1109/ACCESS.2021.3096749

Torky, M., & Chow, T. P. (2024). Determination of the effective critical breakdown field for Si, wide, and extreme bandgap semiconductor superjunction devices. *AIP Advances*, *14*(5), 055001. Advance online publication. DOI: 10.1063/5.0211696

Totten, G. E. (2016). *Heat treating of nonferrous alloys. Volume 4E.* ASM Handbook. ASM International., Retrieved from https://books.google.co.in/books?id=TtSZwgEACAAJ DOI: 10.31399/asm.hb.v04e.9781627081696

Trenfield, S. J., Madla, C. M., Basit, A. W., & Gaisford, S. (2018). Binder jet printing in pharmaceutical manufacturing. In A. W. Basit & S. Gaisford (Eds.), *3D printing of pharmaceuticals* (pp. 41-54): Springer, Cham. DOI: 10.1007/978-3-319-90755-0_3

Trevisan, F., Calignano, F., Aversa, A., Marchese, G., Lombardi, M., Biamino, S., Ugues, D., & Manfredi, D. (2018). Additive manufacturing of titanium alloys in the biomedical field: Processes, properties and applications. *Journal of Applied Biomaterials & Functional Materials*, *16*(2), 57–67. DOI: 10.5301/jabfm.5000371

Tsai, F.-C., Yan, B.-H., Kuan, C.-Y., Hsu, R.-T., & Hung, J.-C. (2009). An investigation into superficial embedment in mirror-like machining using abrasive jet polishing. *International Journal of Advanced Manufacturing Technology*, *43*(5-6), 500–512. DOI: 10.1007/s00170-008-1734-8

Tyagi, L. K., Gupta, A., & Sisodia, V. S. (2025). Machine and deep learning classifications for IoT-enabled healthcare devices. In *2025 International Conference on Pervasive Computational Technologies (ICPCT)* (pp. 582–587). DOI: 10.1109/ICPCT64145.2025.10940415

Ulery, B. D., Nair, L. S., & Laurencin, C. T. (2011). Biomedical applications of biodegradable polymers. *Journal of Polymer Science. Part B, Polymer Physics*, *49*(12), 832–864. DOI: 10.1002/polb.22259

Umezawa, H. (2018). Recent advances in diamond power semiconductor devices. *Materials Science in Semiconductor Processing*, *78*, 147–156. DOI: 10.1016/j.mssp.2018.01.007

Vallone, M., Damasio, M., Atzeni, E., Poggi, M., & Salmi, A. (2023). *Thermo-mechanical Modelling of the Directed Energy Deposition (DED)*. Process for the Optimization of Deposition Strategies., DOI: 10.1007/978-3-031-28547-9_23

Vanmore, V. V., & Dabade, U. A. (2018). Experimental and parametric optimization of micro holes on ANSI 304 using micro abrasive jet machining (MAJM). *Procedia Computer Science, 133*, 620–626. DOI: 10.1016/j.procs.2018.07.096

Varatharajulu, M., Duraiselvam, M., Pradeep, G., & Jagadeesh, B. (2025). Exploring Epitaxial Grain Growth and Marangoni-Induced Conversion in Columnar Dendrites: A Study of Inconel 718 Additive Manufacturing via Direct Laser Energy Deposition. *Surface Review and Letters, 32*(1), 2450102–2450115. DOI: 10.1142/S0218625X24501026

Varley, J. B., Shen, B., & Higashiwaki, M. (2022). Wide bandgap semiconductor materials and devices. *Journal of Applied Physics, 131*(23), 230401. Advance online publication. DOI: 10.1063/5.0100601

Vashista, M., & Paul, S. (2012). Correlation between full width at half maximum (FWHM) of XRD peak with residual stress on ground surfaces. *Philosophical Magazine, 92*(33), 4194–4204. DOI: 10.1080/14786435.2012.704429

Veerabomma, S. K. S., Rajeswari, G., Murthy, B. S., & Selvan, R. S. (2024). Optimising IoT-healthcare systems with deep learning: Sensors data fusion, prognostic modelling, and smart management. In *2024 Second International Conference Computational and Characterization Techniques in Engineering & Sciences (IC3TES)* (pp. 1–6). DOI: 10.1109/IC3TES62412.2024.10877558

Velmurugan, V., Kumar, D. D., & Thanikaikarasan, S. (2020). Experimental evaluation of mechanical properties of natural fibre reinforced polymer composites. *Materials Today: Proceedings, 33*, 3383-3388. DOI: 10.1016/j.matpr.2020.05.190

Venkatesh, V. S. S., Rao, G. P., Patnaik, L., Gupta, N., Kumar, S., Saxena, K. K., Sunil, B. D. Y., Eldin, S. M., & Kutham Al- kafaji, F. H. (2023). Processing and evaluation of nano SiC reinforced aluminium composite synthesized through ultrasonically assisted stir casting process. *Journal of Materials Research and Technology, 24*, 7394–7408. DOI: 10.1016/j.jmrt.2023.05.030

Verdhan, N., Bhende, D. D., Kapoor, R., & Charavarty, J. K. (2015). Effect of microstructure on the fatigue crack growth behaviour of a near-α Ti alloy. *International Journal of Fatigue, 74*, 46–54. DOI: 10.1016/j.ijfatigue.2014.12.013

Verma, A. P., & Lal, G. K. (1984). An experimental study of abrasive jet machining. *International Journal of Machine Tool Research*, *24*(1), 19–29. DOI: 10.1016/0020-7357(84)90043-X

Verma, R., Kumar, P., Jayaganthan, R., & Pathak, H. (2022). Extended Finite Element Simulation on Tensile, Fracture Toughness and Fatigue Crack Growth Behaviour of Additively Manufactured Ti–6Al–4V Alloy. *Theoretical and Applied Fracture Mechanics*, *117*, 103163. DOI: 10.1016/j.tafmec.2021.103163

Vetelino, J. F., & Reghu, A. (2017). *Introduction to sensors* (2nd ed.). CRC Press., DOI: 10.1201/9781315218274

Villalón Fornés, I., Vaičiukynienė, D., Nizevičienė, D., Tamošaitis, G., & Pupeikis, D. (2022). The improvement of the thermal and acoustic insulation properties of phosphogypsum specimens by adding waste wood fibre. *Construction & Building Materials*, *331*, 127341. DOI: 10.1016/j.conbuildmat.2022.127341

Vivas, J., Aldanondo, E., Dufour, P., Delgado, C., & Fernández-Calvo, A. I. (2024). Laser Beam Welding of T Joints for Aluminum-Copper-Lithium Aircraft Panels: Effect of Filler Wire on Recyclability and Weldability. *Journal of Materials Research and Technology*, *33*, 5809–5820. DOI: 10.1016/j.jmrt.2024.10.223

Vorobyova, M., Biffoli, F., Giurlani, W., Martinuzzi, S. M., Linser, M., Caneschi, A., & Innocenti, M. (2023). PVD for decorative applications: A review. *Materials (Basel)*, *16*(14), 4919. Advance online publication. DOI: 10.3390/ma16144919 PMID: 37512195

Vrancken, B., Buls, S., Kruth, J., & Van Humbeeck, J. (2016). Preheating of Selective Laser Melted Ti6Al4V: Microstructure and Mechanical Properties. In *Proceedings of the 13th World Conference on Titanium* (pp. 1269–1277). Wiley. https://doi.org/ DOI: 10.1002/9781119296126.ch215

Wagner, L., & Wollmann, M. (2008). Shot peening of non-ferrous alloys to enhance fatigue performance. *ICSP-10, Tokyo, Japan*.

Wang, Y., & Zhao, G. (2020). Hot Extrusion Processing of Al–Li Alloy Profiles and Related Issues: A Review. In *Chinese Journal of Mechanical Engineering (English Edition)* (Vol. 33, Issue 1). Springer. DOI: 10.1186/s10033-020-00479-7

Wang, Z., Xing, Z., Lei, Y., Hu, B., Hu, J., Li, H., Feng, X., & Kang, G. (2025). A review on cyclic plasticity, damage, and fatigue failure of magnesium alloys. *Journal of Materials Science & Technology*. https://doi.org/https://doi.org/10.1016/j.jmst.2025.02.038

Wang, C., Su, Y., & Xie, J. (2024). Advances in electrospun nanofibers: Versatile materials and diverse biomedical applications. *Accounts of Materials Research*, *5*(8), 987–999. DOI: 10.1021/accountsmr.4c00145

Wang, F., Chen, W., Wang, D., Hou, H., & Zhao, Y. (2024). Phase-field modeling and Experimental investigation for rapid solidification in wire and arc additive manufacturing. *Journal of Materials Research and Technology*, *28*, 4585–4599. DOI: 10.1016/j.jmrt.2024.01.021

Wang, G., Yao, S., Chi, Y., Zhang, C., Wang, N., Chen, Y., Lu, R., Li, Z., & Zhang, X. (2024). Improvement of Titanium Alloy TA19 Fatigue Life by Submerged Abrasive Waterjet Peening: Correlation of Its Process Parameters with Surface Integrity and Fatigue Performance. *Chinese Journal of Aeronautics*, *37*(1), 377–390. DOI: 10.1016/j.cja.2023.03.033

Wang, H., Duan, Z., Guo, Q., Zhang, Y., & Zhao, Y. (2023). Machine Learning Design of Aluminum-Lithium Alloys with High Strength. *Computers, Materials & Continua*, *77*(2), 1393–1409. DOI: 10.32604/cmc.2023.045871

Wang, H., Lai, P., Islam, M. Z., Hasan, A. S. M. K., Di Mauro, A., Anika, N. E., Russell, R., Feng, Z., Chen, K., Faruque, A., White, T., Chen, Z., & Mantooth, H. A. (2024). A review of silicon carbide CMOS technology for harsh environments. *Materials Science in Semiconductor Processing*, *178*, 108422. Advance online publication. DOI: 10.1016/j.mssp.2024.108422

Wang, J.-X., Fan, Y.-B., Gao, Y., Hu, Q.-H., & Wang, T.-C. (2009). TiO2 nanoparticles translocation and potential toxicological effect in rats after intraarticular injection. *Biomaterials*, *30*(27), 4590–4600. DOI: 10.1016/j.biomaterials.2009.05.008

Wang, J., Ye, L., Wang, X., Zhang, H., Li, L., Kong, C., & Li, W. (2019). High transmittance β-Ga_2O_3 thin films deposited by magnetron sputtering and post-annealing for solar-blind ultraviolet photodetector. *Journal of Alloys and Compounds*, *803*, 9–15. DOI: 10.1016/j.jallcom.2019.06.224

Wang, L., Li, Y., Zhou, L., Lou, Y., Liu, S., Zheng, D., & Yi, M. (2023). Progress in Additive Manufacturing, Additive Repair and Fatigue Evaluation of Aviation Titanium Alloy Blades. *Materials Research Letters*, *11*(12), 973–1012. DOI: 10.1080/21663831.2023.2275599

Wang, S. C., & Starink, M. J. (2005). Precipitates and intermetallic phases in precipitation hardening Al–Cu–Mg–(Li) based alloys. *International Materials Reviews*, *50*(4), 193–215. DOI: 10.1179/174328005X14357

Wang, W., Biermann, D., Aßmuth, R., Arif, A. F. M., & Veldhuis, S. C. (2020). Effects on tool performance of cutting edge prepared by pressurized air wet abrasive jet machining (PAWAJM). *Journal of Materials Processing Technology*, *277*, 116456. DOI: 10.1016/j.jmatprotec.2019.116456

Wang, X., Shi, Q., Wang, X., & Zhang, Z. (2010). The influences of precrack orientations in welded joint of Ti–6Al–4V on fatigue crack growth. *Materials Science and Engineering A*, *527*(4-5), 1008–1015. DOI: 10.1016/j.msea.2009.09.031

Wang, X., Sridar, S., Klecka, M., & Xiong, W. (2025). Rapid data acquisition and machine learning-assisted composition design of functionally graded alloys via wire arc additive manufacturing. npj. *Advances in Manufacturing*, *2*(1), 17. Advance online publication. DOI: 10.1038/s44334-025-00028-x

Wang, X., Xu, S., Zhou, S., Xu, W., Leary, M., Choong, P., Qian, M., Brandt, M., & Xie, Y. M. (2016). Topological design and additive manufacturing of porous metals for bone scaffolds and orthopaedic implants: A review. *Biomaterials*, *83*, 127–141. DOI: 10.1016/j.biomaterials.2016.01.012

Wang, Y. B., Xu, W. H., You, T. G., Mu, F. W., Hu, H. D., Liu, Y., Huang, H., Suga, T., Han, G. Q., Ou, X., & Hao, Y. (2020). β-Ga_2O_3 MOSFETs on the Si substrate fabricated by the ion-cutting process. *Science China. Physics, Mechanics & Astronomy*, *63*(7), 277311. Advance online publication. DOI: 10.1007/s11433-020-1533-0

Wang, Z., Wu, S., Kang, G., Li, H., Wu, Z., Fu, Y., & Withers, P. J. (2021). In-situ synchrotron X-ray tomography investigation of damage mechanism of an extruded magnesium alloy in uniaxial low-cycle fatigue with ratchetting. *Acta Materialia*, *211*, 116881. DOI: 10.1016/j.actamat.2021.116881

Wanhill, R. J. H. (2014). Aerospace Applications of Aluminum-Lithium Alloys. In Prasad, N. E., Gokhale, A. A., & Wanhill, R. J. H. (Eds.), *Aluminum-Lithium Alloys: Processing, Properties, and Applications* (pp. 503–535). Elsevier., DOI: 10.1016/B978-0-12-401698-9.00015-X

Wankhade, S., Dabade, B., & Khond, M. (2016). Abrasive assisted electrical discharge machining: An innovative hybrid approach. *Proceedings of the International Conference on Communication and Signal Processing* (pp. 290-297). Lonere, Raigad (MS), India: Atlantis Press. DOI: 10.2991/iccasp-16.2017.46

Wankhade, L. N., Rathod, D., Nukulwar, M. R., Agrawal, E. S., & Chavhan, G. R. (2021). Characterization of aluminium-silicon carbide metal matrix composites. *Materials Today: Proceedings*, *44*, 2740–2747. DOI: 10.1016/j.matpr.2020.12.699

Wasa, K. (2012). *Sputtering systems* (2nd ed.). Elsevier. DOI: 10.1016/B978-1-4377-3483-6.00003-6

Weckend, S., Wade, A., & Heath, H. (2016). End-Of-Life Management Of Solar Panels. In *IRENA*. DOI: 10.1109/NRSC58893.2023.10152958

Weerasinghe, R. P. N. P., Yang, R. J., Wakefield, R., Too, E., Le, T., Corkish, R., Chen, S., & Wang, C. (2021). Economic viability of building integrated photovoltaics: A review of forty-five (45) non-domestic buildings in twelve (12) western countries. *Renewable & Sustainable Energy Reviews*, *137*, 110622. DOI: 10.1016/j.rser.2020.110622

Weiss, I., & Semiatin, S. L. (1999). Thermomechanical processing of alpha titanium alloys – an overview. *Materials Science and Engineering A*, *263*(2), 243–256. DOI: 10.1016/S0921-5093(98)01155-1

Westengen, H., & Rashed, H. (2016). Magnesium Alloys: Alloy and Temper Designation System. *Reference Module in Materials Science and Materials Engineering*, 1–3.

Winiczenko, R., Skibicki, A., & Skoczylas, P. (2024). The Experimental and FEM Studies of Friction Welding Process of Tungsten Heavy Alloy with Aluminium Alloy. *Applied Sciences (Basel, Switzerland)*, *14*(5), 2038. DOI: 10.3390/app14052038

Wittenauer, J., & Sherby, O. D. (1987). *Factors influencing the fatigue behavior of ferrous laminates.*

Wolf, M. T., Vodovotz, Y., Tottey, S., Brown, B. N., & Badylak, S. F. (2014). PredictingIn VivoResponses to biomaterials via CombinedIn VitroandIn SilicoAnalysis. *Tissue Engineering. Part C, Methods*, *21*(2), 148–159. DOI: 10.1089/ten.tec.2014.0167

Wonga, C. C., Hartawana, A., & Teo, W. K. (2014). Deep Cold Rolling of Features on Aero-Engine Components. *Procedia CIRP*, *13*, 350–354. DOI: 10.1016/j.procir.2014.04.059

Wong, K. V., & Hernandez, A. (2012). A Review of Additive Manufacturing. *ISRN Mechanical Engineering*, *2012*, 1–10. DOI: 10.5402/2012/208760

Woo-Jin, K., Chang-Young, H., & Ho-Kyung, K. (2006). Fatigue strength of ultrafine-grained pure Ti after severe plastic deformation. *Scripta Materialia*, *54*(10), 1745–1750. DOI: 10.1016/j.scriptamat.2006.01.042

Wu, Y., He, W., Ma, H., Nie, X., Liang, X., Pan, J., Wang, S., Shang, M., & Cheng, L. (2024). Titanium Alloy Materials with Very High Cycle Fatigue: A Review. In *Materials* (Vol. 17, Issue 12). Multidisciplinary Digital Publishing Institute (MDPI). DOI: 10.3390/ma17122987

Wu, C., Zhu, Z., Li, J., Tu, C., Lv, P., & Wang, Y. (2024). Fabrication of h-BN solar-blind ultraviolet detectors by RF magnetron sputtering. *Next Materials*, *3*, 100075. Advance online publication. DOI: 10.1016/j.nxmate.2023.100075

Wu, H., Zhang, C., Ding, K., Wang, L., Gao, Y., & Yang, J. (2017). Efficient planar heterojunction perovskite solar cells fabricated by in-situ thermal-annealing doctor blading in ambient condition. *Organic Electronics*, *45*, 302–307. DOI: 10.1016/j.orgel.2017.03.017

Wu, L., Liu, J., Ren, Y., Yang, Z., Wang, G., Sun, J., Song, J., Xu, W., & Liu, X. (2020). 3D FEM simulation of chip breakage in turning AISI1045 with complicate-grooved insert. *International Journal of Advanced Manufacturing Technology*, *108*(5–6), 1331–1341. DOI: 10.1007/s00170-020-05460-1

Wu, L., Li, X., & Wang, H. (2012). The effect of major constituents on microstructure and mechanical properties of cast Al-Li-Cu-Zr alloy. *Materials Characterization*, *171*, 110800. DOI: 10.1016/j.matchar.2020.110800

Wu, S., Yuan, D., Lü, S., Hu, K., & An, P. (2018). Nano-SiCP particles distribution and mechanical properties of Al-matrix composites prepared by stir casting and ultrasonic treatment. *China Foundry*, *15*(3), 203–209. DOI: 10.1007/s41230-018-8009-2

Wu, Y. (2022). Application of Aluminum Alloy in Aircraft. *Journal of Physics: Conference Series*, *2228*(1), 012024. DOI: 10.1088/1742-6596/2228/1/012024

Wycisk, E., Siddique, S., Herzog, D., Walther, F., & Emmelmann, C. (2015). Fatigue Performance of Laser Additive Manufactured Ti–6Al–4V in Very High Cycle Fatigue Regime up to 10^9 Cycles. *Frontiers in Materials*, *2*, 72. DOI: 10.3389/fmats.2015.00072

Wypij, M., Trzcińska-Wencel, J., Golińska, P., Avila-Quezada, G. D., Ingle, A. P., & Rai, M. (2023). The strategic applications of natural polymer nanocomposites in food packaging and agriculture: Chances, challenges, and consumers' perception. *Frontiers in Chemistry*, *10*, 1106230. Advance online publication. DOI: 10.3389/fchem.2022.1106230

Xiao-Bing, S., Xin-Yi, L., & Ting, Z. (2010). Research progress in fretting-resistance technology of titanium alloys. *Heat Treatment of Metals*, *35*(3), 83–86.

Xiao, G., An, R., Yang, J., & Ma, J. (2024). Research Progress of New Generation of Aluminum-Lithium Alloys Alloying. *Advanced Engineering Materials*, *26*(12), 2301639. DOI: 10.1002/adem.202301639

Xin, W., Wang, S., & Yan, Z. (2025). Study on the high temperature fatigue performance of AZ31B magnesium alloy. *Materials Today. Communications*, *43*, 111584. https://doi.org/https://doi.org/10.1016/j.mtcomm.2025.111584. DOI: 10.1016/j.mtcomm.2025.111584

Xiong, J., Li, Y., Li, R., & Yin, Z. (2018). Influences of process parameters on surface roughness of multi-layer single-pass thin-walled parts in GMAW-based additive manufacturing. *Journal of Materials Processing Technology*, *252*, 128–136. DOI: 10.1016/j.jmatprotec.2017.09.020

Xiong, J., Yin, Z., & Zhang, W. (2016). Forming appearance control of arc striking and extinguishing area in multi-layer single-pass GMAW-based additive manufacturing. *International Journal of Advanced Manufacturing Technology*, *87*(1–4), 579–586. DOI: 10.1007/s00170-016-8543-2

Xue, W., Krishna, B. V., Bandyopadhyay, A., & Bose, S. (2007). Processing and biocompatibility evaluation of laser processed porous titanium. *Acta Biomaterialia*, *3*(6), 1007–1018. DOI: 10.1016/j.actbio.2007.05.009

Xu, J., Gu, X., Ding, D., Pan, Z., & Chen, K. (2018). A review of slicing methods for directed energy deposition based additive manufacturing. *Rapid Prototyping Journal*, *24*(6), 1012–1025. DOI: 10.1108/RPJ-10-2017-0196

Xu, L.-Y., Xi, X.-C., Li, Z.-L., Ma, J.-Y., Gao, Q., & Zhao, W.-S. (2022). Wire breakage prevention for reciprocated traveling WEDM based on discharge location detection. *International Journal of Advanced Manufacturing Technology*, *123*(5–6), 1859–1875. DOI: 10.1007/s00170-022-10240-0

Xu, L.-Y., Xi, X.-C., Zhang, Y.-O., Li, Z.-L., Ma, J.-Y., Gao, Q., & Zhao, W.-S. (2022). Online workpiece height estimation method based on discharge location detection for reciprocated traveling WEDM. *Journal of Manufacturing Systems*, *65*, 317–329. DOI: 10.1016/j.jmsy.2022.09.018

Xu, Z., Tang, G., Tian, S., Ding, F., & Tian, H. (2007). Research of electroplastic rolling of AZ31 Mg alloy strip. *Journal of Materials Processing Technology*, *182*(1-3), 128–133. DOI: 10.1016/j.jmatprotec.2006.07.019

Yablonovitch, E., & Cody, G. D. (1982). Intensity Enhancement in Textured Optical Sheets for Solar Cells. *IEEE Transactions on Electron Devices*, *29*(2), 300–305. DOI: 10.1109/T-ED.1982.20700

Yadav, J., Gangil, N., Siddiquee, A. N., Ali Mohsin, M. E., Alnamasi, K., Moinuddin, S. Q., Khan, M. F., & Maheshwari, S. (2025). Effect of Tool Pin Profiles on Defects Formation during Friction Stir Welding of AA 6082-T6 Straight-Pipes-Butt-Weld Joints. *Journal of Materials Research and Technology*, *36*, 1645–1657. Advance online publication. DOI: 10.1016/j.jmrt.2025.03.185

Yamanoglu, R., Efendi, E., Kolayli, F., Uzuner, H., & Daoud, I. (2018). Production and mechanical properties of Ti–5Al–2.5Fe– x Cu alloys for biomedical applications. *Biomedical Materials (Bristol, England)*, *13*(2), 025013. DOI: 10.1088/1748-605X/aa957d

Yan, M., & Yu, P. (2015). An Overview of Densification, Microstructure and Mechanical Property of Additively Manufactured Ti-6Al-4V — Comparison among Selective Laser Melting, Electron Beam Melting, Laser Metal Deposition and Selective Laser Sintering, and with Conventional Powder. In *Sintering Techniques of Materials*. InTech. DOI: 10.5772/59275

Yan, C., Hao, L., Hussein, A., & Raymont, D. (2012). Evaluations of cellular lattice structures manufactured using selective laser melting. *International Journal of Machine Tools & Manufacture*, *62*, 32–38. DOI: 10.1016/j.ijmachtools.2012.06.002

Yan, C., Hao, L., Hussein, A., & Young, P. (2015). Ti–6Al–4V triply periodic minimal surface structures for bone implants fabricated via selective laser melting. *Journal of the Mechanical Behavior of Biomedical Materials*, *51*, 61–73. DOI: 10.1016/j.jmbbm.2015.06.024

Yánez, A., Herrera, A., Martel, O., Monopoli, D., & Afonso, H. (2016). Compressive behaviour of gyroid lattice structures for human cancellous bone implant applications. *Materials Science and Engineering C*, *68*, 445–448. DOI: 10.1016/j.msec.2016.06.016

Yang, C., Hu, W., Liu, J., Han, C., Gao, Q., Mei, A., Zhou, Y., Guo, F., & Han, H. (2024). Achievements, challenges, and future prospects for industrialization of perovskite solar cells. *Light: Science & Applications 2024 13:1*, *13*(1), 1–48. DOI: 10.1038/s41377-024-01461-x

Yang, C.-S., Smith, L. L., Arthur, C. B., & Parsons, G. N. (2000). Stability of low-temperature amorphous silicon thin film transistors formed on glass and transparent plastic substrates. *Journal of Vacuum Science & Technology. B, Microelectronics and Nanometer Structures : Processing, Measurement, and Phenomena : An Official Journal of the American Vacuum Society*, *18*(2), 683–689. DOI: 10.1116/1.591259

Yang, Y., Li, M., & Li, K. R. (2014). Comparison and analysis of main effect elements of machining distortion for aluminum alloy and titanium alloy aircraft monolithic component. *International Journal of Advanced Manufacturing Technology*, *70*(9-12), 1803–1811. DOI: 10.1007/s00170-013-5431-x

Yan, R., Luo, D., Huang, H., Li, R., Yu, N., Liu, C., Hu, M., & Rong, Q. (2018). Electron beam melting in the fabrication of three-dimensional mesh titanium mandibular prosthesis scaffold. *Scientific Reports*, *8*(1), 750. DOI: 10.1038/s41598-017-15564-6

Yashpal, K., Jawwalkar, C. S., & Kant, S. (2015). A review on use of aluminium alloys in aircraft components, *i-manager's. Journal of Materials Science*, *3*(3), 33–38. DOI: 10.26634/jms.3.3.3673

Yazawa, K., Malay, A. D., Masunaga, H., Norma-Rashid, Y., & Numata, K. (2020). Simultaneous effect of strain rate and humidity on the structure and mechanical behavior of spider silk. *Communications Materials*, *1*(1), 10. DOI: 10.1038/s43246-020-0011-8

Yen, C. C., Singh, A. K., Wu, P. W., Chou, H. Y., & Wuu, D. S. (2023). Interface engineering in epitaxial growth of sputtered β-Ga_2O_3 films on Si substrates via TiN (111) buffer layer for Schottky barrier diodes. *Materials Today. Advances*, *17*, 100348. Advance online publication. DOI: 10.1016/j.mtadv.2023.100348

Yıldız, A., Kurt, A., & Yağmur, S. (2020). Finite element simulation of drilling operation and theoretical analysis of drill stresses with the deform-3D. *Simulation Modelling Practice and Theory*, *104*, 102153. DOI: 10.1016/j.simpat.2020.102153

Yin, J., & Luan, S. (2016). Opportunities and challenges for the development of polymer-based biomaterials and medical devices. *Regenerative Biomaterials*, *3*(2), 129–135. DOI: 10.1093/rb/rbw008

Yoon, Y., Hwang, W. S., & Shin, M. (2022). Solar-blind ultrathin Sn-doped polycrystalline Ga_2O_3 UV phototransistor for normally off operation. *Advanced Photonics Research*, *3*(5), 2100316. Advance online publication. DOI: 10.1002/adpr.202100316

Yorseng, K., Rangappa, S. M., Pulikkalparambil, H., Siengchin, S., & Parameswaranpillai, J. (2020). Accelerated weathering studies of kenaf/sisal fiber fabric reinforced fully biobased hybrid bioepoxy composites for semi-structural applications: Morphology, thermo-mechanical, water absorption behavior and surface hydrophobicity. *Construction and Building Materials*, *235*, 117464. DOI: 10.1016/j.conbuildmat.2019.117464

Yoshikawa, A., Matsunami, H., & Nanishi, Y. (2007). Development and applications of wide bandgap semiconductors. In *Wide bandgap semiconductors: Fundamental properties and modern photonic and electronic devices* (pp. 1–24). Springer., DOI: 10.1007/978-3-540-47235-3_1

Yoshimura, R., Konno, T. J., Abe, E., & Hiraga, E. (2003a). Transmission electron microscopy study of the early stage of precipitates in aged Al–Li–Cu alloys. *Acta Materialia, 51*(10), 2891–2903. DOI: 10.1016/S1359-6454(03)00104-6

Yoshimura, R., Konno, T. J., Abe, E., & Hiraga, K. (2003b). Transmission electron microscopy study of the evolution of precipitates in aged Al–Li–Cu alloys: The θ′ and T1 phases. *Acta Materialia, 51*(14), 4251–4266. DOI: 10.1016/S1359-6454(03)00253-2

Yu, F.-P., Ou, S.-L., & Wuu, D.-S. (2015). Pulsed laser deposition of gallium oxide films for high performance solar-blind photodetectors. *Optical Materials Express, 5*(5), 1240–1249. DOI: 10.1364/OME.5.001240

Yuvaraj, N., Koli, Y., Vedabouriswaran, G., & Issar, P. (2023). Mechanical and Tribological Properties of AA6061/SiC/Aloe Vera Powder Hybrid Al Composites Fabricated by Stir Casting. *Silicon, 15*(5), 2451–2465. DOI: 10.1007/s12633-022-02168-7

Yu, Y., Yu, T., Mao, Y., Yang, Y., & Liang, S. (2025). Key Factors affecting Over-expanded Flow Separation in Design of Large Expansion Ratio Single Expansion Ramp Nozzle. *Aeronautical Journal, 129*(1333), 717–736. DOI: 10.1017/aer.2024.70

Zampaloni, M., Pourboghrat, F., Yankovich, S. A., Rodgers, B. N., Moore, J., Drzal, L. T., Mohanty, A.K., & Misra, M. J. (2007). Kenaf natural fiber reinforced polypropylene composites: A discussion on manufacturing problems and solutions. *Composites Part A: Applied Science and Manufacturing, 38*(6), 1569-1580. DOI: 10.1016/j.compositesa.2007.01.001

Zhang, C., Zou, D., Mazur, M., Mo, J. P., Li, G., & Ding, S. (2023). The state of the art in machining additively manufactured titanium alloy Ti-6Al-4V. *Materials (Basel), 16*(7), 2583. DOI: 10.3390/ma16072583

Zhang, H., Lin, X., Cao, X., Wang, Y., Wang, J., & Zhao, Y. (2023). Developing natural polymers for skin wound healing. *Bioactive Materials, 33*, 355–376. DOI: 10.1016/j.bioactmat.2023.11.012

Zhang, J., Wang, B., & Tian, Y. (2022). Study of Pre-Strain before Artificial Aging on Mechanical Property and Stress Corrosion Behavior of 2297 Al-Cu-Li Alloy. *Materials Science and Engineering A, 831*, 142174. DOI: 10.1016/j.msea.2021.142174

Zhang, L., & Chen, L. (2019). A Review on Biomedical Titanium Alloys: Recent Progress and Prospect. *Advanced Engineering Materials*, *21*(4), 1801215. Advance online publication. DOI: 10.1002/adem.201801215

Zhang, L., Kuriyagawa, T., Yasutomi, Y., & Zhao, J. (2005). Investigation into micro abrasive intermittent jet machining. *International Journal of Machine Tools & Manufacture*, *45*(7-8), 873–879. DOI: 10.1016/j.ijmachtools.2004.11.003

Zhang, X., Wang, L., Wang, X., Chen, Y., Shao, Q., Wu, G., Wang, X., Lin, T., Shen, H., Wang, J., Meng, X., & Chu, J. (2020). High-performance β-Ga_2O_3 thickness dependent solar blind photodetector. *Optics Express*, *28*(3), 4169–4177. DOI: 10.1364/OE.385470 PMID: 32122074

Zhang, Y., Wu, L., Guo, X., Kane, S., Deng, Y., Jung, Y.-G., Lee, J.-H., & Zhang, J. (2018). Additive Manufacturing of Metallic Materials: A Review. *Journal of Materials Engineering and Performance*, *27*(1), 1–13. DOI: 10.1007/s11665-017-2747-y

Zhang, Z., Ao, Y., Li, S., & Gu, G. X. (2024). An Adaptive Machine Learning-based Optimization Method in the Aerodynamic Analysis of a Finite Wing under Various Cruise Conditions. *Theoretical & Applied Mechanics Letters*, *14*(1), 100489. DOI: 10.1016/j.taml.2023.100489

Zhao, B., Wang, Y., Peng, J., Wang, X., Ding, W., Lei, X., Wu, B., Zhang, M., Xu, J., Zhang, L., & Das, R. (2024). Overcoming Challenges: Advancements in Cutting Techniques for High Strength-Toughness Alloys in Aero-Engines. *International Journal of Extreme Manufacturing*, *6*(6), 062012. DOI: 10.1088/2631-7990/ad8117

Zhao, Y., Zhao, X., Chen, X., Song, J., Cui, Y., Niu, F., Du, X., Lu, W., & Zuo, D. (2024). Modeling and analysis of plastic abrasive jet flow characteristics. *Journal of Manufacturing Processes*, *123*, 96–111. DOI: 10.1016/j.jmapro.2024.05.064

Zheng, J., Zhang, M., Lau, C. F. J., Deng, X., Kim, J., Ma, Q., Chen, C., Green, M. A., Huang, S., & Ho-Baillie, A. W. Y. (2017). Spin-coating free fabrication for highly efficient perovskite solar cells. *Solar Energy Materials and Solar Cells*, *168*, 165–171. DOI: 10.1016/j.solmat.2017.04.029

Zheng, W., Cui, M., Song, Q., & Yuan, W. (2025). Effect of Low-Temperature Deformation and Deep-Cryogenic Treatment on the Mechanical and Corrosion Resistance of 2050 Aluminum-Lithium Alloy. *Journal of Alloys and Compounds*, *1034*, 181347. DOI: 10.1016/j.jallcom.2025.181347

Zhou, B., Liu, B., & Zhang, S. (2021). The Advancement of 7XXX Series Aluminum Alloys for Aircraft Structures: A Review. *Metals*, *11*(5), 718. DOI: 10.3390/met11050718

Zhu, D., Mobasher, B., & Rajan, S. D. (2011). Dynamic tensile testing of Kevlar 49 fabrics. *Journal of Materials in Civil Engineering, 23*(3). DOI: 10.1061/(ASCE)MT.1943-5533.0000156

Zhu, J., Jie, Z., Chen, C., Zheng, H., & Wang, W. (2023). Fatigue Crack Propagation of Corroded High-Strength Steel Wires Using the XFEM and the EIFS. *Materials (Basel), 16*(13), 4738. Advance online publication. DOI: 10.3390/ma16134738

Zhu, J., Xu, Z., Ha, S., Li, D., Zhang, K., Zhang, H., & Feng, J. (2022). Gallium oxide for gas sensor applications: A comprehensive review. *Materials (Basel), 15*(20), 7339. Advance online publication. DOI: 10.3390/ma15207339 PMID: 36295403

Zhu, Y. S., Wu, J., Lu, W. Z., Zuo, D. W., Xiao, H. P., Cao, D. W., & Ko, T. J. (2022). Surface formation mechanics and its microstructural characteristics of AAJP of aluminum alloy by using amino thermosetting plastic abrasive. *International Journal of Precision Engineering and Manufacturing-Green Technology, 9*(1), 59–72. DOI: 10.1007/s40684-020-00284-6

Ziaee, M., & Crane, N. B. (2019). Binder jetting: A review of process, materials, and methods. *Additive Manufacturing, 28*, 781–801. DOI: 10.1016/j.addma.2019.05.031

Zuccarello, B., Bongiorno, F., & Militello, C. (2022). Basalt fiber hybridization effects on high-performance sisal-reinforced biocomposites. *Polymers, 14*(7), 1457. DOI: 10.3390/polym14071457

Zuo, J. H., Wang, Z. G., & Han, E. H. (2008). Effect of microstructure on ultra-high cycle fatigue behavior of Ti–6Al–4V. *Materials Science and Engineering A, 473*(1-2), 147–152. DOI: 10.1016/j.msea.2007.04.062

About the Contributors

Deepanraj Balakrishnan is currently working as Assistant Research Profesor in the Department of Mechanical Engineering, Prince Mohammad Bin Fahd University, Saudi Arabia. He has 13 years of working experience, which include both teaching and research. Dr. Deepanraj has published more than 200+ research articles in peer reviewed international journals and conferences and holds 4 Indian patents. He is serving as guest editor, editorial board member and reviewer for many international peer reviewed journals including well known publishers like Elsevier, Springer and Inderscience. His area of interest includes Renewable Energy Utilization; Materials Engineering, Waste Management and Sustainability.

Ratna Sunil received his bachelor's degree in mechanical engineering from Andhra University, India, in 2004 and Master's degree in Mechanical Engineering from Osmania University, India in 2007. Dr Ratna Sunil earned his Ph.D. from the Indian Institute of Technology Madras, India, in 2014 in the field of Biomaterials. Dr. Sunil has 15 years of experience in teaching and research in the areas of Materials Engineering, Manufacturing Processes and Surface Engineering. Dr. Sunil has been listed in world top 2% scientists announced by Stanford University, USA, consecutively for five years (from 2020 to 2024). Dr. Sunil has published more than 140 scientific papers in international peer-reviewed journals and conference proceedings and authored a book titled "Surface Engineering by Friction-Assisted Processes: Methods, Materials, and Applications" available with CRC Press, USA. From the research findings, Dr. Sunil has filed 20 Indian patents.

Wattala Rohan Fernando holds the position of Lecturer in Mechanical and Industrial Engineering at the School of Science and Engineering (SSEN), and he is also a lecturer at The Centre for Energy, Petroleum, and Mineral Law and Policy (CEPMLP), where he teaches energy-related modules associated with SSEN. After completing his studies at Northumbria University Newcastle Upon Tyne, UK, in 2011, Fernando gained significant experience through roles with Ericson, Vodafone, and Huawei, contributing to various projects, including the London Olympic Project. In 2015, he completed his MSc in Energy, Water, and Environmental Management at Abertay University Dundee, UK. The practical experience acquired at Ericson, Vodafone, and Huawei proved invaluable, particularly in power and energy engineering, facilitating his MSc studies. In 2020, he successfully earned his PhD in energy engineering from Edinburgh Napier University. Concurrently, during his PhD, Fernando completed six European Credit Transfer and Accumulation System (ECTS) PhD courses at Aalborg University in Aalborg, Denmark. Before joining the University of Dundee in January 2023, Fernando held positions as a research officer at Hydrogen Accelerator, St Andrews University, a research and teaching assistant at Edinburgh Napier University, and a visiting lecturer at the University of Applied Management Studies, Mannheim, Germany, and German Business School, the University of Tunis, for an MBA in Energy Management. With more than 12 years of experience, he actively participated in national and international industry projects in electronics, telecommunications, and power engineering before transitioning to academia.

Abdul Aziz Afzal earned his B. Sc. in Mechanical Engineering from King Fahd University of Petroleum & Minerals (KFUPM), KSA in 2017 and his M. Sc. in Mechanical Engineering with specialization in Manufacturing & Applied Materials from KFUPM, KSA in 2019. During his time at KFUPM, he worked as a Research Assistant. He was involved in funded research projects and provided mentorship to undergraduate and graduate students working on design-based projects. His research interests include application-focused material design of alloys and life cycle assessment of manufacturing processes, focusing primarily on welding processes. He is a co-author of two journal publications and has received a US Patent

Hameed AlRashidi holds a Ph.D. in Renewable Energy. His research focuses on energy systems, particularly photovoltaic technologies. He serves as an engineering advisor for the water and energy sectors at the Kuwait Fund for Arab Economic Development, combining academic expertise with practical experience in sustainable development.

Thanikasalam Arumugam is an Associate Professor in the Department of Marine Engineering at AMET Deemed to be University, Chennai, with 23+ years of academic and research expertise. Holding a Ph.D. (2020) and M.E. (2006) in Mechanical Engineering from Anna University, and a B.E. (2000) from Madurai Kamaraj University, he specializes in deep-hole drilling, composite materials, cryogenic drilling, and tool condition monitoring. A prolific researcher, he has authored 15+ international journal papers, five book chapters, and co-authored the textbook Composite Materials Design. He holds 4 patents (2 Indian, 2 UK) and has organized Faculty Development Programs funded by Anna University. Recognized for bridging industry and academia, he reviews for reputed journals and is a life member of ISTE, AMM, and PMAI. His career underscores innovation in engineering education and advancing sustainable technological solutions.

Prashant A. Athavale, Teaching the UG students in Electrical Engineering since 2009, at BMS Institute of Technology and Management, Yelahanka, Bengaluru. Interested in Signal Processing and allied areas of research.

Jagadeesh Baskaran is currently working as a Faculty member in the Department of Mechanical Engineering, PSG Institute of Technology and Applied Research. He obtained his B. Tech in Mechanical Engineering and M.Tech in Advanced Manufacturing from SASTRA University. For his performance during the Master's program, he received the best outgoing M. Tech student award, recognising his academic record and research publication from his M.Tech thesis. He has two years of work experience in SAP ERP software consulting and one year of teaching experience in Mechanical and Manufacturing Engineering. During his Ph.D. work at NIT Trichy, he got the DORA plus scholarship to undergo research training in the Mechanical and Industrial Engineering department of Tallinn University of Technology, Estonia. His research interests include Metal 3D printing, Design for Additive Manufacturing, Laser Materials Processing, unconventional machining techniques and materials management. He has authored multiple international journal Publications in metal 3D printing, out of which four are the outcomes of his PhD thesis. One of the presentations from his PhD work earned a best paper award in the ICPPM conference (March 2023).

Bárbara Bermúdez-Reyes: Received a Ph.D. in Metallurgy and Materials Science in 2009. She is a Professor at the High Technology Unit of the National University Autonomous of Mexico. Previously, she was a Research Professor at different universities. She was General Secretary of the University Space Engineering Consortium-Mexico Chapter for 2014–2019. She is a member of the Space Science and Technology Network. She has collaborated on projects with the Mex-ican Air Force and has evaluated projects of the Mexican Naval Research Institute. She is part of the National Research System with level I. Her research lines are designing and processing aerospace ceramic-ceramic and metal-ceramic composite materials, satellite structures, and severe plastic deformation to form aluminum alloy for high-velocity aircraft.

Susanta Das is a seasoned educator with a proven record of teaching and mentoring diverse students, working at multiple institutions, and providing service to the various initiatives of organizations. He received his Ph.D. and M.A. degrees from Western Michigan University (WMU), USA, and M.Sc. degree from Banaras Hindu University (BHU), India all in Physics. He continued his research as a Marie-Curie post-doctoral fellow at Stockholm University, Stockholm, Sweden on beam diagnostics for the DESIREE (Double ElectroStatic Ion Ring ExpEriment) in the project DITANET (Diagnostic Techniques for particle Accelerators – a Marie-Curie initial training NETwork), at the Indian Institute of Science Education and Research-Kolkata, India on high-pressure physics as a Project Scientist-B, and at the University of Electro-Communications, Tokyo, Japan on ion-surface interactions as a Post-doctoral fellow. Throughout his career, Dr. Das worked and collaborated with many researchers, Ph.D., master, and visiting students from different countries. He further visited many countries to discuss research and an international conference participant (UK, Italy, Germany, Greece, Belgium, Bulgaria, Romania, Brazil etc.). He worked at several institutions before assuming his current role at Ajeenkya D Y Patil University. He received the Marie-Curie post-doctoral fellowship in Sweden, Science Academies' Summer Research Fellowship in India, Gwen Frostic Doctoral Fellowship, Department Graduate Research and Creative Scholar Award by WMU, and Leo R. Parpart Doctoral Fellowship by Dept. of Physics, WMU, among many others, throughout his academic journey. He co-authored several research articles in WoS/Scopus indexed international journals, conference proceedings, and scholarly book chapters and co-edited several books published by renowned international publishing houses (IGI-Global, Springer Nature, Nova Science Publishers, Wiley, & Scrivener-Wiley, Notion Press). At present, Dr. Das continues to delve into cutting-edge fields, with a keen interest in nanotechnology, UN SDGs, and data science. His multifaceted contributions, spanning teaching, research, and administrative leadership, showcase a dedicated professional who is instrumental in advancing the vision and mission of the institutions he serves.

Saurav Datta is presently working as Professor in the Department of Mechanical Engineering, National Institute of Technology, Rourkela (Odisha). After receiving B.E. (Honors) in Mechanical Engineering from Jadavpur University, Kolkata in 2003, he was awarded Ph.D. (Engineering) by Jadavpur University in December 2008. He joined the Department of Mechanical Engineering of NIT Rourkela in August 2008. His domain of research includes Manufacturing, Quality Optimization, Statistical Modeling/ Optimization of manufacturing Processes, Industrial Decision Making, Traditional and Non-traditional machining, and Metal Additive Manufacturing. He is Life Associate Member of Indian Institute of Welding. He has published more than 200 articles in International Journals, National Journals of repute, and proceedings of different International and National level Conferences. To date, he mentored 45 M. Tech. projects and supervised a total number of 11 PhDs with the capacity of

Sole/ Principal Supervisor. He executed two sponsored projects funded by SERB and CSIR, with capacity of Principal Investigator and Co-Investigator, respectively. He is currently executing 02 more sponsored projects on Metal Additive Manufacturing funded by the Science and Technology Department, Govt. of Odisha and CSIR, respectively. Recently, his name has been included within the Top 2% of the most Influential Scientists – 2023, by the Stanford University list.

BalakrishnanDeepanraj is currently working as Research Faculty in the Department of Mechanical Engineering, College of Engineering at Prince Mohammad Bin Fahd University, Saudi Arabia. Dr. Deepanraj received his Bachelor's degree in Mechanical Engineering and Master's Degree in Thermal Engineering from Anna University, Chennai and Ph.D degree in Mechanical Engineering from National Institute of Technology, Calicut. He also completed his MBA in Energy Management from Jaipur National University. He has more than10 years of working experience, which include both teaching and research. Dr. Deepanraj has published more than 100 research articles in peer reviewed international journals and conferences and holds 2 Indian patent

C. Dhanesh was born on 18th April 1977, in Tirunelveli District, Tamil Nadu, India. He completed his Bachelor's Degree in Mechanical Engineering in 2000 and then completed his Master's Degree in Thermal Engineering in 2009. He is currently pursuing his Ph.D. Degree in Anna University and his area of research is focused towards Nano Materials, Alternate Fuel Technology, IC Engines. Currently he is serving as Associate Professor and Head of the Department, Mechanical Engineering, S.A. Engineering College (Autonomous), Chennai. Under his leadership as head of the department, he not only shares expertise with dedication but also guides and empowers aspiring engineers and innovators, driving advancement in mechanical engineering education and technological progress

C. K. Dhinakarraj was born on June 5, 1972, in Chengalpattu, India. With a deep-rooted passion for mechanical systems and innovation, he pursued his undergraduate degree in Mechanical Engineering, followed by a postgraduate specialization in Manufacturing Engineering. Driven by a commitment to research and development in advanced materials, he earned his Ph.D. focusing on the characterization of magnesium metal matrix composites and the optimization of machining parameters; an area of increasing relevance in lightweight structural applications. Currently, Dr. Dhinakarraj is a Professor in the Department of Mechanical Engineering, where he also holds the Head of the Department (HOD) position. In this leadership role, he imparts knowledge and mentors budding engineers and researchers, shaping the future of mechanical engineering education and research.

Atef El Khatib is a researcher and educator with a Ph.D. in Mechanical Engineering, specializing in thermo-fluids and renewable energy systems. His work spans energy and aerodynamic systems, with publications in areas such as flow-induced vibrations and photovoltaic technologies. Alongside, he coordinates educational programs and contributes to academic development efforts.

Emad Elsehly, Acquire a rich knowledge in solid state physics theories and material science applications, peruse my postdoctoral research in thermoelectric properties, participate in a variety of water treatment projects by different membranes based on carbon nanotubes and have a list of scientific research publications in the field of Physics. In general, interested in nanostructure applications.

Arzu Eren Şenaras is a distinguished academic specializing in system dynamics, simulation modeling, and metaheuristic optimization methods. Her research focuses on applying dynamic modeling techniques to complex systems, including healthcare resource management, logistics, and industrial systems. She has published extensively in journals and has contributed to several academic book chapters in her field. Her work includes the application of metaheuristic algorithms—such as simulated annealing and genetic algorithms—for solving complex optimization problems. As a faculty member at Uludağ University, she is recognized for her interdisciplinary research, academic leadership, and commitment to innovation in education.

Abdul Gafoor is a student in the aeronautical engineering department at MLR Institute of Technology. With a strong background in composite structures and CFD, he has developed a keen interest in hybrid composites research. Abdul Gafoor has been actively involved in hybrid composites projects.

Dankan Gowda V is currently working as an Assistant Professor in the Department of Electronics and Communication Engineering at BMS Institute of Technology and Management in Bangalore. Previously, he worked as a Research Fellow at ADA DRDO and as a Software Engineer at Robert Bosch. With a total experience of 14 years, including teaching and industry, Dr. Gowda has made significant contributions to both academia and research. He has published over 120 research papers in renowned international journals and conferences. In recognition of his innovative work, Dr. Gowda has been granted six patents, including four from Indian authorities and two international patents. His research interests primarily lie in the fields of IoT and Signal Processing, where he has conducted workshops and handled industry projects. Dr. Gowda is passionate about teaching and strives to create an engaging and effective learning environment for his students. He consistently seeks opportunities to enhance his knowledge and stay updated with the latest advancements in his field.

Samrat Hazra is a PhD scholar at the National Institute of Technology (NIT) in Rourkela, Odisha, India. He completed his Diploma in Mechanical Engineering from the West Bengal State Council of Technical Education (WBSCTE), followed by a B.Tech in Mechanical Engineering from Maulana Abul Kalam Azad University of Technology (MAKAUT), West Bengal, India. He pursued an M.Tech in Manufacturing Technology at the National Institute of Technical Teachers' Training and Research (NITTTR), Kolkata, West Bengal, India. His research interests include advanced manufacturing processes, composite materials, and additive manufacturing.

Anoop Johny is presently working as Associate professor at KMCT College of Engineering, Kozhikode, Kerala. Dr. Anoop did his BE Production Engineering at Karunya Institute of Technology Coimbatore during 2003-07. While working at Venture Lighting India Limited as a production engineer, he did his MTech in Industrial engineering at Dr MGR University, Chennai during 2010-13. Later he got associated with Saveetha University as assistant professor during 2013-19 and pursued his PhD in mechanical engineering and was awarded Ph.D. degree in 2022. Dr Anoop's core research area is non-traditional machining, quality management and industrial engineering.

Mandeep Kaur is working as Assistant Professor in Department of Electronics and Communication engineering, Punjabi University Patiala. She has thirteen years of teaching experience. She has published more than 60 research papers in SCI and Scopus journal and International/National conferences. She has guided 25 M.Tech students. Her area of interest is VLSI design, VLSI DSP, optimization and machine learning.

Muhammad Azhar Ali Khan received his PhD degree in Mechanical Engineering from King Fahd University of Petroleum and Minerals, Saudi Arabia in 2018. He did his MS from Middle East Technical University, Turkey in 2013 and BE from NED University of Engineering and Technology, Pakistan in 2009. During his stay at KFUPM and METU, he gained six years of experience in teaching undergraduate courses. He also assisted a range of mechanical engineering courses on mechanics, materials, and manufacturing. His research is focused on utilization of computational methods and tools for mold design optimization, performance evaluation and reliability assessment of cast parts. He is an author of the book

"Evolution of Metal Casting Technologies: A Historical Perspective", in addition to his other technical publications in reputable international conferences and scientific journals.

Kaushik Kumar holds a Ph.D. in Engineering from Jadavpur University, India, an MBA in Marketing Management from Indira Gandhi National Open University, India and a Bachelor of Technology from Regional Engineering College (Now National Institute of Technology), Warangal, India. For 11 years, he worked in a manufacturing unit of Global repute. He is currently working as a Professor in the Department of Mechanical Engineering, Birla Institute of Technology, Mesra, Ranchi, India. He has 24 Years of Teaching and Research Experience. His research interests include Composites, Optimization, Non-Conventional Machining, CAD / CAM, Rapid Prototyping and Quality Management Systems towards product development for societal and industrial usage and has received 50+ patents for them, many of which are under consideration for donation to Start-ups and Industrial Institutions through Jharkhand Small Industries Association (JSIA). He has published 65+ Books (They are referred as Text Books and Reference Books by 60+ Universities / Institutes in their academic curriculum), 100+ Book Chapters and 200+ Research Papers in peer-reviewed reputed national and international journals. Kaushik has also served as Editor–in–Chief, Series Editor, Guest Editor, Editor, Editorial Board Member and Reviewers for International and National Journals. He has been felicitated with many awards and honours including Distinguished Alumnus Award for Professional Excellence 2023 under Academic and Research from National Institute of Technology, Warangal, India. He has also received Sponsored Research and Consultancy Projects of more than 1 Crore from Govt. of India and abroad. Kaushik has delivered expert lectures as Keynote Speaker at International & National Conferences, Resource Person at Various Workshops, FDP's and Short-Term Courses. He has guided many students of Doctoral, Masters and Under Graduate programmes of his home and other institutions in India and abroad. He also has served as Reviewer and Examiner of Doctoral and Masters dissertation for institutes in India and abroad.

Saad Patel Md is a researcher with expertise in advanced alloy development, additive manufacturing, and microstructural characterization. His research encompasses magnesium based alloys/composites and additive manufacturing of NiTi shape memory alloys, and sustainable manufacturing strategies. He has contributed to high impact journal publications and book chapters, focusing on improving mechanical performance, corrosion resistance, and surface properties in structural materials. His work integrates experimental techniques such as friction stir processing, laser shock peening, and monitoring of additive processes to address critical challenges in aerospace, automotive, and green manufacturing applications.

PalanivendhanMurugadoss is an Assistant Professor in the Department of Automobile Engineering at the School of Mechanical Engineering, SRM Institute of Science and Technology (SRMIST), Kattankulathur, Tamil Nadu. With a cumulative academic experience of nearly 12 years, he is a committed educator, researcher, and mentor with deep expertise in metal matrix composites, tribology, and advanced manufacturing. Dr. Palanivendhan has authored over 40 Scopus-indexed research articles and holds 7 patents, including 5 granted design patents. His scholarly contributions emphasize innovative composite material development using agro-industrial waste reinforcements, particularly ADC-12 aluminum alloys, with practical industrial applications. His collaborative projects with companies like MK Trons Pvt. Ltd. and Brakes India Pvt. Ltd. demonstrate the real-world impact of his research

Gheorghe Nagit is a professor at "Gheorghe Asachi" Technical University of Iasi, Romania. He is a graduate of the Faculty of Machine Manufacturing Engineering since 1991. From the same year he is a professor at the same university. He finalized his doctoral thesis in 1997 with a paper about superficial plastic deformation by vibro-burnishing. He is the author of over 200 scientific papers. The main research areas addressed were: - Cold plastic deformation technologies, - Non-traditional technologies, - Fundamentals of Technical Creation and Technological Innovation, - Experimental planning and modelling of technological processes, - Finishing technologies by cutting and electrophysical methods. Gheorghe Nagit is the director of the Department of Machine Manufacturing Engineering and is a former Dean of the Faculty of Machine Manufacturing.

Ramesh NaiduPalamaneni is currently working as a associate professor of Computer Science and Engineering at Nitte Meenakshi Institute of Technology, Bengaluru. He has total 19 years of teaching experience and 1 year Industry experience. His research area is cloud computing, web technologies and block chain technologies. He has published total 40 international, national journals and 11 patents. He is a life member of the Indian Society for Technical Education (ISTE) and life of Institute of Engineer's. He can be contacted at email: ramesh.naidu@nmit.ac.in

Anand Pandey is an applied physicist specializing in energy materials, optoelectronic devices, and advanced material characterization. He earned his Ph.D. from the University of Allahabad, India, and has published over 15 research articles in reputable journals. His work has received three Best Paper Awards at international conferences and two "Outstanding Reviewer of the Year" awards from IOP Publishing. He is currently a researcher at Linköping University, Sweden, developing electronically steerable guidewires for medical diagnostics. Previously, he worked as a Research Associate at the Indian Institute of Technology Guwahati, focusing on the development of a reliability framework, measurements, and degradation mechanisms in energy devices. Dr. Pandey is affiliated with IOP, ACS, and ECS, and is a lifetime member of the Indian Science Congress Association and the Society for Materials Chemistry. He also serves as a reviewer for journals published by IOP, Elsevier, and AIP.

Prudvi Reddy Paresi is an accomplished mechanical engineering academic and researcher with expertise in sustainable machining, advanced manufacturing, and numerical modelling of high-speed machining. He currently serves as a Lecturer and Program Coordinator at Federation University, Australia, where he teaches and coordinates multiple engineering courses in Diploma of Engineering (Higher Education). Dr. Paresi has a strong research background in the machining of difficult-to-cut materials, and presently working on projects focusing on enhancing machining efficiency through innovative cooling, lubrication, and tool design strategies. His research explores cryogenic cooling, minimum quantity lubrication (MQL), and hybrid cooling techniques to improve machining performance and reduce environmental impact. He has conducted extensive experimental and numerical investigations on the machining of aerospace materials, and difficult-to-machine materials like Inconel alloys and composite materials, using finite element modelling (FEM) and digital simulation techniques. Beyond machining, he has also worked on government-funded projects related to carbon accounting and sustainable design in the signage industry. As a member of Engineers Australia, Dr. Paresi also plays a key role in the peer-review process for international journals and conferences. He serves as a reviewer for high-impact journals, including the International Journal on Interactive Design and Manufacturing, Cogent Engineering, Heliyon, and Advances in Materials Processing Technologies. Dr. Paresi has published in leading international journals and conferences and has successfully secured competitive research grant. He actively collaborates with industry and academic institutions to advance research and teaching methodologies in engineering.

KDV Prasad working as a Faculty (Research) with the Symbiosis Institute of Business Management, Hyderabad; Symbiosis International (Deemed University), Pune India. Dr KDV Prasad possess Masters in Computers Applications, Masters in Software Engineering, PhD in Business Administration and Management. Dr Prasad authored 3 books and published over 200 papers in reputed national and international journals. His interests are machine learning. deep learning, Generate AI, behavioral finance and marketing and Human Resource management

Francisco Reyes Calderón: Francisco Reyes is a full-time research professor at the National Technological Institute of Mexico, in Morelia's Technological Institute. He's PhD in Metallurgy and Materials Science from the UMSNH, a Master in Materials Science and Mechanical Engineer from the same institute. Also, a major in Industrial Welding from CBTis 149. Since 2009, he has been involved in the development of high-strength steels currently used in the automotive industry and he was part of a pioneering research group studying TWIP steels (Twinning Induced Plasticity - AHSS) in Mexico. He is also a co-founder of the Mexican Association of Welding, Metallurgy, and Engineering Sciences, A.C. He holds two industrial patents. He is the organizer and founder of the International Conference on Joining Processes and Technologies in Materials (CIPTUM). His original research focuses on the use and control of different modes of mass and energy transfer in conventional welding processes, additive manufacturing using semi-automatic welding processes, and the introduction of laser welding in the state of Michoacán for various companies.

Emmanuel Reyes-Gordillo, PhD in Metallurgy and Materials with over 8 years of experience in process optimization, additive manufacturing, and applied research for the automotive and advanced manufacturing sectors. As a Six Sigma Black Belt, I have led projects generating savings exceeding $100K USD/year through DMAIC methodologies, focusing on quality enhancement and cost reduction. My expertise includes WAAM (Wire Arc Additive Manufacturing) for strategic components, microstructural characterization, and cross-functional team leadership. I have contributed to patents and publications in indexed journals, taught engineering courses, and driven OEE improvements in industrial settings. Proficient in process simulation, solid-gas kinetics, and casting process of ferrous and no ferrous materials, I am committed to advancing sustainable metallurgical processes and R&D collaborations with industrial impact. My work bridges academic rigor and practical innovation.

N. Senthilkumar is a gold medalist and one of the world's top 2% scientists, listed by Stanford University. He had completed his B.E. (Mechanical) degree from Government Engineering College, Salem, affiliated with the University of Madras in 1998 with I Class. He received his M.E. (Manufacturing) degree in 2007 from Madras Institute of Technology, Chennai, a constituent college of Anna University, Chennai, with I Class and Distinction and a gold medal. Dr. Nsk received his doctoral degree from Anna University, Chennai, in 2014 for his work on the effect of cutting tool geometries in turning. So, he has published 200 research articles in reputed peer-reviewed international and national journals, presented 50 articles at international and national conferences, and has also published 4 textbooks. His core research areas are metal cutting, composite materials, optimization, machining of newer materials, and finite element analysis. He has a total teaching experience of 25 years and industrial experience of 3 years. Under his guidance, 5 scholars have completed doctoral degrees so far, and he is presently guiding 3 research scholars. He acted as lead guest editor and guest editor for several special issues and is acting as a reviewer for 90 international journals, including Elsevier, Springer, SAGE publications, Hindawi, IGI Global, Taylor & Francis, and an editorial member for 3 international journals. Presently, he is working as a professor in the Department of Mechanical Engineering, SIMATS School of Engineering, Saveetha Institute of Medical and Technical Sciences, Chennai, Tamil Nadu, India.

Aditya Shrivastav is currently studying B.Tech. in Computer Engineering at the School of Engineering, Ajeenkya DY Patil University. His research interests are centered around nanoparticle biosynthesis, exploring ionic liquids' use in various applications, and applying Artificial Intelligence to different fields. Aditya is particularly focused on understanding how these technologies can be utilized together to solve practical problems. He aims to advance research in these areas and contribute to innovative solutions that address real-world challenges. He recently participated at the Global Hyperloop Competition (GHC) at IIT Madras as a member of the team Vac-Vectoor. The event took place from 21st February through 24th February 2025. He contributed significantly to the performance of his team. The team Vac-Vectoor achieved 2nd position in Business and Socio-Economic Impact of Hyperloop in the Innoquest category and Special Mention Certificate in Traction sub-system under Design X category.

C. Thiagarajan is a Professor in the Institute of Automobile Engineering. He has completed his doctorate in the field of Manufacturing Engineering in MIT Campus, Anna University, Chennai, India. He has had more than 25 years of experience in Teaching and Industry. His academic and research interests include Manufacturing engineering and Composites machining. He has received three awards in academics and has more than 35 publications to his credit. Under his guidance two research scholars got awarded Ph.D. degree.

Rafael Vargas-Bernal received a bachelor's degree in Communications and Electronics Engineering from the University of Guanajuato in 1995 Mexico, and the degrees of Master of Science and Doctorate in Science with a Specialty in Electronics from the National Institute of Astrophysics, Optics and Electronics (INAOE) in

1997 and 2000 from Mexico, respectively. Since September 2000, he has been a professor- researcher at the Polytechnical University of Guanajuato, Technological University of Salamanca, University of Guanajuato, University La Salle Bajío, and the Higher Technological Institute of Irapuato (ITESI), Mexico, where he has established himself as a researcher in electronics and materials engineering. He has authored 22 articles in journals, 61 chapters in books, and more than 100 conference articles. He has edited three books for IntechOpen. He is a member of the National System of Researchers (SNII- Mexico). He regularly serves as a reviewer of scientific articles in RSC, Elsevier, MDPI, Wiley, Taylor & Francis, IEEE, World Scientific Publishing, and ACS, as well as Standards in Semiconductor Equipment and Materials International (SEMI). His research interests include two-dimensional materials, nanomaterials, space materials, composite materials, gas sensors, and biosensors.

Index

A

B

C

D

F

G

H

I

K

L

M

N

O

www.ingramcontent.com/pod-product-compliance
Lightning Source LLC
LaVergne TN
LVHW082051011026
849162LV00019B/773

* 9 7 9 8 3 3 7 3 9 4 5 4 1 *